International Symposium on Materials Science and Innovation for Sustainable Society – Eco-Materials and Eco-Innovation for Global Sustainability – The 21st Iketani Conference 2011

(ECO-MATES 2011)

Journal of Physics: Conference Series Volume 379

Osaka, Japan
28-30 November 2011

ISBN: 978-1-62276-846-2
ISSN: 1742-6588

Printed from e-media with permission by:

Curran Associates, Inc.
57 Morehouse Lane
Red Hook, NY 12571

Some format issues inherent in the e-media version may also appear in this print version.

Printed by Curran Associates, Inc. (2013)

Institute of Physics
Dirac House, Temple Back
Bristol BS1 6BE UK

Phone: 44 1 17 929 7481
Fax: 44 1 17 920 0979

techtracking@iop.org

Additional copies of this publication are available from:

Curran Associates, Inc.
57 Morehouse Lane
Red Hook, NY 12571 USA
Phone: 845-758-0400
Fax: 845-758-2634
Email: curran@proceedings.com
Web: www.proceedings.com

TABLE OF CONTENTS

RENEWABLE ENERGY

012001 The Photovoltaic Power Generation Era is Coming (Global Energy Network Equipped with Solar Cells and International Superconductor Grids: Genesis) ... 1
Y Kuwano
012002 Unique Features of Cylindrical Type Solar-module Contrasted with Plane Or Conventional Type Ones ... 10
H Hiraki, A Hiraki, M Maeda, Y Takahashi
012003 Local Photovoltaic Characterization for Silicon Thin Film Solar Cells Using a Scanning Probe Microscope ... 15
T Itoh, H Natsuhara, S Nonomura

ENERGY MATERIALS

012004 Synthesis and Thermoelectric Properties of Sintered Skutterudite $CoSb_3$ with a Bimodal Distribution of Crystal Grains ... 21
S Katsuyama, F Maezawa, T Tanaka
012005 Magnetron Sputtering of (TiZr)NiSn Thin Films on Different Substrates for Thermoelectric Applications ... 28
W Wunderlich, Y Shinohara, Y Matsumura
012006 Thermoelectric Properties of Nitrogen-doped TiO_{2-x} Compounds ... 33
M Mikami, K Ozaki
012007 Influence of Preparation Conditions on Thermoelectric Properties of $Ba_8Ga_{16}Si_{30}$ Clathrate by Combining Arc Melting and Spark Plasma Sintering Methods ... 39
H Anno, H Yamada, T Nakabayashi, M Hokazono, R Shirataki
012008 Thermoelectric Properties of Single Crystalline $Ba_8Al_xSi_{46-x}$ Clathrate by Using Flux Czochralski Method ... 47
Y Nagatomo, N Mugita, Y Nakakohara, M Saisho, M Tajiri, R Teranishi, S Munetoh
012009 Synthesis and Thermoelectric Properties of P-type $Ba_8Au_xSi_{46-x}$ Clathrate ... 53
M Saisho, L Bin, Y Nagatomo, Y Nakakohara, R Teranishi, S Munetoh

FUNCTIONAL MATERIALS

012010 The Healing Effect of Stearic Acid Applied on Amorphous Carbon Film with Dispersed Nanodiamonds ... 59
T Yamazaki, K Shoji, T-T Ikeshoji, A Suzumura
012011 Room-temperature Fast Deposition and Characterization of Nanocrystalline $Bi_{0.4}Sb_{1.6}Te_3$ Thick Films by Aerosol Deposition ... 69
S Baba, L Huang, H Sato, R Funahashi, J Akedo
012012 First-principles Study on Oxygen Ion Conduction of La_2GeO_5 Based on the Density Functional Theory ... 76
M Sakaue, W T Cahyanto, W T D Kencana, S M Aspera, K Miwa, H Kishi, S Kunikata, H Nakanishi, W A Diño, H Kasai, T Ishihara
012013 Ferromagnetic Nanostructures of Oxygen on Ag(111) ... 84
Y Kunisada, H Nakanishi, W A Diño, H Kasai
012014 Dispersion State and Oxygen Storage Capacity Properties of Ceria and Zirconia Nanoparticles Supported on Alumina by the Impregnation Process ... 90
M Ozawa, K Matuda, Y Miura, M Hattori, T Okouchi, H Yuzuriha, T Sugimoto, M Haneda
012015 Luminescence Properties of $M_2Si_5N_8:Ce^{3+}$ (M = Ca, Sr, Ba) Mixed Nitrides Prepared by Metal Hydrides as Starting Materials ... 95
D Kuramoto, H S Kim, T Horikawa, M Itoh, K Machida
012016 Luminescence Properties of $CaAlSiN_3:Eu^{2+}$ Mixed Nitrides Prepared by Carbothermal Process ... 100
H S Kim, T Horikawa, H Hanzawa, K Machida

ELECTRONICS

012017 Discrimination of Wireless Electromagnetic Signals by Electro-optic Modulators Using an Array of Patch Antennas Embedded with Orthogonal Gaps 104
Y N Wijayanto, H Murata, Y Okamura

012018 Phase Analysis and Optical Evaluation of Ceria-zirconia-terbia Prepared by Coprecipitation Method 114
M Amimoto, M Haneda, M Ozawa

012019 Clock Field in Arrayed Magnetic Logic Gates with a Magnetic Force Microscope Tip 118
H Nomura, Y Imanaga, Y Hiratsuka, R Nakatani

012020 Effect of Interfacial Reaction on Electrical Conduction Across the Interface Between N-type Gallium Nitride and Contact Materials 122
M Maeda, T Yamasaki, Y Takahashi

012021 Electrical Properties and Structure of Contact Interface Between Ti_3SiC_2 and P-type GaN 130
A B M Halil, M Maeda, Y Takahashi

012022 Optimization of Control Parameters of a Boost Converter for Energy Harvesting 136
M Tanaka, P Hemthavy, K Takahashi

012023 Optimum Condition of Boost Switching Regulator for Charging Tiny Electric Energy to Capacitor 142
Y Honda, M Tanaka, Y Sekiguchi, P Hemthavy, K Takahashi

ELECTRONICS PACKAGING AND MICRO JOINING

012024 Effect of PvP on the Low Temperature Bonding Process Using Polyol Prepared Ag Nanoparticle Paste for Electronic Packaging Application 147
J Yan, G Zou, A Wu, J Ren, J Yan, A Hu, L Liu, Y N Zhou

012025 Study on Erosion Resistance Characteristics of Fe-MWCNT Composite Plating with Respect to Lead-free Solder 153
J Watanabe, N Sekimori, K Hatsuzawa, T Uetani, I Shohji

012026 Bonding of Copper to Silicon Chips Using Vapor-deposited Tin Film 163
T Fujimoto, S Fukumoto, T Miyazaki, Y Kashiba, K Shiotani, K Fujimoto

012027 Microstructural Development at Weld Interface Between Zr-based Glassy Alloy and Stainless Steel by Resistance Microwelding 170
S Fukumoto, M Minami, A Soeda, M Matsushima, M Takahashi, Y Yokoyama, K Fujimoto

012028 Numerical Analysis of Interfacial Deformation and Temperature Rise During Ultrasonic Al Ribbon Bonding 178
Y Takahashi, S Suzuki, Y Ohyama, M Maeda

012029 A Crack Initiation and Propagation Simulation and the Fatigue Characteristics of Solder Joints Considering the Material Property Changes 189
M Matsushima, Y Shishihara, H Matsunami, S Fukumoto, K Fujimoto

SMART PROCESSING

012030 Study on the Low Energy & Pollution Manufacturing of Micro Cutting Tools by Powder Injection Molding Process 197
Y-M Heo, G-H Kim, S-H Chang, G-A Lee

012031 Characterization of Inductively Coupled RF Plasmas for Plasma-assisted Mist CVD of ZnO Films 206
K Takenaka, Y Okumura, Y Setsuhara

012032 Temperature Dependence for Nitridation of Aluminum Films by Sub-atmospheric Pressure Discharge 212
A Saikubo, Y Ohtsuka, Y Ueda

012033 Laser Processing for Strengthening of the Self-restoring Metal-elastomer Interface on a Silicone Sheet 218
K Yasuda

012034 Greener Manufacturing: Superplastic-like Forming 224
J Liu, M-J Tan, A E W Jarfors, S C V Lim, K-S Fong, S Castagne

012035 Laser Micro-machinability of Borosilicate Glass Surface-modified by Electric Field-assisted Ion-exchange Method 228
S Matsusaka, T Kobayakawa, H Hidai, N Morita

APPLIED PHYSICS AND CHEMISTRY

012036 Reactivity of CO_2 and H_2O on TiO_2 Catalysts Studied by Gas Phase FT-IR Method and Deactivation Mechanism 234
Y Y Maruo, T Yamada, M Tsuda

012037 Kinetic study on CO_2 photoreduction by Re complexes 243
Y Ono, M Tsuda, Y Y Maruo, J Nakamura

012038 CO_2 Decomposition Using Electrochemical Process in Molten Salts 250
K Otake, H Kinoshita, T Kikuchi, R O Suzuki

012039 Development of a Selective Electroforming Process for Micro-sized Probe Tips Used in LCD Inspection Machines 258
J S Kim, I K Min, J D Kim

012040 Cluster Model Investigation of Chemical Bonding of Elements in Silica and Water 265
W Takahara

012041 Separation Mechanisms of Reversible Adhesive Joining Using an Elastic Body with Variable Elastic Modulus 269
Y Sekiguchi, P Hemthavy, S Saito, K Takahashi

012042 Time Dependence of Adhesion Contact Between a Rigid Sphere and an Elastic Body 278
P Hemthavy, M F B Samir, S Saito, K Takahashi

BASE MATERIALS

012043 Influence of Scale on Hot Rolling Characteristics of Si-bearing Steel Sheets 285
K Hara, H Utsunomiya, R Matsumoto, T Sakai

012044 Improvement in Rolling Workability of Fe3Al by High-speed Rolling 291
A Muraoka, H Utsunomiya, R Matsumoto, T Sakai

012045 Evaluation of R-value of Steels Using Vickers Hardness Test 297
K Ohashi, H Utsunomiya, R Matsumoto

WELDING AND JOINING

012046 Effect of Columnar Grain Shape Modeling on the Microscopic Stress Distribution at the Clad Weld Metal Surface of Ni Base Alloy 303
T Sakai, Y Mikami, M Mochizuki

012047 Dissimilar Joint Characteristics of SiC and WC-Co Alloy by Laser Brazing 309
K Nagatsuka, Y Sechi, K Nakata

012048 Dissimilar Laser Brazing of h-Bn and WC-Co Alloy in Ar Atmosphere Without Evacuation Process 318
Y Sechi, K Nagatsuka, K Nakata

012049 a Study for High Accuracy Measurement of Residual Stress by Deep Hole Drilling Technique 328
H Kitano, S Okano, M Mochizuki

012050 Residual Stress Variation Due to Piping Processes of Austenitic Stainless Steel 338
R Ihara, T Hashimoto, M Mochizuki

012051 Effect of Electrode Force Condition on Nugget Diameter and Residual Stress in Resistance Spot Welded High-strength Steel Sheets 346
M Iyota, Y Mikami, T Hashimoto, K Taniguchi, R Ikeda, M Mochizuki

012052 Surface Modification of Ferritic Stainless Steel by Active Screen Plasma Nitriding 355
H Nii, A Nishimoto

012053 Comparison of Weld Residual Stress Measurement Results in Low Alloy Welds Between X-ray Diffraction and Stress Relief Methods 362
A Sugahara, S Okano, T Hashimoto, M Mochizuki

012054 Numerical Simulation of the Effect of Low-temperature Transformation Expansion on Residual Stress in Cold Cracking Test Specimens of Different Restraint Factors 369
N Kubota, Y Mikami, M Mochizuki, K Hiraoka

012055 Investigation of Factors Influencing Welding Deformation of Ship Block by Inherent Strain Analysis Using Idealized Explicit FEM 376
M Hata, S Itoh, T Sugihiro, Y Yamashita, A Kamio, M Shibahara, M Mochizuki
Author Index

ECO-MATES 2011

International Symposium on Materials Science and Innovation for Sustainable Society
Eco-Materials and Eco-Innovation for Global Sustainability
The 21st Iketani Conference 2011

Preface

The 21st century has been called the century of environmental revolution. Green innovations and environmentally friendly production systems based on physics, chemistry, materials science, and electronic engineering will be indispensable for ensuring renewable energy and establishing a sustainable society. In particular, production design, materials processing, and fabrication technologies such as welding and joining will be very important components of such green innovations.

For these reasons, the International Symposium on Materials Science and Innovation for Sustainable Society – eco-materials and eco-innovation for global sustainability – (ECO-MATES 2011) was organized by the Joining and Welding Research Institute (JWRI) and the Center of Environmental Innovation Design for Sustainability (CEIDS), Osaka University.

ECO-MATES 2011 was held at Hotel Hankyu Expo Park, Osaka, Japan from 28–30 November 2011. 435 participants from 20 countries around the world attended the symposium. 149 oral presentations including 60 invited talks and 160 posters were presented at the symposium to discuss the latest research and developments in green innovations in relation to environmental issues.

The topics of the symposium covered all environmentally related fields including renewable energy, energy-materials, environment and resources, waste and biomass, power electronics, semiconductor, rare-earth metals, functional materials, organic electronics materials, electronics packaging, smart processing, joining and welding, eco-efficient processes, and green applied physics and chemistry.

Therefore, 55 full papers concerning green innovations and environmentally benign production were selected and approved by the editorial board and the program committee of ECO-MATES 2011. All papers were accepted through peer review processes. I believe that all the papers have many informative contents.

On behalf of the steering committee of the symposium, I would like to express my sincere appreciation to all the committees and secretariats, authors, participants of ECO-MATES 2011, and everybody involved in the publication of this special issue. It is a great honor for me that the special issue of *Journal of Physics: Conference Series* will contribute to establishing green innovations and a sustainable society.

Yasuo Takahashi
Chairman of ECO-MATES 2011

Journal of Physics: Conference Series **379** (2012) 011001 doi:10.1088/1742-6596/379/1/011001

ECO-MATES 2011
November 28-30, 2011
Venue: Hotel Hankyu Expo Park, Osaka, Japan

ECO-MATES 2011 ORGANIZATION

Conference Chairman

Yasuo Takahashi (Osaka University)

Organized by

- Joining and Welding Research Institute (JWRI), Osaka University
- Center of Environmental Innovation Design for Sustainability (CEIDS), Osaka University

Co-organized by

- Global COE Program (Center of Excellence for Advanced Structural and Functional Materials Design), Osaka University
- Global COE Program (Center for Electronic Devices Innovation), Osaka University
- Division of Materials and Manufacturing Science, Graduate School of Engineering, Osaka University
- Institute of Scientific and Industrial Research, Osaka University
- Renovation Center of Instruments for Science and Technology, Osaka University
- Photon Pioneers Center, Osaka University
- Innovation Division, Office for University- Industry Collaboration, Osaka University
- Institute of Multidisciplinary Research for Advanced Materials, Tohoku University
- Iketani Science and Technology Foundation (ISTF)

Cooperated by

- IEEE CPMT Society Japan Chapter Japan Institute of Electronics Packaging (JIEP)
- The Ecomaterials Forum, The Society of Non-Traditional Technology

- The Japan Institute of Metals
- The Iron and Steel Institute of Japan
- The Japan Welding Society
- The Thermoelectrics Society of Japan
- High Temperatures Society of Japan
- The Japan Welding Engineering Society
- The Institute of Electronics, Information and Communication Engineers
- The Ceramic Society of Japan
- The Japan Society of Applied Physics
- The Rare Earth Society of Japan

(in no special order)

Steering Committee and Secretariats

Chair:
Yasuo Takahashi (Osaka Univ., Japan)

Vice Chair:
Shinji Fujimoto (Osaka Univ., Japan)
Shun-Ichiro Tanaka (Tohoku Univ., Japan)
Tadatomo Suga (The Univ. of Tokyo, Japan)

Secretariat General:
Masakatsu Maeda (Osaka Univ., Japan)

Planning Advisor:
Toshimi Yamane (Osaka Univ., Japan)

Organizing Committee:

M. Enoki (The Univ. of Tokyo)
K. Fujimoto (Osaka Univ.)
S. Fujimoto (Vice Chair, Osaka Univ.)
Y. Fujiwara (Osaka Univ.)
M. Fukumoto (Toyohashi Univ. of Technol.)
H. Fukuyama (Tohoku Univ.)
Y. Fukuzawa (Nagaoka Univ. of Technol.)
T. Funaki (Osaka Univ.)
K. Harada (National Institute for Materials Science)
Y. Hirata (Osaka Univ.)
A. Hirose (Osaka Univ.)
Y. Horikoshi (ISTF)
M. Iketani (ISTF)
H. Inui (Kyoto Univ.)
T. Kakeshita (Osaka Univ.)
S. Katayama (Osaka Univ.)
R. Kodama (Osaka Univ.)
M. Kohyama (AIST)
H. Kokawa (Tohoku Univ.)
T. Koseki (The Univ. of Tokyo)
K. Machida (Osaka Univ.)
N. Miyazaki (Kyoto Univ.)
Y. Miyazawa (Tokai Univ.)
Y. Mori (Osaka Univ.)
J. Morimoto (Kinki Univ.)
H. Murakawa (Osaka Univ.)
T. Nagasaka (Tohoku Univ.)
M. Naito (Osaka Univ.)
H. Nakajima (Osaka Univ.)
K. Nakata (Osaka Univ.)
Y. Ohmori (Osaka Univ.)
M. Ozaki (Osaka Univ.)
Y. Shimoda (Osaka Univ.)
K. Shinozaki (Hiroshima Univ.)
T. Suga (Vice Chair, The Univ. of Tokyo)
K. Suganuma (Osaka Univ.)
R. Suzuki (Hokkaido Univ.)
A. Suzumura (Tokyo Institute of Technol.)
Y. Takahashi (Chair, Osaka Univ.)
J. Takeya (Osaka Univ.)
T. Tanaka (Osaka Univ.)
S. Tanaka (Vice Chair, Tohoku Univ.)
K. Taniguchi (Nara National College of Technol.)
T. Tomimura (Kumamoto Univ.)
Y. Tsuda (ISTF)
Y. Umeda (Osaka Univ.)
K. Yamamoto (Osaka Univ.)
T. Yamane (Osaka Univ.)

T. Yoshida (The Univ. of Tokyo)

Executive Committee:

N. Abe (Osaka Univ.)
H. Fujii (Osaka Univ.)
K. Hara (Osaka Univ.)
T. Ikeshoji (Tokyo Institute of Technol.)
M. Ito (Osaka Univ.)
Y. Iwata (Osaka Univ.)
H. Kajii (Osaka Univ.)
S. Katsuyama (Osaka Univ.)
Y. Kawahito (Osaka Univ.)
S. Kirihara (Osaka Univ.)
T. Machimura (Osaka Univ.)
M. Maeda (Osaka Univ.)
S. Matsusaka (Chiba Univ.)
Y. Miura (Osaka Univ.)
Y. Mori (Osaka Univ.)
H. Nishikawa (Osaka Univ.)
A. Nishimoto (Kansai Univ.)
K. Saida (Osaka Univ.)
T. Sano (Osaka Univ.)
T. Shibayanagi (Osaka Univ.)
K. Takahashi (Tokyo Institute of Technol.)
Y. Takahashi (Chair, Osaka Univ.)
M. Tanaka (Osaka Univ.)
S. Tashiro (Osaka Univ.)
H. Terasaki (Osaka Univ.)
H. Tsuchiya (Osaka Univ.)
M. Tsukamoto (Osaka Univ.)
T. Tsumura (Osaka Univ.)
J. Umeda (Osaka Univ.)
M. Uwasu (Osaka Univ.)
Y. Yamaguchi (Osaka Univ.)
K. Yasuda (Nagoya Univ.)

National Advisory Board:

E. Arai (Osaka Univ.)
T. Araki (Ehime Univ.)
H. Araki (Osaka Univ.)
T. Ise (Osaka Univ.)
K. Itoh (Osaka Univ.)
Y. Kanematsu (Osaka Univ.)
E. Kasai (Tohoku Univ.)
M. Kato (Tokyo Institute of Technol.)
Y. Komizo (Osaka Univ.)
K. Kondo (Osaka Univ.)
T. Masaki (Osaka Univ.)
F. Minami (Osaka Univ.)
Y. Minamino (Osaka Univ.)
M. Mochizuki (Osaka Univ.)
K. Nishimoto (Osaka Univ.)
F. Saito (Tohoku Univ.)
R. Satoh (Osaka Univ.)
J. Suzuki (Mie Univ.)
J. Takeda (Osaka Univ.)
T. Takemoto (Osaka Univ.)
A. Tokai (Osaka Univ.)
T. Usui (Fukui Univ. of Technol.)
T. Watanabe (Niigata Univ.)
S. Yamanaka (Osaka Univ.)
H. Yamashita (Osaka Univ.)

International Advisory Board:

H. K. D. H. Bhadeshia (Univ. of Cambridge, UK)
W. S. Chang (Research Institute of Industrial Sci. & Technol., Korea)
B. Chang (Tsinghua Univ., China)
U. Dilthey (RWTH Aachen Univ., Germany)
N. Ekere (Univ. of Greenwich, UK)
J. C. Feng (Harbin Institute of Technol., China)
T. Gang (Harbin Institute of Technol., China)
M. Goosey (Loughborough Univ., UK)
M. Howarth (Sheffield Hallam Univ., UK)

J. E. Indacochea (Univ. Illinois, Chicago, USA)
S. B. Jung (Sungkyunkwan Univ., Korea)
J. P. Jung (The Univ. of Seoul, Korea)
C. Kang (Pusan National Univ., Korea)
Y. S. Kim (Korea Maritime Univ., Korea)
C. W. Lee (Korea Institute of Industrial Technol., Korea)
Y. C. Lee (Univ. Colorado, Boulder, USA)
C. A. Lewinsohn (Ceramatec, USA)
C. Li (Xi'an Jiao Tong Univ., China)
M. Mayer (Univ. Waterloo, Canada)
J. U. Park (Chosun Univ., Korea)
J. C. Schuster (Univ. of Wien, Austria)
W. C. Seo (Pukyong National Univ., Korea)
Y. Shin (Chung-Ang Univ., Korea)
Y. B. Sun (Kyonggi Univ., Korea)
W. Tillmann (Institute of Materials Eng., Germany)
D. Zhang (Shanghai Jiao Tong Univ., China)
K. Zhang (Hanan Univ. of Sci. and Technol., China)
N. Y. Zhou (Univ. Waterloo, Canada)
Z. Sun (Singapore Institute of Manufacturing Technol., Singapore)

Program Committee:

T. Ikeshoji (Tokyo Institute of Technol.)
M. Itoh (Osaka Univ.)
H. Kajii (Osaka Univ.)
S. Katsuyama (Osaka Univ.)
T. Machimura (Osaka Univ.)
M. Maeda (Chair, Osaka Univ.)
Y. Miura (Osaka Univ.)
Y. Takahashi (Osaka Univ.)
S. Tashiro (Tohoku Univ.)
H. Tsuchiya (Osaka Univ.)
J. Umeda (Osaka Univ.)
K. Yasuda (Vice Chair, Nagoya Univ.)

Conference Secretariats:

Y. Nakagawa (Osaka Univ.)
A. Ou (Osaka Univ.)

Editorial Board:

N. Abe (Osaka Univ.)
H. Fujii (Osaka Univ.)
S. Fujimoto (Vice Chair, Osaka Univ.)
Y. Fujiwara (Osaka Univ.)
M. Fukumoto (Toyohashi Univ. of Technol.)
S. Fukumoto (Osaka Univ.)
T. Funaki (Osaka Univ.)
K. Hara (Osaka Univ.)
A. Hiraki (Osaka Univ.)
A. Hirose (Osaka Univ.)
T. Ikeshoji (Tokyo Institute of Technol.)
M. Ito (Osaka Univ.)
Y. Iwata (Osaka Univ.)
H. Kajii (Osaka Univ.)
S. Katsuyama (Osaka Univ.)
Y. Kawahito (Osaka Univ.)
S. Kirihara (Osaka Univ.)
M. Kohyama (AIST)
K. Kondoh (Osaka Univ.)
K. Machida (Osaka Univ.)
T. Machimura (Osaka Univ.)
M. Maeda (Osaka Univ.)
K. Matsu (Tokyo Braze Co., Ltd.)
S. Matsusaka (Chiba Univ.)
Y. Miura (Osaka Univ.)
Y. Miyazawa (Tokai Univ.)
Y. Mori (Osaka Univ.)
H. Nakajima (Osaka Univ.)
H. Nishikawa (Osaka Univ.)
A. Nishimoto (Kansai Univ.)
Y. Ohmori (Osaka Univ.)

Journal of Physics: Conference Series **379** (2012) 011001 doi:10.1088/1742-6596/379/1/011001

K. Saida (Osaka Univ.)
T. Sano (Osaka Univ.)
T. Shibayanagi (Osaka Univ.)
K. Suzuki (Tokyo Univ. of Science, Yamaguchi)
R. Suzuki (Hokkaido Univ.)
W. Takahara (Osaka Univ.)
K. Takahashi (Tokyo Institute of Technol.)
Y. Takahashi (Chair, Osaka Univ.)
M. Tanaka (Osaka Univ.)
S. Tanaka (Vice Chair, Tohoku Univ.)
T. Tanaka (Osaka Univ.)
S. Tashiro (Osaka Univ.)
H. Terasaki (Osaka Univ.)
T. Tomimura (Kumamoto Univ.)
H. Tsuchiya (Osaka Univ.)
M. Tsukamoto (Osaka Univ.)
T. Tsumura (Osaka Univ.)
J. Umeda (Osaka Univ.)
M. Uwasu (Osaka Univ.)
Y. Yamaguchi (Osaka Univ.)
K. Yasuda (Nagoya Univ.)

Peer review statement

All papers published in this volume of *Journal of Physics: Conference Series* have been peer reviewed through processes administered by the proceedings Editors. Reviews were conducted by expert referees to the professional and scientific standards expected of a proceedings journal published by IOP Publishing.

The Photovoltaic power generation era is coming (Global energy network equipped with solar cells and international superconductor grids: GENESIS)

Yukinori Kuwano

Photovoltaic Power Generation Technology Research Association

Oshima Building 5th Floor, 4-7-13 Kudan-Minami, Chiyoda-ku, Tokyo 102-0074, Japan

E-mail: pvtec-jimu@pvtec.or.jp

Abstract. Solar cells are considered a critical technology for overcoming global environmental problems and energy problems. This paper reviews the development history, current status, and application technologies of solar cells. The paper concludes with a discussion on the prospects of a future global-scale energy supply system based on solar cells.

1. Introduction

Solar energy is an ideal form of energy because it is clean, inexhaustible, and available everywhere in the world. The Earth receives 170 billion MW of solar energy from the Sun. Among the methods available for utilizing the Sun's energy, the most prominent method is solar cells, which use the photovoltaic (PV) effect of semiconductors to convert solar light energy to electrical energy. In this paper, the author will review the steps and status of solar cell development before going into solar cell applications. Then the author will cover future prospects by talking about a global energy supply system comprised of solar cells

2. Need for clean energy

2.1. Dwindling energy resources

The consumption of fossil fuels has reached a turning point from the standpoint of confirmed global energy reserves, as shown in Figure 1.

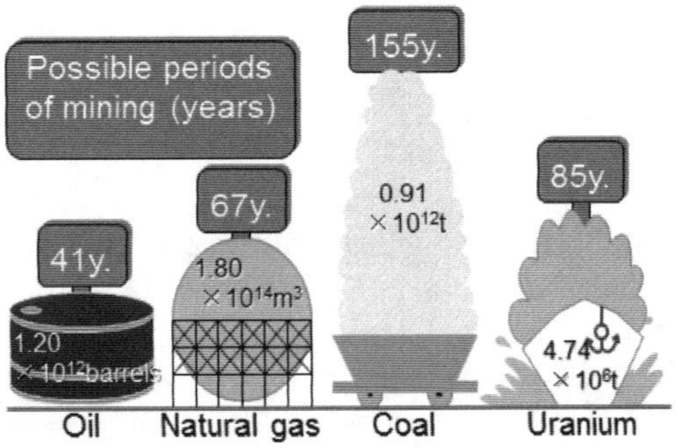

Figure 1. Reserve of various energy sources.

If we continue to consume energy at current levels, we will exhaust all fossil fuel resources on the Earth within 200 years. Not only is there a limit to the ultimate amount of reserves, but predictions hold that the future population will increase, as shown in Figure 2. Therefore, we can expect an explosive increase in the amount of energy consumed, shown in Figure 3, as the population of the world and energy consumption per capita increase. In 2020–2030, the gap between the energy required for humankind and the production volume of fossil fuels (called "energy gap") will grow. As a result, we need new energy sources.

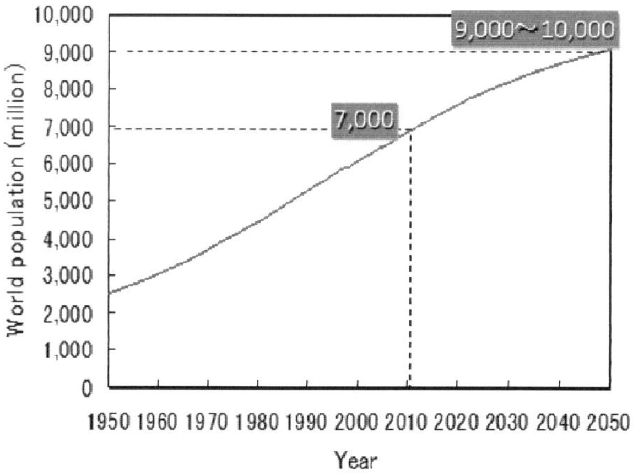

Figure 2. Predicted future world population.

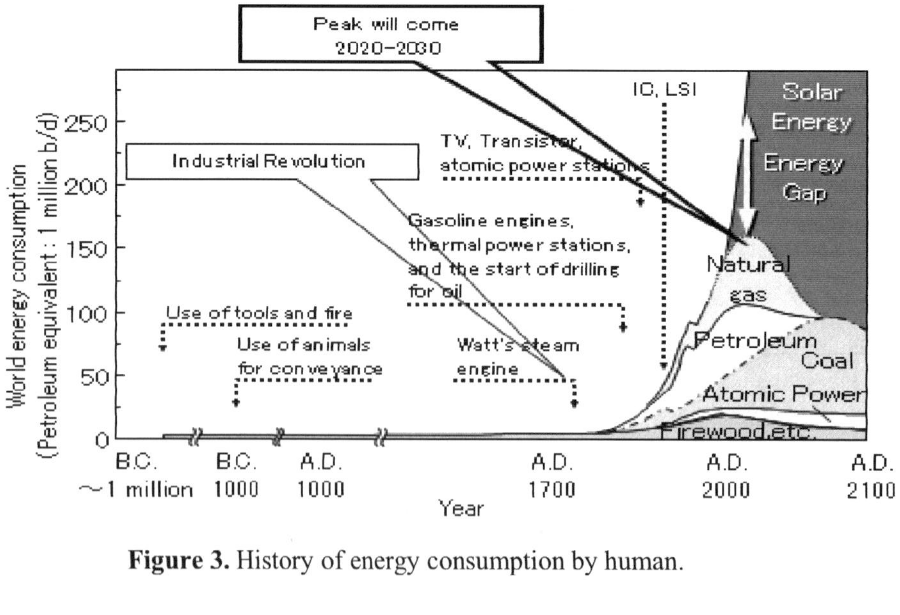

Figure 3. History of energy consumption by human.

2.2. Solar energy as a new source of clean energy

In view of global environmental problems and dwindling energy resources, we must develop new energy sources that are both abundant and safe as a substitute for fossil fuels. In this respect, solar energy is an ideal form of energy because it is clean, inexhaustible and available everywhere in the world. The Earth receives 170 billion MW of solar energy, as shown in Figure 4. This quantity is enormous that one hour's worth of solar energy could provide the annual energy needed for the entire world.

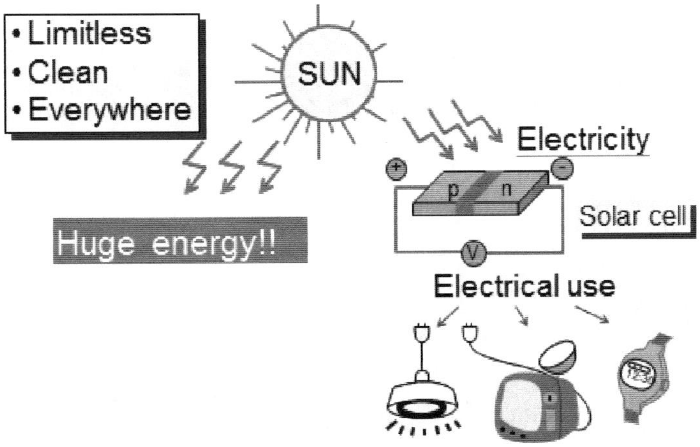

Figure 4. Uses of solar energy.

Among the methods available for utilizing solar energy, the most prominent method is solar cells, which use the photovoltaic effect (PV effect) of semiconductors to convert solar light energy to electrical energy, as shown in Figure 4. When sunlight enters a semiconductor which has a p-n junction, a hole with a positive charge and an electron with a negative charge are produced. These are separated along the p-n junction, and the positive and negative charges collect at both electrodes. When these two electrodes are connected, electric current flows and work is performed. Since sunlight is used as an energy source, the result is a power generating element that does not require fossil fuels, does not produce exhaust gas, and does not have moving parts. In addition to providing the many features of solar energy, solar cells also offer the following.

①They provide electrical energy directly.

②The conversion efficiency is the same irrelevant of the scale of power generation, that is, whether it is 1 W or 1 MW.

③The service life is basically semipermanent because there are no movable parts.

Finally, silicon—the main material comprising solar cells—is the second most abundant element on the Earth, so there is absolutely no problem in resource availability.

2.3. The recent massive earthquake and PV

Recently, a massive earthquake occurred in the sea off East Japan. The earthquake and ensuing tsunami damaged nuclear power plants, resulting in serious accidents. We must propose to mankind a new energy source that will replace nuclear power. Solar cells can surely be considered a critical technology for overcoming global environmental problems and energy problems.

3. Progress in solar cells

3.1. History of solar cells

The history of solar cells began with Pearson in the United States in 1954. 1n 1958, solar cells were sent up in a satellite. Since then, they have been used in lighthouses and other locations, but their high price prohibits them from achieving widespread use. Following the so-called “oil shocks” of 1973, however, the superior characteristics of solar cells received attention as a substitute source of energy. Technical development was pursued in the United States and in each country of the EU under the supervision of the National Renewable Energy Laboratory in Japan as part of the Sunshine Project. The history of solar cells is shown in Figure 5. The first step is called "the dawn of solar power generation." The second step is called "advances to supply electric power." The third step is called "world-scale spread of solar power generation." We are currently in the world-scale PV era.

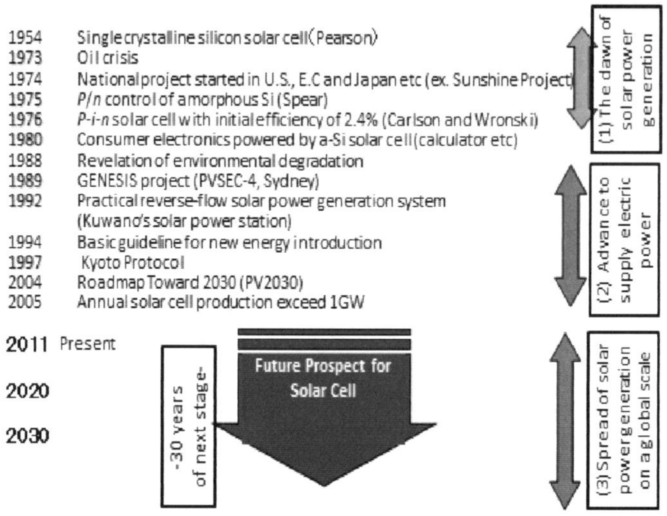

Figure 5. History of the solar cell.

3.2. Status of solar cells

3.2.1. Types of solar cells There are many different types of solar cells depending on the material used (silicon, compound semiconductors, organic semiconductors, etc.) and the shape of the crystals of the material (single crystal, polycrystalline, amorphous, etc., or a combination). Silicon is the typical material use for solar cells. Some representative methods for manufacturing Si solar cells are shown in Figure 6.

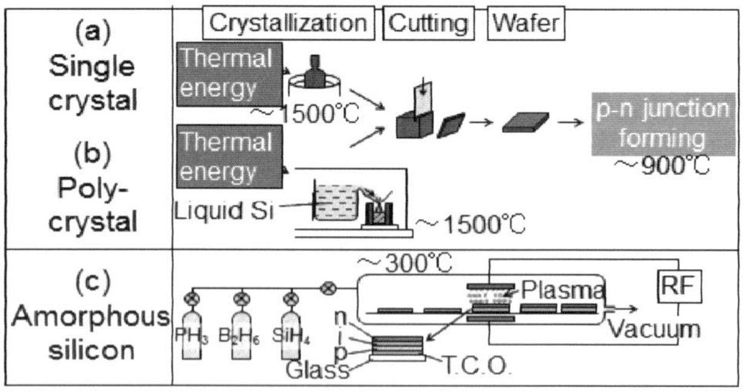

Figure 6. Comparison of fabrication methods of various solar cells.

3.2.2. Improvements in conversion efficiency The most important factor determining the performance of solar cells is the efficiency of energy conversion. In the beginning, the efficiency of solar cells was low, but conversion efficiency was improved by the efforts of many people, as shown in Figure 7. This data is based on the data of the National Renewable Energy Laboratory. The practical module used as electricity was improved to 13%–20% in the crystalline Si module, 6%–11% in the film Si module, 8%–12% in the compound (copper indium gallium selenide: CIGS) module.

3.2.3. Production volume and cost The world production volume of solar cells increased to approximately 24 GW in 2010 (Figure 8)[1)]. The world PV cumulative capacity was approximately 35 GW in 2010.

The sharp decrease in cost is a consequence of the increased production volume and manufacturing technology developments (Figure 9).

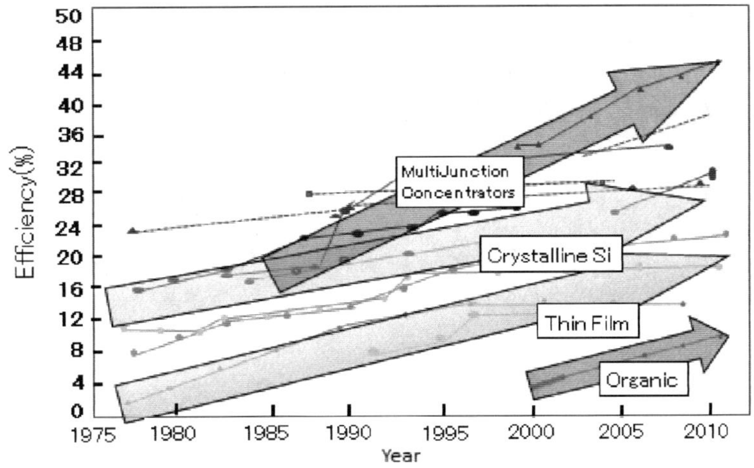

Figure 7. Progress in conversion efficiency.

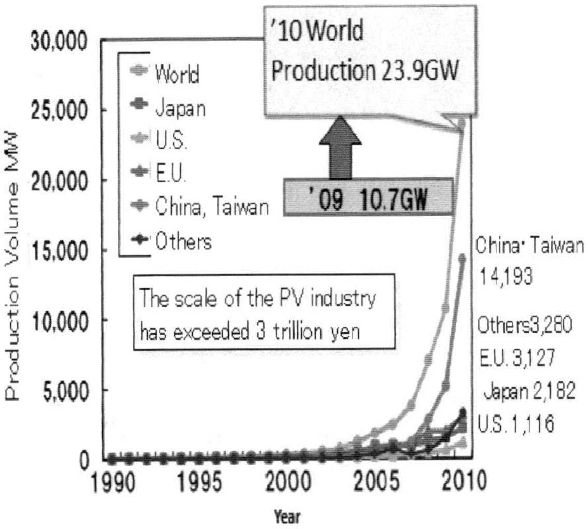

Figure 8. Progress in production volume of solar cells.

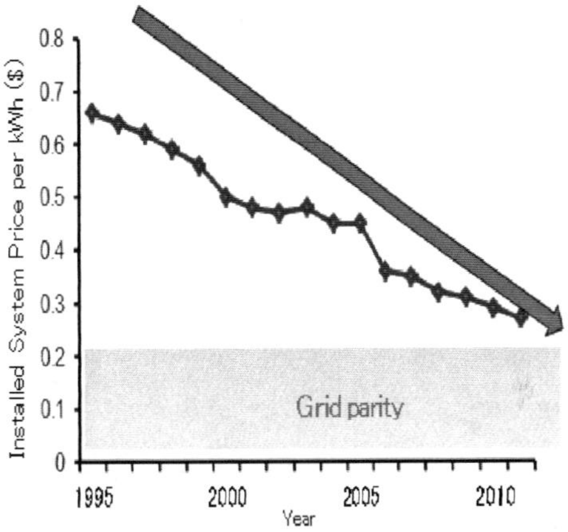

Figure 9. Tendency of installed system prices.

4. Applications and systems using solar cells

4.1. Use of solar cells in consumer electronics

First, solar cells were used as a remote power supply, such as power supplies for satellites and lighthouses. Then, solar cells were applied as a power supply for various kinds of electric products, such as electronic calculators, wristwatches and radios by Japanese electric companies. Figure 10 shows various types of electronic products powered by a-Si solar cells. These solar powered pocket calculators were put on the market in 1980. They were the world's first practical use of the amorphous silicon solar cells.

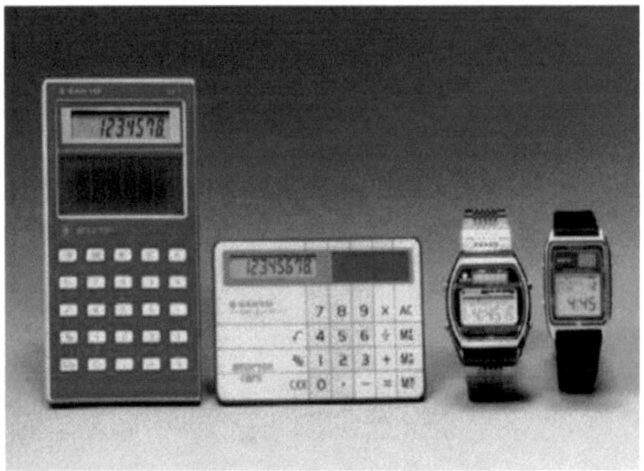

Figure 10. Solar powered calculators and wristwatches.

Figure 11. First practical use of a reverse-flow PV system in a Japanese house.

4.2. Application to power generation systems

In 1992, the regulations for grid connections of solar power generation systems were changed in Japan. PV system owners could sell the electricity produced by the PV system back to the electric power company through the utility grid. Figure 11 shows the first practical use of a reverse-flow PV system in an actual house (the author's house). Reverse flow means that surplus power is fed back to the power system. Figure 12 shows the operation data for the PV system from 1992 to last year (2011), a period of more than 19 years. The peak electricity is stable, and the system still operates well.

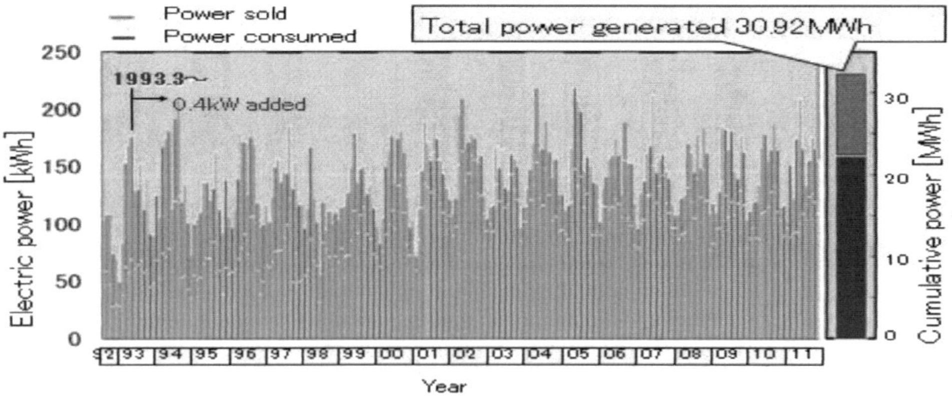

Figure 12. Operating data of the PV System in Figure 11.

Environmental problems have prompted governments around the world to expand plans to promote the PV. A new system called a Feed-in Tariff was established in Germany. It buys electricity generated by

Figure 13. 24 MW PV system in Germany.

PV systems at a high price (3 times the normal price). Similar systems are also spreading to other countries. Large-scale PV generation systems have been installed all over the world. Figure 13 shows a large-scale PV plant in Germany.

4.3. Effects of PV

4.3.1. Energy Payback Time (EPT) One of the most important concepts when evaluating solar cells as a new energy source is energy payback time (*EPT*). *EPT* is the number of years required for solar cell modules to generate the same amount of electric power as was consumed in their fabrication. ETP depends on the conversion efficiency and the production volume of solar cells.

$$EPT = E_0/E_g \, , \qquad (1)$$

where E_0 is the energy needed to produce PV, E_g is the annual energy generated by PV.

The *EPT* is estimated to be approximately 1 year for a-Si solar cells, and 1.5-2 years for poly-Si and single crystal Si solar cells[2)]. In any case, the *EPT*s of these solar cells are much shorter than their lifetime, which is estimated to be longer than 20 years, that is, the solar cells can possibly self-multiply, as shown in Figure 14. Thus, solar cells are quite suitable as a new energy source.

Figure 14. Self-multiplication of solar cells with solar energy.

4.3.2. Reduction of CO_2 emission The PV of 1 KW reduces CO_2 emission to approximately 0.57 ton in one year. Table 1 shows the quantity of reduction of the CO_2 emission according to the scale of the PV system[2)].

Table 1. Dependence of CO_2 reduction on the PV system scale.

Scale of the PV System	Quantity of Reduction of CO_2 Emission (Unit ton)
1KW	0.56
1MW	5.6×10^2
1GW	5.6×10^5

5. What are the challenges for PV systems?

Approximately 60 years have passed since the development of Si solar cells. During this time, their efficiency has increased 4 to 10 times. Their cost has decreased to approximately 1/100. This proves that PV systems can act as an alternative energy source.

However, PV is not yet sufficient to become a basic energy source for people.

Future problems include the following:
(1) The generation cost of the PV system needs to be equivalent to the conventional electricity cost, which is called "grid parity." It is necessary to cut the cost of the PV system by a half or a quarter.
(2) The life of the PV system must be much longer. It should be extended from 20 to 40 years.

6. Future prospects for the GENESIS Project

The author proposed a global energy system for the future of PV was proposed at the 4th PVSEC (held in Sydney, Australia) in 1989[3). The project was named the GENESIS Project. GENESIS stands for a Global Energy Network Equipped with Solar cells and International Superconductor Grids. The project proposes PV systems to be built all around the world, especially in desert areas, and connected to one another by superconductive cables without transmission loss, as shown in Figure 15.

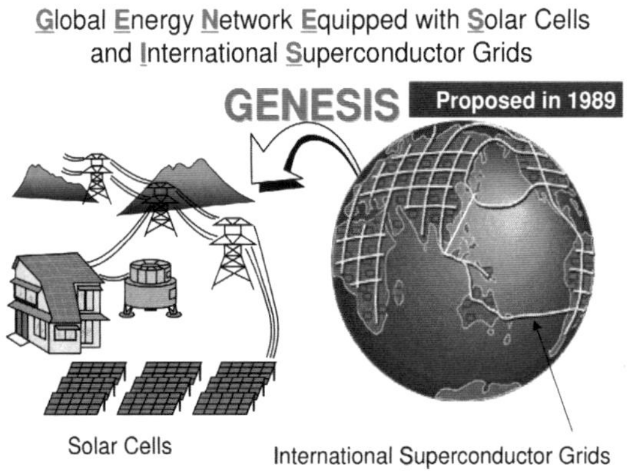

Figure 15. The GENESIS Project.

The shortcoming of the PV system is that it does not generate power at night. However, because the sun is always shining somewhere in the world, the electric power can be transported from the day-time areas to the night-time areas by an international network. Table 2 shows the world's energy consumption and the required PV system area. It was estimated that in 2010, the world's energy consumption will require approximately 14 billion kl of oil per year. Assuming 10% PV system efficiency, an area approximately 800 km square would be required. This area is not very large and equals to only 4% of the world's total desert area.

Table 2. World energy consumption and the required solar cell system area.

Year	Energy consumption (billion kl/y.) :oil equivalent	System efficiency (%)	System area (km square)
2000	11	10	729
2010	14	10	802
2050	35	15	1,030
2100	111	15	1,850

Implementing this project is not unrealistic. The three steps shown in Figure 16 reflect our view for realizing this project.

Step 1: PV systems are installed in homes and factories, and these are connected to an electric power grid. 35 GW of the global PV systems are connected to the utility grid. Note that the first step of the GENESIS Project has already started.

Step 2: Each country's transmission lines are subsequently connected to create an international network. In Europe and other continental regions, electric power supply grids have already been connected to those in other countries.

Step 3: Many large-scale PV plants are built in the desert to expand the international country network, and thus creating a global network.

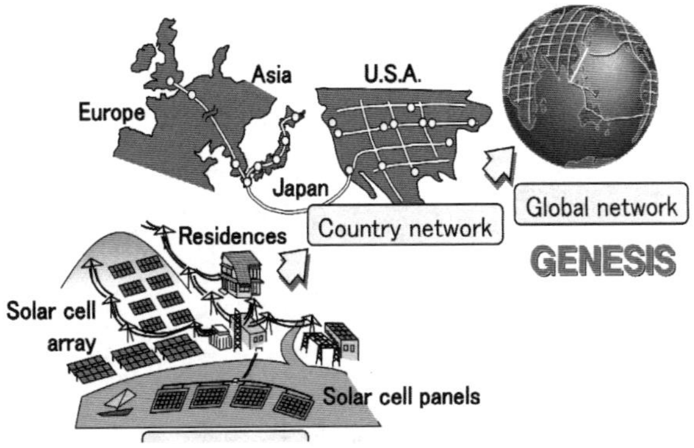

Figure 16. Steps in the GENESIS project.

Not only the solar cells, but also superconductive cables are important for the GENESIS project. A breakthrough occurred 5 years ago in this field when Bi-based high-temperature superconducting wires were developed by Sumitomo Electric Industries, Limited[4)].

These wires have been drastically improved since then. The critical current exceeds 250 A, and the cable length exceeds 1 km. In fact, in 2006, a superconductive cable was connected to a practical operating power grid in the United States.

7. Conclusions

The serious nuclear power plant accident has prompted people to reaffirm the need for safe, clean energy. PV systems are the leading candidate as a new, clean energy source. In order to resolve our energy problems, we must build a global PV system.

References

[1] Krulewitz A 2011 *PVNEWS* **30**, Number 5, 5
[2] Kuwano.Y 2011 *The development history of solar cells* (Tokyo, Ohomsha) p384
[3] Kuwano Y 1989 *Proc. 4th PVSEC Sydney*, II 40－47
[4] Sato K 2007 *TEION KOGAKU* **42** 338

Unique features of cylindrical type solar-module contrasted with plane or conventional type ones

Hirohisa Hiraki[1], Akio Hiraki[2], Masakatsu Maeda[1] and Yasuo Takahashi[1]

[1]Osaka University JWRI, Incubation Building, 2-1 Yamadaoka, Suita, Osaka 565-0871, Japan

[2]Osaka University, Incubation Building, 2-1 Yamadaoka, Suita, Osaka 565-0871, Japan

E-mail: hirahirohisa@yahoo.co.jp

Abstract. Due to their shape and construction, Cu(In,Ga)Se_2 [CIGS] cylindrical photovoltaic [PV] panels have unique features that planar or conventional PV panels do not have. For example, a) they capture sunlight over an angular range of 360°, b) they are self-ballasting (no roof penetrations, no attachments required) and c) they are high-reliability hermetically sealed cylindrical packages. In field tests in Japan, cylindrical PV panels have proven their durability against typhoons, and also their excellent electrical properties. Finally, as a new application of cylindrical PV panels, we suggest the new concept of fusion or combination of PV generation and agriculture.

1. Introduction

PV generation is an especially promising technology among renewable energies. Although crystalline silicon PV cells are currently the dominant technology, CIGS PV cells have been researched and commercialized in many countries like Germany, USA and Japan. This is because CIGS PV cells show the highest energy conversion efficiency of thin film PV cells, and the production costs of CIGS PV cells can be lower than those of crystalline silicon PV cells in principle. The strong points of CIGS PV cells are as follows: (a) high power output, (b) high reliability (no light-induced degradation), (c) wide spectral response, (d) strong spectral absorption, (e) possibility of low production cost.

Solyndra, the US solar panel maker, has adopted CIGS cylindrical type PV modules and succeeded in commercializing them. Solyndra's PV panels are widespread in European countries and the USA, but are just entering the market in Japan. As depreciation in the price of PV systems has accelerated recently, it is difficult to spread PV systems without any unique features. To spread CIGS cylindrical PV panels in Japan, it seems the following 3 points are needed:

(1) Having incomparable unique features.
(2) Positive proof through field tests in Japan.
(3) A new concept other PV do not have.

This paper describes the above-mentioned items (1)-(3).

Published under licence by IOP Publishing Ltd

2. Unique features of cylindrical PV panel

Figure 1 shows the appearance of Solyndra's cylindrical PV panel. Due to its shape and construction, the cylindrical PV panel has many unique features that planar or conventional PV panels do not have. The main features of the cylindrical PV panel are as follows.

(a) Captures sunlight from an angular range of 360°

Cylindrical modules capture sunlight across a 360° photovoltaic surface capable of converting direct, diffuse and reflected sunlight into electricity as shown in Figure 2. Due to the cylindrical shape, the photovoltaic surface area can be larger than that of a planar module, so it can be expected to capture more sunlight and to generate more electric power than a planar module by optimum design of the panel structure. Also, cylindrical PV modules have functions of self tracking and can capture more light early and late in the day.

(b) Self-ballasting

The cylindrical construction allows wind to flow through the panel and as a result no additional ballast, attachments or roof penetrations are required in wind up to 57.8 m/s.

(c) High reliability of hermetically sealed cylindrical package

Isolating CIGS films from moisture is very important for long-term stability. Cylindrical modules easily achieve high reliability by hermetic seals at the ends of the cylindrical packages.

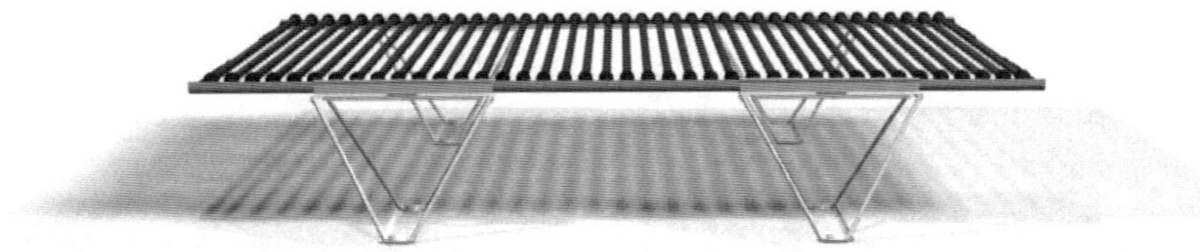

Figure 1. Appearance of the cylindrical PV panel.

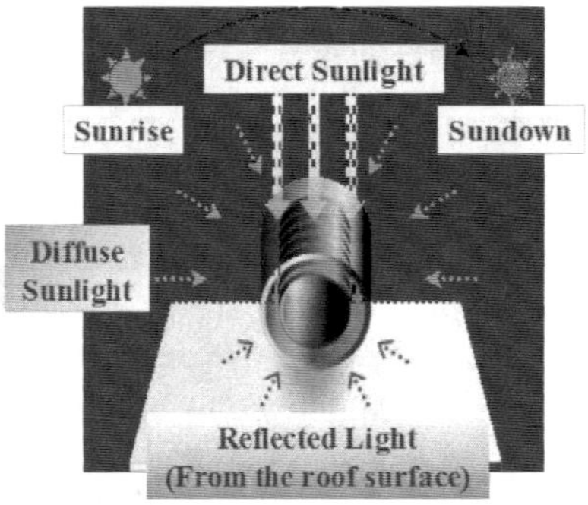

Figure 2. Collection of sunlight.

3. Field tests in Japan

As PV generation needs to match the natural environment, field tests are indispensable. Currently, field tests have been done in some areas in Japan and a large amount of data will be acquired from now on. In the case of Japan, most customers worry about the reliability of self-ballasted installation. At first a wind tunnel test was done in Okinawa in order to win their confidence. We installed cylindrical PV panels when typhoon No. 9 (in 2011) struck Okinawa and checked whether the panels had moved or not. The panels had not moved at all, even though the maximum instantaneous wind speed of typhoon No. 9 was around 50 m/s. Cylindrical PV panels demonstrated positive proof of their durability against wind of 50 m/s through this field test. Figures 3-5 show the appearance of the field test in Okinawa.

Figure 3. Preparation for the field test.

Figure 4. Typhoon No. 9 was passing over Okinawa.

Figure 5. Panels did not move at all.

Next, we report the result of power generation tests at Osaka. Figure 6 shows the correlation between power and irradiance. The nominal power of the cylindrical PV panels is 1 kW. From the test data, these panels generated 650 W when the irradiance was 700 W/m^2. This result was about 95% of the nominal power, so it was a fairly good result.

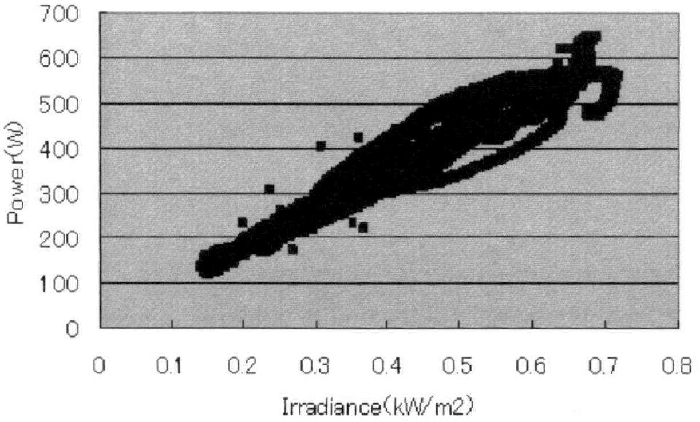

Figure 6. Power vs. Irradiance.

(Term 9:20 a.m ~14:30 p.m. 2011.11.15)

We have also done a comparative study between a cylindrical PV module and a conventional CIGS PV module. Both modules were installed on a flat rooftop at an inclination of 0°. Figure 7 shows the correlation between power and irradiance for the two modules. For the conventional CIGS PV module (yellow dots), there are two curves for the same irradiance from 0.1 kW/m^2 to 0.6 kW/m^2 (this will be discussed later). And can also be seen, the slope of the curve for the cylindrical PV module and the conventional CIGS PV module are slightly different. The slope for the cylindrical one is higher than that for the conventional module. That means the cylindrical PV module can generate higher electric power than the conventional CIGS PV module.

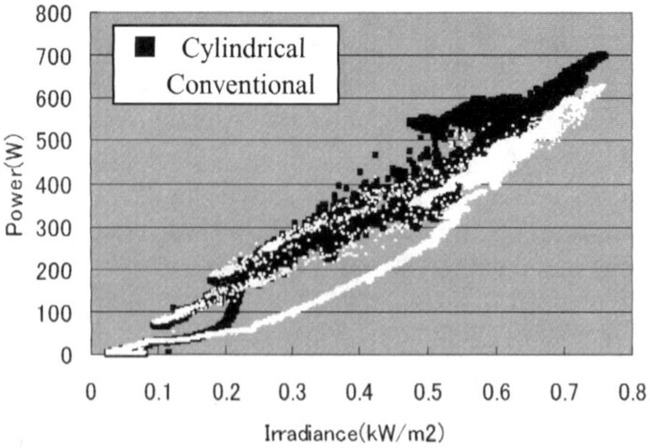

Figure 7. A comparative study between a cylindrical PV module and a conventional CIGS PV module.

4. A new concept of a fusion of PV generation and agriculture

Concerns for PV generation in Japan have jumped after the Great East Japan Earthquake, while the decrease in the agricultural population is getting worse. Creating a new technology that allows farmers to earn sustainable income seems to be one solution to the decrease in the agricultural population. As shown in Figure 8, we suggest a new concept of fusion of PV generation and agriculture by using cylindrical PV panels that do not fully shut out the sunlight, wind and rain that are needed for plant growth. This concept can produce dual functions by using sunlight, that is, harvesting crops and converting sunlight into electricity. This attempt has the potential to relieve the decrease in the agricultural population.

We have already started a demonstration test for this concept at a small test field at Aioi city, Hyogo prefecture (Figure 9). And we have already harvested some crops and generated electric power. But, after this, demonstration tests for many farm products will be needed.

Figure 8. A new concept of PV generation and agriculture.

Figure 9. A small test field at Aioi city.

5. Discussion

As mentioned above, we acquired strange data from a comparative study between a cylindrical PV module and a conventional CIGS PV module. For the conventional CIGS PV module, there are two data patterns at the same irradiances from 0.1 kW/m^2 to 0.6 kW/m^2, while the data for the cylindrical PV module does not show such a strange pattern.

During our speculation about this reason, we paid attention to the difference of module shapes. One module has a cylindrical shape and the other is flat. The cylindrical shape has a function of self tracking and can capture more light early and late in the day. So we expected the power at early times to be different between the two. Figure 10 shows the correlation between power and irradiance from 7:30 a.m. to 9:30 a.m. The power of the cylindrical module is obviously higher than that of the conventional CIGS PV module. Figure 10 supports our speculation. In the case of sunlight from a high angle (from around 10:00 a.m. to 14:30 p.m.), the flat module can also capture sunlight easily, so the conventional module would have two data patterns at the same irradiance from 0.1 kW/m^2 to 0.6 kW/m^2.

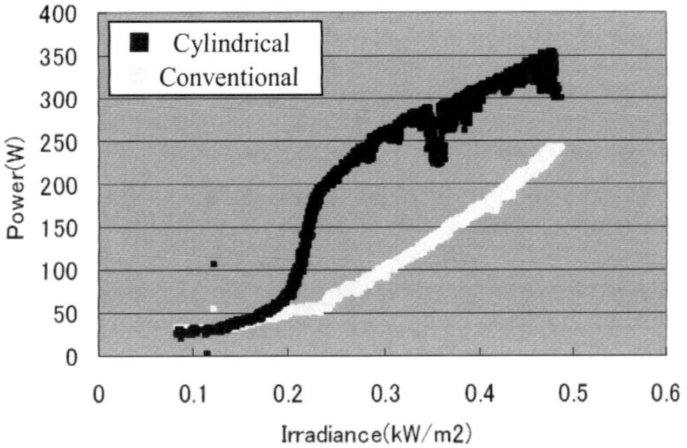

Figure 10. Correlation between power and irradiance from 7:30 a.m. to 9:30a.m.

Journal of Physics: Conference Series **379** (2012) 012002 doi:10.1088/1742-6596/379/1/012002

6. Conclusions

(a) The cylindrical PV panel has many unique features compared to conventional plane PV panels.
(b) The cylindrical PV panel has its own applications such as agricultural fields.
(c) The abilities of the cylindrical PV panel have been proven through field tests in Japan.

References

[1] Kunioka A, M.Mizutani, Hagiwara Y and Nakada T 2001 *Solar Energy Materials and Solar Cells* **67** 255-260.
[2] Morad R US Patent No.US2011/0045674 A1 Pub.Date: Feb.24.2011 Method and Apparatus for Inline Deposition of Materials on A Non-Planar Surface,
[3] Diaz J C and Fitzgerald M 2011 Tubular Glass Solar Energy Advancement *Ceramic Industry* **161** 812-13

Local photovoltaic characterization for silicon thin film solar cells using a scanning probe microscope

T Itoh[1], H Natsuhara[1] and S Nonomura[2]

[1] Department of Electrical and Electronic Engineering, Gifu University, 1-1 Yanagido, Gifu 501-1193, Japan
[2] Environment Renewable Energy System Division, Gifu University, 1-1 Yanagido, Gifu 501-1193, Japan

E-mail: itoh@gifu-u.ac.jp

Abstract. The photovoltaic characterization of silicon thin film solar cells with nano-scale spatial resolution is demonstrated by a new characterization method developed using a scanning microscope. In the surface topography of p-i-n type amorphous silicon thin film solar cells, large and small convex grains corresponding to the textured surface of an Asahi U-type substrate and the crystalline grains in the n-type microcrystalline silicon layer are found. In the local photo-current image, distribution in local photo-current correlated with the structure of the n-type microcrystalline silicon layer is observed in the p-i-n type amorphous silicon (a-Si:H) thin film solar cells. The local surface potential in amorphous silicon thin film solar cells with light irradiation is also evaluated. In the local surface potential image of the p-i-n a-Si:H thin film solar cells without light irradiation, the local surface potential on the large convex grains found in the surface topography is smaller than that in the concave region between the large convex grains. Similar distribution of the local surface potential is observed in the local surface potential images with light irradiation. The local surface potential difference between the large convex grains and the concave region decreases with increasing power of the irradiation light.

1. Introduction

Microcrystalline materials such as hydrogenated microcrystalline silicon (μc-Si:H) are used as a photo-absorption layer and doped layer materials in Si thin film solar cells such as μc-Si:H and amorphous Si (a-Si:H) thin film solar cells. The microcrystalline materials have a hetero-structure composed of an amorphous phase, nano-sized crystalline grains, and grain boundaries. Therefore, local properties in the nano-scale range are different. Typical Si thin film solar cells are deposited on glass substrates coated with textured transparent conductive oxide (TCO). The properties of a-Si:H and μc-Si:H films are affected by the preparation on the textured TCO [1, 2]. Based on these points, the local photovoltaic performance of Si thin film solar cells in the nano-scale range would be different. Consequently, the obtained efficiency is the combination of the local photovoltaic performances. In order to improve the photovoltaic performance of silicon thin film solar cells, understanding the local properties is important. However, the photovoltaic performance for the Si thin film solar cells has been mostly evaluated by macroscopic measurements.

Light-beam-induced current (LBIC) is used in the measurement of local characterization in the polycrystalline silicon solar cells [3]. Compared with the EBIC, higher spatial resolution is necessary in the local characterization of Si thin film solar cells because microcrystalline materials with nano-scale structure are used as photo-absorption and doped layer materials in the Si thin film solar cells. An atomic force microscope (AFM) is a scanning probe microscope (SPM) which is an imaging technique at nano-scale spatial resolution. Using the method based on the AFM technique, local electrical measurement with nano-scale spatial resolution was conducted [4-6]. However, local electrical characterization in only single films was carried without light irradiation. Therefore, we have developed a local photovoltaic characterization method using the AFM technique.

In this report, the newly developed local photovoltaic characterization method is explained. Using the developed method, the first measurement of the local surface potential in Si thin film solar cells with light irradiation was conducted in this work.

2. Development of local photovoltaic characterization method

At first, a local photo-current measurement system was developed using the scanning near-field microscope (SNOM) technique [7]. In p-i-n type Si thin film solar cells, light is irradiated from the p-layer side. Using the system based on the SNOM technique, the light is irradiated from the n-layer side in the p-i-n type solar cells and the result is not obtained in actual behaviour. Therefore, we developed a new local photo-current measurement system in which light could be irradiated from the p-layer side [8].

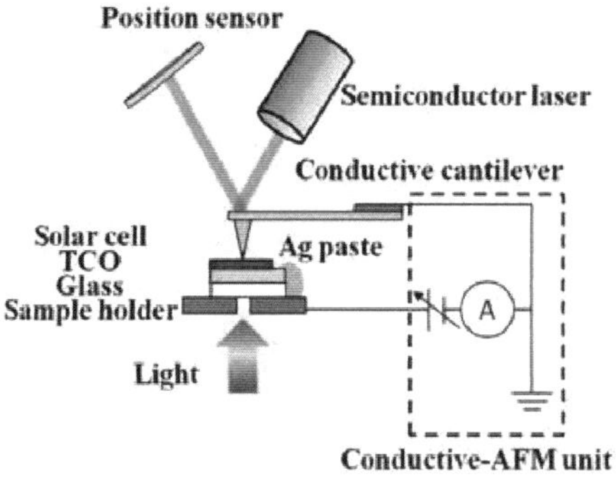

Figure 1. Schematic illustration of a local photo-current measurement system based on conductive-AFM.

Figure 1 shows the schematic illustration of the local photo-current measurement system. This system is based on conductive-AFM (Seiko Instruments Nano-Technology, SPI3800N-SPA400). In order to measure the local current under light irradiation from the p-layer side of p-i-n type solar cells, a new attachment of a sample holder with a light source was developed. A halogen lamp (white light) and LED light (wavelength of 625 nm or white light) were used as light sources. Spot size of the light at the surface of solar cells is less than 2 mm in diameter. The scanning frequency is typically 0.5 Hz. The conductive cantilever and sample holder were connected to a pico-ampere current meter and an electrical source of a sink current type which were embedded systems in the AFM. In this measurement, the electrical contact property between the cantilever and solar cell is important. In the early stage, Rh coated Si cantilevers (Seiko Instruments, SI-DF3-R) with a spring constant of 1.6 N/m were used as a probe. However, the contact property between the Rh coated cantilever and solar cell was not good, since a shottky barrier would form between the cantilever and the solar cell. The electrical contact property between the conductive cantilever and solar cells was improved by coating with a defective ZnO thin film deposited by DC-sputtering at room temperature [8]. Using this system, local photo-current - voltage characteristics and photo-current mappings can be measured. However, the local oxidation led to the formation of a SiO_2 thin film on the surface of Si thin film solar cells, because this measurement was conducted at room temperature in air [9]. We proposed topographical and Current-Voltage Imaging (TCVI) as a technique to avoid local oxidation [9, 10]. Using the TCVI, the topography and local photo-current – voltage characteristic can be obtained simultaneously in only a single scan. Therefore, the local photo-current – voltage characteristics and local photo-current images can be obtained without local oxidation.

An example of the results measured using the developed system is shown. Figure 2 shows (a) the surface topography and (b) photo-current image in the p-i-n type a-Si:H thin film solar cell. Here, the structure of the a-Si:H thin film solar cell was an Asahi U-type substrate (textured SnO_2 on glass)/p-type a-$Si_{1-X}C_X$:H (15 nm)/i-type a-Si:H (450 nm)/n-type μc-Si:H (47 nm) and the conversion efficiency obtained by macroscopic measurement was 7.1%. LED light with a wavelength of 625 nm was used as the irradiation light and the photon energy of the irradiation light was larger than the band gap energy of i-type a-Si:H used as a photo-absorption layer material. As the local photo-current image was measured, a bias voltage of -2 V was applied to the sample holder connected to TCO using a silver paste. The brighter regions in the surface topography and photo-current image indicate higher surface height and larger current flow, respectively. In the surface topography, large and small convex grains were found. In the surface topography of the Asahi U-type substrate, large convex grains were found [10]. In the surface topography of the n-type μc-Si:H film deposited on the glass substrate, small convex grains were found [11]. The size of the large convex grains found in the surface topography of the Asahi U-type substrate and that of the n-type μc-Si:H film deposited on the glass substrate was almost the same as that of the p-i-n type a-Si:H thin film solar cell, respectively. Therefore, the large and small convex grains correspond to the textured surface of the Asahi U-type substrate and crystalline Si grains, respectively. Comparing the two images, the photo-current on the crystalline Si grains was larger than that in the grain boundaries, and the photo-current was very small in the region between the large convex grains. This result suggests that photo-current flows mainly through the crystalline Si grains in the n-type μc-Si:H layer.

Based on this result, the local photo-current image with nano-scale resolution can be measured in Si thin film solar cells using this system.

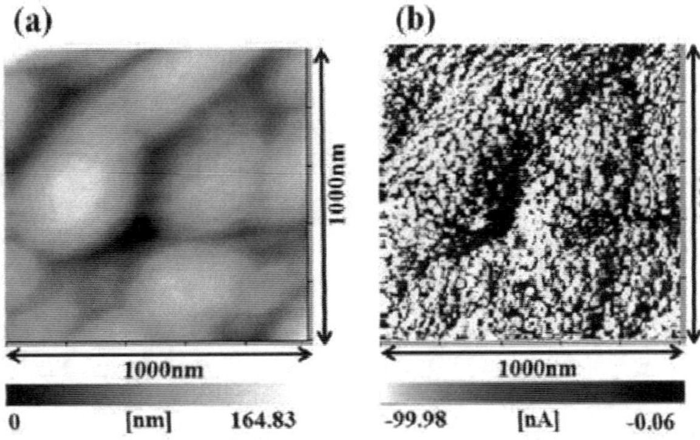

Figure 2. (a) Surface topography and (b) local photo-current image of the a-Si:H thin film solar cell.

3. Local surface potential in a-Si:H thin film solar cells with light irradiation

In the developed system, local surface potential images with light irradiation were also measured using a kelvin force microscope (KFM) unit instead of a conductive-AFM unit. In this measurement, the surface topography and photovoltaic properties can be obtained simultaneously in a single scan. Using this system, we measured the local surface potential in a-Si:H thin film solar cells. Au coated Si cantilevers (Seiko Instruments, SI-DF3-A) with a spring constant of 1.6 N/m was used as a probe and the scanning frequency was 0.4 Hz. The structure of the a-Si:H thin film solar cell used in this work was the Asahi U-type substrate/p-type a-$Si_{1-X}C_X$:H (15nm)/i-type a-Si:H (450nm)/n-type μc-Si:H (45nm) and the conversion efficiency obtained by macroscopic measurement was 6.4%.

Figure 3 shows (a) the surface topography and (b) the local surface potential image of the a-Si:H thin film solar cell without light irradiation. The brighter regions in the surface topography and the local surface potential images indicate higher surface height and larger surface potential, respectively. In the surface topography, an uneven surface was observed. This large convex grain is related to the textured surface of the Asahi U-type substrate as explained above. Comparing the two images, the surface potential on the large grains was smaller than that in the region between the large convex grains. In n-type μc-Si:H films deposited on Asahi U-type substrates, similar tendency is obtained [11]. Therefore, this surface potential distribution observed in the a-Si:H thin film solar cell would be related to that in the n-type μc-Si:H layer. On the growth of the μc-Si:H film on the Asahi U-type substrate, the collision of crystalline columns occurs, resulting in the formation of grain boundaries [1]. In the local surface potential image of the n-type μc-Si:H film on the glass substrate, the local surface potential on the small convex grains found in the surface topography was smaller than that in the region between the small convex grains, i.e., grain boundary. Therefore, the surface potential distribution observed in the surface potential images of the n-type μc-Si:H films deposited on the Asahi U-type films would be related to the grain boundary [11].

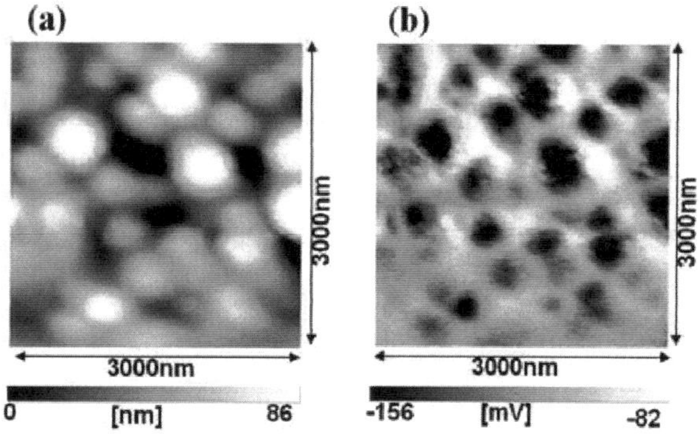

Figure 3. (a) Surface topography and (b) local surface potential image of the a-Si:H thin film solar cell without light irradiation.

Next, we measured the local surface potential images with light irradiation. Here, the white light LED (OSRAM, LCW G6Sp-CBEB-409Q) was used as the irradiation light and the power of the irradiation light was varied from 0 to 2.24 mW/cm^2. Figure 4 shows (a) the surface topography and (b) the local surface potential image of the a-Si:H thin film solar cell with light irradiation. Here, the power of the irradiation was 2.24 mW/cm^2. Comparing Figures 3 and 4, almost the same surface potential distribution was observed with/without light irradiation. However, the numerical value was changed by light irradiation. Variation in the average of the local surface potential in the scanning area corresponds to that of the macroscopic surface potential by light irradiation. In order to understand the change of the surface potential by light irradiation, the average of the local surface potential was first obtained.

Figure 5 shows the dependence of the average of the local surface potential on the power of the irradiation light. In Figure 5, the average of surface potential decreased with increasing power of the irradiation light. This result could be understood as follows. In a-Si:H thin film solar cells, photo-generated electrons are collected into the n-layer. In this measurement, the potential on the p-layer is connected to the conductive cantilever. Therefore, the potential on the n-layer decreases by light irradiation. As a result, the average of local surface potential decreases with increasing power of the irradiation light. In Figures 3 and 4, the potential difference was found between the large convex grain and concave region, i.e., grain boundary. The potential difference $\Delta\phi_{ave}$ between the large convex grain and grain boundary was estimated as:

$$\Delta\phi_{\text{ave}} = \phi_{\text{GBave}} - \phi_{\text{CGave}} \tag{1}$$

Here, ϕ_{CGave} and ϕ_{GBave} were averages of the potential on the top of the large convex grains and at the bottom of the concave region between the large convex grains, i.e., grain boundary, respectively. Figure 6 shows the dependence of the potential difference $\Delta\phi_{\text{ave}}$ between the large convex grain and grain boundary on the power of the irradiation light. In Figure 6, the potential difference $\Delta\phi_{\text{ave}}$ decreased with increasing power of irradiation. One possible cause is the potential difference in the n-layer. In the n-type μc-Si:H films deposited on the Asahi U-type substrates, the local potential on the large convex grain is smaller than that in the grain boundary. Therefore, electrons collected into the n-layer would move from the large grains to the grain boundary. Thus, potential in the grain boundary decreases more than that on the large convex grain. As a result, the potential difference decreases with increasing power of irradiation. These results suggested that the grain boundaries in the n-type μc-Si:H layer might be one of the factors which causes decreasing efficiency in a-Si:H thin film solar cells. At present, however, the relation between the efficiency and the change of the $\Delta\phi_{\text{ave}}$ is not clear. Therefore, further investigation using a-Si:H thin film solar cells with different efficiencies is necessary.

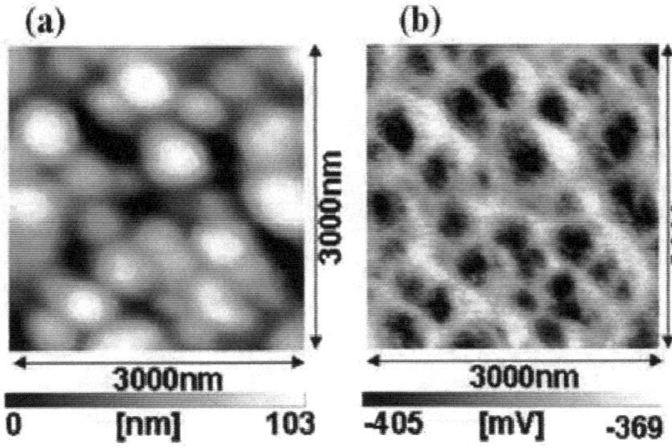

Figure 4. (a) Surface topography and (b) local surface potential image of the a-Si:H thin film solar cell with light irradiation. The power of irradiation was 2.24 mW/cm^2.

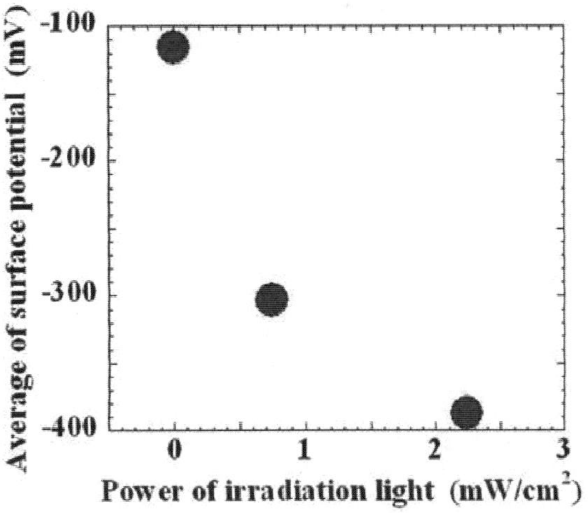

Figure 5. Dependence of the average of the local surface potential on the power of irradiation.

4. Summary

We developed a new local photovoltaic characterization method for Si thin film solar cells based on AFM and demonstrated local photovoltaic characterization with nano-scale spatial resolution using this method. Local photo-current and local surface potential images were observed in a-Si:H thin film solar cells. These results suggested that the local photovoltaic performance with nano-scale spatial resolution can be evaluated using this method. Based on these results, this method is useful for the evaluation of not only Si thin film solar cells, but also multi-crystalline solar cells such as CuInGaSe solar cells.

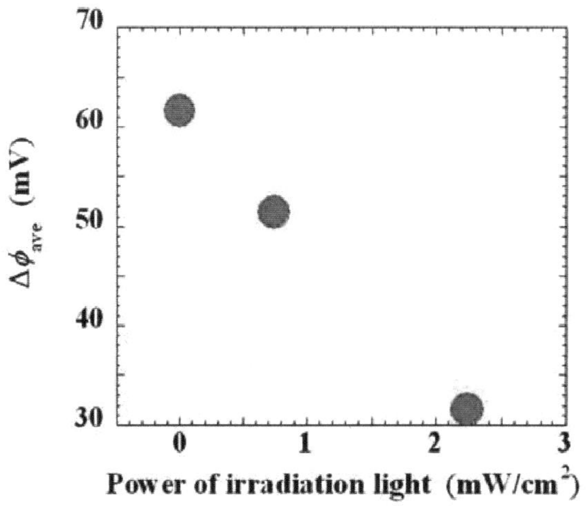

Figure 6. Dependence of the potential difference $\Delta\phi_{ave}$ between the large convex grain and grain boundary on the power of the irradiation light.

References

[1] Nasuno Y, Kondo M and Matsuda A 2001 *Jpn. J. Appl. Phys.* **40** L303
[2] Sugita K, Koyama K, Ohdaira K and Matsumura H 2007 *Jpn. J. Appl. Phys.* **46** 2852
[3] Nishioka K, Yagi T, Uraoka Y and Fuyuki T 2007 *Sol. Energy Mater. Sol. Cells* **91** 1.
[4] Rezek B, Stuchlik J, Fejfar A Kocka J 2002 *J. Appl. Phys.* **92** 578
[5] Fejfar A, Kocka J 2004 *J. Non-Cryst. Solids* **338** 303
[6] Gotoh T, Sugawara K and Tanaka K 2004 *Jpn. J. Appl. Phys.* **43** L818
[7] Gotoh T, Yamamoto Y, Shen Z, Ogawa S, Yoshida N, Itoh T and Nonomura S 2009 *Jpn. J. Appl. Phys.* **48** 2009
[8] Kawai M, Kawakami T, Inaba T, Ohashi F, Natsuhara H, Itoh T and Nonomura S 2010 *Current Appl. Phys.* **10** S392
[9] Shen Z, Eguchi M, Gotoh T, Yoshida N, Itoh T and Nonomura S 2008 *Thin Solid Films* **516** 588
[10] Shen Z, Gotoh T, Eguchi M, Yoshida N, Itoh T and Nonomura S 2007 *Jpn. J. Appl. Phys.* **46** 2858
[11] Itoh T, Ito T, Sakai T, Moriya Y, Kuriyama H, Natsuhara H and Nonomura S 2011 *Technical Digest 21st Int. Photovoltaic Science and Engineering Conf.* (Fukuoka) 3D-2P-12

Synthesis and thermoelectric properties of sintered skutterudite $CoSb_3$ with a bimodal distribution of crystal grains

S Katsuyama[1], F Maezawa[1] and T Tanaka[1]

[1] Division of Materials and Manufacturing Science, Graduate School of Engineering, Osaka University, 2-1 Yamadaoka, Suita, Osaka 565-0871, Japan

E-mail: katsuyama@mat.eng.osaka-u.ac.jp

Abstract. We have synthesized sintered skutterudite $CoSb_3$ composed of large and small crystal grains. The sintered materials with larger crystal grains tend to show lower electrical resistivity, while those with smaller grains possess lower thermal conductivity. The possibility of enhancing thermoelectric performance by producing sintered materials with a bimodal distribution of crystal grain sizes so that they possess both low electrical resistivity and thermal conductivity is discussed. The electrical resistivity and Seebeck coefficient generally increase as the ratio of small particles increases. The sample with a mass ratio of small particles of 30% possesses the largest power factor of the sample. The thermal conductivity decreases as the ratio of small particles increases. As a result, the samples containing a higher ratio of smaller particles show higher *ZT*.

1. Introduction

In the face of impending environmental and energy problems, thermoelectric energy conversion of heat to electricity has attracted the attention of many researchers. Skutterudite compounds are known as potential thermoelectric materials [1-3]. Skutterudite compounds can be represented as MX_3, where M is Co, Rh or Ir and X is P, As or Sb. Most binary skutterudite compounds possess a reasonably large Seebeck coefficient (S) and good electrical conductivity (σ), resulting in a large power factor ($S^2\times\sigma$) comparable to state-of-the-art thermoelectric materials. However, their thermal conductivity (κ) remains too high to make skutterudite compounds efficient thermoelectric materials [3].

The total thermal conductivity κ_{total} of a material consists of a lattice κ_{ph} and carrier contributions κ_{car}. κ_{car} generally increases as the number of carriers in a material increases. κ_{ph} decreases when phonons are scattered by disorder in a material, for example, by grain boundaries. The degree of phonon scattering depends on the density of the disorder; the more disorder present in a material, the more strongly phonons are scattered. The minute grains in sintered materials effectively improve phonon scattering at grain boundaries, resulting in reduced thermal conductivity [4].

On the other hand, electron or hole carriers are scattered at grain boundaries, so sintered materials with smaller crystal grains generally have larger electrical resistivity. As a result, it is difficult to realize both low electrical resistivity and thermal conductivity in sintered materials composed of crystal grains of a single size, *i.e.*, with a single-modal distribution of crystal grains. In sintered materials composed of a mixture of large and small crystal grains, electron or hole carriers can pass through the large crystal grains without being disturbed, while phonons are scattered at the grain boundaries between small crystal grains because phonons do not follow a path like carriers. Therefore, both low electrical resistivity and thermal conductivity can be attained in sintered materials with a bimodal distribution of crystal grains.

From this viewpoint, Zhao *et al.* investigated the thermoelectric properties of Bi_2Te_3 containing nano- and microsized crystal grains, which were synthesized by mechanical milling for different times and spark plasma sintering [5]. Mechanical milling grinds alloys or intermetallic compounds into fine powders using a ball or vibration mill. In this study, we have synthesized sintered $CoSb_3$ with a bimodal distribution of crystal grains by mechanical milling and hot pressing. The thermoelectric properties of the resulting $CoSb_3$ compounds have been investigated.

2. Experiment

The method used to prepare the samples is summarized in Figure 1. $CoSb_3$ was synthesized by direct reaction of the constituent elements. Co (Rare Metallic Co., Ltd., Japan, 99.9 at.% purity) and Sb (Rare Metallic Co., Ltd., Japan, 99.9999% at.% purity) powders were mixed in the desired ratio, evacuated in a silica ampoule and then heated at 1073 K for 24 h. After the reaction was completed, the ampoule was opened and the products were ground with a mortar and pestle into fine powders. For large particles of $CoSb_3$, the obtained powders were classified by a sieve into particles with diameters between 32 and 63 μm. For small particles of $CoSb_3$, the obtained powders were mechanically milled. $CoSb_3$ powder (about 10 g) was placed in a zirconium pot with a volume of about 45×10^{-6} m^3 with 10 zirconium balls with a diameter of 10 mm. The pot was filled with Ar gas. Mechanical milling was carried out for 20 h using a planetary ball mill (P-6, Fritsch, Germany). The obtained large and small particles were weighed with the desired mass ratios of large particles to small particles of 100-x : x, where x=0, 30, 70 and 100. The weighed powders were agitated ultrasonically in ethanol for 10 min. Dry mixed powers were obtained after the evaporation of ethanol. The mixed powders were packed into graphite dies and hot-pressed at 923 K for 1 h under a vacuum of 10^{-3} Pa and pressure of 50 MPa.

Phase identification of the obtained samples was performed by X-ray diffraction at room temperature. The Seebeck coefficient and electrical resistivity of the sintered samples were simultaneously measured by the standard four probe dc method in under a flow of Ar gas for the temperature range from room temperature to about 773 K. The thermal diffusion coefficient (λ) of the samples was measured by the laser flash method using a thermal constant analyzer (TC7000, ULVAC, Japan). The specific heat (C_p) of the samples was measured by a differential scanning calorimeter (DSC-50, Shimadzu, Japan). The density (d) of the samples was measured by the Archimedes method.

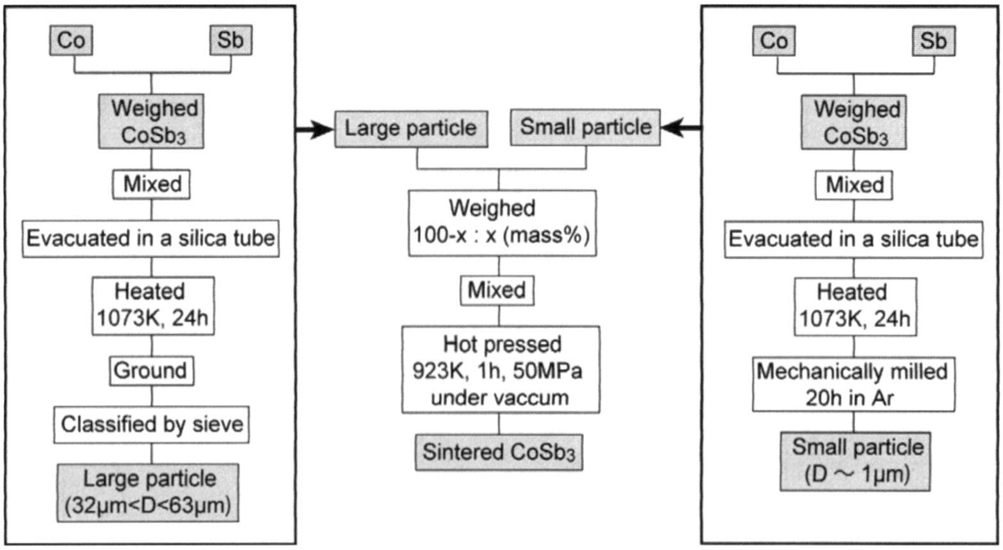

Figure 1. Flow chart showing the procedure followed to prepare samples.

The thermal conductivity (κ) was calculated from the measured λ, C_p and d using the relationship $\kappa = \lambda C_p d$.

3. Results and discussion

The X-ray diffraction patterns for the sintered samples with various ratios of particle sizes contained only diffraction peaks consistent with $CoSb_3$ with cubic skutterudite structure (data not shown).

Figure 2 shows SEM images of the sintered $CoSb_3$ with various ratios of particle sizes. As shown in Figure 2(a), the sample containing only large particles possesses more pores than the other samples. The relative density of the sample composed of only large particles is about 92%, while the relative densities of the other samples are between 96 and 98%. The crystal grain size is several tens of μm, which shows that the crystal grains do not grow during sintering. As the ratio of small particles is increased, the area occupied by the small particles increases, but the small particles gather together and so are not evenly dispersed. The sample containing only small particles exhibits a dense and homogeneous texture (Figure 2(d)), and some of the crystal grains grow during sintering.

The temperature dependence of the Seebeck coefficient for $CoSb_3$ with various ratios of particle sizes is shown in Figure 3. The Seebeck coefficient of all of the samples is positive, and it generally increases as the ratio of small particles in the samples increases. Previously, we reported that the Seebeck coefficient of half-Heusler $ZrNiSn_{0.98}Sb_{0.02}$ is increased by mechanical milling, which is because of the smaller grain size in the mechanically milled samples [4]. Yang *et al.* reported that $CoSb_3$ with small grains shows a larger Seebeck coefficient than that with larger grains, regardless of

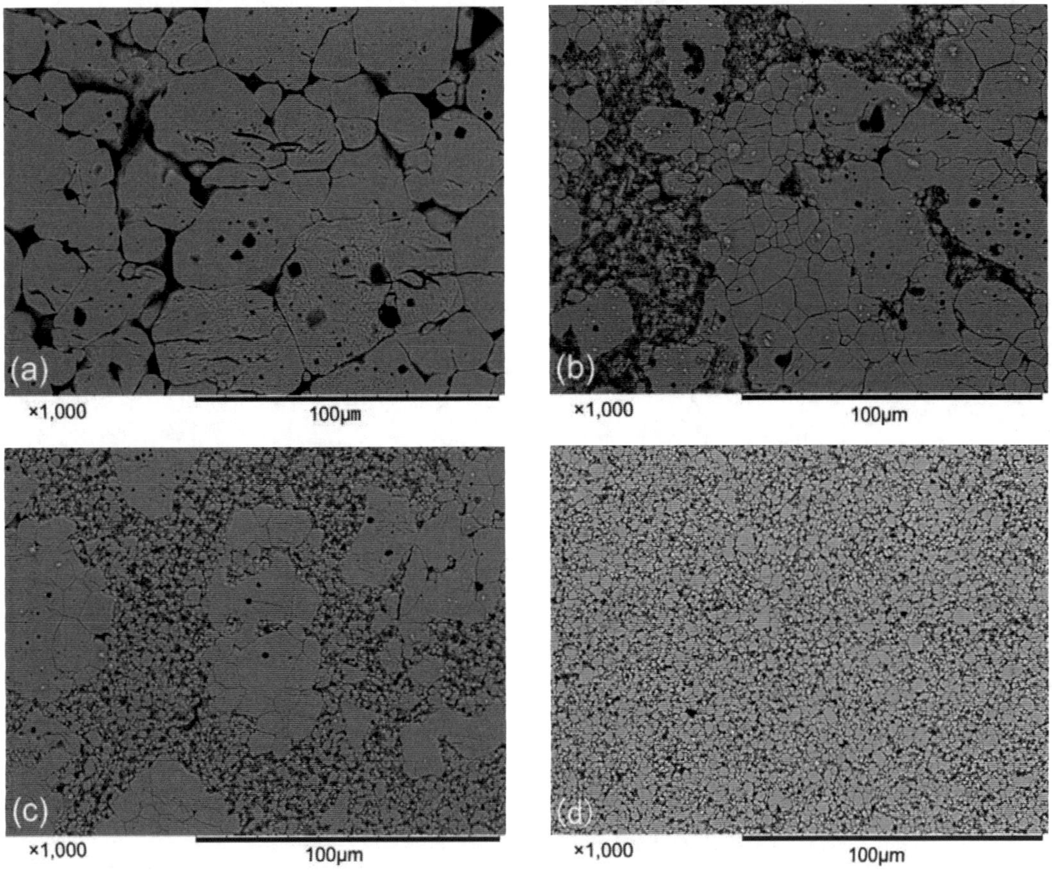

Figure 2. SEM images for sintered $CoSb_3$ with ratios of small particles of (a) 0%, (b) 30%, (c) 70% and (d) 100%.

their similar carrier densities [6]. They considered that the increase in the Seebeck coefficient is mainly caused by potential barrier scattering. A potential barrier should form at grain boundaries when the typical grain size is sufficiently small. Carriers with high enough energy will overcome the potential barrier, while those of lower energy will be stopped or strongly scattered by the potential barrier at grain boundaries [7,8].

Figure 4 shows the temperature dependence of the electrical resistivity ρ for $CoSb_3$. The electrical resistivity decreases with temperature, so these samples show semiconducting behavior. The electrical resistivities of these samples are larger than those of a previous report; Sharp *et al.* reported that the electrical resistivity of $CoSb_3$ at room temperature was 40 μΩm [3]. The relative density of their sample is 90%, which is lower than that of our samples, so the larger electrical resistivity of our samples cannot be ascribed to their density. There may be a slight difference in the chemical compositions of the samples, which would affect their carrier density. The electrical resistivity generally increased as the ratio of small particles increased. In the samples containing higher ratios of

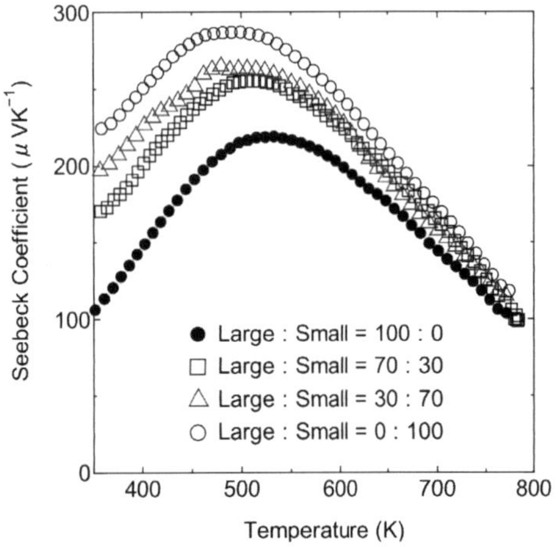

Figure 3. Temperature dependence of the Seebeck coefficient for $CoSb_3$ samples containing various ratios of small particles.

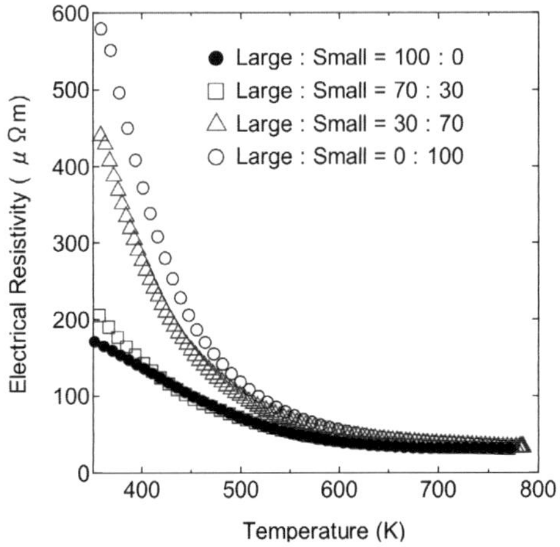

Figure 4. Temperature dependence of the electrical resistivity for $CoSb_3$ samples containing various ratios of small particles.

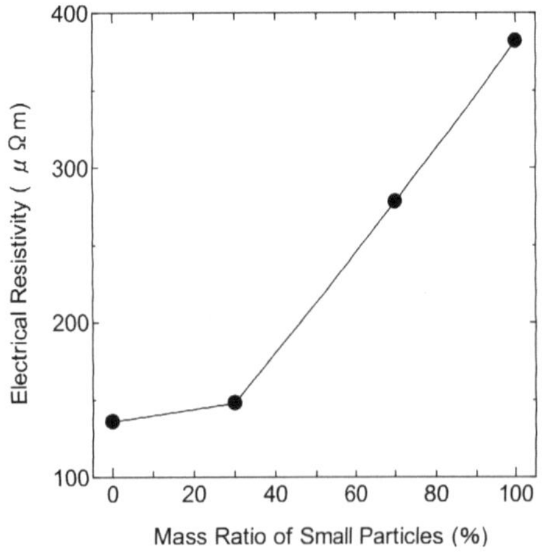

Figure 5. Electrical resistivity for $CoSb_3$ samples as a function of mass ratio of small particles.

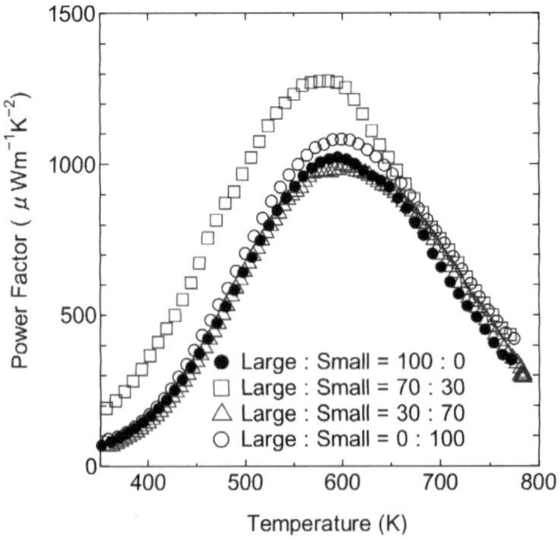

Figure 6. Temperature dependence of the power factor for $CoSb_3$ samples containing various ratios of small particles.

small particles, the density of grain boundaries is higher, so this behavior is considered to be caused by carrier scattering at grain boundaries.

Figure 5 shows the electrical resistivity at 400 K as a function of the mass ratio of small particles in the samples. The sample with a ratio of small particles of 30% shows relatively low electrical resistivity, while for those containing >30% small particles, the electrical resistivity increases linearly with the ratio of small particles. Zhao *et al.* observed similar behavior for Bi_2Te_3 samples [5]. The electrical resistivity of their samples rapidly increased when the ratio of small particles exceeded about 60%, which was attributed to an effect similar to percolation.

The temperature dependence of the power factor for $CoSb_3$ samples containing various ratios of small particles is shown in Figure 6. The sample with a ratio of small particles of 30% shows the largest power factor of the samples over the whole temperature range. As mentioned above, as the ratio of small particles increases, both the Seebeck coefficient and electrical resistivity increase. This result shows that the sample with a ratio of small particles of 30% has relatively low electrical resistivity.

The temperature dependence of the thermal conductivity κ for $CoSb_3$ samples containing various ratios of small particles is presented in Figure 7. The thermal conductivity also generally decreases as the ratio of small particles increases. As already mentioned, the total thermal conductivity κ_{total} of the material consists of κ_{ph} and κ_{car}. To determine whether the lattice or carrier conductivity governs the thermal conductivity of these samples, κ_{ph} and κ_{car} were estimated by the Wiedemann-Franz relationship. According to this relationship, $\kappa_{car} = LT/\rho$, where L is the Lorentz number. We calculated κ_{car} using a theoretical value of the Lorentz number for a metal with free electrons of 2.45×10^{-8} $W\Omega K^{-2}$ [9] . κ_{ph} was obtained by subtracting κ_{car} from κ_{total}.

Figure 8(a) and (b) show the temperature dependence of κ_{ph} and κ_{car} for $CoSb_3$ samples containing various ratios of small particles, respectively. The values of κ_{ph} are much larger than those of κ_{car}, which shows that the thermal conductivity of this system is mainly dominated by the lattice. Both κ_{ph}

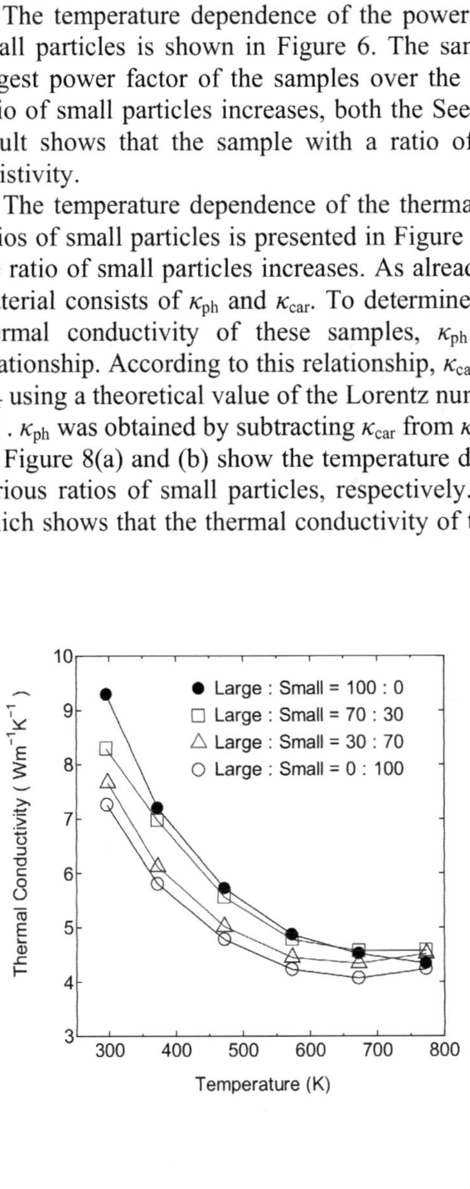

Figure 7. Temperature dependence of the thermal conductivity for $CoSb_3$ samples containing various ratios of small particles.

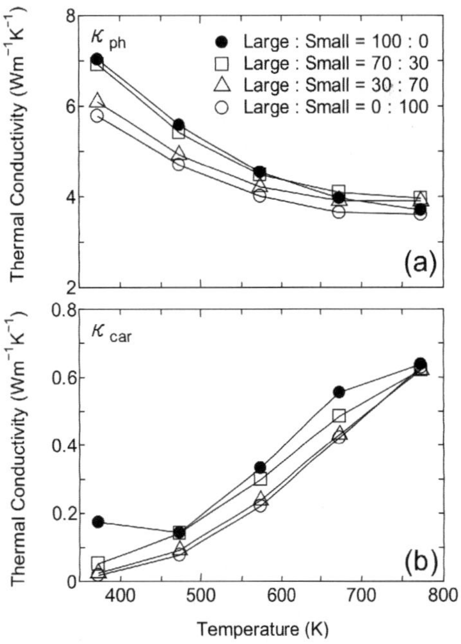

Figure 8. Temperature dependence of (a) κ_{ph} and (b) κ_{car} for $CoSb_3$ samples containing various ratios of small particles.

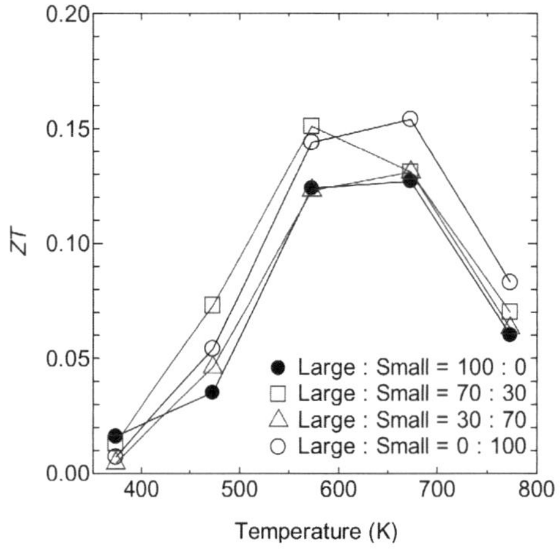

Figure 9. Temperature dependence of *ZT* for $CoSb_3$ samples containing various ratios of small particles.

and κ_{car} generally decrease as the ratio of the small particles increases, but considering the scale of the vertical axis, κ_{car} can be regarded as almost constant independent of the ratio of small particles. The decrease of κ_{ph} shows that phonons are efficiently scattered at the grain boundaries in the samples with a higher ratio of small particles. κ_{car} gradually increases with temperature. This shows that the number of carriers increases with temperature, which coincides with the results for the electrical resistivity of the samples.

The dimensionless figure of merit *ZT* was calculated from the Seebeck coefficient, electrical resistivity and thermal conductivity of the samples. Figure 9 shows the temperature dependence of *ZT* for $CoSb_3$ samples containing various ratios of small particles. The sample with a ratio of small particles of 30% possesses the largest power factor of the samples (Figure 6). On the other hand, the higher the ratio of small particles in the sample, the lower its thermal conductivity (Figure 7). As a result, the samples with ratios of small particles of 70% and 100% show higher *ZT*. The maximum *ZT* obtained in this study is 0.154 at 673 K for the sample containing only small particles.

4. Conclusions

To realize both low electrical resistivity and thermal conductivity, sintered skutterudite $CoSb_3$ with a bimodal distribution of crystal grain sizes was synthesized and its thermoelectric properties were investigated. The electrical resistivity generally increases as the ratio of small particles in the samples is increased, which is considered to be a result of carrier scattering at grain boundaries. The Seebeck coefficient also increases as the ratio of small particles is increased, while the thermal conductivity decreases. The sample with a ratio of small particles of 30% shows the largest power factor of those examined. As a result, the sample composed entirely of small particles shows the highest *ZT*.

The sample with a ratio of small particles of 30% shows relatively low electrical resistivity regardless of the small particles, but a clear influence of the bimodal distribution of crystal grain size on the thermoelectric performance of $CoSb_3$ was not observed in this study. SEM observations showed that the small particles in the sintered materials tended to aggregate. Improved dispersion of the small particles in the matrix may be necessary to enhance the thermoelectric performance of sintered materials with a bimodal distribution of crystal grain sizes.

On the other hand, Yang *et al.* reported that $CoSb_3$ nano/microcomposites, which are composed of nanosized $CoSb_3$ particles obtained by the polyol method and microsized $CoSb_3$ particles obtained by solid state reaction, exhibit *ZT* almost twice that of the sample without nanosized particles [6]. The carrier density of nanosized $CoSb_3$ produced by polyol method is larger than that of microsized $CoSb_3$ formed by solid state reaction because the nanosized $CoSb_3$ contains excess Sb, so the electrical resistivity of the composites is much smaller than that of $CoSb_3$ without nanoparticles, regardless of its lower thermal conductivity. This result shows that control of the carrier density may be an effective way to enhance the thermoelectric performance of sintered materials with a bimodal distribution of crystal grain sizes.

Acknowledgement
This work was supported by an Advanced Low Carbon Technology and Development Program (ALCA) (no. 11102875) from Japan Science and Technology Agency (JST), Japan.

References

[1] Caillat T, Borshevsky A and Fleurial J P 1993 *Proc. 12th Int. Conf.* (Yokohama, Japan) p132
[2] Slack G A and Tsoukala V G 1994 *J. Appl. Phys.* **76** 1665
[3] Sharp J W, Jones E C, Williams R K, Martin P H and Sales B C 1995 *J. Appl. Phys.* **78** 1013
[4] Katsuyama S and Kobayashi T 2010 *Mat. Sci. Eng. B* **166** 99
[5] Zhao L D, Zhand B P, Liu W S and Li J F 2009 *J. Appl. Phys.* **105** 023704
[6] Yang L, Hng H H, Li D, Yan Q Y, Ma J, Zhu T J, Zhao X B and Huang H 2009 *J. Appl. Phys.* **106** 013705
[7] Broido D A and Mingo N 2006 *Phys. Rev. B* **74** 195325
[8] Lin Y M and Dresselhaus M S 2003 *Phys. Rev. B* **68** 075304
[9] Franz R and Wiedemann G 1853 *Annalen der Physik* 497

Magnetron sputtering of (TiZr)NiSn thin films on different substrates for thermoelectric applications

Wilfried Wunderlich[1], Yoshiaki Shinohara[2] and Yoshito Matsumura[2]

[1] Tokai University, Faculty of Engineering, Material Science Department, 4-1-1 Kitakaname, Hiratsuka, Kanagawa-ken 259-1292, Japan
[2] Tokai University, Faculty of Engineering, Energy Department, 4-1-1Kitakaname, Hiratsuka, Kanagawa-ken 259-1292, Japan

Email: wi-wunder@rocketmail.com

Abstract. Half-Heusler alloys are considered to be high-efficiency thermoelectric materials. Such thin film devices are expected to have a wide range of applications. In the present study, a dense 1μm thin films of (TiZr)NiSn were deposited by magnetron sputtering on different substrates including Ni-, Cu-, Ti-metals, or Si wafers. SEM-EDX observations confirmed a homogenous element distribution. X-ray diffraction measurements indicated a hexagonal P -6 2 m space group (#189). All thin films showed reasonable Seebeck coefficients with the highest values obtained for (TiZr)NiSn thin films deposited on Cu or Ti, as explained by the values of the work function. As the Seebeck coefficient is significantly different from the values measured for bulk specimens, this research opens a new dimension for the development of thermoelectric materials.

1. Introduction

Intermetallic phases composed of $(Ti_xZr_{1-x})Ni_{1+y}Sn_{1-y}$ crystallize with a half-Heusler (HH) structure and are considered to be effective thermoelectric (TE) materials [1-3] with a maximum Seebeck coefficient for $x = 0.4$, $y = 0$ [2], where Ti and Zr commonly share one of the three available crystallographic sites in the MgAgAs structure. In contrast to bulk material, thin film composites integrated into Si technology are expected to be of large interest for practical applications in small devices. This paper describes magnetron sputtering of (TiZr)NiSn on different substrates, including Si wafers, for the first time. Deposition of such films on MgO led to a slightly smaller Seebeck coefficient than bulk specimens [4], but our intention is to measure the difference when deposited on metallic substrates.

The conventional design of thermoelectric devices has a Π-shape in which one leg is an n-type, and the other is a p-type semiconductor [5]. Modifications have been reported such as an arch-shape for utilizing the heat of water pipes [6], or a Y-shape when composites are used [5]. In the latter geometry the interface between different materials lies parallel to the heat gradient and is designed to reduce their mutual thermal expansion stresses. This idea has even been brought to an extreme by placing a layered structure with an interface on the hot and cold ends of a thermoelectric device [7, 8]. This design provides an easy and cheap way to utilize interface effects in thermoelectric materials and, hence, it is used in the experiments presented in this paper. After explaining the experimental details and microstructure observations, this paper describes the thermoelectric performance of such layered structures under a large temperature gradient [2, 9, 10] and compares their properties to bulk samples.

2. Experimental

After preparing bulk specimens, typically 2 mm in thickness and 15 mm in diameter, by arc melting from powder pellets [2], a sputtering target was manufactured by attaching about ten such sliced specimens on top of a brass disc, 60 mm in diameter, which was covered by a 1-mm-thick Ti sheet and glued with silver paste. The substrate consists of 10x8x1 mm pieces of Si wafer (5" n-type, Sumco, Finebrand, Japan) with polished (111) or (100) surfaces and pieces of Ni, Cu and Ti pure metals (High Purity Chemicals Ltd, Japan). The pieces were mounted on the substrate holder by conducting tape.

The sputtering conditions were 50 W plasma power, target-to-substrate distance of 60 mm, temperature 200 °C, and sputtering time of 30 min. As shown in table 1, two sputtering experiments were performed using sputtering targets of different compositions. Specimen characterization was performed using a Hitachi SEM system operated at 30kV, equipped with a Noran EDX instrument, and a Rigaku Miniflex XRD system with a Cu Kα beam. The Seebeck voltage was measured by attached Ni wires, as shown in the insets of Figure s. 4 and 5. One Ni wire was attached to the cold side and the other to the hot side of the specimen, which was heated by a ceramic heater (Sawaguchi

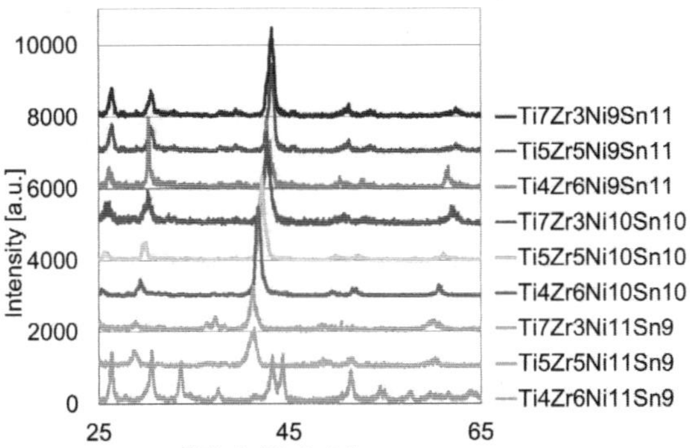

Figure 1. XRD of (TiZr)NiSn bulk specimens with the different compositions shown.

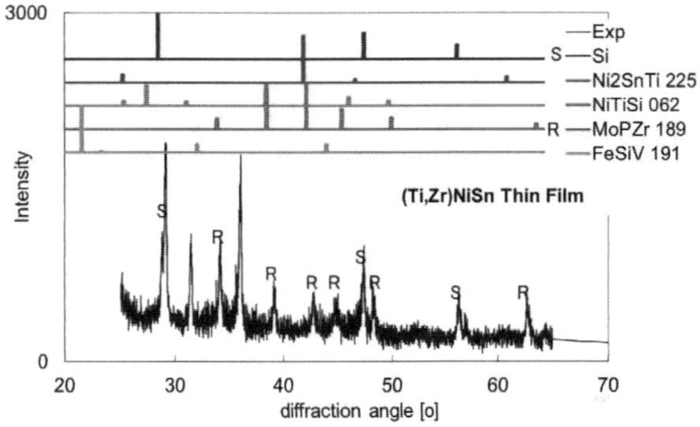

Figure 2. XRD of (TiZr)NiSn thin film on Si (111) wafers with peaks marked with "S". The results match best to the MoPZr (189) phase, whose peaks are marked with "R".

Table 1. SEM-EDX results for the composition of thin film and bulk, as well as the Seebeck coefficient *S* of thin films on Si(111) wafers for two sputtering experiments.

Compo-sition [at%]	Bulk		Thin Film	
	1.Exp.	2.Exp.	1.Exp.	2.Exp.
Ti	16	16	18	12
Zr	16	13	23	7
Ni	33	22	35	18
Sn	33	49	20	63
S [mV/K]	0.075	0.01	-0.05	-0.02

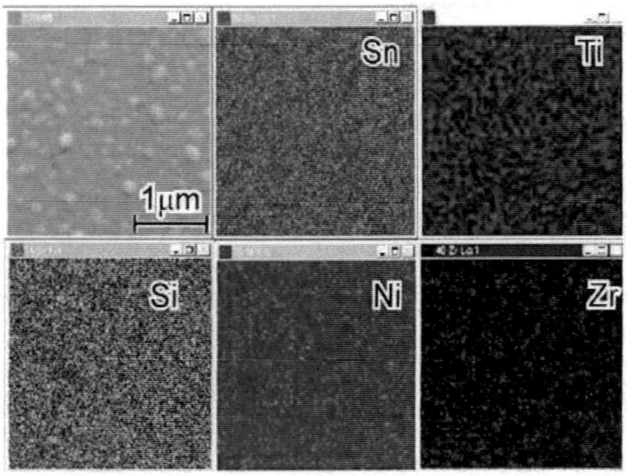

Figure 3. SEM-EDX chemical mapping of the thin film deposited on Si (111) wafer.

Journal of Physics: Conference Series **379** (2012) 012005 doi:10.1088/1742-6596/379/1/012005

MS1000, Japan); details have been described previously [2, 9, 10]. The measurements were performed under a high temperature gradient up to ΔT = 650 K, which results in slightly larger Seebeck coefficients than that usually measured under small temperature gradients [2]. By closing the circuit between the hot and cold end with different resistors (1 Ω, 100 Ω, 1 kΩ, 100 kΩ, 1 MΩ), the electric current versus voltage (*I-U-* characteristics) can be measured [2].

3. Results and discussion

Figure 1 shows the XRD results for bulk specimens with different compositions. The bulk specimens were confirmed to possess the half-Heusler crystal structure phase with a systematic decrease in diffraction angle, and hence increase of the lattice parameter, when the Ni content is increased (y= -0.1, 0, +0.1). The only exception is the $(Ti_{0.4}Zr_{0.6})Ni_{1.1}Sn_{0.9}$ specimen, which consists of several phases obviously beyond the half-Heusler stability limit. Concerning the XRD data for the sputtered thin film on the Si(111) wafer (Figure 2), the simulated XRD patterns for several crystal structures are drawn in the upper part of Figure 2, including the Si substrate marked with "S". The best fit is achieved to a phase, in literature referred to as ZrNiAl, MoPZr or CrNiAs prototypes, and also called the ordered Fe_2P phase, with space group of P -6_2 m number 189 and lattice constants of 0.629 and 0.376 nm. The simulated XRD peaks are marked with the letter "R". This phase occurs in Pettifor maps [2] in regions neighbouring the full- and half-Heusler phases. The XRD results for the thin films in both sputtering experiments agree with each other, and the same XRD patterns are observed for thin films on the Cu, Ni and Ti- substrates, indicating that the thin film growth is not limited by epitaxial strain originating from the substrate, but is mainly driven by the thermodynamic stability of its crystallographic phase. The SEM-EDX chemical maps confirmed the film homogeneity (Figure 3).

Thermoelectric measurements of the bulk specimens show a linear increase of the Seebeck voltage with increasing temperature difference, as indicated by an orange line and arrows in Figure 4, while the blue line indicates cooling in each figure in this paper. This behaviour indicates a p-type semiconductor with a maximum Seebeck voltage of 53 mV yielding a Seebeck coefficient S = 0.07 mV/K (dotted line in Figure 4). The Seebeck voltage of the thin film was measured in the geometry shown in the inset of Figure 5. Two main differences can be recognized. The Seebeck voltage reverses its sign yielding n-type behaviour and the maximum appears at around ΔT = 500K.

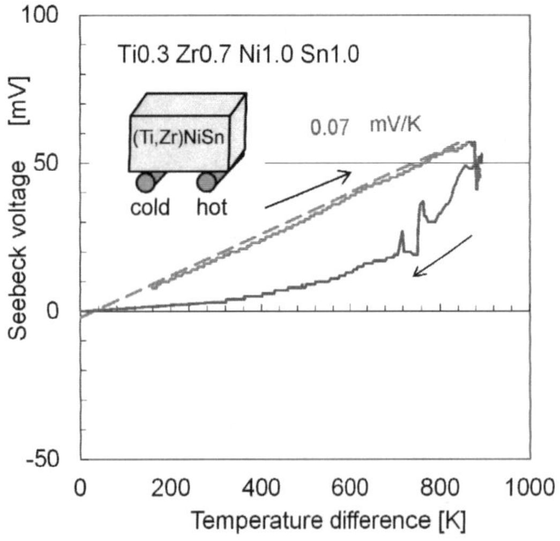

Figure 4. Seebeck voltage as a function of temperature difference for the $(Ti_{0.3}Zr_{0.7})Ni_{1.0}Sn_{1.0}$ bulk specimen.

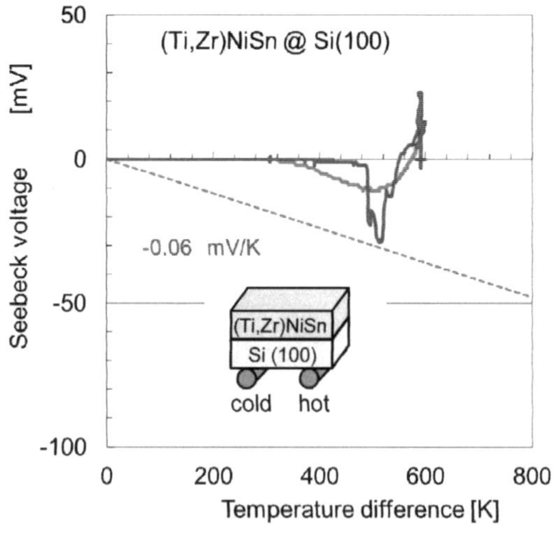

Figure 5. Seebeck voltage as a function of temperature difference for (TiZr)NiSn deposited on Si(100) wafer in the geometry shown in the inset.

The Seebeck coefficient is reduced to less than S = -0.06 mV/K. When the specimen is put upside down, and the Ni wires are attached on the side of the substrate, quite similar values are obtained.

Measurements on the bulk specimens were extended compared to a previous study [2] over a large range of composition and are shown in Figure 6. When the Ni content increased to y = 0.1, or decreased to y < -0.1, the Seebeck voltage drops drastically. The corresponding closed circuit current shows a behavior similar to that reported previously; here we show the current-voltage characterization for $(Ti_xZr_{1-x})Ni_1Sn_1$ specimens for the first time (Figure 8). The current is very high with a maximum of 410 μA for the specimen with x = 0.2 (filled squares in Figure 8), and the Seebeck voltage remains almost constant at 45mV, when loaded with different resistors or even short-circuited.

For the sputtered (TiZr)NiSn thin films (Table 1, conditions of experiment #2) on different metallic substrates negative Seebeck voltages were also measured, as shown in Figure 7. The Seebeck coefficient for films on Ni substrates is about the same as that on Si (Figure 5), S = -0.06 mV/K shown as the dotted line in Figures 5 and 7. On the Cu substrate, however, it is much higher, S = -0.25 mV/K with saturation occurring for ΔT > 400 K. In the case of Ti substrates a Seebeck voltage of more than -250 mV is measured between ΔT = 200 K and 300 K with a decrease at higher temperature differences.

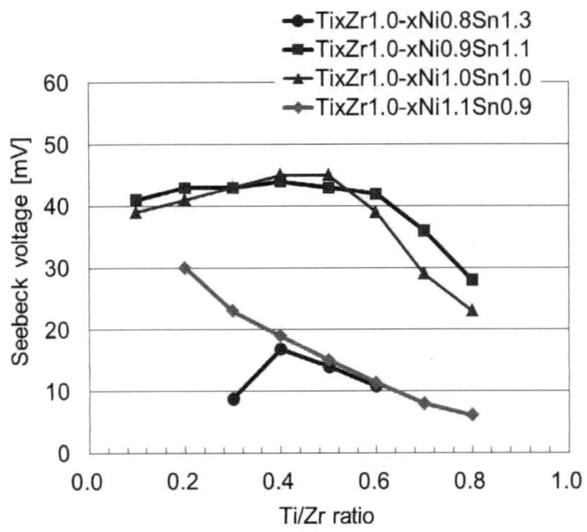

Figure 6. Seebeck voltage as a function of the Ti/Zr ratio x for several bulk specimens with the Ni and Sn compositions shown.

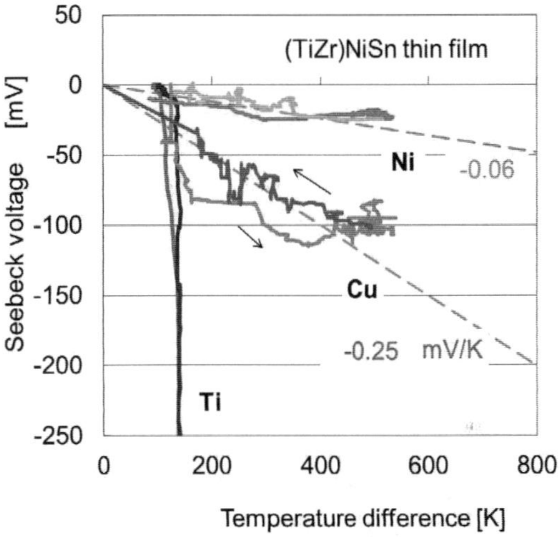

Figure 7. Seebeck voltage as a function of temperature difference for (TiZr)NiSn thin films deposited on Ni, Cu and Ti.

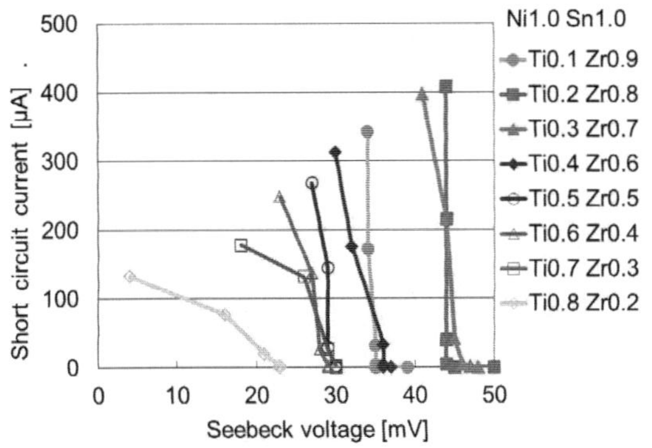

Figure 8. Current-voltage characteristic of $(Ti_xZr_{1-x})Ni_{1.0}Sn_{1.0}$ bulk specimens as a function of Ti/Zr ratios x, as indicated on the right side.

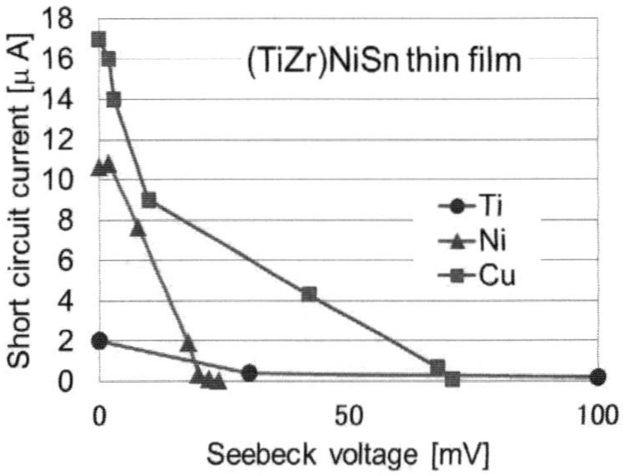

Figure 9. Absolute values of the electric current as a function of Seebeck voltage for (TiZr)NiSn thin films deposited on Ni, Cu, and Ti.

This large Seebeck voltage was confirmed for two specimens and also for the reversed orientation, when the substrate is lying on the wires opposite to the case shown in the inset of Figure 5. When the electric circuit is closed, bulk specimens show currents up to 400 μA (Figure 8), while for thin films deposited on Ti the electric current is only -2 μA, much lower than the case for (TiZr)NiSn on Cu or Ni (Figure 9) with a maximum value of -16 μA for Cu. The difference from bulk specimens can be explained by the far lower thickness of 1 μm compared to 2 mm for bulk specimens.

The results showed that depositing (TiZr)NiSn thin films on different substrates led to different Seebeck voltages and *I-U*-characteristics. The large Seebeck voltage on Ti seems to be contrary to the large current on Cu -substrates or bulk specimens. The change from the p-type behaviour of bulk specimens to the n-type behaviour of the thin films can be explained by the large amount of negative charge carriers provided by the metal. For the same reason, the thin film on n-doped Si has a reversed sign of the Seebeck voltage. The sequence Ti, Cu, Si, Ni, for the value of the Seebeck coefficient can be explained by the sequence of values of the work functions Ti 4.33 eV, Cu 4.65 eV, Si 4.85 eV, and Ni 5.15 eV. Assuming that the largest Seebeck coefficient is expected for a composite with a large difference in work function, it can be deduced that the work function of a (TiZr)NiSn thin film has a value around 5.2 eV.

The results of thermoelectric measurements presented in this paper suggest that a layered structure composed of metallic substrate and (TiZr)NiSn thin film allows the tuning of both, the Seebeck voltage and the current and opens a new dimension for development of thermoelectrics. .

4. Conclusions

The following results have been obtained.

(1) The (TiZr)NiSn bulk specimens consisted of half-Heusler crystals, with decreasing lattice parameter when the Ni content is increased.
(2) (TiZr)NiSn thin films deposited on Si(111) consists of a crystalline ZrNiAl structure with hexagonal P-6 2 m space group number 189, a phase which appears in other ternary intermetallic com-pounds as well.
(3) The thin film showed a deficit of Sn compared to the bulk, which can be explained by its large vapor pressure.
(4) The Seebeck coefficient shows a remarkable high value, but becomes negative, when the (TiZr)NiSn thin film is deposited on a Cu substrate. Thin films deposited on Ti show a very low electric current, and those on Ni or Si show a very low Seebeck voltage.
(5) The obtained Seebeck coefficients for thin films on four different substrates follow the order of the difference in work functions between the thin film and substrate.

References

[1] Culp S R, Poon S J, Hickman N, Tritt T M and Blumm J 2006 *Appl. Phys. Lett.* **88** 042106
[2] Wunderlich W, Motoyama Y, Sugisawa S and Matsumura Y 2011 *J. Electr. Mat.* **40** 583
[3] Wunderlich W and Motoyama Y 2009 *MRS Symp. Proc.* **1128-U01-10** 1
[4] Jaeger T, Mix C, Schwall M, Kozina X, Barth J, Balke B, Finsterbusch M, Idzerda Y U, Felser C and Jakob G 2011 *Thin Solid Films* **520** 1010
[5] Crane D T and Bell L E 2006 *IEEE Proc. ICT2006* p 11, doi: 10.1109/ICT.2006.331259
[6] Funahashi R, Mihara T, Urata S, Hisazumi Y and Kegasa A 2006 *IEEE Proc. ICT2006* 58, doi: 10.1109/ICT.2006.331269
[7] Span G, Wagner M, Holzer S and Grasser T 2006 *IEEE Proc. ICT2006* 23, doi: 10.1109/ICT.2006.331261
[8] Span G, Wagner M, Grasser T and Holmgren L 2007 *Phys. stat. sol. (RRL)* **1** 241 doi: 10.1002/pssr.200701171
[9] Wunderlich W and Soga S 2010 *J.Cer.Proc. Res.* **11** 233 http://jcpr.kbs-lab.co.kr/english/journal/JCPR/vol11num2/233.pdf
[10] Wunderlich W 2009 *J. Nucl. Mat.* **389** 57

Thermoelectric properties of nitrogen-doped TiO_{2-x} compounds

M Mikami and K Ozaki

National Institute of Advanced Industrial Science and Technology, 2266-98 Anagahora, Shimo-Shidami, Moriyama, Nagoya, Aichi 463-8560, Japan

E-mail: m-mikami@aist.go.jp

Abstract. The thermoelectric properties of nitrogen-doped TiO_{2-x} containing oxygen defects have been investigated. The oxygen-defective TiO_{2-x} was synthesized by a solid-state reaction of TiO_2 and titanium mononitride, TiN. By inducing a sufficient amount of oxygen defects, a sintered sample consisting of the Magneli phase titanium oxide Ti_nO_{2n-1} was obtained. While TiO_2 is an intrinsically insulating material, the electrical resistivity was reduced by the formation of oxygen defects, which introduce electron carriers. Although the Seebeck coefficient also decreased with an increase in the amount of oxygen defects, the thermoelectric power factor was improved by the further reduction of electrical resistivity. In addition, the thermal conductivity was strongly reduced by phonon scattering at nanometer-spaced periodic planar defects of the Magneli phase crystallographic structure, resulting in the nitrogen-doped TiO_{2-x} exhibiting good n-type performance at high temperatures.

1. Introduction

Thermoelectric devices have recently attracted renewed interest for their potential application in clean energy-conversion systems. Especially, vast amounts of waste heat are targeted for thermoelectric power generation in order to enhance energy utilization efficiency. The conversion efficiency of a thermoelectric device depends mainly on the material's thermoelectric properties, which are evaluated using the thermoelectric figure of merit $Z = S^2/\rho\kappa$, where S is the Seebeck coefficient, ρ is the electrical resistivity, and κ is the thermal conductivity. Thus, the exploration of materials having a high Z value is indispensable for the practical use of thermoelectric technology as an eco-friendly energy recovery system.

Recent discovery of large S coexisting with low ρ in layered cobaltites has opened a route to the exploration of oxides as thermoelectric materials [1–4]. The cobaltites, such as Na_xCoO_2 and $[Ca_2CoO_3]_pCoO_2$, which is known as $Ca_3Co_4O_9$, exhibit p-type conduction and the ZT (where T is the absolute temperature) values are greater than 1 at 1000 K [2-4], which is the required criterion for practical application. In addition, because of their high thermal stability, high tolerance to oxidation, and low toxicity, these oxides are promising candidates for practical use for thermoelectric power generation in air at high temperatures.

On the other hand, in order to construct a thermoelectric device, an n-type thermoelectric oxide material is required. Recently, it has been reported that titanium oxides, such as Nb-doped $SrTiO_3$ and reduced TiO_2, show good n-type thermoelectric performance [5-7]. Although these titanium oxides are intrinsically insulating materials, ρ was reduced by the electron doping resulting from the Nb doping or the reduction of oxygen content. In addition, a large thermoelectric power factor (PF=S^2/ρ) can be obtained by the adjustment of the doping level to keep large S. In addition, κ should be reduced to obtain a large Z value. Since κ in titanium oxide is mainly governed by the lattice thermal conductivity, the enhancement of phonon scattering by the introduction of crystal lattice defects or grain boundaries is effective to reduce the total κ. For titanium dioxide, a relatively wide range of oxygen content can be allowed such as from TiO_2 to TiO. Therefore, an oxygen deficiency and corresponding crystal lattice defects can be easily introduced. Moreover, depending on the oxygen content, the crystallographic structure is deformed. For instance, rutile type TiO_2 was changed to a homologous series of the Magneli phase expressed as Ti_nO_{2n-1} (n=4~10) when a sufficient amount of oxygen defects was introduced [8]. Since the crystallographic structure of the Magneli phase can be described as a shear structure of rutile type TiO_2 [9,10], nanometer-spaced periodic planar defects can be naturally introduced. Then, a strong phonon scattering effect can be expected. In this study, the amount of oxygen defects and the crystallographic phase of titanium oxide were controlled by the solid-state reaction of TiO_2 and titanium mononitride, TiN, and the effect on the thermoelectric properties was investigated.

2. Experimental

The oxygen-defective titanium oxide was synthesized by a solid-state reaction of TiO_2 and TiN. The mixing ratios of these two powders were adjusted to coincide with the chemical compositions of the Magneli phase Ti_nO_{2n-1}, as listed in table 1. For instance, aiming to form the Magneli phase of Ti_8O_{15}, the mixing ratio was fixed at TiO_2:TiN=7:1. These raw powders were mixed with ethanol as a lubricant by planetary ball milling at 250 rpm for 20 h. After drying, the mixed powder was sintered at 1473 K for 10 min under a uniaxial pressure of 50 MPa by electric current sintering using a graphite mold. The heating and cooling rates were about 100 K/min. For comparison, a sintered sample made only from TiO_2 powder was also prepared to investigate the oxygen deficiency effect on the thermoelectric properties. The obtained bulk samples, with a typical size of 10 mm in diameter and 2 mm in thickness, were used for measurement of κ and then cut into a bar shape for measurement of ρ and S.

Table 1. Mixing ratio of raw powders and identified phases by XRD measurement.

Nominal Composition	Mixing ratio TiO_2:TiN	Identified phase
TiO_2	1:0	Rutile
$Ti_9(O,N)_{17}$	8:1	Ti_8O_{15}
$Ti_8(O,N)_{15}$	7:1	Ti_7O_{13}, Ti_8O_{15}
$Ti_7(O,N)_{13}$	6:1	Ti_6O_{11}, Ti_7O_{13}
$Ti_6(O,N)_{11}$	5:1	Ti_5O_9, Ti_6O_{11}

Crystallographic structure analysis was performed using X-ray diffraction (XRD) with Cu Kα radiation. The value of ρ was measured in a He atmosphere using a conventional four-probe DC technique. The value of S was calculated from a plot of the thermoelectric voltage against the temperature difference between each end of the bar shaped sample. Two Pt-Pt/Rh (R-type) thermocouples were attached to the sample ends and the Pt wires of the thermocouples were used as voltage terminals. Measured S values were corrected by S for the Pt wires to obtain the net S values of the samples. These measurements of ρ and S were simultaneously conducted using a RZ2001i (Ozawa Science Co. Ltd.) in the temperature range from 350 K to 1150 K. κ was evaluated from the density (D), thermal diffusivity (α), and heat capacity (C_p) with the relationship $\kappa=D\times\alpha\times C_p$. Of those parameters, D was measured using the Archimedes method; α and C_p were evaluated using the laser flash method in the temperature range from 300 K to 550 K.

3. Results and discussion

3.1. Phase identification

XRD measurements were performed on the polished cross section of each sintered sample. As shown in Fig. 1(a), the XRD pattern of the sample made only from TiO_2 powder agreed with the reported data for the rutile TiO_2 structure, with the exception of the weak diffraction peak (marked with ●) from a secondary phase around 2θ=26.5°. Since the sample was sintered in the reductive ambient of a graphite mold, some amount of oxygen should be reduced. In fact, the color of the sample turned to dark blue and ρ was significantly reduced compared to the intrinsically insulating TiO_2, as described below. Therefore, the secondary phase can be considered to be a reductive phase of titanium oxide such as the Magneli phase, which gives a strong diffraction peak around the 2θ angle of the additional peak.

The XRD patterns of the samples made from the mixture of TiO_2 and TiN are clearly different to the rutile phase and agree with the reported data for Magneli phase titanium oxide, with the exception of a weak diffraction peak (marked with ♦) from a secondary phase around 2θ=20.7°. This additional peak could not be identified at present. Although the identified majority phase is indicated in Fig. 1, a detailed phase identification confirmed that these samples consisted of more than one Magneli phase, as listed in table 1. It should be noted that the identified phases are described according to the XRD pattern database. These chemical formulas then just express the crystallographic phase, not the actual chemical composition. Since the sintering condition was a reductive atmosphere and titanium oxide is easily reduced at high temperature, the crystallographic phase was formed under unstable conditions. Therefore, samples consisting of a single intended Magneli phase could not be obtained by our sample preparation method. In addition, the n index of the Magneli phase Ti_nO_{2n-1} of the identified phase is almost one lower than the nominal composition of the starting mixed powders. This result is also attributed to the reductive reaction during sintering.

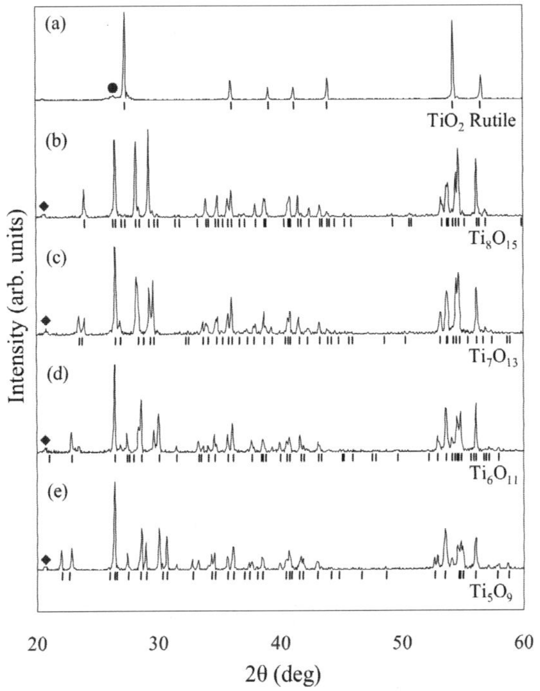

Figure 1. XRD pattern (Cu Kα radiation) from titanium oxide sintered samples with a nominal composition of a) TiO_2, b) $Ti_9(O,N)_{17}$, c) $Ti_8(O,N)_{15}$, d) $Ti_7(O,N)_{13}$, e) $Ti_6(O,N)_{11}$. The identified main phase is indicated below each XRD pattern. Peaks not identified are marked with ● or ♦.

In a preliminary composition analysis, a reduction of oxygen content during sintering was detected, while the nitrogen loss was relatively small. The difference in decomposition behavior between oxygen and nitrogen during sintering seems to be attributed to the difference in chemical stability between TiO_2 and TiN in the reductive atmosphere at high temperatures. A precise composition analysis will be conducted for further investigation of the nitrogen doping effect on the thermoelectric properties of titanium oxide. Although the chemical composition and the crystallographic phase could not be precisely controlled, the results of XRD measurements confirmed that a series of Magneli phase titanium oxides $Ti_n(O,N)_{2n-1}$ having a different n index from 8 to 5 can be synthesized by a solid-state reaction of TiO_2 and TiN. In addition, this result also indicates that the amount of oxygen deficiency can be roughly controlled by the mixing ratio of the raw powders of TiO_2 and TiN.

3.2. Electrical properties

Figure 2 shows the temperature dependence of ρ for the samples incorporating different amounts of oxygen defects. All the samples exhibit a semiconducting temperature dependence. The reduction rate of ρ with increasing temperature is reduced by an increase in the amount of oxygen deficiency. This result implies that additional electronic states were introduced into the band gap of the original rutile TiO_2 by the oxygen defects. For the sample made only from TiO_2 powder, the ρ value is significantly lower than that for the intrinsically insulating rutile TiO_2 material. This large reduction of ρ can be attributed to a reduction of the oxygen content during sintering. For the samples consisting of the Magneli phase, a further reduction of ρ was observed. Since the amount of electron doping is directly related to the oxygen deficiency and indirectly related to the n index of the Magneli phase, the monotonic decrease in ρ with decreasing n index is consistent.

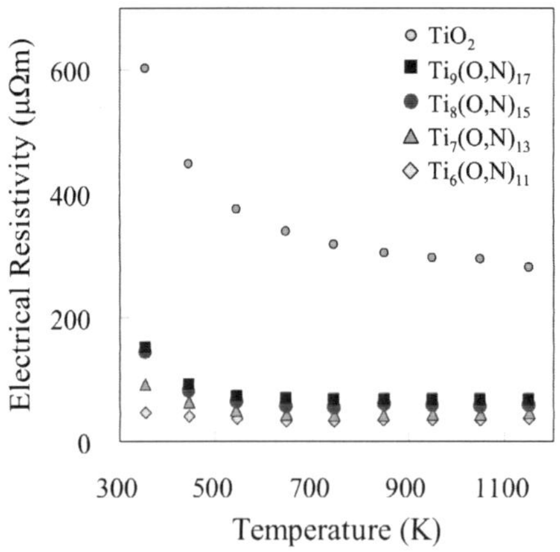

Figure 2. Temperature dependence of electrical resistivity in titanium oxide sintered samples.

Figure 3 shows the temperature dependence of S. All the samples exhibit a negative S value, resulting from the electron doping effect by the introduction of oxygen defects. The absolute S value, which is generally inversely proportional to the carrier density, decreases with increasing oxygen deficiency because the amount of electron doping is increased by the increase in oxygen reduction level. Although electron doping diminishes S, the further enhancement of electrical conductivity can enhance the thermoelectric power factor PF ($=S^2/\rho$), especially at high temperatures, as shown in Fig. 4. All the samples consisting of the Magneli phase exhibit a higher PF value compared to the rutile phase titanium oxide. The PF value monotonically increases with temperature mainly due to the increase in S at high temperatures. This temperature dependence indicates that the reduced titanium oxide can be a good thermoelectric material at high temperatures.

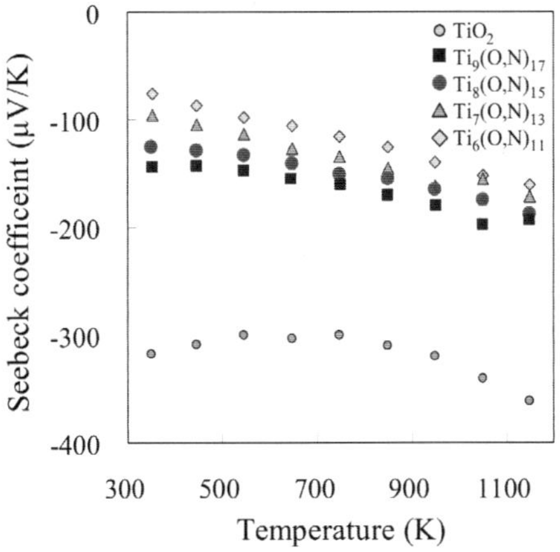

Figure 3. Temperature dependence of Seebeck coefficient in titanium oxide sintered samples.

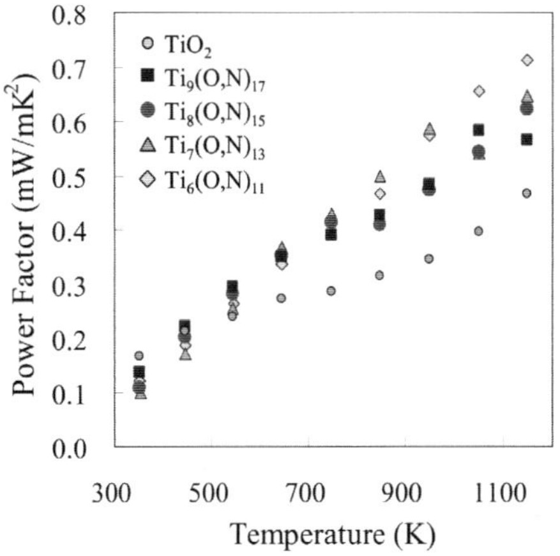

Figure 4. Temperature dependence of thermoelectric power factor in titanium oxide sintered samples.

3.3. Thermal conductivity and thermoelectric performance

In order to estimate the thermoelectric performance of the reductive titanium oxide, the κ value was measured as shown in Fig. 5. The κ value of the rutile phase titanium oxide sample is much lower than the reported κ value in single crystals (7-11 W/mK at 300 K) [9]. This reduction of κ can be attributed to the enhancement of phonon scattering at grain boundaries and point defects at oxygen sites, because the sintered sample consists of grains having sizes of several micrometers and contain a certain amount of oxygen defects as mentioned above.

For the sample consisting of the Magneli phase, a further reduction of κ was observed. For the sample mainly consisting of the Ti_7O_{13} Magneli phase (indicated by a nominal composition of $Ti_8(O,N)_{15}$ and marked with • in Fig. 5), the κ value is almost half that for the rutile phase sample. The crystallographic structure of the Magneli phase can be described as a shear structure of the original rutile type TiO_2 structure. For instance, for the Ti_8O_{15} (n=8) phase, one planar defect is introduced into every eight pure rutile type titanium dioxide stacking units. This means that nanometer-spaced periodic planar defects can be naturally introduced by the formation of the Magneli phase. Since the distance between the planer defects is the same order as the phonon mean free path in the original rutile TiO_2, the formation of the Magneli phase structure should have a strong impact on the enhancement of the phonon scattering effect. In fact, the Magneli phase samples exhibit relatively temperature independent κ values, as shown in Fig. 5. This result implies that the temperature independent phonon scattering effect of structural defects is much larger than the temperature dependent phonon-phonon scattering effect. Therefore, the large reduction of κ can be attributed to the formation of the nanometer-spaced periodic planar defect structure of the Magneli phase.

Finally, the ZT value of the sintered samples was calculated as shown in Fig. 6. Since the lattice part of κ and the electrical conductivity of the samples are almost independent of temperature above 600 K, ZT values above 600 K (indicated by open symbols) were estimated using the κ value of 550 K. The estimated ZT values increase with increasing temperature mainly due to the enhancement of S at high temperatures. All the samples consisting of the Magneli phase exhibit much larger ZT values compared to the rutile phase sample. The large reduction of κ resulting from the formation of the Magneli phase can contribute to the improvement of the ZT value.

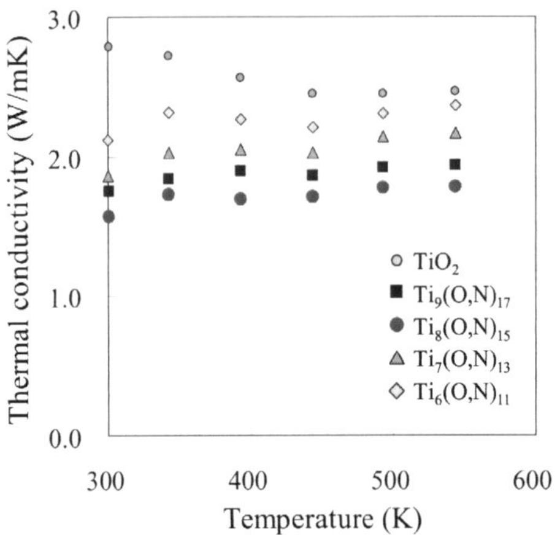

Figure 5. Temperature dependence of thermal conductivity in titanium oxide sintered samples.

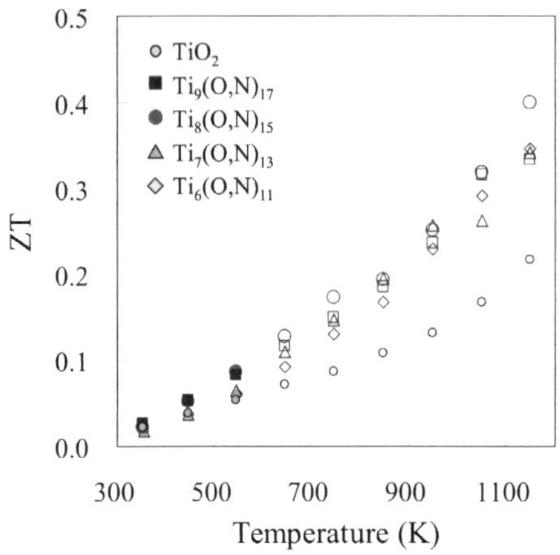

Figure 6. Temperature dependence of dimensionless thermoelectric figure of merit in titanium oxide sintered samples. Above 600 K, the κ value at 550 K was used for the estimation (marked with open symbols).

4. Conclusions

The thermoelectric properties of rutile type TiO_2 having oxygen defects and the Magneli phase $Ti_n(O,N)_{2n-1}$ were investigated. The amount of oxygen deficiency and the n index of the Magneli phase can roughly be controlled by the solid-state reaction of TiO_2 and TiN. The introduction of oxygen deficiency is effective for electron doping, resulting in a significant reduction of ρ. Although the electron doping diminishes S, a relatively high PF value was obtained at high temperatures due to the increase in S with temperature. The nanometer-order periodic planar defect structure of the Magneli phase can contribute to the strong enhancement of phonon scattering, resulting in a large reduction of κ. The estimated ZT value reached 0.4 at 1150 K in the sample mainly consisting of the Ti_7O_{13} Magneli phase. This ZT value is as high as the reported value for n-type thermoelectric oxide material.

Acknowledgements
This work was partly supported by a Grant-in-Aid for Young Scientists (B) 21760571 from the Ministry of Education, Culture, Sports, Science and Technology (MEXT), Japan.

References

[1] Terasaki I, Sasago Y and Uchinokura K 1997 *Phys. Rev. B* **56** 12685
[2] Fujita K, Mochida T and Nakamura K 2001 *Jpn. J. Appl. Phys.* **40** 4644
[3] Funahashi R, Matsubara I, Ikuta H, Takeuchi T, Mizutani U and Sodeoka S 2000 *Jpn. J. Appl. Phys.* **39** L1127
[4] Shikano M and Funahashi R 2003 *Appl. Phys. Lett.* **82** 1851
[5] Ohta S, Nomura T, Ohta H, Hirano M, Hosono H and Koumoto K 2005 *Appl. Phys. Lett.* **87** 092108
[6] Lu Y, Hirohashi M and Sato K 2006 *Mater. Trans.* **47** 1449
[7] Harada S, Tanaka K and Inui H 2010 *J. Appl. Phys.* **108** 083703
[8] Andreson S, Collen B, Kuylenstierna U and Magnel A 1957 *Acta Chem. Scand.* **11** 1641
[9] Andreson J S and Hyde B G 1967 *J. Phys. Chem. Solids* **28** 1393
[10] Liborio L and Harrison N 2008 *Phys. Rev. B* **77** 104104
[11] Thurber W and Mante A 1965 *Phys. Rev.* **139** A1655

Influence of preparation conditions on thermoelectric properties of $Ba_8Ga_{16}Si_{30}$ clathrate by combining arc melting and spark plasma sintering methods

Hiroaki Anno[1, 2], Hiroki Yamada[1], Takahiro Nakabayashi[1], Masahiro Hokazono[1] and Ritsuko Shirataki[1, 2]

[1] Department of Electrical Engineering, Tokyo University of Science, Yamaguchi, 1-1-1 Daigaku-Dori, Sanyoonoda, Yamaguchi 756-0884, Japan
[2] JST, CREST, 5 Sanbancho, Chiyoda-ku, Tokyo 102-0075, Japan

E-mail: anno@rs.tus.ac.jp

Abstract. We have investigated the thermoelectric properties of clathrate samples with a nominal $Ba_8Ga_{16}Si_{30}$ composition, prepared by combining arc melting and spark plasma sintering methods under different conditions. The carrier concentration of the samples greatly decreases from the order of 10^{21} cm^{-3} to 10^{20} cm^{-3} due to an increase in Ga content by changing the arc melting conditions. Consequently, the Seebeck coefficient increases and the thermal conductivity decreases while the electrical conductivity decreases. The effective mass, carrier mobility and lattice thermal conductivity are determined from an analysis using a single parabolic band model with alloy disorder and phonon scattering mechanisms. The thermoelectric figure of merit *ZT* significantly increases up to about 0.55 (900 K) for a carrier concentration of about 6×10^{20} cm^{-3} (RT). The simulation predicts that *ZT* can reach 0.8 (900 K) at the optimum carrier concentration (about 2×10^{20} cm^{-3}) in the $Ba_8Ga_{16}Si_{30}$ system.

1. Introduction

Inorganic clathrate systems, a class of semiconductors, have received much attention as high temperature thermoelectric materials based on the phonon-glass electron-crystal concept [1–5]. Among silicon clathrate systems, $Ba_8Ga_{16}Si_{30}$ clathrate has been well characterized and demonstrated to be a potential material for high temperature applications by several researchers [6–11]. In Si clathrate systems, a deviation of the chemical composition from the Zintl condition, however, easily occurs and yields excess carriers, which degrade the thermoelectric properties due to a shift to the metallic region [12–14]. The control of the chemical composition of the $Ba_8Ga_{16}Si_{30}$ system is one of key factors for improving the thermoelectric properties. The crystallographic properties of the $Ba_8Ga_{16}Si_{30}$ system, including the chemical composition, defects, and homogeneity are significantly affected by the synthesis method. We have reported the thermoelectric properties of several polycrystalline Ge- and Si-based clathrates prepared by combining arc melting and spark plasma sintering (SPS) methods [15–20]. Most recently, Tsujii et al. [14] reported the synthesis of polycrystalline $Ba_8Al_xSi_{46-x}$ clathrates without a framework deficiency by a similar approach and details of their thermoelectric properties. These results support the advantage of the arc melting method for preparation of polycrystalline Si clathrates. Thus, the influence of the synthesis conditions on the chemical composition and the thermoelectric properties is one of the most important issues to be investigated in order to determine the ideal synthesis conditions for $Ba_8Ga_{16}Si_{30}$ clathrate. Therefore, in the present study, we prepared polycrystalline $Ba_8Ga_{16}Si_{30}$ (nominal composition) clathrate samples by combining arc melting and SPS methods under different conditions, and investigated their thermoelectric properties.

2. Experimental

Ba ingots (3N), Ga shots (5N) and Si grains (5N) were used as starting materials. $Ba_8Ga_{16}Si_{30}$ clathrate samples were first synthesized by arc melting of the constituent elements with the Ba-rich composition $Ba_{8.2}Ga_{16}Si_{30}$ at different arc melting times and arc currents of A1 (short time: 4.4 min×5 times, 150 A), A2 (short time: 4.6 min×5 times, 250 A), B1 (long time: 16.4 min×5 times, 150 A) and B2 (long time: 22.6 min×5 times, 200 A) in an argon atmosphere. Then, the arc ingots were ground to powder (< 90 μm) and the materials were sintered at 30 or 60 MPa and 1113–1273 K for 60–120 min in an argon atmosphere by using a SPS technique to obtain dense samples. Powder x-ray diffraction (XRD) measurements were performed at room temperature in the 2θ range of 10–80 deg using Cu*K*α radiation (40 kV and 40 mA) with a Rigaku Ultima-III diffractometer. The chemical compositions were determined from the polished surface of the sintered samples by field-emission scanning electron microscopy (FE-SEM, Hitachi High Technologies S-4800) combined with energy-dispersive x-ray spectroscopy (EDS, EDAX Inc).

Measurements of the electrical conductivity σ and the Seebeck coefficient S were performed in the temperature range of 300–900 K. Hall measurements were conducted at room temperature by applying an external magnetic field of 1 T using the van der Pauw method. The Hall carrier concentration n was determined from the Hall coefficient R_H using the relation $R_H = 1/en$, where e is the electron charge. The Hall mobility μ was determined from the electrical conductivity σ and the Hall coefficient R_H using the relation $\mu = \sigma R_H$. The thermal diffusivity α and the specific heat C_p were measured by the flash method (Xe flash analyzer, Netzsch LFA447). The thermal conductivity κ was calculated by using the relation $\kappa = \rho \alpha C_p$, where ρ is the density of the sample.

3. Results and discussion

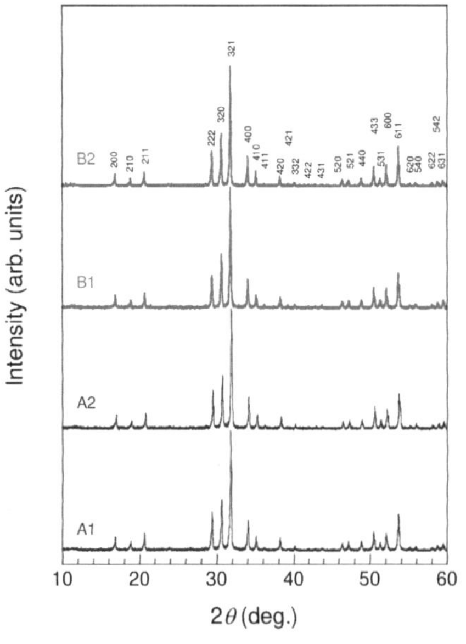

Figure 1. X-ray diffraction patterns for samples prepared under different conditions.

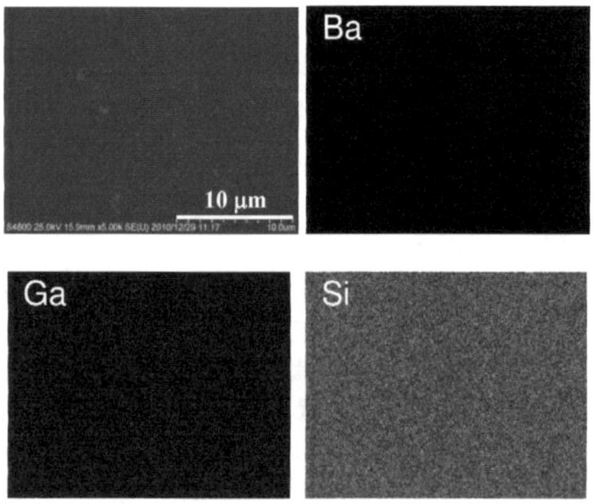

Figure 2. SEM image and elemental mapping for selected sample (B1 condition).

Table 1. Chemical composition, lattice constant a (nm), density ρ (g/cm^3), carrier concentration n (10^{20} cm^{-3}), Seebeck coefficient S (μV/K), electrical conductivity σ (S/cm) and thermal conductivity κ (Wm^{-1}K^{-1}) at room temperature for samples prepared under different conditions.

Conditions	EDS composition			a	ρ	n	S	σ	κ
	Ba	Ga	Si						
A1	7.65	14.42	31.93	1.0539	4.287	13.5	–41.6	1280	2.19
A2	7.72	13.96	32.32	1.0540	4.236	13.6	–42.2	1150	2.11
B1	7.65	14.51	31.84	1.0545	4.306	5.57	–54.6	947	1.75
B2	7.77	14.55	31.69	1.0544	4.279	9.52	–49.2	1090	1.74

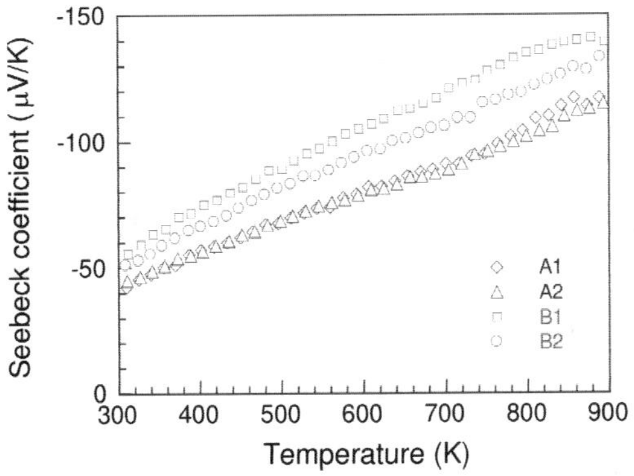

Figure 3. Temperature dependence of Seebeck coefficient for samples prepared under different conditions.

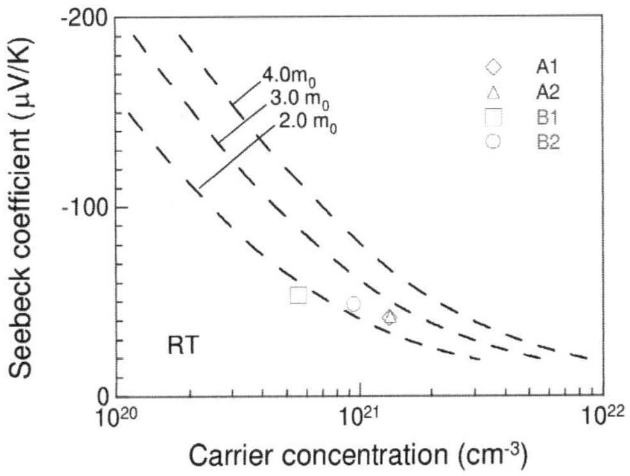

Figure 4. Seebeck coefficient as a function of carrier concentration for samples prepared under different conditions.

Figure 1 shows the powder XRD patterns for samples prepared under different conditions. The XRD patterns indicate that the main phase of all samples is the type-I clathrate structure (cubic unit cell, space group *Pm_3n*). FE-SEM observation and EDS measurements at several positions on each sample also confirmed that the chemical compositions of the samples were homogeneous. Figure 2 shows a SEM image and the distribution of Ba, Ga and Si elements for a selected sample (B1 condition). The chemical compositions determined by EDS measurements and other properties of the samples are tabulated in Table 1. The chemical compositions of the samples shift from the nominal composition to Si-rich ones. The samples for long arc conditions (B1 and B2) have slightly higher Ga contents than the samples for short arc conditions (A1 and A2). The long melting time and low arc current condition (B1) seems to be better than the others at increasing the actual Ga content, resulting in a decrease in the carrier concentration n. The Seebeck coefficient S and the electrical conductivity σ at room temperature vary reasonably with the change in the carrier concentration n. Although a deviation of the chemical composition occurs in all samples, reflections from a Ga phase are not observed in the diffraction patterns (figure 1) and no Ga phase is detected in the SEM observations with EDS (figure 2). It is inferred that most of the excess Ga was removed through the high pressure process of SPS.

Figure 3 shows the temperature dependence of the Seebeck coefficient for samples prepared under different conditions. The absolute values of the Seebeck coefficient increase with decreasing carrier concentration, as expected. The Seebeck coefficient increases with increasing temperature below the intrinsic region. The behavior is typical for a degenerate semiconductor. The room temperature

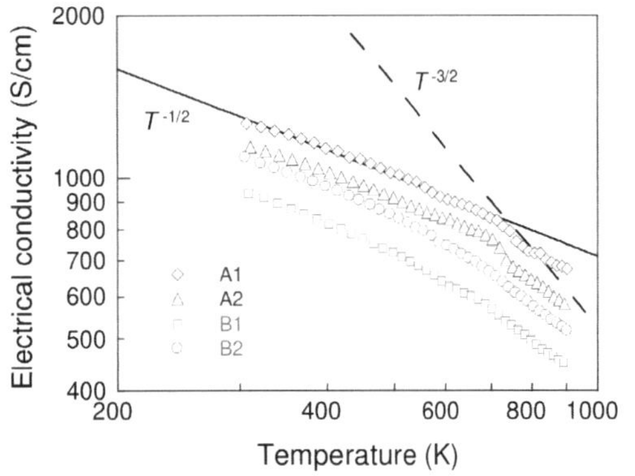

Figure 5. Temperature dependence of electrical conductivity for samples prepared under different conditions.

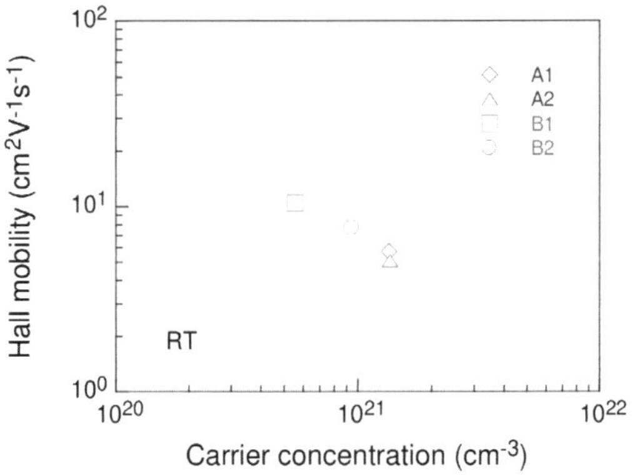

Figure 6. Hall mobility as a function of carrier concentration for samples prepared under different conditions.

Seebeck coefficient of the samples prepared under different conditions is shown in figure 4 as a function of the carrier concentration. As mentioned above, the absolute value of the Seebeck coefficient increases with decreasing carrier concentration. The dashed curves indicate the calculated Seebeck coefficient values at different effective mass m^* values using the Boltzmann transport equation. The calculation of the Seebeck coefficient as well as the estimation of the effective mass will be described in detail later. The Seebeck coefficient values vary with the carrier concentration, and it seems that the effective mass m^* of the samples is around $2.0m_0$ (m_0: the free electron mass).

Figure 5 shows a log-log plot of the temperature dependence of the electrical conductivity for samples prepared under different conditions. The temperature dependence indicates the saturation region and the electrical conductivity decreases with increasing temperature according to the scattering mechanism of carriers. We first examine the transport mechanism in the samples. In the saturation region, assuming the carrier concentration is constant, the temperature dependence of the electrical conductivity depends on the temperature dependence of the carrier mobility. The solid and dashed lines in figure 5 show the relation $\sigma \propto \mu \propto T^{-r}$ for $r = -1/2$ and $-3/2$ respectively. The temperature dependence below 700 K for all samples is very close to the relation $T^{-1/2}$ for a scattering mechanism due to alloy disorder (energy dependence $E^{-1/2}$) [21, 22]. As one possible explanation, it is likely that the disorder scattering is due to a non-uniform distribution of Ga atoms substituted for Si atoms, as is the case for Ge clathrates such as $Ba_8Cu_xGa_yGe_{46-x-y}$ [17] and $Sr_8Ga_xGe_{46-x}$ [18]. On the other hand, at high temperatures above 700 K, the temperature dependence seems to come close to the relation $T^{-3/2}$ for a scattering mechanism due to acoustic phonon scattering (energy dependence $E^{-1/2}$) [22]. We thus estimated the density-of-states effective mass m^* from the carrier concentration dependence of the Seebeck coefficient at room temperature by assuming a single parabolic band model with alloy disorder scattering and/or acoustic phonon scattering as the dominant scattering mechanism (energy dependence $E^{-1/2}$), as follows [23]. The Seebeck coefficient S is give by

$$S = -\frac{k_B}{e}\left[2\frac{F_1(\eta)}{F_0(\eta)} - \eta\right], \tag{1}$$

where k_B is the Boltzmann constant, η is the reduced Fermi energy and F_n is the Fermi-Dirac integral of n-th order. The carrier concentration n is given by

$$n = \frac{4}{\sqrt{\pi}}\left(\frac{2\pi m^* k_B T}{h^2}\right)^{3/2} \cdot F_{1/2}(\eta), \tag{2}$$

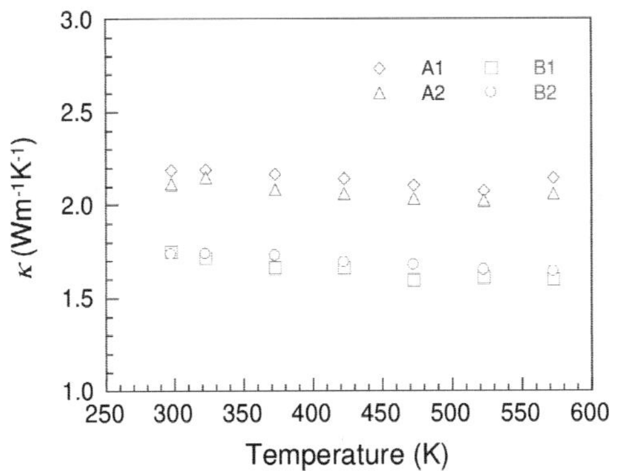

Figure 7. Temperature dependence of thermal conductivity κ for samples prepared under different conditions.

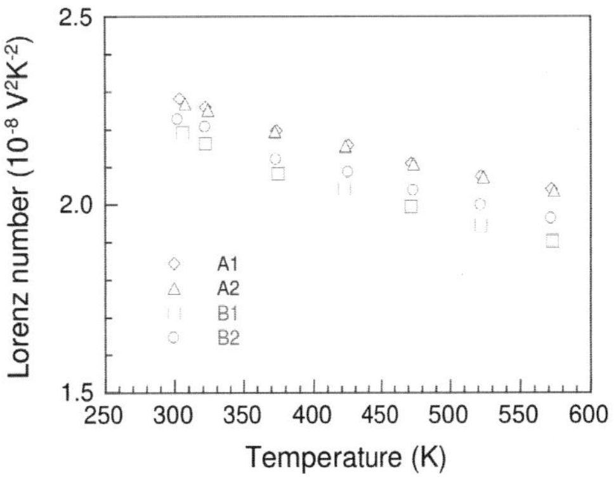

Figure 8. Temperature dependence of Lorenz number for samples prepared under different conditions.

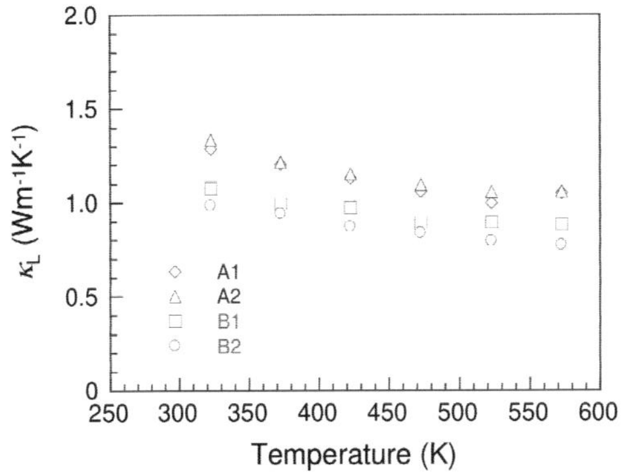

Figure 9. Temperature dependence of lattice thermal conductivity κ_L for samples prepared under different conditions.

where h is the Planck constant. The dashed curves in figure 4 are calculated using equations (1) and (2) as a function of η for different effective mass values. The effective mass m^* for the samples is almost constant at about $2.0m_0$. The effective mass m^* value of the samples agrees well with the value reported ($m^* = 2.2m_0$) by Kuznetsov et al. [7]. The effective mass m^* of the samples is larger than the reported ones for n-type $Ba_8Ga_xGe_{46-x}$ clathrates [16], suggesting that the optimum carrier concentration for the $Ba_8Ga_{16}Si_{30}$ system should be higher than that for $Ba_8Ga_xGe_{46-x}$.

The room temperature Hall mobility of samples prepared under different conditions is plotted in figure 6 as a function of the carrier concentration. The Hall mobility increases gradually with decreasing carrier concentration. The Hall mobility for the samples is comparable to that for $Ba_8Ga_xGe_{46-x}$ clathrate samples [16]. The Hall mobility value is about 10 $cm^2V^{-1}s^{-1}$ in the carrier concentration range of 10^{20} cm^{-3}, which corresponds to the optimum carrier concentration (about 2×10^{20} cm^{-3}), as mentioned later. The relaxation time ($\tau = \mu m^*/e$) of the order of 10^{-14} s is consistent with the typical value obtained in theoretical calculations of the transport properties of some clathrates [8].

Figure 7 shows the temperature dependence of the thermal conductivity of the samples prepared under different conditions. The thermal conductivity κ for the B1 and B2 conditions is smaller than that for the A1 and A2 conditions. One of factors that cause the difference in the thermal conductivity is the difference in the electronic contribution κ_e to κ. The electronic thermal conductivity κ_e was estimated by the Wiedemann-Franz relation $\kappa_e = LT\sigma$, where L is the Lorenz number. L was calculated using the reduced Fermi energy determined from the Seebeck coefficient by the Boltzmann transport

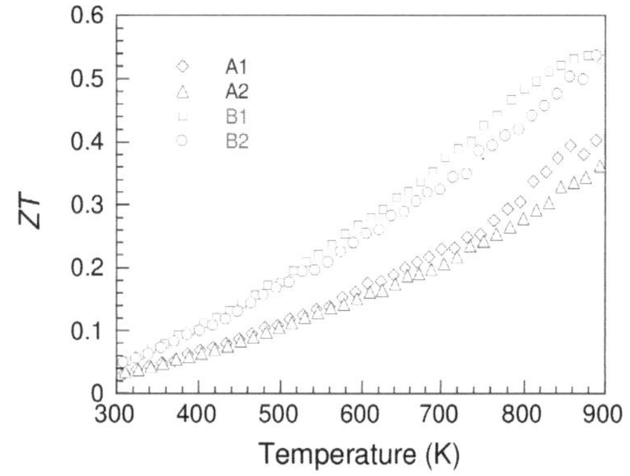

Figure 10. Temperature dependence of the dimensionless thermoelectric figure of merit ZT for samples prepared under different conditions.

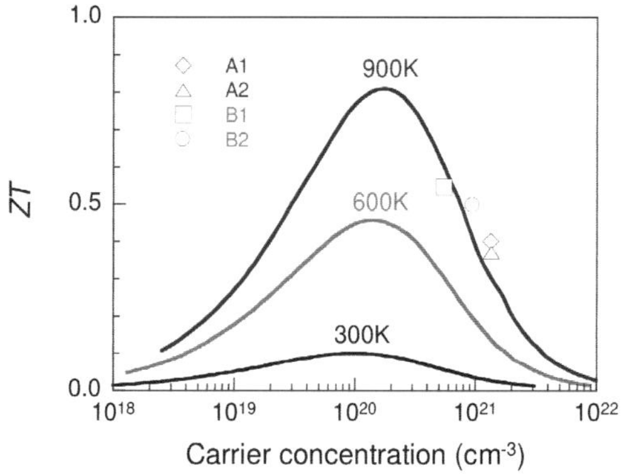

Figure 11. Dimensionless thermoelectric figure of merit ZT as a function of carrier concentration for samples prepared under different conditions.

equation [23]:

$$L = \left(\frac{k_B}{e}\right)^2 \frac{3F_0(\eta)F_2(\eta) - 4F_1(\eta)^2}{F_0(\eta)^2}. \quad (3)$$

The temperature dependence of the Lorenz number L for samples prepared under different conditions is presented in figure 8. The Lorenz number for the B1 and B2 conditions is smaller than that for the A1 and A2 conditions due to the lower carrier concentration. The electronic thermal conductivity κ_e for the B1 and B2 conditions is lower than that for the A1 and A2 conditions due to the smaller L and σ. The lattice thermal conductivity κ_L was thus estimated by subtracting the electronic thermal conductivity κ_e from κ. Figure 9 shows the temperature dependence of the lattice thermal conductivity κ_L for samples prepared under different conditions. The lattice thermal conductivity κ_L for the B1 and B2 conditions is lower than that for the A1 and A2 conditions. The lattice thermal conductivity κ_L for the B1 condition is estimated to be as low as about 1.1 $Wm^{-1}K^{-1}$, which is comparable to the reported value for $Ba_8Ga_{16}Si_{30}$ [4]. The lattice thermal conductivity κ_L decreases with increasing temperature. The temperature behavior is attributed to the Umklapp process as a typical scattering mechanism of phonons at high temperatures [24–27], and is fitted to the relation $\kappa_L \propto T^{-b}$ (b is constant): $0.36 < b < 0.45$, which is in good agreement with the theoretical value ($b = 0.5$).

The dimensionless thermoelectric figure of merit ZT was estimated from the temperature dependence of the Seebeck coefficient S, the electrical conductivity σ and the thermal conductivity κ. The thermal conductivity values at 575 K were used in the estimation of ZT above 575 K. Figure 10 shows the temperature dependence of ZT for samples prepared under different conditions. As shown in Figure 10, the samples prepared by the B1 and B2 conditions show a relatively high ZT of about 0.55 at 900 K. The improvement of ZT is attributed mainly to the decrease in the carrier concentration toward the optimum value (about 2×10^{20} cm^{-3} as mentioned below) by achieving a relatively high Ga content.

Figure 11 shows the carrier concentration dependence of the dimensionless thermoelectric figure of merit ZT for samples prepared under different conditions. ZT was calculated using a modified model that was proposed originally by Slack [1] with the Boltzmann transport equation using the experimental values of the effective mass $m^* = 2.0m_0$, Hall mobility $\mu_0 = 10$ $cm^2V^{-1}s^{-1}$ and lattice thermal conductivity $\kappa_{L0} = 1.1$ $Wm^{-1}K^{-1}$ at room temperature. The model calculation takes into account the temperature dependence as $\mu = \mu_0 (T/T_0)^{-1/2}$ for alloy disorder scattering and $\kappa_L = \kappa_{L0} (T/T_0)^{-0.41}$, where the b value is an average value extracted from the experimental fits, assuming a constant $L = 2.44\times10^{-8}$ V^2/K^2. If the Lorenz number presented in figure 8 is used, an estimation of ZT is obtained that is larger than that shown in figure 11. The calculated curve agrees well with the experimental data. Further, the calculation predicts that the ZT at 900 K is about 0.8 at the optimum carrier concentration of about 2×10^{20} cm^{-3}. Thus, it is likely that the ZT value will be further improved if the carrier concentration can be reduced in the $Ba_8Ga_{16}Si_{30}$ system. In addition, the ZT value will become higher if the carrier mobility is improved by optimization of the synthesis process such as annealing of polycrystalline samples.

4. Conclusions

Preparation conditions and the thermoelectric properties were investigated for polycrystalline samples with a nominal $Ba_8Ga_{16}Si_{30}$ composition by combining arc melting and SPS methods. The chemical composition of polycrystalline samples depends significantly on the synthesis conditions. The adjustment of the arc melting conditions leads to an increase in the actual Ga content in the samples. As a result, the carrier concentration of the samples decreases from the order of 10^{21} cm^{-3} to 10^{20} cm^{-3}. The long arc melting time and low arc current conditions are effective to some degree but still insufficient for obtaining the Zintl condition. The thermoelectric properties of the samples vary reasonably with carrier concentration. The thermoelectric material parameters m^*, μ and κ_L that govern the material efficiency were determined for the polycrystalline samples. The thermoelectric figure of merit ZT was estimated to be about 0.55 at 900 K and a relatively high ZT value of about 0.8 (900 K) is predicted from the material parameters if the carrier concentration is optimized. Further investigation into the optimization of the thermoelectric properties of the $Ba_8Ga_{16}Si_{30}$ system is in progress.

Acknowledgments

The authors would like to thank Professor N. Toshima of Tokyo University of Science, Yamaguchi for use of the Xe flash analyzer. This work was supported by JST, CREST and KAKENHI(C) No.22560708.

References

[1] Slack G A 1995 New materials and performance limits for thermoelectric cooling *CRC Handbook of Thermoelectrics* ed D M Rowe (Boca Raton: CRC Press) pp 407–440
[2] Slack G A 1997 Design concepts for improved thermoelectric materials *Mat. Res. Soc. Symp. Proc.* vol 478, ed T M Tritt, M G Kanatzidis, H B Lyon Jr and G D Mahan (Warrendale, Pennsylvania: MRS Press) pp 47–54
[3] Nolas G, Cohn J L, Slack G A and Schujman S B 1998 *Appl. Phys. Lett.* **73** 178–80
[4] Cohn J L, Nolas G S, Fessatidis V, Metcalf T H and Slack G A 1999 *Phys. Rev. Lett.* **82** 779–82
[5] Nolas G S, Slack G A, Schujman S B, 2001 Semiconductor clathrates: A phonon glass electron crystal material with potential for thermoelectric applications *Recent Trends in Thermoelectric Materials Research I, (Semiconductors and Metals* vol 69*)* ed T M Tritt (San Diego: Academic Press) Chapter 6 pp 255–300
[6] Eisenmann B, Schäfer H and Zagler R 1986 *J. Less-Common Metals* **118** 43–55
[7] Kuznetsov V L, Kuznetsova L A, Kaliazin A E and Rowe D M 2000 *J. Appl. Phys.* **87** 7871–75
[8] Blake N P, Latturner S, Bryan J D, Stucky G D, Metiu H 2001 *J. Chem. Phys.* **115** 8060–73
[9] Mudryk Y, Rogl P, Paul C, Berger S, Bauer E, Hilsher G, Godart C and Noël H 2002 *J. Phys.: Condens. Matter* **14** 7991–8004
[10] Bentien A, Iversen B B, Bryan J D, Stucky G D, Palmqvist A E C, Schultz A J and Henning R W 2002 *J. Appl. Phys.* **91** 5694–99
[11] Bentien A, Nishibori E, Paschen S and Iversen B B 2005 *Phys. Rev. B* **71** 144107
[12] Condron C L, Martin J, Nolas G S, Piccoli P M B, Schultz A J and Kauzlarich S M 2006 *Inorg. Chem.* **45** 9381–86

[13] Mugita N, Nakakohara Y, Motooka T, Teranishi R and Munetoh S 2011 *Materials Science and Engineering* **18** 142007
[14] Tsujii N, Roudebush J H, Zevalkink A, Cox-Uvarov C A, Snyder G J and Kauzlarich S M 2011 *J. Solid. State Chem.* **184** 1293–1303
[15] Anno H, Hokazono M, Kawamura M, Nagao J and Matsubara K 2002 Thermoelectric properties of $Ba_8Ga_xGe_{46-x}$ clathrate compounds *IEEE Proc. XXI Int. Conf. on Thermoelectrics (ICT2002) (Long Beach, CA, USA, 25–29 August 2002)* (Piscataway: IEEE) pp 77–80
[16] Anno H, Hokazono M, Kawamura M and Matsubara K 2003 Effect of transition element substitution on thermoelectric properties of semiconductor clathrate compounds *IEEE Proc. XXII Int. Conf. on Thermoelectrics (ICT2003) (La Grande-Motte, France, 17–21 August 2003)* (Piscataway: IEEE) pp 121–26
[17] Hokazono M, Anno H and Matsubara K 2005 *Mater. Trans.* **46** 1485–89
[18] Fujita I, Kishimoto K, Sato M, Anno H and Koyanagi T 2006 *J. Appl. Phys.* **99** 093707
[19] Koga K, Suzuki K, Fukamoto M, Anno H, Tanaka T and Yamamoto S 2009 *J. Electron. Mater.* **38** 1427–32
[20] Nakabayashi T, Hokazono M, Anno H, Ba Y and Koumoto K 2011 *Materials Science and Engineering* **18** 142008
[21] Krishnamurthy S, Sher A and Chen A 1985 *Appl. Phys. Lett.* **47** 160–162
[22] Vining C B 1991 *J. Appl. Phys.* **69** 331–41
[23] Goldsmid H J 1986 Principles for the selection and optimization of materials *Electronic Refrigeration* (London: Pion Limited) Chapter 3 pp 58–64
[24] Klemens P G 1960 *Phys. Rev.* **119** 507–09
[25] Callaway J and Baeyer H C 1960 *Phys. Rev.* **120** 1149–54
[26] Abeles B 1963 *Phys. Rev.* **131** 1906–11
[27] Yang J, 2004 Theory of thermal conductivity *Thermal Conductivity – Theory, Properties, and Applications – Physics of Solids and Liquids* ed T M Tritt (New York: Kluwer Academic/Plenum Publishers) Chapter1.1 pp 1–20

Thermoelectric properties of single crystalline $Ba_8Al_xSi_{46-x}$ clathrate by using flux Czochralski method

Yuya Nagatomo, Naoki Mugita, Yusuke Nakakohara, Makoto Saisho, Mikihito Tajiri, Ryo Teranishi and Shinji Munetoh

Department of Materials Science and Engineering, Kyushu University, 744 Motooka, Nishi-ku, Fukuoka 819-0395, Japan

E-mail: nagatomo@zaiko14.zaiko.kyushu-u.ac.jp

Abstract. Single crystalline $Ba_8Al_xSi_{46-x}$ clathrates for use as thermoelectric materials were synthesized by the flux Czochralski method. Their aluminum content increased with that of the solution. The Seebeck coefficients and electric specific resistances of the $Ba_8Al_xSi_{46-x}$ clathrates increased with aluminum content, whereas the carrier concentration decreased. The chemical compositions of the single crystalline samples from solutions of $Ba_8Al_{30}Si_{30}$, $Ba_8Al_{40}Si_{30}$ and $Ba_8Al_{50}Si_{30}$ were $Ba_{7.8}Al_{14.1}Si_{31.9}$, and $Ba_{7.8}Al_{14.6}Si_{31.4}$, and $Ba_{7.8}Al_{14.9}Si_{31.1}$, respectively. The power factors of the single crystalline $Ba_{7.8}Al_{14.1}Si_{31.9}$, $Ba_{7.8}Al_{14.6}Si_{31.4}$ and $Ba_{7.8}Al_{14.9}Si_{31.1}$ at 773 K were 6.86×10^{-4}, 7.75×10^{-4} and 8.89×10^{-4} $V^2/K^2\Omega m$. The results indicate that the carrier concentration of the $Ba_8Al_xSi_{46-x}$ clathrates can be controlled by the composition of the solution, and that the flux Czochralski method is effective for synthesizing high performance thermoelectric materials.

1. Introduction

Although fossil fuel is essential in our daily life, there is increasing concern about the increase in waste gas from cars, and the fact that the internal combustion engine exhausts 70 % of total energy as heat. Thermoelectric materials have been studied in recent decades for energy harvesting from waste heat [1, 2]. It is necessary for practical applications for them to have a figure of merit $ZT > 1$ [3, 4]. The figure of merit is defined as $ZT = S^2T/\rho\kappa$, where S is the Seebeck coefficient, T is the absolute temperature, ρ is the specific electric resistance and κ is the thermal conductivity. Several kinds of thermoelectric materials, such as BiTe, PbTe, and SiGe, have been studied, but these materials have problems such as a low melting point, harmfulness, and high cost for practical applications [5, 6, 7]. Recently, some research on other thermoelectric materials has been performed [8, 9]. Clathrate compounds have attracted attention because of their low thermal conductivity [10, 11]. The chemical formula of metal-induced type-I clathrate compounds is $A_8M_xB_{46-x}$ (A is the guest atom, group I or II elements, B is the host atom, group IV elements, M is Al, Ga, Cu, Ni, Ag, Au, etc.). The clathrate structure consists of two pentagonal dodecahedra and six tetrakaidecahedra host cages containing the guest atoms [12, 13]. The metal atoms can substitute for the host atoms. Clathrate compounds have a low thermal conductivity because of the phonon scattering of the rattling motion, which is the vibration of the guest atoms inside the cages [14, 15, 16]. It was reported that $Ba_8Ga_xGe_{46-x}$ and $Sr_8Ga_xGe_{46-x}$ clathrates

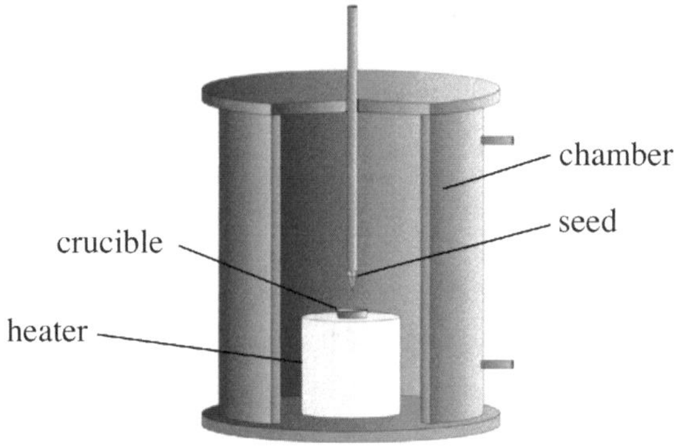

Figure 1. Schematic diagram of the CZ method.

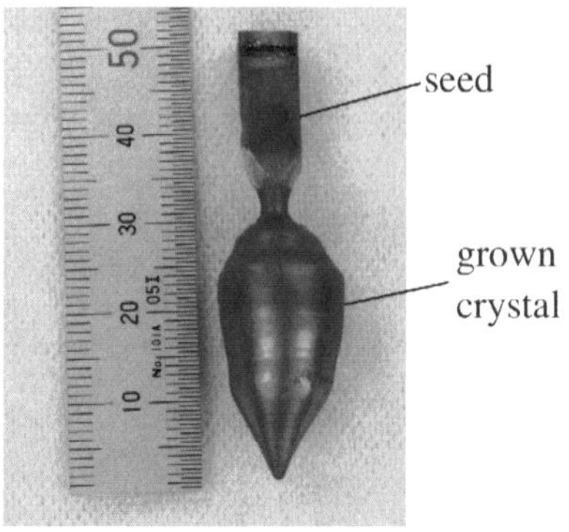

Figure 2. Typical image of crystal grown by the CZ method.

showed high thermoelectric performance [17, 18, 19, 20]. We have investigated $Ba_8Al_xSi_{46-x}$ clathrates which consists of the same group elements as $Ba_8Ga_xGe_{46-x}$ and $Sr_8Ga_xGe_{46-x}$ clathrates. In a previous study[21], we synthesized a single crystalline $Ba_8Al_xSi_{46-x}$ clathrate by the Czochralski method and reported that single crystalline samples showed high thermoelectric performance as compared to polycrystalline samples because of their low specific electric resistance. However, the chemical composition of the single crystal, $Ba_{8.0}Al_{12.0}Si_{34.0}$, was different to the stoichiometric composition, $Ba_8Al_{16}Si_{30}$. In this system, the carrier type can be controlled by adjusting the concentrations of electron donors and acceptors [22]. The Ba and Al elements act as a two-electron donor and a one-electron acceptor, respectively, and the sample with $x < 16$ shows n-type conduction in the $Ba_8Al_xSi_{46-x}$ clathrate system. The carrier concentration decreases as x approaches 16. It is well known that the Seebeck coefficient increases with decreasing carrier concentration. In the present study, single crystalline $Ba_8Al_xSi_{46-x}$ clathrates with high aluminum compositions were synthesized by the Czochralski method from a solution rich in aluminum, called the flux CZ method, in order to control the carrier concentration.

2. Experiment

To prepare the solution for the CZ method, polycrystalline $Ba_8Al_{16}Si_{30}$, $Ba_8Al_{30}Si_{30}$, $Ba_8Al_{40}Si_{30}$ and $Ba_8Al_{50}Si_{30}$ clathrates were formed by arc melting of pure elements, Ba (99%), Al (99.99%) and Si (99.999%). The samples prepared by arc melting consisted of a mixed phase of the clathrate and an Al rich phase. The polycrystalline samples were crushed, put in the crucible and melted at 950 ~ 1000 °C in an argon atmosphere. The argon gas flow was at 1.0 L/min. A schematic diagram of the CZ method is shown in Figure 1. A [001] orientation controlled single-crystalline clathrate was used as a seed crystal for the CZ method. The seed crystal was dipped into the solution and pulled up at a pulling speed of 1 mm/h and a rotation speed of 10 rpm. The crystal structures and lattice parameters of the samples were analyzed by powder X-ray diffraction (XRD) with Cu Kα radiation. Orientation Imaging

Table 1. Chemical composition of the CZ samples by WDX.

Nominal composition of melt	Chemical composition of CZ sample
$Ba_8Al_{16}Si_{30}$	$Ba_{8.0}Al_{12.0}Si_{34.0}$
$Ba_8Al_{30}Si_{30}$	$Ba_{7.8}Al_{14.1}Si_{31.9}$
$Ba_8Al_{40}Si_{30}$	$Ba_{7.8}Al_{14.6}Si_{31.4}$
$Ba_8Al_{50}Si_{30}$	$Ba_{7.8}Al_{14.9}Si_{31.1}$

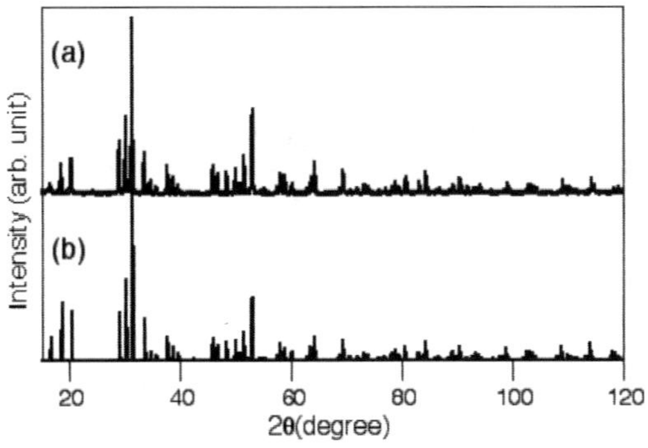

Figure 3. XRD patterns of (a) CZ sample of $Ba_{7.8}Al_{14.1}Si_{31.9}$ and (b) simulation result.

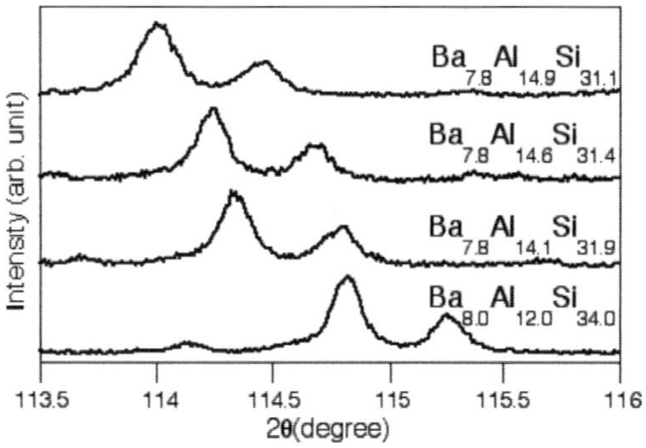

Figure 4. XRD patterns near (2 3 11) peak of CZ samples.

Microscopy images of the samples were obtained by Electron Backscatter Diffraction (EBSD). The Seebeck coefficient was determined by the relationship between the voltage and temperature difference between the two ends of the samples. The electric specific resistance was measured by the four probe method. Single crystalline samples were cut into disks perpendicular to the growth direction, and Seebeck coefficient and electric specific resistance measurements were performed along the [001] direction of each sample.

3. Results and discussion

A typical crystal grown by the CZ method is shown in Figure 2. A 30 mm long and 10 mm diameter BaAlSi crystal was pulled from the solution by using the CZ method. The chemical compositions of the CZ samples are shown in Table 1. The composition is determined by assuming that the total number of aluminum and silicon atoms is 46. The chemical compositions of the samples were different from the compositions of the solution. The barium content of the samples was slightly lower than the nominal compositions. This suggests that the barium atoms were released from the silicon cages during the crystal growth. The aluminum content of all samples was under 16, and $Ba_{8.0}Al_{12.0}Si_{34.0}$ was pulled from the stoichiometric solution ($Ba_8Al_{16}Si_{30}$). The aluminum content of the samples gradually increased with increasing aluminum content of the solution and $Ba_{7.8}Al_{14.9}Si_{31.1}$ was pulled from the solution of $Ba_8Al_{50}Si_{30}$. The results of powder XRD measurements are shown in Figure 3. The patterns for all the samples were in good agreement with the simulation result for a type-I clathrate and no peaks due to other phases were observed. The diffraction patterns near the (2 3 11) peak of the CZ samples are shown Figure 4. The (2 3 11) peak shifted to lower angle with increasing aluminum content. The lattice parameters of the $Ba_{8.0}Al_{12.0}Si_{34.0}$, $Ba_{7.8}Al_{14.1}Si_{31.9}$, $Ba_{7.8}Al_{14.6}Si_{31.4}$, and $Ba_{7.8}Al_{14.9}Si_{31.1}$ samples were 10.584, 10.612, 10.618 and 10.632 Å, respectively, and increased with the aluminum content of

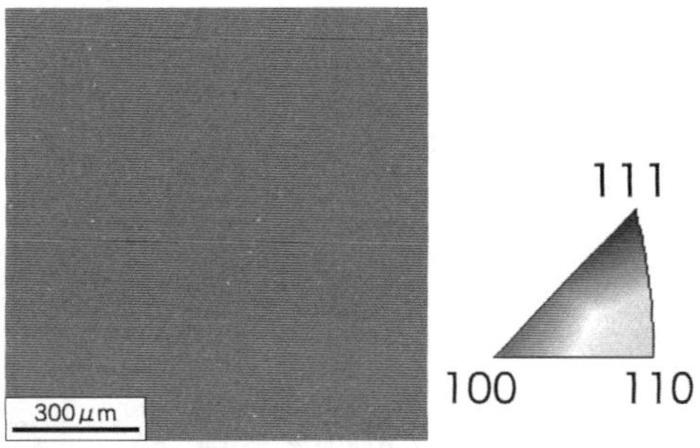

Figure 5. Orientation Imaging Microscopy image of CZ sample.

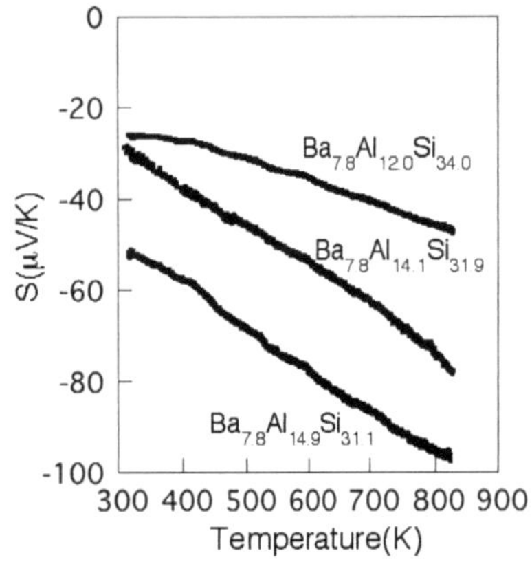

Figure 6. Seebeck coefficient of CZ sample.

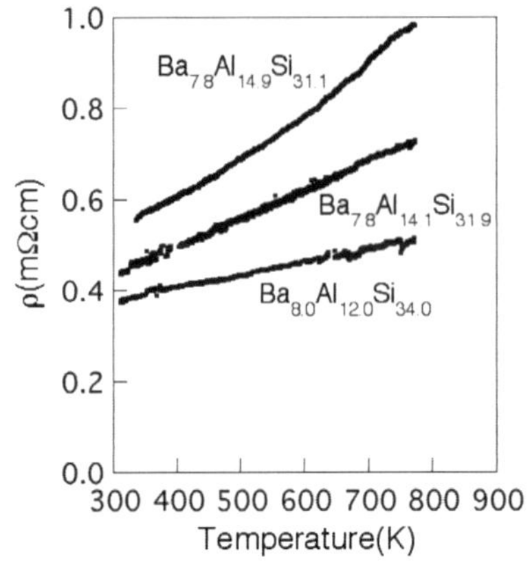

Figure 7. Electric specific resistances of CZ sample.

the samples. This suggests that the aluminum atoms substituted for the silicon atoms in the cages of the clathrate structure because the atomic size of aluminum is larger than silicon. An Orientation Imaging Microscopy image of the $Ba_{7.8}Al_{14.1}Si_{31.9}$ sample pulled from the solution of $Ba_8Al_{30}Si_{30}$ clathrate is shown in Figure 5. The color shows the directions of the sample. The image shows an almost uniform color corresponding to [001]. This indicates that a single crystal was successfully synthesized by the CZ method and the growth direction was controlled to [001]. The results for the Seebeck coefficients are shown in Figure 6. All the samples showed n-type conduction, and the absolute values of the Seebeck coefficients increase with temperature and also with the aluminum content of the samples. The Seebeck coefficients of the samples show low values of -26.0 ~ -51.7 μV/K at low temperature and the values for the $Ba_{7.8}Al_{14.1}Si_{31.9}$, $Ba_{7.8}Al_{14.6}Si_{31.4}$, and $Ba_{7.8}Al_{14.9}Si_{31.1}$ single crystalline clathrates were -71.4, -81.3 and -93.5 μV/K at 773 K, respectively. Kuznetsov *et al.* reported that the Seebeck coefficients of a polycrystalline BaGaSi clathrate was around -120 μV/K at 773 K [23]. Uemura *et al.* reported that the value was around -90 μV/K at 773 K for a polycrystalline BaAlGe clathrate [24]. In Figure 6, the slope of the curves becomes steeper at high temperature for all samples. The slopes also show a slight dependence on the aluminum content of the samples. This suggests that the electronic structure and the carrier concentration change by the substitution of aluminum. It is known that the Seebeck coefficient increases with decreasing carrier concentration and our results suggest that the carrier concentration of the single crystalline clathrate decreases with the aluminum content of the sample. Figure 7 shows the results for the electric specific resistances. The electric specific resistance

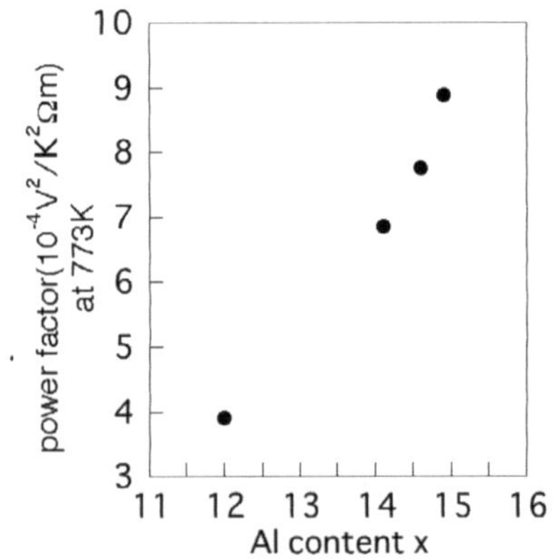

Figure 8. power factor of the CZ sample at 773K.

increases with temperature, indicating a typical heavily doped semiconductor behavior. The electric specific resistance of the samples also increases with the aluminum content of the samples. The values for the $Ba_{7.8}Al_{14.1}Si_{31.9}$, $Ba_{7.8}Al_{14.6}Si_{31.4}$ and $Ba_{7.8}Al_{14.9}Si_{31.1}$ single crystalline clathrates were 0.725, 0.865 and 0.990 mΩcm at 773 K, respectively. These results suggest that the carrier concentration of the single crystalline clathrates decreases with the aluminum content of the samples. The power factors of the single crystalline clathrates at 773 K are shown in Figure 8. In our samples, although the electric specific resistance increased with the aluminum content, the Seebeck coefficient has more effect on the power factor (S^2/ρ) than the electric specific resistance. As a result of the Seebeck coefficient and the electric specific resistance, the power factor increased with the aluminum content. The maximum value was 8.89×10^{-4} $V^2/K^2\Omega m$ at 773 K for the $Ba_{7.8}Al_{14.9}Si_{31.1}$ single crystalline clathrates. The thermoelectric properties of the $Ba_8Al_xSi_{46-x}$ clathrate improved as the aluminum content approached 16 because of the high Seebeck coefficient due to the low carrier concentration. In this study, although the $Ba_{7.8}Al_{14.9}Si_{31.1}$ single crystalline clathrate was succesfully synthesized by using the CZ method, the carrier concentration of the sample seems to be high, of the order of 10^{21} cm^{-3}. It is expected that the power factor of the $Ba_8Al_xSi_{46-x}$ clathrates can be further improved by decreasing the carrier concentration.

4. Conclusions

Single crystalline $Ba_8Al_xSi_{46-x}$ clathrates were synthesized by the flux Czochralski method from solutions with various chemical compositions. The chemical composition of the CZ samples were different from the composition of the solution in that the aluminum content of the samples was less than that of the solution. The aluminum content of the samples increased with the aluminum content of the solution. The Seebeck coefficient and the electric specific resistance increased with the aluminum content due to a decrease in the carrier concentration. The value of the Seebeck coefficient at 773 K was -93.5 μV/K for the $Ba_{7.8}Al_{14.9}Si_{31.1}$ sample. As a result, the power factor increased with the aluminum content because of the increase in the Seebeck coefficient, and the value for $Ba_{7.8}Al_{14.9}Si_{31.1}$ was 8.89×10^{-4} $V^2/K^2\Omega m$ at 773 K. These results suggest that the flux Czochralski method can control the carrier concentration of $Ba_8Al_xSi_{46-x}$ clathrates, and is effective for synthesizing high performance thermoelectric materials.

References

[1] Rama Venkatasubramanian, Edward Siivola, Thomas Colpitts and Brooks O Quinn 2001 *NATURE* **413** 597

[2] Sales B C, Mandrus D and Williams R K 1996 *SCIENCE* **272** 1325

[3] Bed Poudel, Qing Hao, Yi Ma, Yucheng Lan, Austin Minnich, Bo Yu, Xiao Yan, Dezhi Wang, Andrew Muto, Daryoosh Vashaee, Xiaoyuan Chen, Junming Liu, Mildred S Dresselhaus, Gang Chen and Zhifeng Ren 2008 *SCIENCE* **320** 634

[4] Lon E Bell 2008 *SCIENCE* **321** 1457

[5] Li Bassi A, Bailini A, Casari C S, Donati F, Mantegazza A, Passoni M, Russo V and Bottani C E 2009 *J. Appl. Phys.* **105** 124307

[6] Taichao Su, Xiaopeng Jia, Hongan Ma, Jianggang Guo, Yiping Jiang, Nan Dong, Le Deng, Xinbing Zhao, Tiejun Zhu, Cui Wei 2009 *J. Alloys Compd.* **468** 410

[7] Giri Joshi, Hohyun Lee, Yucheng Lan, Xiaowei Wang, Gaohua Zhu, Dezhi Wang, Ryan W Gould, Diana C Cuff, Ming Y Tang, Mildred S Dresselhaus, Gang Chen and Zhifeng Ren 2008 *Nano Lett.* **8** 4670

[8] Yoshiyuki Kawaharad, Ken Kurosaki, Masayoshi Uno and Shinsuke Yamanaka 2001 *J. Alloys Compd.* **315** 193

[9] Park K, Jang K U, Kwon H C, Kimc J G and Chod W S 2006 *J. Alloys Compd.* **419** 213

[10] Jung Hwan Kim, Norihiko L Okamoto, Kyosuke Kishida, Katsushi Tanaka and Haruyuki Inui

2006 *Acta Mater.* **54** 2057
[11] Huo D, Sakata T, Sasakawa T, Avila M A, Tsubota M, Iga F, Fukuoka H, Yamanaka S, Aoyagi S, and Takabatake T 2005 *Phys. Rev. B* **71** 075113
[12] Norihiko L Okamoto, Katsushi Tanaka and Haruyuki Inui 2006 *J. Appl. Phys.* **100** 073504
[13] Norihiko L Okamoto, Katsushi Tanaka and Haruyuki Inui 2006 *Acta Mater.* **54** 173
[14] Charles W Myles, Koushik Biswas and Emmanuel Nenghabi 2007 *Phys. B* **401-402** 695
[15] Jianjun Dong, Otto F Sankey, Ganesh K Ramachandran and Paul F McMillan 2000 J. Appl. Phys. **87** 7726
[16] Takasu Y, Hasegawa T, Ogita N, Udagawa M, Avila M A, Suekuni K, and Takabatake T 2010 *Phys. Rev. B* **82** 134302
[17] Daniel Bryan J, Nick P Blake, Horia Metiu and Galen D Stucky, Bo B Iversen, Rasmus D Poulsen and Anders Bentien 2002 *J. Appl. Phys.* **92** 7281
[18] Eric S Toberer, Christensen M, Iversen B B, and Jeffrey Snyder G 2008 *Phys. Rev. B* **77** 075203
[19] Nolas G S, Cohn J L, Slack G A and Schujman S B 1998 *Appl. Phys. Lett.* **73** 178
[20] Isao Fujita, Kengo Kishimoto, Masaya Sato, Hiroaki Anno and Tsuyoshi Koyanagi 2006 *J. Appl. Phys.* **99** 093707
[21] Mugita N, Nakakohara Y, Teranishi R and Munetoh S 2002 *J. Matt. Res* **26** 1857
[22] Avila M A, Huo D, Sakata T, Suekuni K and Takabatake T 2006 *J. Phys. Condens. Matter* **18** 1585
[23] Kuznetsov V L, Kuznetsova L A, Kaliazin A E, and Rowe D M 2000 *J.Appl. Phys.* **87** 7871
[24] Takashi Uemura, Koji Akai, Kenji Koga, Terumitsu Tanaka, Hiroki Kurisu, Setsuo Yamamoto, Kengo Kishimoto, Tsuyoshi Koyanagi, and Mitsuru Matsuura 2008 *J. Appl. Phys.* **104** 013702

Synthesis and thermoelectric properties of p-type $Ba_8Au_xSi_{46-x}$ clathrate

Makoto Saisho, Liu Bin, Yuya Nagatomo, Yusuke Nakakohara, Ryo Teranishi and Shinji Munetoh

Department of Materials Science and Engineering, Kyushu University, 744 Motooka, Nishi-ku, Fukuoka 819-0395, Japan

E-mail: saisyo@zaiko14.zaiko.kyushu-u.ac.jp

Abstract. Poly-crystalline $Ba_8Au_xSi_{46-x}$ clathrates were synthesized by the arc melting method and a subsequent annealing treatment. The as-synthesized sample contained some precipitates rich in Au at the grain boundaries, and the precipitates were reduced by annealing at 1000 °C. The Au content of the clathrate phase for the as-synthesized sample, $Ba_{7.4}Au_{4.7}Si_{41.3}$, was slightly lower than that for the stoichiometric composition, $Ba_8Au_6Si_{40}$. In the annealed samples, the Au content of the clathrate phase increased with annealing time. These results indicated that Au can diffuse from Au-rich precipitates to the clathrate phase by annealing. The Seebeck coefficient of the as-synthesized and 1 hour annealed sample, $Ba_{7.5}Au_{5.7}Si_{40.3}$, were -80 and +100 μV/K at 500 °C, respectively. This suggested that the carrier type was changed by the Au content in the clathrate. In the samples annealed for longer than 1 hour, the Seebeck coefficient decreased with annealing time. This indicated that the hole carrier concentration increased in excess because of overdoping due to the diffusion of Au.

1. Introduction

The energy conversion efficiency of a typical internal combustion engine is around 30% and the remainder is exhausted as waste heat. Thermoelectric materials have been studied for several decades as power generators to convert waste heat into electric power [1, 2, 3]. The dimensionless figure of merit ($ZT = \alpha^2T/\rho\kappa$) is an index of the performance of a thermoelectric material, where α is the Seebeck coefficient, ρ is the electrical resistivity, κ is the thermal conductivity, and T is the absolute temperature. The ZT value needs to be larger than 1 for practical applications [4]. For thermoelectric materials like BiTe, PbTe and SiGe compounds [5,6,7], the charge carrier concentration can be controlled by the substitution and doping of impurities in order to tune the thermoelectric properties. However, these compounds have some problems such as a low melting point, harmfulness and high cost elements. Many other compounds have been studied as thermoelectric materials [8,9,10]. Si-based clathrate compounds have also attracted attention as high-performance thermoelectric materials [11,12]. These compounds are harmless and their melting temperature is high, at around 1000 °C. Type-I Si clathrates consist of two regular dodecahedra and six 14-faced polyhedral Si cages. The guest atoms, alkaline earths like Ba, are enclosed in each Si cage. The clathrate compounds have a low thermal conductivity caused by the rattling vibration of the guest atoms [13,14]. Many studies of high-performance Si clathrates have been performed by controlling the chemical composition. However, it is reported that most Si-based clathrate compounds show n-type conduction [15,16] and there are few reports of p-type Si-based clathrates [17]. The p-type material is necessary in order to manufacture the n-p modules of thermoelectric generators. Transition elements, like Au, Ag and Cu, can be substituted for Si atoms as electron accepters. Jaussaud et al. [17] have reported that $Ba_8Au_xSi_{46-x}$ clathrates with x = 5.4 and 5.9 show n- and p-type conduction, respectively. The Seebeck coefficients for the $Ba_8Au_xSi_{46-x}$ clathrates with x = 5.4 and 5.9 are around -7.0×10^{-6} and $+2.8 \times 10^{-5}$ V/K at 250 K, respectively. However, no thermoelectric properties at high temperature, around 500 °C, have been reported. In this study, we synthesized the p-type $Ba_8Au_xSi_{46-x}$ clathrate by arc melting and subsequent annealing and measured the thermoelectric properties at high temperature.

2. Experimental

Pure Ba (99%), Au (99.9%) and Si (99.999%) were prepared as the starting materials. These elements were weighed to the composition of $Ba_8Au_6Si_{40}$ because Au atoms substitute for Si in the 6c site of the space group $Pm\bar{3}n$. The poly-crystalline samples were synthesized by the arc melting method. These samples were enclosed in quartz tubes in vacuum and annealed at 1000°C for five different times (10 minutes, 1, 5, 10 and 24 hours). The heating rate during annealing was set to 10 °C/min. The annealed samples were cut into specimens of $7 \times 5 \times 1.5$ mm^3 for the measurements of the thermoelectric properties. The crystal structure of these samples was examined by powder X-ray diffraction (XRD). The chemical composition of each sample was determined by EPMA with a wavelength dispersive X-ray spectrometer (WDX). The electrical resistivity was measured by the four- probe method and the Seebeck coefficient was determined by the relationship of the voltage and the temperature difference between the two ends of the samples from room temperature to 500 °C, with a heating rate of 10 °C/min. The temperature difference between the two ends of the sample was set at 20 °C.

3. Results and discussion

An X-ray diffraction pattern for the as-synthesized sample and the simulation result are shown in Figure 1. The simulation result was calculated assuming that the 6c site was fully occupied by Au atoms. These peaks are indexed to the cubic unit cell corresponding to the clathrate structure. The lattice parameter of the as-synthesized sample was 10.36 Å. This value is slightly smaller than 10.42 Å for $Ba_8Au_{5.9}Si_{40.1}$ reported by Jaussaud et al. Figure 2 shows back-scattered electron (BSE) images of the as-synthesized and annealed samples. The grain size of the as-synthesized sample was around 50 μm. Many precipitates were observed at the crystal grain boundaries for the as-synthesized sample. The grain size increased to around 200 μm by annealing. The 10 minutes annealed sample contains light gray regions around the precipitates. These light gray regions expanded with annealing time and the whole grain was occupied by them after 5 hours of annealing.

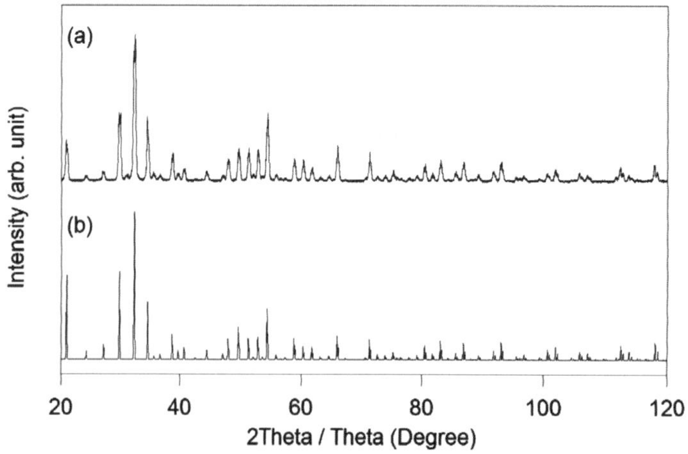

Figure 1. XRD spectra: (a) as-synthesized sample and (b) simulation.

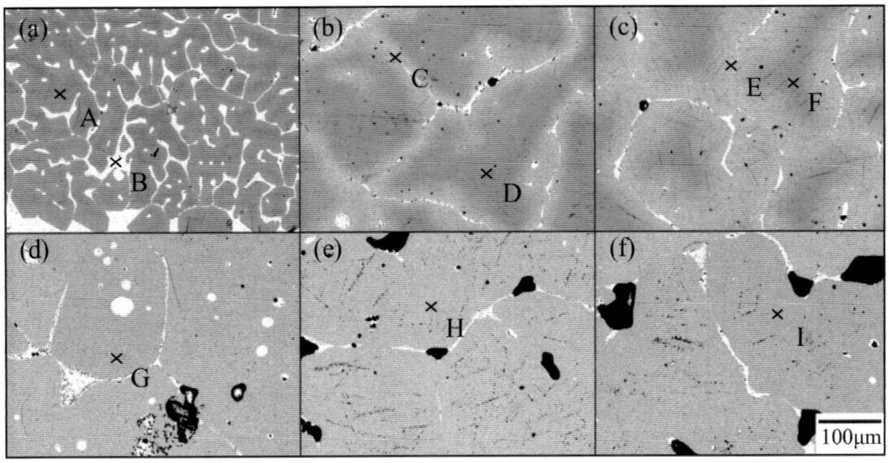

Figure 2. BSE images of (a) as-synthesized and (b) 10 minutes, (c) 1 hour, (d) 5 hours, (e) 10 hours and (f) 24 hours annealed samples.

Table 1 shows the chemical compositions determined by WDX at the points indicated in Figure 2. The values of the WDX results are normalized by assuming that the total number of Au and Si atoms is 46 with the exception of the point B. It could be confirmed that the white and the dark gray regions in Figure 2(a) are precipitates rich in Au and the $Ba_8Au_xSi_{46-x}$ clathrate phases by the results of WDX, respectively. The Au content in the clathrate phase is lower than that of the nominal composition, x = 6, because the precipitates contain a large amount of Au. The Au content for the light gray regions around the precipitates, like points C, E and G in Figure 2(b) and (c) is slightly larger than that for the dark gray $Ba_8Au_xSi_{46-x}$ clathrate phases. Based on the results of the chemical compositions by WDX, it is considered that these light gray regions were also $Ba_8Au_xSi_{46-x}$ clathrate phases. The volume ratio of the light gray $Ba_8Au_xSi_{46-x}$ clathrate phase increases with annealing time. It is suggested that Au can diffuse to the $Ba_8Au_xSi_{46-x}$ clathrate phases from the precipitates rich in Au during the annealing. For the samples annealed over 5 hours, in Figure 2(d), (e) and (f), the chemical compositions of the $Ba_8Au_xSi_{46-x}$ clathrate phases were almost uniform and their Au contents gradually increase with annealing time. Figure 3 shows the temperature dependence of the Seebeck coefficient. Although the as-synthesized sample shows n-type conduction, all annealed samples containing the $Ba_8Au_xSi_{46-x}$ clathrate phases rich in Au show p-type conduction. These results can be explained by the charge valance. In the $Ba_8Au_xSi_{46-x}$ clathrate system, Ba behaves as a double electron donor and Au behaves

Table 1. Chemical compositions by WDX for the points (A ~ I) of Figure 2.

The point of Figure 2	Chemical Composition		
	Ba	Au	Si
A	7.4	4.7	41.3
B	18	55	27
C	7.3	5.2	40.8
D	7.4	4.7	41.3
E	7.5	5.4	40.6
F	7.4	4.5	41.5
G	7.4	5.3	40.70
H	7.5	5.5	40.5
I	7.5	5.7	40.3

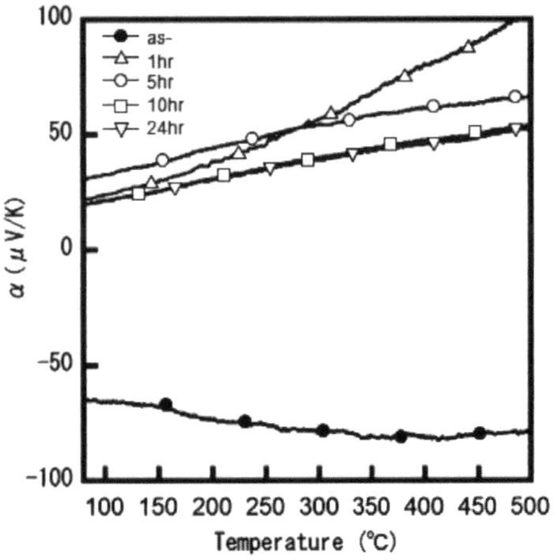

Figure 3. Temperature dependence of the Seebeck coefficient.

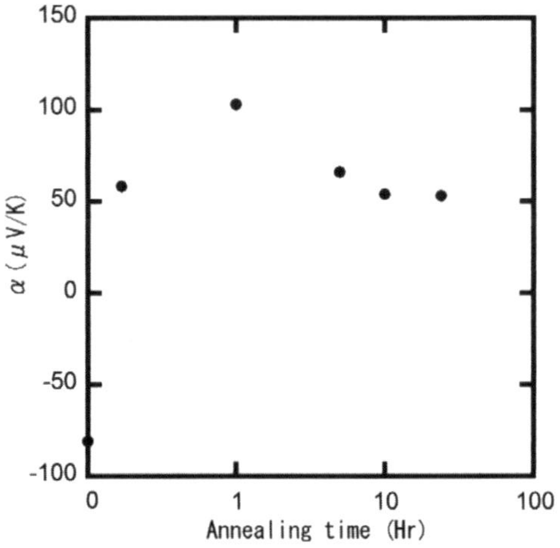

Figure 4. Annealing time dependence of the Seebeck coefficient at 500 ˚C.

as a triple electron acceptor. In this system, the Au content required for exact electron compensation is x = 5.33 and then the content ratio of Ba/Au is 1.5. A sample with a Ba/Au content ratio smaller than 1.5 should exhibit p-type conduction. The Ba/Au content ratio for the as-synthesized sample was 1.57. This causes n-type conduction. For the annealed samples containing the $Ba_8Au_xSi_{46-x}$ clathrate phases rich in Au, the Ba/Au content ratio is smaller than 1.5. These $Ba_8Au_xSi_{46-x}$ clathrate phases cause the p-type conduction.

Figure 4 shows the annealing time dependence of the Seebeck coefficient at 500 ˚C. The Seebeck coefficient has a maximum point at 1 hours annealing and this value is around +100 μV/K. In the range of annealing time from 0 to 1 hour, the Seebeck coefficient increased with annealing time. In this range, the volume ratio of the $Ba_8Au_xSi_{46-x}$ clathrate phases rich in Au increases with annealing time by diffusion of Au from the precipitates. The absolute value of the Seebeck coefficient of the samples increases with the volume ratio of the $Ba_8Au_xSi_{46-x}$ clathrate phases rich in Au. However, the Seebeck coefficient decreased with the increase of the annealing time for the samples annealed over 1 hour. This indicated that Au was doped in excess in the clathrate phases and the carrier concentration of the clathrate phase increased by the annealing due to the diffusion of Au from the precipitates rich in Au to the $Ba_8Au_xSi_{46-x}$ clathrate phases.

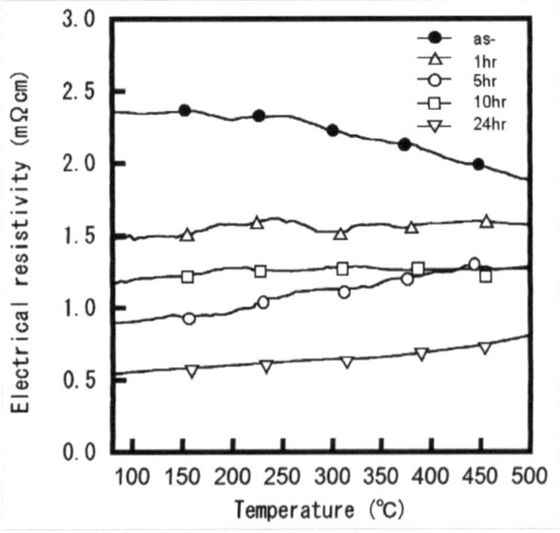

Figure 5. Temperature dependence of the electrical resistivity.

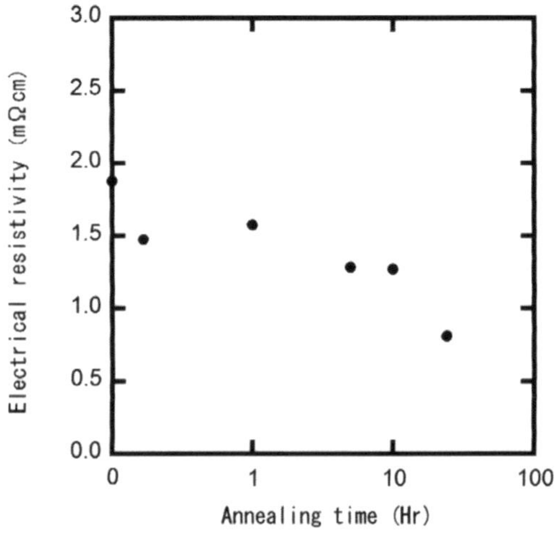

Figure 6. Annealing time dependence of the electrical resistivity at 500 ˚C.

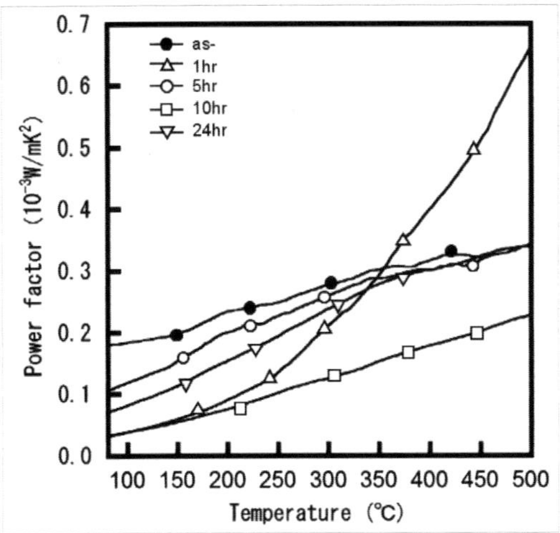

Figure 7. Temperature dependence of the power factor.

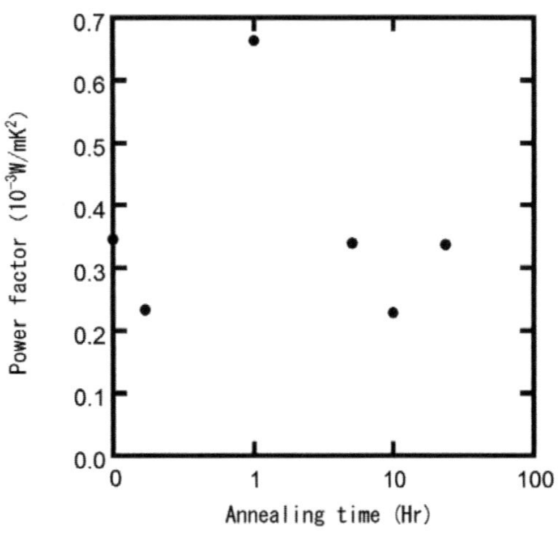

Figure 8. Annealing time dependence of the power factor at 500 ˚C.

Figure 5 shows the temperature dependence of electrical resistivity and Figure 6 shows the annealing time dependence of the electrical resistivity at 500 ˚C. The value for the as-synthesized sample is higher than that for all the annealed samples because of the difference of the crystal grain size. As the crystal grain size increases, the electron mobility of the annealed samples becomes higher than that of as-synthesized sample. The specific electric resistances for the over 1 hour annealed samples decrease with the increase of the annealing time because the carrier concentrations of the clathrate phase increase with increasing annealing time by the diffusion of Au from the precipitates to the $Ba_8Au_xSi_{46-x}$ clathrate phases. Figure 7 shows the temperature dependence of the power factor and Figure 8 shows the annealing time dependence of the power factor at 500 ˚C. The maximum power factor was around 0.7×10^{-3} W/mK^2 for the 1 hour annealed sample. This value is smaller than the power factor for a poly-crystalline sample $Ba_8Au_{5.33}Ge_{40.67}$. Zhang et al. has reported this value to be 1.2×10^{-3} W/mK^2 [18]. More improvements of the thermoelectric properties of the p-type $Ba_8Au_xSi_{46-x}$ clathrate are necessary for practical applications.

4. Conclusions

Poly-crystalline $Ba_8Au_xSi_{46-x}$ clathrates were synthesized by arc melting and subsequent annealing. The annealing times were 10 minutes, 1, 5, 10 and 24 hours. The as-synthesized sample had the $Ba_8Au_xSi_{46-x}$ clathrate phase poor in Au and showed n-type conduction. All annealed samples contained the $Ba_8Au_xSi_{46-x}$ clathrate phases rich in Au and showed p-type conduction. The Seebeck coefficient of the sample annealed for 1 hour reached around 100 μV/K at 500 ˚C. In the annealing time range from 0 to 1 hour, the Seebeck coefficient increased with annealing time because the volume ratio of the clathrate phase, whose Ba/Au content ratio is smaller than 1.5, increases with annealing time by diffusion of Au from the precipitate to the clathrate phase. For the over 1 hour annealed samples, the Seebeck coefficient decreases with increasing annealing time because the hall carrier concentration of these samples increases in excess by overdoping with Au. The resistivity decreases with increasing annealing time, for the over 1 hour annealed samples, because the carrier concentration increases by diffusion of Au. The maximum value of the power factor was for the 1 hour annealed sample and that value is 0.66×10^{-3} W/mK^2.

References

[1] Disalvo F J 1999 *SCIENCE.* **285** 703
[2] Mahan G, Sales B and Sharp J 1997 *Phys. Today.* **50** 42
[3] Wood C 1988 *Rep. Pro. Phys.* **51** 459

[4] Bed Poudel, Qing Hao, Yi Ma, Yucheng Lan, Austin Minnich, Bo Yu, Xiao Yan, Dezhi Wang, Andrew Muto, Daryoosh Vashaee, Xiaoyuan Chen, Junming Liu, Mildred S Dresselhaus,Gang Chen and Zhifeng Ren 2008 *SCIENCE* **320** 634

[5] Brostow W, Menard K P and White J P 2001 *Materials Research Society Symposium Proceedings.* **691** 425

[6] Terasaki I, Sasago Y and Uchinokura K 1997 *Phys. Rev.* **56** 12685

[7] Giri Joshi, Hohyun Lee, Yucheng Lan, Xiaowei Wang, Gaohua Zhu, Dezhi Wang, Ryan W Gould, Diana C. Cuff, Ming Y. Tang, Mildred S. Dresselhaus, Gang Chen and Zhifeng Ren 2008 *Nano Lett.* **8** 4670

[8] H Hohl, A P Ramirez, C Goldmann, G Ernst, B Wolfing and E bucher 1999 *J. Phys. condens matter* **11** 1697

[9] H Anno, K Ashida, K Matsubara, G S Nolas, K Akai, M Matsuura and J Nagao 2002 *Mat. Res. Soc. Symp. Proc.* **691** 49

[10] Y Isoda, N Shioda, H Fuziu, Y Imai, Y Shinohara 2004 *Proc. The 6th Inter. Conf. On Ecobalance* 649

[11] Bentien A, Iversen B B, Bryan J D, Stucky G D, Palmqvist A E C, Schultz A J and Henning R W 2002 *J. Appl. Phys.***91** 5694

[12] D Huo, T Sakata, T Sasakawa, M A Avila, M Tsubota, F Iga, H Fukuoka, S Yamanaka, S Aoyagi, and T Takabatake 2005 *Phys. Rev. B* **71** 075113

[13] Nolas G S, Slacl G A and Schujman S B 2001 *Recent Trends in Thermoelectric Materials Research I.* **69** 255

[14] Y Takasu, T Hasegawa, N Ogita, M Udagawa, M A Avila, K Suekuni, and T Takabatake 2010 *Phys. Rev. B* **82** 134302

[15] Simon J, Anders B, Georg K H M and B B Iversen 2006 *Chem. Mater.* **18** 4633

[16] N Mugita, Y Nakakohara, R Teranishi and S Munetoh 2002 *J. Matt. Res* **26** 1857

[17] Nicolas J, Pierre G, Stanislas P, Bernard C, Michel M, Patrice D, Rodolphe D, Graziella G, Christian C and Michel P 2005 *C. R. Chimie.* **8** 39

[18] Hui Z, Horst B, Niels O, Christophe C, Walter S, Marcus S, Ulrich B, Michael B, Jing-Tai Z and Yuri G 2011 *Inorg. Chem.* **50** 1250

The healing effect of stearic acid applied on amorphous carbon film with dispersed nanodiamonds

T Yamazaki[1], K Shoji[2], T-T Ikeshoji[1] and A Suzumura[1]

[1]Graduate School of Science and Engineering, Tokyo Institute of Technology, 2-12-3 I1-31 O-okayama, Meguro-ku, Tokyo 152-8552, Japan
[2]Graduate student of Graduate School, Tokyo Institute of Technology, ABB 26-2 Cerulean Tower Sakuragaoka, Shibuya-ku, Tokyo 150-8512, Japan

E-mail: yamazaki.t.ae@m.titech.ac.jp

Abstract. Diamond-like carbon (DLC) films have been shown to improve the friction and wear properties of applications with severe loads. Hydrogenated amorphous carbon films (a-C:H), a type of DLC film, were coated onto 440C stainless steel disk surfaces. The plasma enhanced chemical vapour deposition method was used as a coating process, and nanodiamond (ND) particles were dispersed into the film during the deposition process. Stearic acid was expected to affect the lubricating properties of the film because of its reported effectiveness on diamond surfaces. A lubricated reciprocal sliding friction test with 10,000 cycles confirmed that stearic acid improved the friction and wear properties of the ND dispersed a-C:H film. Furthermore, film wear caused by a sliding test without stearic acid could be healed by sliding with stearic acid as long as the level of critical damage was not reached. Raman spectra for the worn tracks revealed the healing effect of the stearic acid on the film because the spectra for the worn tracks after 10,000 cycles were identical to those of the control film, which was not worn. In contrast, the spectra for the worn tracks or debris showed a broad D band peak and no G peak, which was consistent with disordered amorphous carbon.

1. Introduction

Hard coatings, such as diamond-like carbon (DLC), applied on sliding surfaces, have been utilized in machines to suppress wear and friction [1]. However, the hard coating could become worn with use and unable to perform properly. The *in situ* healing of the film can sustain low friction and extended life. Thus, DLC films have been extensively utilized in the industry due to their ability to heal. For example, these films have been used in oil on valve lifters in gasoline engines, where the contact pressure can be as high as several GPa, as energy saving eco-materials. The term "diamond-like carbon" describes a diverse group of materials with different properties, and they can be used in a large number of applications. Donnet et al. investigated hydrogen containing-a-C:H films and DLC films containing transition metals with doped hydrogen [2], whereas Voevodin et al. investigated DLC films with embedded tungsten carbide (WC) and tungsten disulfide (WS_2) nanograins for aerospace use [3]. Nanocluster diamond (ND) powder is composed of particles with a diameter of 5 nm. Dispersed ND powders combined with a stearine lubricant can operate very quietly on sliding surfaces [4]. Hydrogenated amorphous carbon can produce smooth surfaces on a nanometer scale [5, 6]. Films with ND can operate under high contact pressure. Stearic acid acts as an adhesive for carbon films due to its large number of end groups [7]; thus, it has been applied to films for its healing ability to improve the recovery of the surfaces and the friction coefficient in oil.

The difference between films with and without stearic acid after a reciprocal sliding test was investigated by Raman spectroscopic analysis. The Raman spectra of disordered and amorphous carbon are given in reference [8]. Performing X-ray photoelectron spectroscopy (XPS) for the film surface before and after the scratch test confirmed structural differences between using hydrocarbon, graphite and diamond. The XPS result was used to consider the differences between surfaces produced in a previous severe scratch test [9] and the current test near the conduction band.

2. Experimental procedures

2.1. Fabrication of ND/a-C:H Film

Substrate disks of 440C stainless steel were heat-treated for surface hardening and for precipitation of carbon to enable the adherence of the ND/a-C:H film. The disks were subsequently polished to produce a smooth surface. The average Vickers hardness of the disks was 640. The ND powder, which was washed in an H_2SO_4 and HNO_3 acid solution at an elevated temperature and rinsed with distilled water, was obtained from Tokyo Diamond Tools Mfg. Co., Ltd [4, 10]. Approximately 0.5 mg of the ND powder was spread on a pure graphite lower electrode with a diameter of 80 mm, as shown in Figure 1. The films were deposited on the disks by radio frequency (RF) plasma-enhanced chemical vapour deposition in an atmosphere of methane, argon and hydrogen gas. The stainless steel disk was fixed on the water-cooled upper electrode. The disk was cleaned by argon ion sputtering for 10 minutes under a direct current (DC) high voltage of 2 kV after the chamber was evacuated to 10^{-3} Pa. A gas mixture of argon with 10% methane and hydrogen gas was injected into the chamber at the same flow rate of 0.83×10^{-6} m^3/s (50×10^{-3} l/min) after the chamber was evacuated to 10^{-3} Pa again. The mixture gas pressure in the chamber was set to approximately 1 Pa. The distance between the electrodes was set to a maximum of 52 mm.

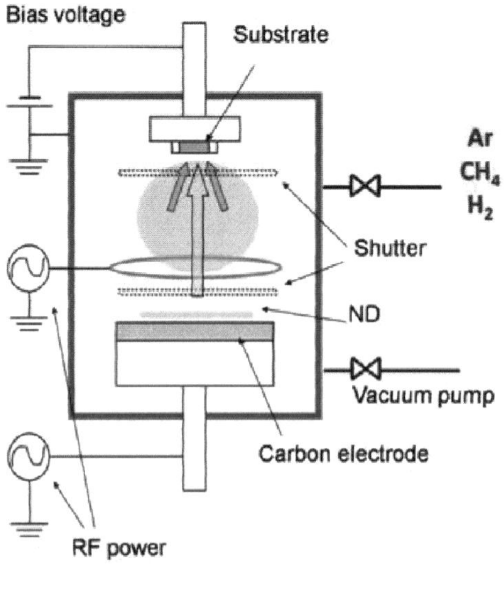

Figure 1. RF plasma-enhanced chemical vapour deposition apparatus with parallel plates.

Deposition was performed under the following conditions to produce an a-C:H film with 46 at% hydrogen [2]. An RF power of 220 W was applied to the lower electrode. Another RF power source was connected to the middle coil to maintain the plasma. A negative DC bias voltage (under –100 V) was applied to the disk. The highest amount of ND particles dropped out from the deposited film when no voltage was applied. Therefore, the bias voltage was set to -100 V to secure the ND powder in the film. This plasma deposition process was maintained for one hour to deposit a 200 nm thick ND/a-C:H film on the disk. The deposited film was checked by measuring the electrical resistance over a length of about 10 mm at the surface. The threshold values were 10^4 Ω for an a-C:H film, 10^6 Ω for a ta-C film and 10^2 Ω for a passive oxide film on stainless steel 440C.

Visual observations revealed transparent films with a lead-like surface lustre, which was in contrast to most DLC films, which are brown or black. The films adhered well to the 440C stainless steel disks. Figure 2 shows a photograph of a film deposited on a stainless steel sample. The upper corners were masked during deposition, so that the difference between the heights of the masked and unmasked areas can be used to determine the thickness of the film. An optical microscopy image of the film surface is shown in Figure 3. Several particles approximately 1 μm in diameter can be observed. These particles were observed when the ND particles were mounted on the graphite electrode in the deposition process.

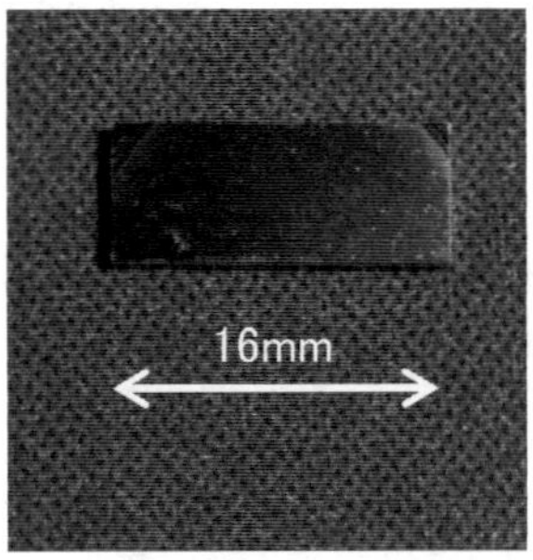

Figure 2. Photograph of film deposited on quenched stainless steel 440C.

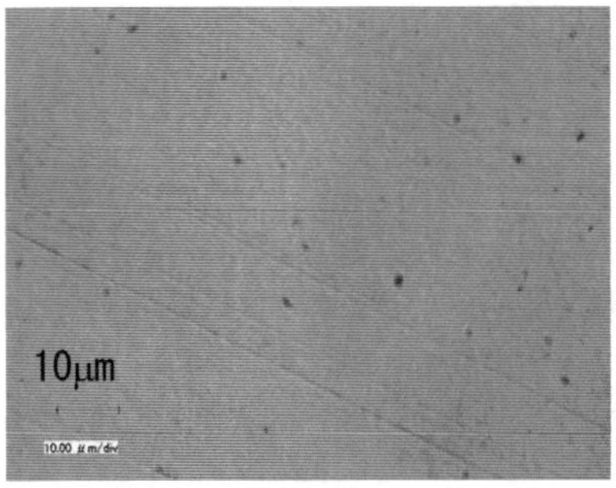

Figure 3. Optical microscopy image of the film surface.

The surface roughness R_{max} of the film was about 10 nm according to the image shown in Figure 4, acquired by AFM in non-contact mode (NC-AFM). The grayscale in the figure indicates a range from 1 to 10 nm. The roughness probably resulted from the 5-nm-diameter ND particles that impinged on the film during the deposition. The small roughness value indicates that the ND powder particles were individually dispersed into the film during plasma deposition. Peng et al. investigated the relationship between the surface roughness and the ion impingement energy [5]. According to their results, a gas pressure of 1 Pa and a -100 V bias should produce a smooth DLC film with a surface roughness of R_q < 0.1 nm. The ND particles that precipitated electrostatically due to the low bias voltage remained on the film because of their mass, whereas the hydrocarbon ions accelerated by the voltage penetrated into the film. The experiment showed that the size of the ND powder particles influenced the surface roughness, which was consistent with reference [5].

2.2. Reciprocal sliding test

A reciprocal sliding test sensor was developed to apply a severe load to the deposited films at high contact pressure without the testing tip jumping on the substrate disk. Figure 5 shows the hemispherical shell structure of the sensor. A silicon nitride ball with a 1.7 mm diameter was mounted on the stainless steel tube plunger as a tip for the pin that contacted the film. The tip was fixed on the stainless steel hemispherical shell sensor. Thus, the tip produced a high contact pressure on the film. A friction test was performed following deposition of oil droplets on the film.

Reciprocal sliding motion was produced by a stepping motor and a linear guide gear fixed on the stage. The specimen disk was fixed on the X-axis stage, and the vertical load was applied by moving the Z-axis stage downward. The distortion of the hemispherical shell was detected by two strain gauges that were adhered to the shell.

The test conditions were as follows: the sliding length was 1 mm, the sliding velocity was 1 mm/s and the lubricant oil was paraffin-based oil VG32 with a dynamic viscosity of 28.8-35.2 mm^2/s at 313 K (40°C). A stearine coat was applied to dissolve the stearic acid powder into the oil.

When the lubricant was tested, contact first occurred at a sliding velocity of zero, and was maintained to both ends of the reciprocal sliding test strip. The lubricant test was conducted with oil alone or oil with dissolved stearic acid.

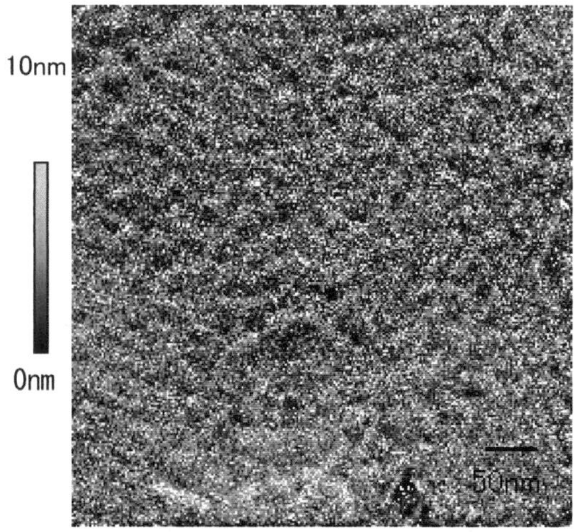

Figure 4. Topograph of the deposited film surface acquired by non-contact mode atomic force microscopy.

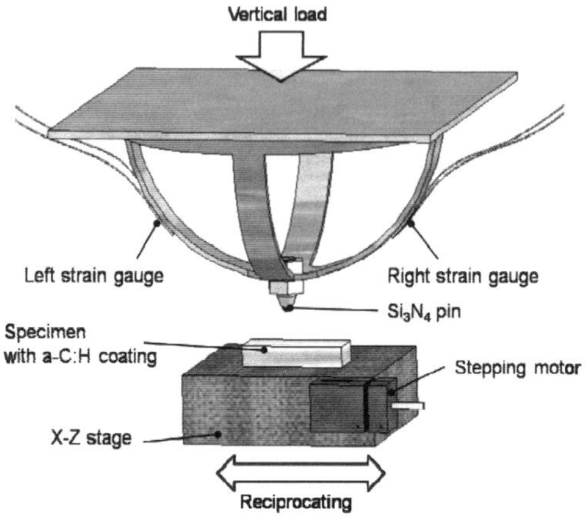

Figure 5. High-stiffness reciprocal sliding test instrument.

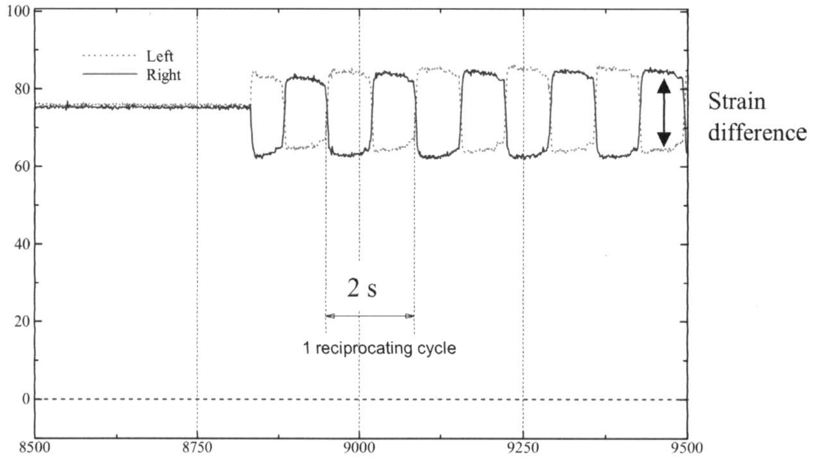

Figure 6. Output signal from strain gauges. Sliding velocity: 1.0 mm/s.

The output signals of the right and left strain gauges in the cases of sliding velocities of 1.0 mm/s are shown in Figure 6. The output signals reflected the sliding motion of the tip. The output signal from the left strain gauge was just the inverse of that from the right strain gauge during sliding motion. Therefore, the normal force was measured by reading the same value of the strain gauges at the intersection of the output signals, and the horizontal force with sliding was measured by reading the difference between the values from both strain gauges during sliding motion. In Figure 6, the normal load was calculated from the strain values. On the vertical axis, the value was 75, which corresponded to 7.5 με, and the load calculated from the linear equation was 0.98 N. On the other hand, the force related to friction was estimated from the strain difference between the output signals during sliding. The value was 20, which corresponded to 2.0 με; thus the friction force was estimated from the proportional relationship between the strain difference and lateral force to be 0.087 N. The friction coefficient was calculated to be 0.089. The change in the friction coefficient of the film revealed the effect of adding stearic acid to the oil within the first 10 sliding cycles. The friction coefficient decreased from 0.12 to 0.08 with friction. The structure on the scratched surface in the high contact region withstood the compressive stress and provided wear resistance, similar to the thin diamond film in reference [12], which also exhibited a buckling pattern produced by compressive stress. The small deformation represented by the arcs of the surface structure fixed on the ND in reference [9] indicated a low friction coefficient because a large surface deformation increases the surface roughness and prevents the pin from sliding smoothly.

In the case that the sliding velocity was 0.5 mm/s, the output signal showed a spike in response to the sticking motion of the pin at the initiation of reciprocal sliding. When the pin was stuck, the tip of the pin jumped up from the film. Thus, it was impossible to evaluate the damage to the film and its healing. The sliding velocity was set to 1.0 mm/s to prevent sticking.

When the load reached 3.93 N, the developed pin-on-disk sensor indicated severe abrasive wear conditions of the deposited film on stainless steel disk. The maximum Hertzian contact pressure, P_{max} was estimated to be 3.8 GPa from equations (1, 2) [13].

$$P_{max} = \frac{3W}{2\pi a^2} \tag{1}$$

$$a = \left(\frac{RW}{K}\right)^{1/3} \quad ; \frac{1}{R} = \frac{1}{R_1} + \frac{1}{R_2}, \; \frac{1}{K} = \frac{3}{4}\left(\frac{1-\nu_1^2}{E_1} + \frac{1-\nu_2^2}{E_2}\right) \tag{2}$$

In equation (1), W is the load, and a is the radius of the contact area in the Hertz contact model. In equation (2), two elastic spherical bodies are denoted by the subscripts 1 and 2. Number 1 represents the Si_3N_4 ball and Number 2 represents the ND/a-C:H film. R is the radius of the body, E is Young's modulus, and ν is the Poisson ratio. For the Si_3N_4 ball, E_1 is 280 GPa and ν_1 is 0.25. For amorphous carbon, R_2 is infinity, E_2 is 1076 GPa, which is almost identical with the value acquired by a nano-indentation test for ND/a-C:H, and ν_2 is 0.2 [14] ,.

3. Results and discussion

3.1. Number of critical sliding cycles and value of critical contact pressure

Figure 7 shows the traces of the friction coefficients with sliding cycles. The test conditions were as follows: a contact pressure of 2.49 GPa without stearic acid or 2.37 GPa with stearic acid. Figure 8 shows optical microscopy images of the wear tracks after 10,000 cycles of sliding; these correspond to the plots shown in Figure 7.

When the contact pressure was 2.0 GPa, the widths of the wear tracks after 10,000 cycles of sliding with or without stearic acid were both approximately 20 μm. The estimated value from equation (2) was 22.7 μm. These traces were related to the elastic deformation of the silicon nitride balls, stainless steel substrates and ND/a-C:H films and indicated a small amount of damage to the film. For the case without stearic acid shown in Figure 8(a), wear tracks were seen as a 50 μm wide ribbon, although the contact width was estimated to be a 28 μm wide band with the maximum contact pressure in the center of the mated silicon nitride ball. This was related to deep wear of the film. The friction coefficient of 0.13 at 4,000 cycles of sliding gradually decreased with sliding. This was related to a decrease of the maximum contact pressure as the contact area enlarged by wear. The sample with stearic acid shown in Figure 8(b) showed few wear tracks and had a friction efficient of 0.08 from Figure 7(b), and the wear tracks became wider than 28 μm. These images show deformation and wear of the films as dark particles. In the case that the contact pressure was approximately 3 GPa, the widths of wear tracks became wider, due to large deformation of the films and deep wear. Based on these wear tracks, the critical number of reciprocal sliding cycles was set to 10,000 for the remaining sliding tests on the films at a contact pressure of 2.5 GPa.

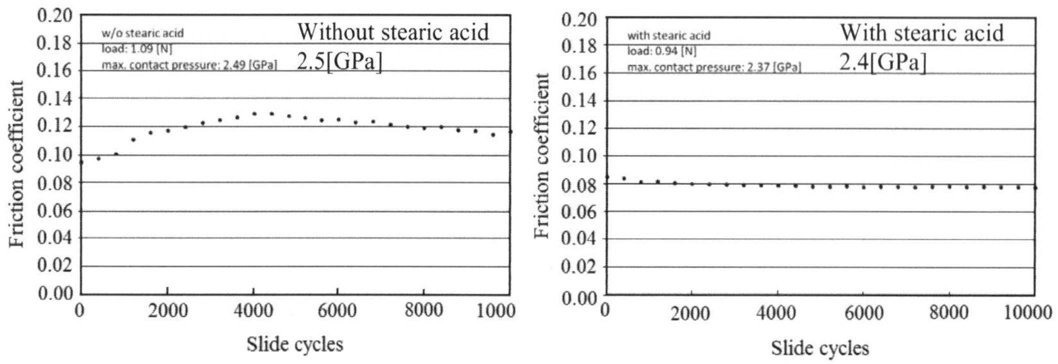

(a) Without stearic acid (b)With stearic acid

Figure 7. Plots of average friction coefficient every 400 cycles at contact pressure of about 2.5 GPa with or without stearic acid.

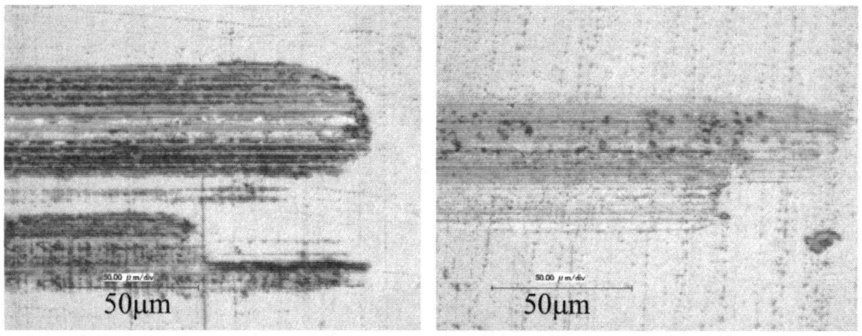

(a) Without strearic acid (b) With strearic acid

Figure 8. Wear tracks after 10,000 cycles of sliding.

3.2. Raman spectra for wear tracks on the films

Raman spectra (laser input wavelength of 514.5 nm) from the ND/a-C:H film are shown in Figure 9. These spectra were acquired from the specimens after washing in an ultrasonic bath filled with acetone. The spectra differed from the reported spectra for amorphous carbon films. A sharp G peak was observed, but no D peak was seen. This was likely due to the higher optical band gap of ta-C:H [8]. The Raman peak indicated a Stokes shift related to the conduction band. The formation of a diamond structure with a tetrahedral arrangement of carbon atoms decreased the amount of conductive carbon.

As shown in Figure 10, this phenomenon was also observed in XPS results for the valence electron energy of the ND/a-C:H film before and after a scratch test in a previous study [9]. The scratch test was performed using the silicon cantilever of an atomic force microscope. Peaks from the benzene ring were observed for the scratched sample, but no peak was detected for the as-deposited sample. The results indicate the scratched surface was composed of benzene rings, similar to the reconstructed diamond surface [12, 15, 16]. Furthermore, the deposited surface was composed of disordered carbon containing hydrogenated carbon atoms. The author considers this crystallization mechanism to be similar to crystallization due to high internal stresses in amorphous material, because bombardment by high energy ions of the ND/a-C:H film increases the number of sp^3 bonds [17-20].

The Raman spectra for the tracks after 10,000 cycles of the reciprocal sliding test without stearic acid exhibited a broad D peak and a diminished sharp G peak, as shown in the middle panel of Figure 9. In contrast, the Raman spectra for the tracks with stearic acid resembled the spectra of the ND/a-C:H control film, and a sharp G peak was observed. The wear track became wider than the elastically deformed silicon nitride ball, as mentioned in section 3.1, and the friction coefficient and Raman spectrum indicated an undamaged film. The friction coefficient of 0.08 is similar to that for a hydrogenated amorphous carbon film [20] and it indicated that the stearic acid protected against the release of hydrogen atoms from the friction surface, with the stearic acid molecules adhered to the wear track or the silicon nitride ball.

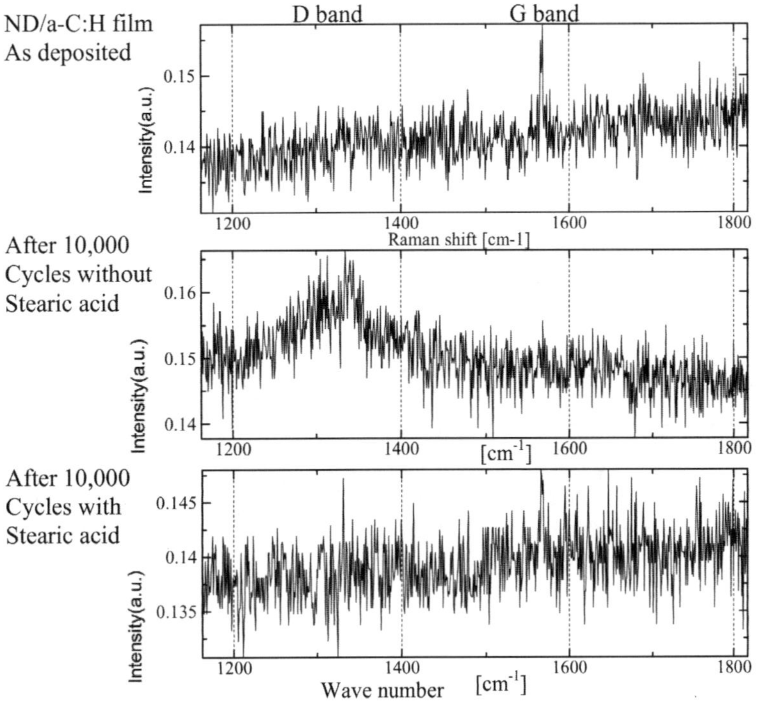

Figure 9. Raman spectra from wear tracks with an input laser wavelength of 514.50 nm.

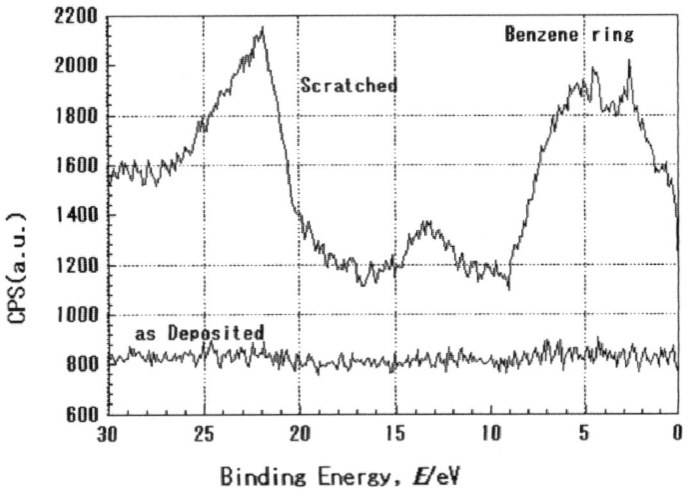

Figure 10. XPS analysis of valence electrons in ND/a-C:H film.

3.3. Healing effect of stearic acid on ND/a-C:H film

Figure 11 shows the change in friction coefficient due to the healing effect of applying stearic acid to a ND/a-C:H film after the film was damaged by sliding at a contact pressure of 2.5 GPa without stearic acid. Upon the application of stearic acid after 4,000 cycles, the friction coefficient decreased below 0.08, similar to the friction coefficient of the film with stearic acid applied at the beginning of the test. As shown in Figure 12, the friction coefficient reached 0.15 following 4,000 sliding cycles before stearic acid application. The healing effect was confirmed when the track, damaged after 4,000 cycles without stearic acid, recovered as stearic acid was applied. As a result, the friction coefficient decreased from 0.13 to 0.08. When the stearic acid was applied to the worn track after 10,000 cycles, the friction coefficient decreased to 0.10.

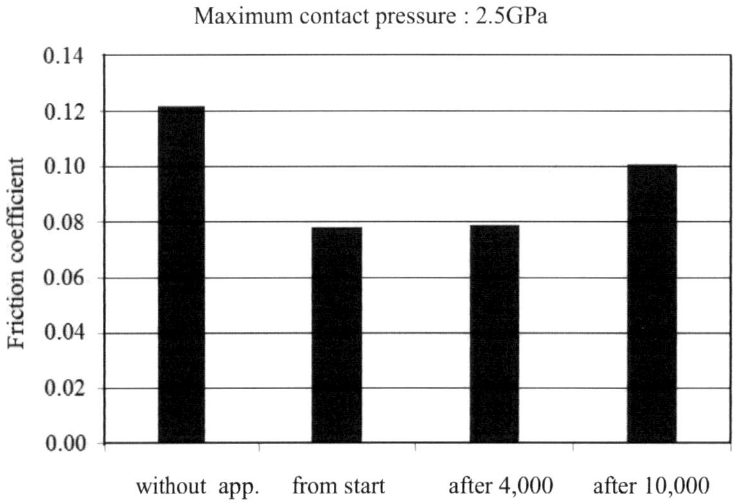

Figure 11. Change in friction coefficient of worn film after stearic acid application.

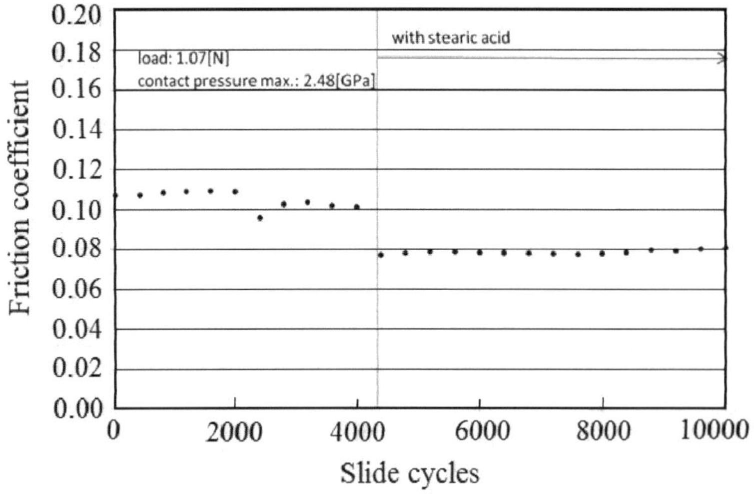

Figure 12. Healing effect of stearic acid on friction coefficient of film with contact pressure of about 2.5 GPa.

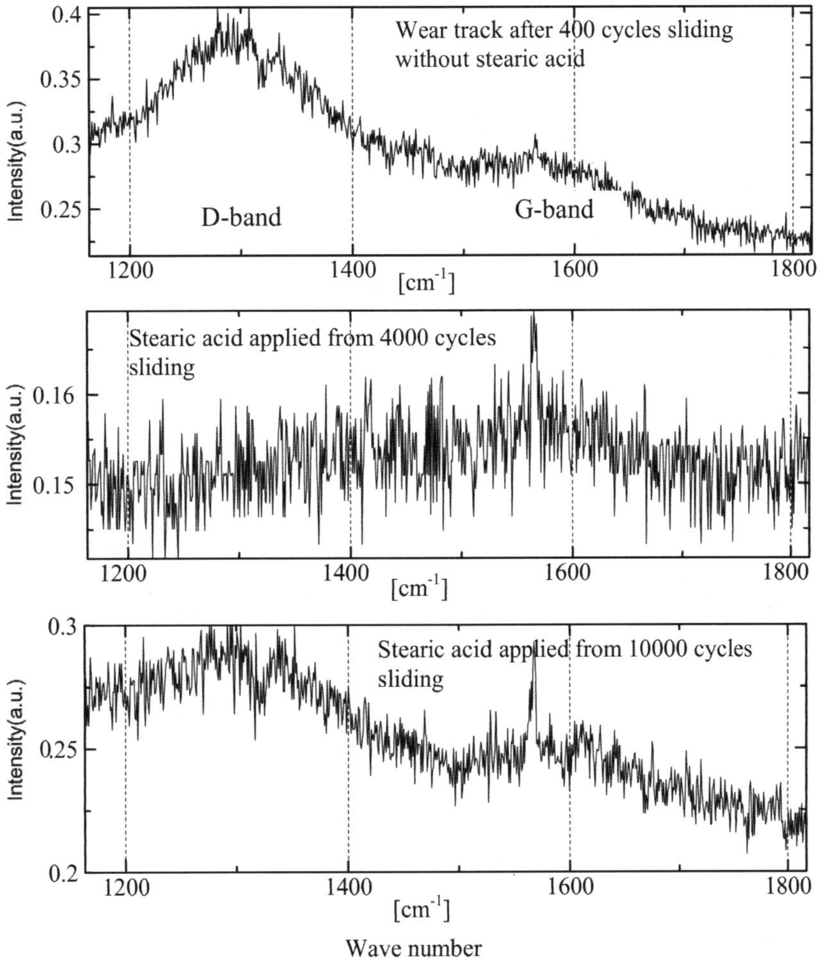

Figure 13. Raman spectra from wear tracks.

In the Raman spectrum in the top panel of Figure 13, a broad D peak is observed for the wear track for which the friction coefficient increased up to 0.15 with 400 cycles of sliding without stearic acid application. It resembles the Raman spectra for the specimen after the 10,000 cycles sliding test without stearic acid in Figure 9. The Raman spectrum for the healed specimen in the middle panel is the same as that in the bottom panel with stearic acid in Figure 9. It is clear that the stearic acid healed the worn ND/a-C:H film, and that the friction coefficient recovered to 0.08. Upon the application of stearic acid after 10,000 cycles, the decrease in the friction coefficient was smaller, becoming 0.10. The Raman spectrum in the bottom panel of Figure 13 shows a broad D peak, indicating that the severe damage to the film did not recover as stearic acid was applied. In this case, the healing effect from the stearic acid decreased the friction coefficient only minimally because the pin traveled across the worn tracks where the dark particles were produced by the pin under the high load.

4. Conclusions

Amorphous carbon films containing hydrogen and dispersed ND powders were produced using RF plasma-enhanced chemical vapour deposition. The deposited surface was very smooth, with a surface roughness within 10 nm. The films were approximately 200 nm thick and transparent with a visible surface lustre.

Friction properties between a silicon nitride tip and the film were investigated in oil using an instrument with high stiffness to withstand severe loads. Stearic acid has been reported to lubricate diamond surfaces, and it was shown that with 10,000 cycles of the reciprocal sliding friction test, the friction and wear properties of ND dispersed a-C:H film improved with the application of stearic acid. Furthermore, wear of the film caused by sliding tests without stearic acid could be healed by sliding with stearic acid, as long as critical damage has not been sustained. Noncritical damage occurred in the film after 4,000 cycles of reciprocal sliding at a contact pressure of 2.5 GPa, and the friction coefficient became 0.13. When stearic acid was applied to the film in oil, the friction coefficient was reduced to 0.08 after 10,000 cycles of reciprocal sliding.

Raman spectra for the worn track revealed the healing effect of the stearic acid on the film because the spectra for the wear tracks after 10,000 cycles were the same as the control ND/a-C:H film with no wear. In contrast, the spectra for the wear tracks or the debris spectra showed a broad D band peak and no G peak, consistent with the spectra for disordered amorphous carbon.

Acknowledgements
The authors thank Mr. M. Mayuzumi of Tokyo Diamond Tools Mfg. Co., Ltd., for discussion about ND powder and Mr. M. Suzuki of the Aerospace Research and Development Directorate, Japan Aerospace Exploration Agency, for his advice on constructing and setting up the deposition system. This test program was supported by TOYOTA MOTOR CORPORATION.

References

[1] Korzec D, Fedosenco G, Georg A and Engemann J 2000 *Surf. Coat. Technol.* **131** 20-5
[2] Donnet C, Fontaine J, Le Mogne T, Belin M, Heau C, Terrat J.P, Vaux F and Pont G 1999 *Surf. Coat. Technol.* **120-121** 548-54.
[3] Voevodin A A, O'Neil J P and Zabinski J S 1999 *Surf. Coat. Technol.* **116-119** 36-45
[4] Mayuzumi M 1994 *SOKEIZAI* 30-6 (in Japanese)
[5] Peng X L, Baeber Z H, and Clyne T W 2001 *Sur. Coat. Technol.*, **138** 23-32
[6] Van der Donck T, Muchlado M, Zein Eddine W, Achanta S, Carvalho N J M, Celis J-P 2009 *Surf. Coat. Technol.* **203** 3472-79.
[7] Kashiwagura Y, Ozawa K, Umezawa T 1994 *Technical Report of IEICE*, MR93-75
[8] Ferrari A C, Robertson 2000 *J. Phys. Rev.* B **61**(20) 14096-107
[9] Yamazaki T, Obara S, Imagawa K 2004 24th International Symposium on Space Technology and Science (Miyazaki, 2004), p.2004-o-1-04v(2004).
[10] Makita H 1997 *New Diamond* **12** 8-13 through NEDO Project Evaluated Report 2001, pp 11-29 (in Japanese)
[11] Pandey K C 1981 *Phys Rev. Lett.* **28** 1913-7
[12] Kern G and Hafner 1998 *J. Phys Rev. B.* **58** 13167-75.
[13] Kato T, Masuko M 2006 *Bases of Tribology* (Japan: Baifukan) (in Japanese)
[14] Kimura Y, Okabe H 1998 *Introduction to Tribology* (Japan: Yokendo) (in Japanese)
[15] Mylvaganam M and Zhang L 2002 *Proc. JSME/ASME Int. Conf. Materials and Processing*, Hawaii, 2002, pp.260-3
[16] McKenzie D R, Mullar D and Paithorpe B A 1991 *Phys. Rev. Lett.*, **67** 773-6
[17] Lifshitz Y, Kasi S R, Rabalais W and Eckstein W 1990 *Phys. Rev. B***41** 10468-80
[18] Iwamura E 2002 *Materia Japan* **41** 635-43. (in Japanese)
[19] Sriraman S, Agarwal S, Aydil E S and Maroudas D 2002 *Nature* **418** 62-5
[20] Fontaine J, Donnet C, Grill A andLeMogne T 2001 *Surface and Coatings Technology* **146 –147** 286–91

Room-temperature fast deposition and characterization of nanocrystalline $Bi_{0.4}Sb_{1.6}Te_3$ thick films by aerosol deposition

S Baba[1], L Huang[1], H Sato[1], R Funahashi[2] and J Akedo[1]
[1]Advanced Manufacturing Research Institute, National Institute of Advanced Industrial Science and Technology, 1-2-1 Namiki, Tsukuba, Ibaraki 305-8564, Japan
[2]Research Institute for Ubiquitous Energy Devices, National Institute of Advanced Industrial Science and Technology, 1-8-31 Midorigaoka, Ikeda, Osaka 563-8577, Japan

E-mail: so-baba@aist.go.jp

Abstract. P-type nanocrystalline $Bi_{0.4}Sb_{1.6}Te_3$ thick films were directly deposited on copper substrates by aerosol deposition (AD) involving shock compaction of the starting powder at room temperature. The film deposition rate was over 700 μm/min. The relative density of the films was above 95%. Films with a thickness of over 2 mm were obtained. They consisted of nanocrystals with grain sizes of several tens of nanometers. The crystal structure of the films was the same as the bulk materials prepared by spark plasma sintering (SPS) using the same starting powder. We measured the electrical conductivity, carrier concentration, Hall mobility, and thermal conductivity of the films and compared them with those of the bulk materials. The thermal conductivity of the AD film was 20–30% lower than that of the bulk materials, which is attributed to the increased number of grain boundaries and some defects in the AD films.

1. Introduction

Thick film thermoelectric devices used for localized and decentralized energy harvesting are important for making compact or large-area thermoelectric generators (TEGs) that operate at very small temperature differences. TEGs have attracted extensive interest for several decades because of the following unique features: no moving parts, quiet operation, low environmental impact, and high reliability [1]. The performance of a thermoelectric material is evaluated by its dimensionless thermoelectric figure of merit (ZT). Here, $Z = S^2\sigma / \kappa$, where S is the Seebeck coefficient, σ is the electrical resistivity, κ is the thermal conductivity, and T is the absolute temperature. Furthermore, κ is the sum of two independent components: an electronic contribution κ_e and a phonon contribution κ_p ($\kappa = \kappa_e + \kappa_p$). To estimate the κ_p value, the κ_e value is often obtained by assuming the validity of the Wiedemann–Franz law $\kappa_e = LT\sigma$, where L is the Lorenz number. Almost all thermoelectric materials are small band gap semiconductors with carrier concentrations of the order of 10^{19}–10^{21} cm^{-3} (depending on the material system); these carrier concentrations are intermediate between those of metals and insulators. For this reason, it has been suggested that the ideal thermoelectric material is amorphous to phonons while maintaining high electron mobility, the so-called 'phonon glass electron crystal' [1]. Until now, bismuth telluride (Bi_2Te_3) and its alloys with antimony telluride (Sb_2Te_3) and bismuth selenide (Bi_2Se_3) have been considered the best thermoelectric materials near room temperature. They can generate power from low-quality energy sources such as low-temperature waste heat. The *ZT* value of these materials is around unity at low temperatures (i.e., between room temperature and about 100°C); no other practical materials that have a *ZT* value significantly above unity have been discovered for almost 50 years [2]. Recently, p-type nanocrystalline bismuth antimony telluride ($Bi_xSb_{2-x}Te_3$, $x = 0.4$ and 0.5) bulk materials fabricated by hot pressing of ball-milled nanopowders with an average size of about 20 nm were found to have a peak *ZT* of 1.4 at 100°C and about 1.2 at room temperature [3]. This high *ZT* is reported due to a low lattice thermal conductivity caused by increased phonon scattering from grain boundaries and defects. While thermoelectric research has focused on the synthesis of novel and efficient bulk materials [4], film manufacture has received less attention.

We have recently been investigating various films of electronic materials such as ceramics, metals, and composites formed by aerosol deposition (AD) [5]. In AD, a powder is mixed with a carrier gas such as nitrogen or air to form an aerosol jet, which is ejected through a nozzle installed in a deposition chamber via a narrow slit onto substrates. AD is based on shock compaction of particles in starting powders at room temperature, known as room-temperature impact consolidation (RTIC). In RTIC, the aerosol jet enables deposition of high-density ceramic, metal, and composite films that are less than 1 μm thick to several hundreds of micrometers thick on various substrates such as glass, metal, ceramic, and plastic in a few minutes at room temperature. We previously found that aerosol-deposited films (AD films) are likely to have low thermal conductivities because they have nanocrystalline grains and defects that increase the phonon scattering [6]. In this study, we deposited $Bi_{0.4}Sb_{1.6}Te_3$ films onto copper (Cu) substrates by AD and investigated their thermoelectric properties (i.e., electrical conductivity, carrier concentration, Hall mobility, and thermal conductivity).

2. Experimental procedure

2.1. Sample preparation

P-type $Bi_{0.4}Sb_{1.6}Te_3$ films were directly deposited on 1-mm-thick Cu substrates by AD at room temperature. $Bi_{0.4}Sb_{1.6}Te_3$ with a purity of over 99.9%, which was obtained from commercial suppliers in powder form (200 mesh or smaller), was used as the starting powder. Figure 1 shows a schematic diagram of the experimental setup. The starting powder was mixed with a carrier gas (nitrogen, N_2) and then an aerosol was formed an aerosol chamber. The aerosol was transported through a tube to a slit nozzle with an orifice size of 20 mm × 0.5 mm installed in a deposition chamber, which was evacuated by a mechanical booster pump and a rotary pump in series. The aerosol jet was ejected from

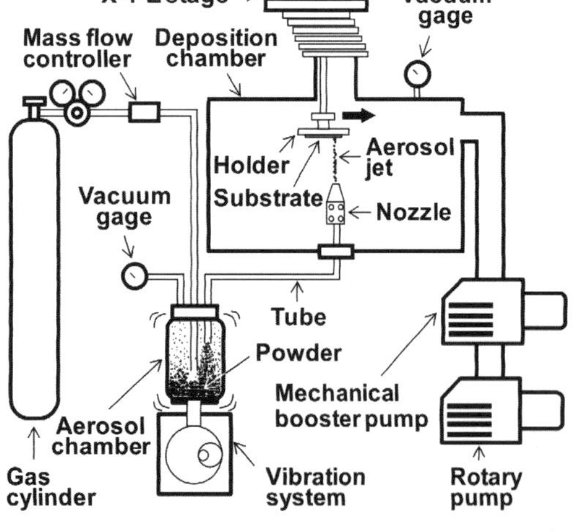

Figure 1. Schematic diagram of the aerosol deposition equipment.

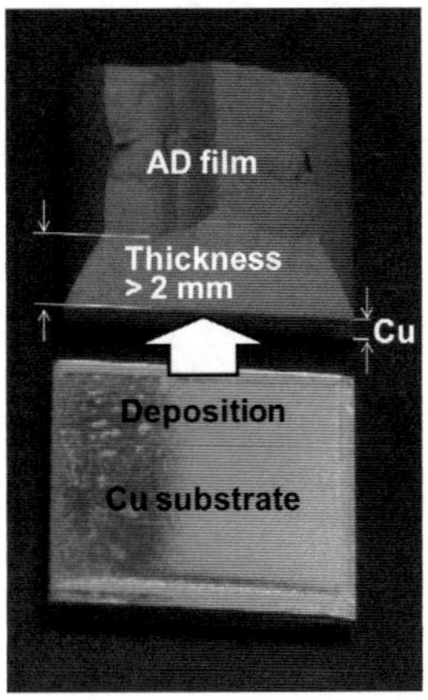

Figure 2. Photograph of the AD $Bi_{0.4}Sb_{1.6}Te_3$ thick film on a Cu substrate.

the slit nozzle due to the pressure differential between the aerosol and the deposition chambers. The Cu substrate, which was located at 10 mm from the nozzle exit, was set on the substrate holder and was then horizontally reciprocated at a scanning speed of 0.1 mm/s. The deposition chamber pressure was approximately 1 kPa during deposition. After deposition, the film surface was mechanically polished using 800–2000 grade emery paper. Figure 2 shows a photograph of an example of a Cu substrate and an as-deposited film. For comparison, bulk materials were sintered from the same starting powder by spark plasma sintering (SPS) at 450°C for 5 min at a pressure of 40 MPa in an argon atmosphere.

2.2 Sample characterization
The film deposition rate was determined by thickness measurements using a surface texture measuring instrument. The sample density was measured using the Archimedes method. The crystallinity of the samples was analyzed by X-ray diffraction (XRD) with Cu Kα radiation in the 2θ range of 10–80°. The electrical conductivity (σ) of the samples was obtained by the direct-current four-probe method. The thermal diffusivity of the samples was measured by the laser flash method. The heat capacity of the samples was measured using a differential scanning calorimeter. The thermal conductivity (κ) of the samples was calculated from the thermal diffusivity (α), the heat capacity (C_p), and the density (ρ) using the equation $\kappa = \alpha C_p \rho$. All measurements were performed at temperatures between room temperature and 300°C. The carrier concentration (n) of the samples was measured by the van der Pauw method at a magnetic field strength of 9 T between room temperature and 127°C. The carrier mobility of the samples was calculated by the equation $\mu = (ne/\sigma)^{-1}$, where e is the electron charge. The microstructures of the samples were examined by transmission electron microscopy (TEM) at an operating voltage of 200 kV.

3. Results and discussion
As shown in Figure 2, $Bi_{0.4}Sb_{1.6}Te_3$ films that had been directly deposited on Cu substrates had a thickness of above 2 mm; the film deposition rate was approximately 700 μm/min. Figure 3 shows a photograph of an example of the appearance of polished film and the bulk material. The surface of the polished film had a metallic luster. This is thought to be because the AD films, which have dense crystal structures consisting of crystal grains of the order of several tens of nanometers, have a surface roughness of a few nanometers without grain dropout by surface polishing. The relative density of the film was 97.6% of the theoretical density. In contrast, the bulk material had a relative density of 90.1%.

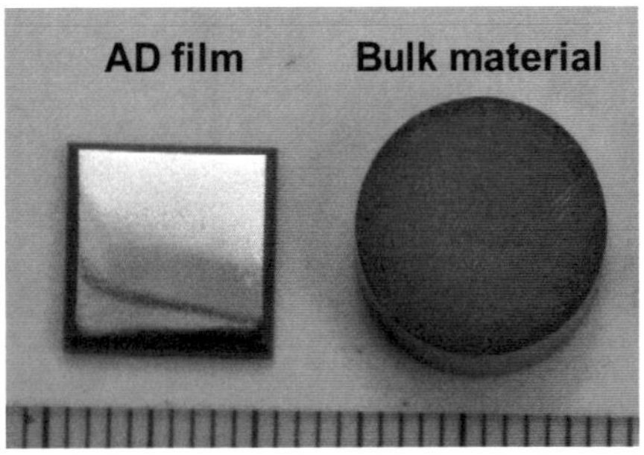

Figure 3. Photograph of the polished $Bi_{0.4}Sb_{1.6}Te_3$ thick film and bulk material.

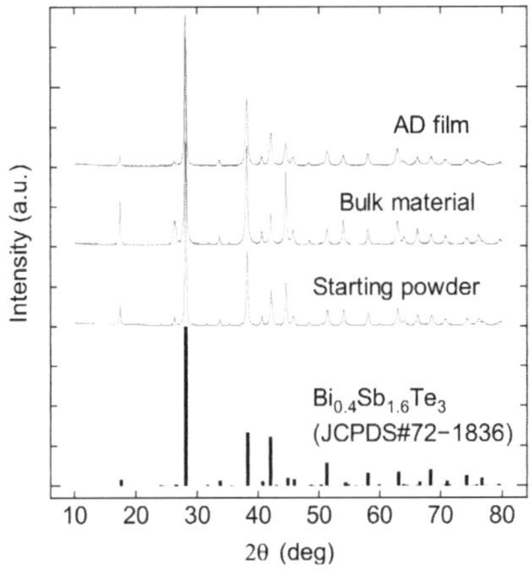

Figure 4. XRD patterns of the obtained $Bi_{0.4}Sb_{1.6}Te_3$ thick film, bulk material, and starting powder. Powder diffraction file is also shown.

Figure 4 shows typical XRD patterns of the AD film and the bulk material. The patterns of the starting powder are also shown. Comparison with the powder diffraction file (#72-1836) shows that the XRD patterns of both the AD film and the bulk material can be attributed to the $Bi_{0.4}Sb_{1.6}Te_3$ phase. The lattice parameter values of the AD film and the bulk material are 3.23 and 3.22 Å, respectively, which are very similar. The crystallite size (D) was evaluated using the Scherrer formula $D = 0.94\lambda/(\beta\cos\theta)$, where λ is the X-ray wavelength, β is the angle in radians at which the intensity is half the maximum intensity (FWHM), and θ is the angle at which the intensity is maximum. The crystallite sizes evaluated using the Scherrer formula for the AD film and the bulk material are 26.8 and 30.0 nm, respectively; hence, the AD film had a smaller crystallite size than the bulk material. This result indicates that the fracture and consolidation of the starting powder resulted in the reduction of the crystallite size in the AD films. These results are consistent with the metallic luster surface of the AD film (see Figure 3).

Figure 5 shows the temperature dependencies of the electrical conductivity of the AD film and the bulk material. The electrical conductivity of the AD film gradually increases with increasing temperature, showing typical semiconductor behaviour, while that of the bulk material decreases with increasing temperature. The electrical conductivity of the AD film was at least one order of magnitude less than that of the bulk material. As is well known, electrical conductivity depends on the carrier concentration and mobility according to $\sigma = ne\mu$. Figure 6 shows the temperature dependence of the carrier concentration and Hall mobility of the AD film and the bulk material. This data reveals that both the AD film and the bulk material are p-type. The carrier concentration of the AD film was comparable to that of the bulk material. This is probably because defects (e.g., point defects and dislocations) increase the carrier concentration. Therefore, the reduction in the electrical conductivity shown in Figure 5 is due to a reduction in the Hall mobility. In both the AD film and the bulk material, the carrier concentration and the Hall mobility are almost temperature independent. Further study is needed to determine more precise temperature dependencies of the carrier concentration and the Hall mobility of the AD film and the bulk material by considering electrode formation for Hall measurements.

Figure 7 shows the temperature dependencies of the total and lattice thermal conductivities for the AD film and the bulk material. The total thermal conductivity (κ) can be written as $\kappa = \kappa_e + \kappa_p$,

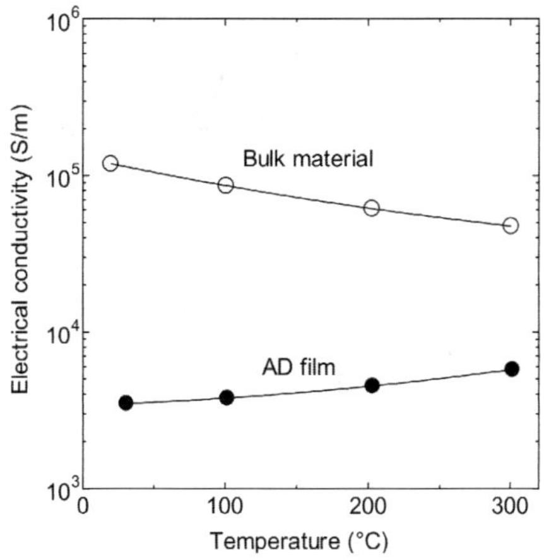

Figure 5. Temperature dependencies of electrical conductivity of the AD film and bulk material.

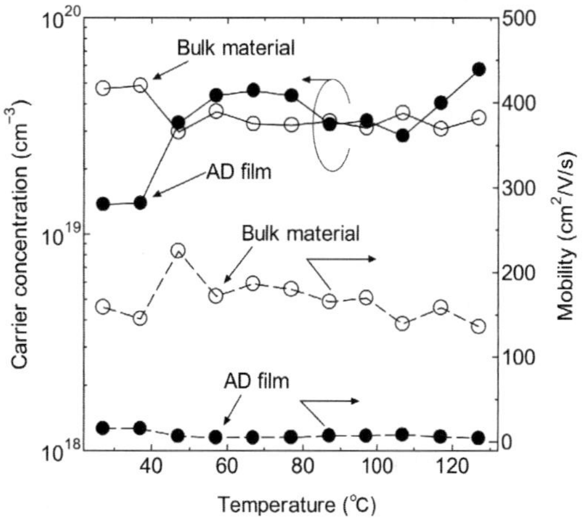

Figure 6. Temperature dependencies of carrier concentration and mobility of the AD film and bulk material.

where κ_e and κ_p are electron and phonon components. The former can be roughly calculated from experimental values of the electrical conductivity using the Wiedemann–Franz law ($\kappa_e = LT\sigma$), where L is the Lorenz number (2.44 × 10^{-8} V^2K^{-2}) and T is the absolute temperature. Both the total and lattice thermal conductivities initially decrease and then increase with increasing temperature. At low temperatures, phonon–phonon scattering is enhanced due to the reduction in the lattice thermal conductivity with increasing temperature as a result of enhanced lattice vibrations. However, with a further increase in temperature, bipolar diffusion of electrons and holes arising from intrinsic radiation will result in an additional thermal conductivity, thereby increasing the lattice thermal conductivity [7]. In the AD film, the lattice thermal conductivity accounted for 89–95% of the total thermal conductivity because of the reduction in the electrical conductivity (see Figure 5).

Figure 8 shows TEM images of the AD film and the bulk material. The AD film had a grain size ranging from a few to several tens of nanometers. In contrast, the grain size of the bulk material was several hundreds of nanometers due to grain growth by Ostwald ripening. The grain size differs from the crystallite size, which was obtained from XRD patterns (see Figure 4), since a single grain can be

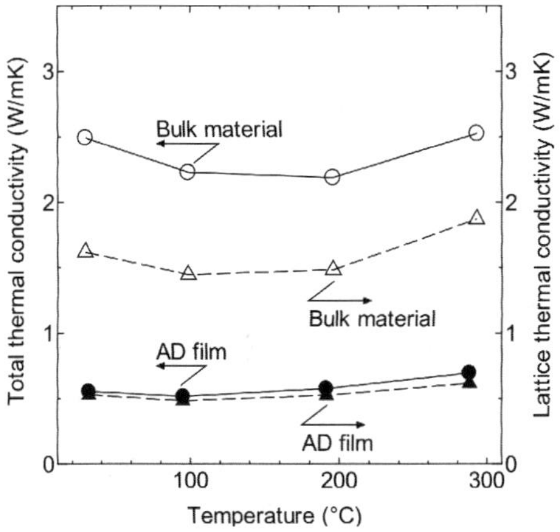

Figure 7. Temperature dependence of total and lattice thermal conductivity of the AD film and bulk material.

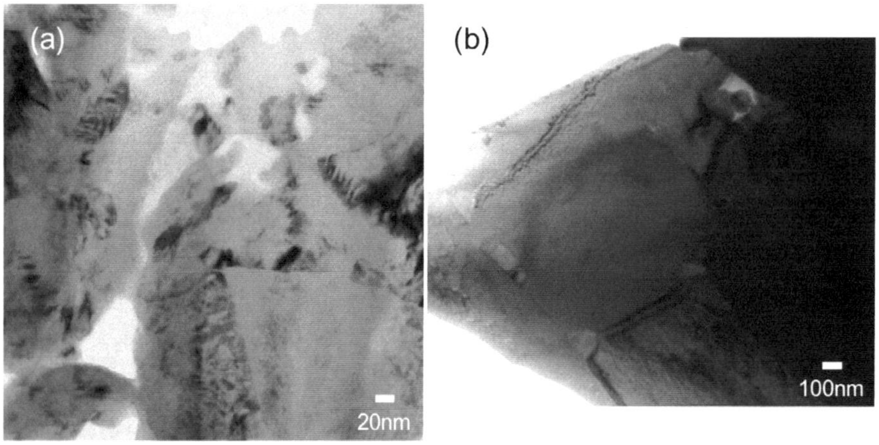

Figure 8. Bright-field TEM images of the (a) AD film and (b) bulk material.

composed of several crystallites. Therefore, the reduction in the electrical conductivity and the thermal conductivity in the AD film may be attributed mainly to increased carrier and phonon scattering, respectively, by the many grain boundaries induced by nanocrystals. We are currently measuring the recovery of the electrical conductivity of the AD film on post-annealing of the film. It has been reported that to minimize carrier scattering by the grain boundaries, nanometer-sized grains with the same phase as bulk materials should be used to reduce the band edge offset [8, 9]. The carrier transport processes in AD films need to be better understood to prevent nanocrystals from increasing the electrical conductivity.

To check the feasibility of the AD films for use in thermoelectric generator modules, we measured their Seebeck coefficient (the thermoelectric power of a material is a measure of the magnitude of an induced thermoelectric voltage in response to a temperature difference across a material). The Seebeck coefficients of the AD film and the bulk material at a temperature of 100°C were about 150 and 200 μV/K, respectively. By assuming a charge-carrier scattering distance of energy, the Seebeck coefficient at a low temperature can be expressed as:

$$S = \frac{8\pi^2 k_B^2}{3eh^2} m^* T \left(\frac{\pi}{3n} \right)^{\frac{2}{3}} \quad , \tag{1}$$

where k_B is the Boltzmann constant, h is the Plank constant, m^* is the effective mass, and n is the carrier concentration. Thus, the Seebeck coefficient decreases with increasing carrier concentration [7]. In general, $Bi_{0.4}Sb_{1.6}Te_3$ materials prepared at room temperature (such as the AD films) have many point defects; hence, a post thermal treatment required to eliminate them [10]. The Seebeck coefficient of the films may be enhanced due to the reduction in the carrier concentration due to there being fewer defects, which act as acceptors [11]. Holes in the AD films are scattered by the grain boundaries and defects; hence, they may be prevent the realization of an improved ZT . To further improve the thermoelectric films produced by AD, it is necessary to control the number of nanocrystals-grains and suppress defects to reduce the thermal conductivity and increase the electrical conductivity and the Seebeck coefficient.

4. Conclusions

To develop a thick-film TEG, we have prepared $Bi_{0.4}Sb_{1.6}Te_3$ films deposited on Cu substrates by AD. The AD films were directly deposited on the substrates at room temperature. They had a thickness of over 2 mm. The deposition rate of over 700 μm/min was obtained. The surface of the polished films, which has a relative density of 97.6%, has a metallic luster. XRD and TEM analyses revealed that the obtained films have dense crystal structures consisting of grains with sizes ranging from a few to several tens of nanometers. Consequently, the thermal conductivity of the film is 20–30% lower than that of the bulk materials. The AD films increase the thermoelectric figure of merit (ZT) by about 70–80%. The electrical conductivity of the films was at least one order of magnitude lower than that of the bulk material. These results suggest that grain boundaries and defects in the AD film affect both phonon conduction and electron conduction. We intend to confirm the recovery of electrical conductivity by heat treating the films.

Acknowledgements
This research was supported by the New Energy and Industrial Technology Development Organization (NEDO) in Japan under a Grant for Industrial Technology Research.

References

[1] Slack G A 1995 *CRC Handbook of Thermoelectrics*, ed D M Rowe (CRC Boca Raton) p 427
[2] Yan X, Poudel B, Ma Y, Liu W S, Joshi G, Wang H, Lan Y, Wang D, Chen G and Ren Z F 2010 *Nano Lett.* **10** 3373
[3] Poudel B, Hao Q, Ma Y, Lan Y C, Minnich A, Yu B, Yan X, Wang D Z, Muto A, Vashaee D, Chen X Y, Liu J M, Dresselhaus M S, Chen G and Ren Z 2008 *Science* **320** 634
[4] Snyder G J and Toberer E S 2008 *Nature Materials* **7** 105
[5] Akedo J 2008 *J. Therm. Spray Technol.* **17** 181
[6] Baba S and Akedo J 2009 *Appl. Sur. Sci.* **255** 9791
[7] Wang S, Xie W, Li H and Tang X 2010 *J. Phys. D: Appl. Phys.* **43** 335404
[8] Fan S, Zhao J, Guo J, Yan Q, Ma J and Hng H H 2010 *Appl. Phys. Lett.* **96** 182104
[9] Minnich A J, Dresselhaus M S, Ren Z F and Chen G 2009 *Energy Environ. Sci.* **2** 466
[10] Liao C N, Liou K M and Chu H S 2008 *Appl. Phys. Lett.* **93** 042103
[11] Takashiri M, Miyazaki K and Tsukamoto H 2008 *Thin Solid Films* **516** 6336

First-principles study on oxygen ion conduction of La_2GeO_5 based on the density functional theory

M Sakaue[1], W T Cahyanto[1], W T D Kencana[1], S M Aspera[1], K Miwa[1], H Kishi[1], S Kunikata[1], H Nakanishi[1], W A Diño[1], H Kasai[1] and T Ishihara[2]

[1] Department of Precision Science & Technology and Applied Physics, Osaka University, 2-1 Yamadaoka, Suita, Osaka 565-0871, Japan

[2] Department of Applied Chemistry, Faculty of Engineering, Kyushu University, Motooka 744, Nishi-Ku, Fukuoka 819-0395, Japan

E-mail: sakaue@dyn.ap.eng.osaka-u.ac.jp

Abstract. We performed first-principles simulations based on the density functional theory for investigations on the atomic geometry of La_2GeO_5, which is a fast oxygen ion conductor applicable to solid oxide fuel cells. While two experimental studies have reported contradicting results about the configuration of GeO_4 tetrahedral substructures, i.e., sp^2- or sp^3-like, we found that only the sp^3-like form is stable. We confirmed that the favorability of oxygen sites for vacancy formation is fundamentally affected by this configuration. The bonding mechanisms between atoms are discussed based on analyses of atomic distances and electronic density of states.

1. Introduction

Development of rechargeable batteries with high energy densities, high capacities and long lifetimes has become an urgent issue for realizing a low-carbon economy based on technologies for stable electricity supplies by renewable energy resources, effective leveling of electric power loads, and replacement of fossil fuel-powered vehicles with electric vehicles. Currently, lithium-ion batteries and lithium-ion polymer batteries have become common commercial products with high energy densities; however, their capacities are still insufficient to meet the demands of a low-carbon economy. Moreover, the use of lithium makes the batteries unstable and unsafe at the high temperatures reachable during practical operation.

One promising replacement is use of solid oxide fuel cells (SOFCs) designed as metal-air rechargeable batteries. The electrolyte membranes and electrodes of SOFCs consist of oxygen ion conductors, which transmit O^{2-} ions between the air and fuel electrodes. In the discharging process, oxygen molecules are dissociated into O^{2-} at the air electrode by cathode reactions, diffuse through the membrane toward the fuel electrode due to the concentration gradient, and finally react by anode reactions with the fuel contained in the storage container connected to the fuel electrode. In contrast, in the charging process, O^{2-} ions are produced from the oxidized fuels in the storage container by cathode reactions at the fuel electrode, diffuse through the membrane toward the air electrode due to the concentration gradient and electrochemical potential gradient, and finally form oxygen molecules at the air electrode by anode reactions. SOFCs are stable, safe and less affected by the atmosphere even at high temperatures of around 1300 K because the devices are ceramic-based and the reactions of fuels take places in the inner part of the devices. Thus, SOFCs can realize high energy densities safely in high temperature ranges for which the carrier mobility is high.

So far, various ceramic materials based on or containing rare-earth elements have been studied as oxygen ion conductors[1]. Yttria-stabilized zirconia and ceria-based ceramics are representative conventional materials for SOFCs. The operating temperatures of SOFCs realized by these materials are approximately in the range from 1073 to 1273 K, which is still extremely high for practical applications in which the lower limit needs to be about 500 K. Therefore, it is crucial to develop new oxygen ion conductors that have higher ionic conductivities at low temperatures.

The mechanisms that determine the oxygen ion conductivity are not understood yet. It had been believed that a high isotropy of the crystal structure gives rise to high ionic conductivity, as for the case of the discovery of $LaGaO_3$ with a pseudocubic perovskite lattice as a fast oxygen ion conductor [2–5]. However, modification of this hypothesis has become necessary due to the discovery that La_2GeO_5 with a highly anisotropic monoclinic lattice also has a high ionic conductivity over a wide range of oxygen partial pressure [6–8].

Usually, oxygen ion conduction take place by migration of lattice defects, which can be classified as either interstitial oxygen atoms or oxygen vacancies; for example, $LaGaO_3$ and La_2GeO_5 belong to the latter category. In oxygen-deficient materials where vacancy propagation accounts for the conduction, identification of oxygen-rich layers that contribute to the conduction is an important subject for clarification of the conduction mechanisms.

The atomic geometry of stoichiometric La_2GeO_5 was first studied by Ishihara *et al.* using the X-ray diffraction method with Rietveld analysis, and was reported to be characterized by distorted GeO_4 tetrahedral substructures [6]. It was also found that the atomic coordination is preserved in nonstoichiometric $La_{2-x}GeO_{5-\delta}$ for $x < 0.33$ [6] while the monoclinic lattice is preserved for $x < 0.45$. This results in two types of well-defined planar layers, i.e., $LaGeO_3$ and LaO_2. According to the atomic distances, three atoms in the $LaGeO_3$ layers are coordinated around a germanium atom in the configuration of sp^2 hybridization. On the other hand, Berastegui *et al.*, who also performed X-ray diffraction measurements, reported that the GeO_4 tetrahedral substructures in the stoichiometric La_2GeO_5 become rather symmetric by undergoing sp^3-like Ge-O coordination [8]. They also found that the lattice structure changes to a hexagonal apatite-type structure for a nonstoichiometric composition of $x = 0.45$, which leads to higher oxygen ion conductivities than the monoclinic case. Thus, before discussing the conduction mechanisms, it is necessary to resolve this experimental contradiction regarding the atomic geometry. However, no theoretical study on the stable geometries of monoclinic La_2GeO_5-based materials has been reported yet.

In this paper, we investigate the stable atomic geometry of stoichiometric La_2GeO_5 with a monoclinic lattice by first-principles calculations based on the density functional theory (DFT). We also carry out an analysis of the favorable sites for oxygen vacancy formation based on total energy calculations. The bonding properties between atoms are discussed with the aid of an analysis of the electronic density of states.

2. Computational methodology

The calculations were performed using the Vienna Ab-initio Simulation Package (VASP) code. We employed the generalized gradient approximation (GGA) within the Perdew-Burke-Ernzerhof (PBE) functional for the exchange-correlation energy. The interactions between ions and electrons were given by the projector augmented-wave (PAW) method. The total energy convergence was confirmed with a cut-off energy of 900 eV and k-point sampling by the Monkhorst-Pack scheme with a $3 \times 3 \times 3$ mesh. It should be noted that the required cut-off energy was much higher than the usually expected value, which is about 550 eV for treatment of the target composition with standard PAW potentials of VASP. The tetrahedron method with Blöchl corrections is employed for total energy calculations.

We employed a monoclinic unit cell containing 32 atoms as the initial structure according to the experimentally obtained data [6, 8] as shown in Table 1. The differences in the volumes and lattice

Table 1. Atomic coordinates of La_2GeO_5 in the initial geometries with the $P2_1/c$ space group given in fractional coordinates for each corresponding lattice with $a = 9.59693$ Å, $b = 7.47359$ Å, $c = 7.10745$ Å and $\beta = 107.5770°$ (485.971 Å^3 volume) for the sp^2-based structure [6] and $a = 9.60405$ Å, $b = 7.46902$ Å, $c = 7.11199$ Å and $\beta = 107.550°$ (486.417 Å^3 volume) for the sp^3-based one [8].

	sp^2-based [6]			sp^3-based [8]		
Site	X	Y	Z	X	Y	Z
La(1)	0.38937	0.14161	0.57941	0.38289	0.14396	0.58432
La(2)	0.02056	0.12360	0.72966	0.02148	0.12451	0.73529
Ge	0.29732	0.58935	0.53976	0.29880	0.58817	0.54080
O(1)	0.32250	0.06691	0.86216	0.2919	0.0748	0.8573
O(2)	0.34794	0.48623	0.75895	0.3695	0.4558	0.7496
O(3)	0.35157	0.77243	0.49523	0.4074	0.7723	0.5523
O(4)	0.04677	0.61215	0.49451	0.1117	0.6371	0.4917
O(5)	0.04310	0.12912	0.45330	0.1140	0.1217	0.4493

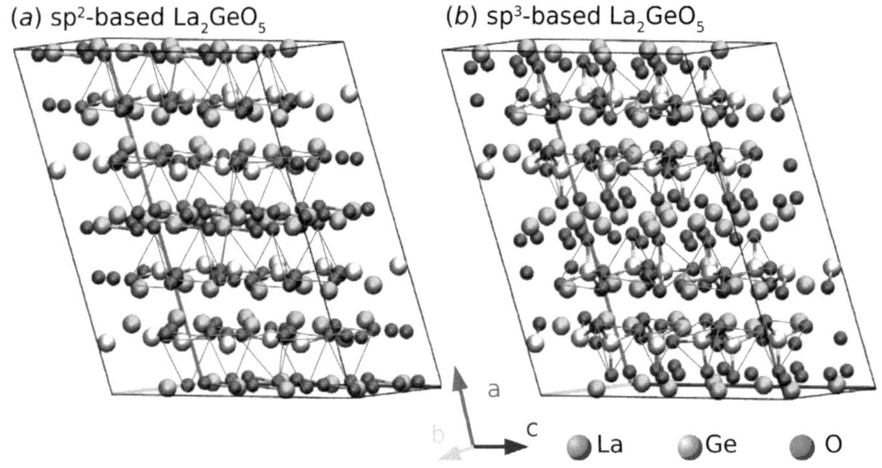

Figure 1. Initial geometries of La_2GeO_5 for (a) the sp^2-based structure and (b) the sp^3-based structure in an expanded cell with $2 \times 2 \times 2$ periods. Green, gray and red balls denote lanthanum, germanium and oxygen atoms, respectively. Arrows indicate a, b and c axes, respectively. Magenta lines indicate GeO_4 tetrahedral substructures.

parameter between the two structures are within 1 %. The corresponding geometries are shown in Figure 1. While the oxygen positions are quite different between the two structures, the positions of lanthanum and germanium are almost equivalent with each other. In both structures, La(1), Ge, O(1), O(2) and O(3) form $LaGeO_3$ layers and La(2), O(4), O(5) form LaO_2 layers. The GeO_4 tetrahedral substructures consist of Ge, O(1), O(2), O(3) and O(4). It is seen that the LaO_2 layers are well-defined in the sp^2-based structure while they are spread in the sp^3-based one.

The Ge-O(1), Ge-O(2) and Ge-O(3) distances in the sp^2-based structure are 1.79, 1.67 and 1.53 Å, respectively. Thus, the Ge-O(1) and Ge-O(2) distances are similar to that for a single bond (1.73 Å) and the Ge-O(3) distance is similar to that for a double bond (1.57 Å), according to Pauling's definition. The Ge-O(4) distance is 2.33 Å and is significantly longer than the covalent bond length. Here, the

La(2)-O(4) (2.24 Å) and La(2)-O(5) (2.04 Å) distances are somewhat shorter. Both the La(2)-O(4) and La(2)-O(5) distances are shorter than the La-O single bond length, 2.24 Å, defined in the same way as Ge-O. The O(4)-O(5) distance is 2.19 Å and is therefore considered to be non-interacting. On the other hand, the Ge-O(1), Ge-O(2), Ge-O(3) and Ge-O(4) distances in the sp^3-based structure are 1.77, 1.74, 1.71 and 1.76 Å, respectively, all of which are similar to the single bond length. Here, the La(2)-O(4), La(2)-O(5) and O(4)-O(5) distances are 2.48, 2.40 and 2.94 Å, respectively, and therefore longer than the covalent bond length.

The lattice parameters and atomic positions were relaxed by the conjugate-gradient algorithm with a force scaling constant of 0.5 fs until the total energy convergence reached less than 0.01 meV. The relaxation calculations were executed in two steps by initializing the electronic basis set after the first convergence in order to reduce the Pulay stress. Then, the optimized electronic properties were obtained by static calculations with the relaxed lattice parameters and atomic positions. Here, we note that some static calculations for unrelaxed geometries were also performed for purposes of reference.

3. Simulation results and discussions

3.1. Geometry

Relaxed geometries were obtained for both structures. However, the result for the sp^2-based structure is unfortunately equivalent to the sp^3-based one while the displacement by relaxation from the sp^3-based structure is almost negligible (< 0.1 Å, and is visually equivalent to Figure 1(b)). The obtained lattice parameters for the sp^2-based structure are a = 9.702 Å, b = 7.504 Å, c = 7.179 Å, α = 90.209°, β = 107.127° and γ = 90.185°, which give a volume of 499.431 Å^3, and those for the sp^3-based structure are a = 9.662 Å, b = 7.516 Å, c = 7.118 Å, α = 89.998°, β = 107.476° and γ = 90.005°, which give a volume of 493.061 Å^3. The changes by relaxation and differences between the two structures are within the error margins expected from the precision of the calculation, i.e., a few percent. Consequently, the difference in the obtained geometries between the calculations for the two structures is found to be negligible since the differences in atomic positions are also within the calculation error margins.

Equivalent results were also found with harder PAW potentials and other functional types, i.e., GGA-PBE revised for solids [9] and the local density approximation (LDA), and with phenomenological van der Waals corrections (see Appendix A). This was also confirmed by distorting the initial geometry so that there was no local energy minimum to support the sp^2-based structure. Thus, it is concluded that the sp^2-based structure is unstable for the experimentally determined composition.

Table 2. Atomic coordinates in the relaxed geometry with standard deviations among equivalent positions projected onto each axis.

Site	X	Y	Z
La(1)	0.3896 ± 0.0133	0.1418 ± 0.0039	0.5800 ± 0.0104
La(2)	0.0322 ± 0.0096	0.1261 ± 0.0008	0.7398 ± 0.0053
Ge	0.3116 ± 0.0097	0.5919 ± 0.0060	0.5364 ± 0.0039
O(1)	0.2974 ± 0.0086	0.0821 ± 0.0005	0.8500 ± 0.0046
O(2)	0.3935 ± 0.0114	0.4583 ± 0.0143	0.7452 ± 0.0074
O(3)	0.4041 ± 0.0067	0.7886 ± 0.0107	0.5138 ± 0.0141
O(4)	0.1211 ± 0.0092	0.6336 ± 0.0010	0.4968 ± 0.0050
O(5)	0.1233 ± 0.0105	0.1199 ± 0.0016	0.4501 ± 0.0066

Table 2 shows the atomic coordinates in the relaxed geometry. It should be noted that calculated values for the sp^2-based structure are employed hereafter. The equivalent positions fit within less

Table 3. Total energies of systems containing one oxygen vacancy per unit cell. The unit is eV.

Vacancy site	sp^2-based	sp^3-based
O(1)	-209.09	-238.67
O(2)	-210.16	-239.26
	-210.18	-239.29
	-210.22	
O(3)	-212.07	-239.88
O(4)	-214.13	-238.87
O(5)	-211.75	-236.62

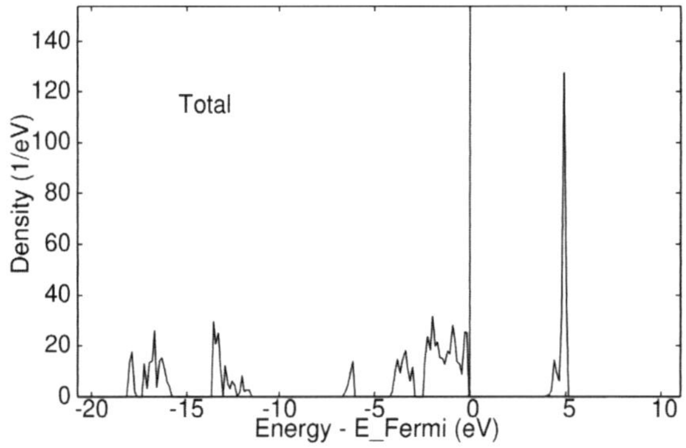

Figure 2. Total electronic density of states for the relaxed geometry. The vertical line at energy 0 indicates the Fermi energy defined as the top of the occupied states.

than 0.015 standard deviations in fractional coordinates. It is understood that the difference from the experimental sp^3-based structure is also within a similar range to the standard deviations except for Z of O(3) which is 0.0385 because of its distorted initial configuration. The obtained Ge-O(1), Ge-O(2), Ge-O(3) and Ge-O(4) distances are 1.83, 1.79, 1.76 and 1.81 Å, respectively, and are classified as single bonds. Here, the La(2)-O(4), La(2)-O(5) and O(4)-O(5) distances become 2.49, 2.38 and 2.96 Å, respectively, by keeping them longer than the covalent bond lengths.

In order to investigate the effects of geometry on the tendency of oxygen vacancy formation, we performed total energy analyses on systems where one oxygen is removed from a unit cell based on static DFT calculations where all atoms are fixed. The analyses were made on the sp^2-based initial geometry and the (sp^3-based) relaxed geometry. It should be noted that the analyzed configurations are for simulating unstable domains to accept oxygen ions. The current analyses were made by assuming a neutral total charge where charge trapping around the vacancies is negligible. The results are shown in Table 3. The energies for equivalent positions of O(2) are spread while the remainder are identical within an error of 0.02 eV. The energy difference of about 30 eV between the two structures is in a similar range to the energy gain by geometric relaxation for the stoichiometric composition, which is 26.64 eV (from -221.84 to -248.48 eV). The lowest energy is found for O(4) in the sp^2-based structure while it is for O(3) in the sp^3-based one. The relative instability of O(5) vacancies suggests a large contribution of O(5) to the formation of LaO_2 layers although the atomic distances and electronic properties do not indicate covalent features. Thus, if the sp^2-based structure is retained for some reason [6], the LaO_2 layers will be favorable for oxygen vacancy formation while the $LaGeO_3$ will be favorable in the sp^3-

based structure. It should be noted that the actual formation energies of oxygen ions should be lower than those evaluated from Table 3, considering the effects of geometric relaxation and charge compensation by electron trapping due to local metal-insulator transitions [10].

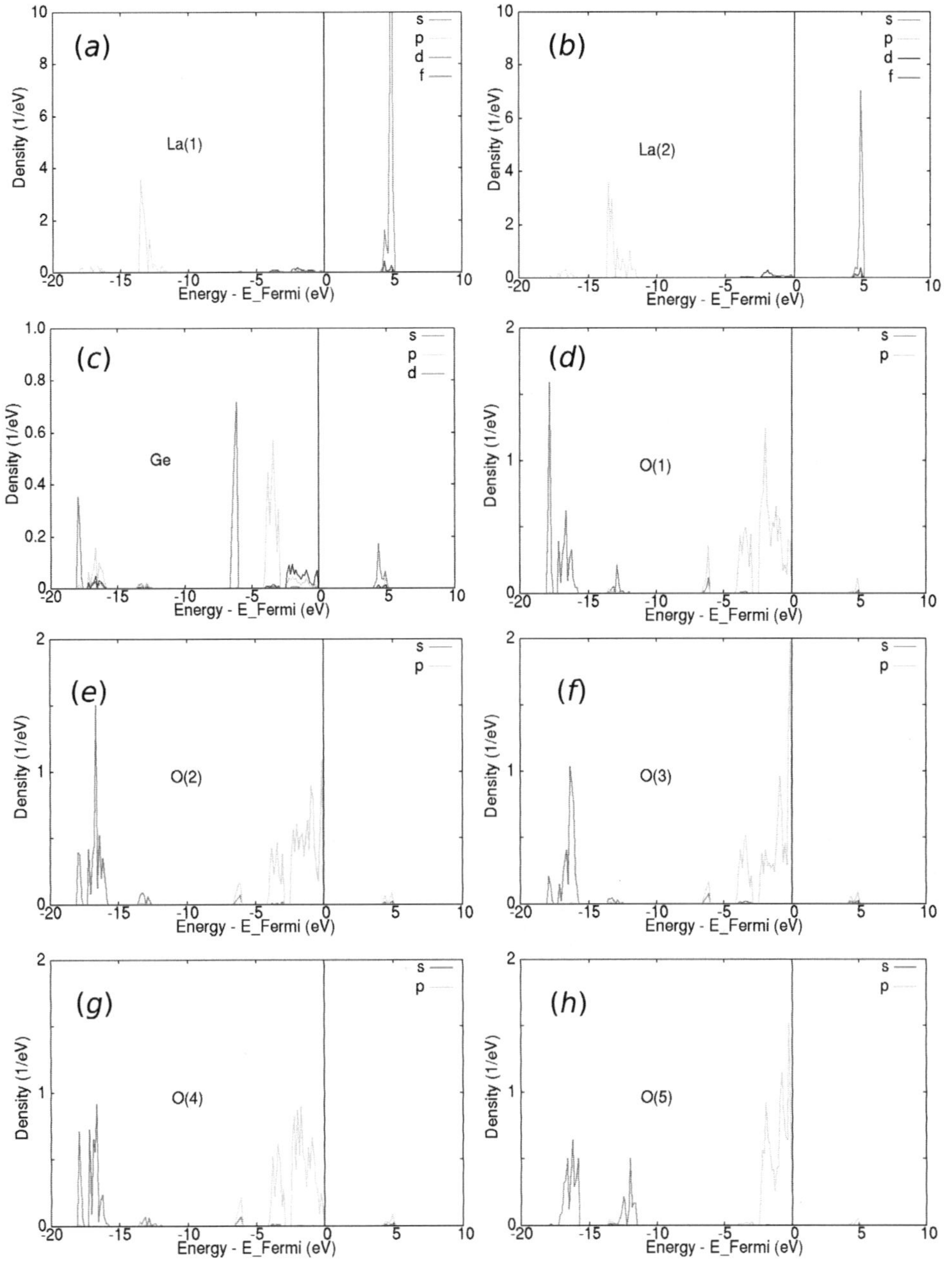

Figure 3. Partial electronic densities of states for the relaxed geometry projected onto valence (s, p, d and f) orbitals at each atomic site: (*a*), (*b*), (*c*), (*d*), (*e*), (*f*), (*g*) and (*h*) are for La(1), La(2), Ge, O(1), O(2), O(3), O(4) and O(5) sites, respectively.

3.2. Electronic density of states

The total electronic density of states obtained for the relaxed geometry was determined to be non-spin-polarized and is shown in Figure 2. The band gap of 4.029 eV indicates that this material is electronically insulating. It should be noted that the actual band gap can be larger since DFT-based calculations tend to underestimate it.

Contributions from each atomic orbital were extracted by obtaining the partial densities of states projected onto valence orbitals at each atomic site, which are summed over the magnetic quantum numbers (see Figure 3). It is understood that the portion from -5 to 0 eV mainly consists of p orbitals of oxygen and also contains small contributions from p orbitals of germanium and much smaller contributions from d orbitals of germanium and lanthanum. The similarity in the features between La(1) and La(2) indicates non-covalent interactions with the surrounding atoms where the details of the atomic coordinates are not significant. The p orbital portions for O(1), O(2), O(3) and O(4) are split into two parts by a gap from -3.0 to -2.5 eV while that for O(5) possesses only the upper part. Thus, the lower p portions for O(1), O(2), O(3) and O(4) form isotropic covalent bonding orbitals with the p orbitals of germanium. The upper p portions for the all oxygen sites overlap with the d orbital components of lanthanum and germanium. It is considered that this p-d interaction accounts for the La(2)-O(5) bonding.

4. Conclusions

It was clarified that stoichiometric La_2GeO_5 contains sp^3-type GeO_4 tetrahedral substructures in its stable atomic geometry and has a tendency to form oxygen vacancies in the $LaGeO_3$ layers. Although the formation mechanism of the distorted sp^2-type GeO_4 substructures is still unclear, it is expected that the oxygen vacancies are more likely to be formed in the LaO_2 layers if the geometry is retained as reported in experiments. The contradiction regarding the geometry between the experiments may be due to the difference in the methods of sample preparation, which can cause unexpected element substitution, e.g., of lanthanum with strontium contained in the ingredients [6, 8].

Acknowledgments

This work was supported in part by MEXT (Ministry of Education, Culture, Sports, Science and Technology) through the G-COE (Special Coordination Funds for the Global Center of Excellence) program "Atomically Controlled Fabrication Technology", Grant-in-Aid for Scientific Research on Innovative Areas Program (2203-22104008) and Scientific Research (c) (22510107) program. It was also supported in part by JST (Japan Science and Technology Agency) through ALCA (Advanced Low Carbon Technology Research and Development) Program "Development of Novel Metal-Air Secondary Battery Based on Fast Oxide Ion Conductor Nano Thickness Film", Elements Science and Technology Project "New development of self-forming nano-particle catalyst without precious metals" and Strategic Japanese-Croatian Cooperative Program on Materials Science "Theoretical modeling and simulations of the structural, electronic and dynamical properties of surfaces and nanostructures in materials science research".

References

[1] Atkinson A, Barnett S, Gorte R J, Irvine J T S, Mcevoy A J, Mogensen M, Singhal S C and Vohs J 2004 *Nat. Mater.* **3** 17–27
[2] Ishihara T, Matsuda H and Takita Y 1995 *Solid State Ionics* **79** 47–151
[3] Ishihara T, Kilner J A, Honda M and Takita Y 1997 *J. Am. Chem. Soc.* **119** 2747–2748
[4] Ishihara T, Matsuda H and Takita Y 1998 *J. Am. Chem. Soc.* **116** 3801–3803
[5] Slater P R, Irvine J T S, Ishihara T and Takita Y 1998 *J. Solid State Chem.* **139** 135–143
[6] Ishihara T, Arikawa H, Akbay T, Nishiguchi H and Takita Y 2001 *J. Am. Chem. Soc.* **123** 203–209
[7] Ishihara T, Arikawa H, Nishiguchi H and Takita Y 2002 *Solid State Ionics* **154–155** 455–460
[8] Berastegui P, Hull S, García F J G and Grins J 2002 *J. Solid State Chem.* **168** 294–305
[9] Perdew J P, Ruzsinszky A, Csonka G I, Vydrov O A, Scuseria G E, Constantin L A, Zhou X and Burke K 2008 *Phys. Rev. Lett.* **100** 136406 (4 pages)

[10] Kishi H, Sarhan A A A, Sakaue M, Aspera S M, David M Y, Nakanishi H, Kasai H, Tamai Y, Ohnishi S and Awaya N 2011 *Jpn. J. Appl. Phys.* **50** 071101 (6 pages)
[11] Grimme S 2006 *J. Comp. Chem.* **27** 1787–1799
[12] Schwerdtfeger P 2011 Table of experimental and calculated static dipole polarizabilities for the electronic ground states of the neutral elements (in atomic units) version 2.8 `http://ctcp.massey.ac.nz/dipole-polarizabilities`
[13] Miller T M 2002 *CRC Handbook of Chemistry and Physics* (New York: CRC Press)
[14] Bondi A 1964 *J. Phys. Chem.* **68** 441–452

Appendix A. Parameters for van der Waals corrections

Phenomenological van der Waals (vdW) corrections based on the Grimme empirical force-field method [11] were made with parameters defined for each element: $C_6 = 185.1$ J nm^6 mol^{-1} and $R_0 = 1.727$ Å for lanthanum, $C_6 = 17.1$ J nm^6 mol^{-1} and $R_0 = 1.727$ Å for germanium, and $C_6 = 0.7$ J nm^6 mol^{-1} and $R_0 = 1.342$ Å for oxygen, where C_6 and R_0 denote the dispersion coefficient and the vdW radius, respectively. The values for germanium and oxygen were based on the defaults of VASP version 5.2.11. C_6 for lanthanum was given by the definition $C_6 = 0.05 \times N I_p \alpha_{\mathrm{D}}$ [11] with an ionization potential $I_p = 538.1$ kJ mol^{-1}, a static dipole polarizability $\alpha_{\mathrm{D}} = 210$ a.u. [12, 13] and a number of electrons of the rare gas element in the same row, i.e., radon, $N = 86$. There is no reliable data for the vdW radii of sixth-row elements in the public databases, however, so we took the value for lanthanum to be the same as germanium by assuming that the radii of the sixth-row transition metals and rare-earth elements are slightly smaller than platinum, whose value is considered to be 1.75 Å [14].

Ferromagnetic nanostructures of oxygen on Ag(111)

Y Kunisada, H Nakanishi, W A Diño and H Kasai

Department of Applied Physics, Osaka University,
2-1 Yamada-oka, Suita, Osaka 565-0871, Japan

Email: kasai@dyn.ap.eng.osaka-u.ac.jp

Abstract. First principles calculations were performed to investigate the robustness of oxygen adsorbed structures that show ferromagnetic properties. These ferromagnetic properties are induced by ferromagnetic superexchange interactions and Ruderman-Kittel-Kasuya-Yosida (RKKY) interactions. Such ferromagnetic oxygen adsorbed structures appear only in the case of coverages higher than 0.5 ML, because the ferromagnetic superexchange interactions and RKKY interactions are appreciable between O-O only for short distances. In this work, we focused on single oxygen atom diffusion parallel and perpendicular to Ag(111) surfaces. The in-plane diffusion could induce associative desorption, which reduces the oxygen coverage. On the other hand, diffusion to the subsurface could quench the magnetic moment. These diffusion paths have effective diffusion barriers of 0.78 eV. Thus, such diffusion rarely occurs in low temperature regions.

1. Introduction

Surface magnetism has attracted a great deal of interest in basic science and application fields. From the viewpoint of basic science, singular behaviours of magnetic impurities [1-8] (and references therein) on surfaces have been investigated, and are caused by the two-dimensionality of the surface. In addition, understanding and controlling these singular behaviours are also very important for realizing high capacitance and robustness of magnetic devices. Moreover, studies suggested that surface magnetic states can be useful for controlling catalytic reactivity [9-13] in fuel cell electrodes. These studies show that the efficacy of surface magnetism extends from spintronics devices to energy-related technologies. Design and control of magnetic surfaces are promising ways for developing new effective devices and catalysts.

Recently, we reported metastable ferromagnetic states of the O/Ag(111) system [14,15] by potential energy surface (PES) studies on O_2 dissociative adsorption on Ag(111) surfaces, and suggested possibilities for controlling the surface magnetic properties in this system. The ferromagnetic oxygen adsorbed structures on Ag(111) surfaces are induced by the ferromagnetic superexchange interactions and Ruderman-Kittel-Kasuya-Yosida (RKKY) interactions. Such ferromagnetic oxygen adsorbed structures appear only in the case of coverages higher than 0.5 ML, because the ferromagnetic superexchange interactions and RKKY interactions are appreciable between O-O only for short distances. In this work, we investigated the robustness of ferromagnetic oxygen adsorbed structures on Ag(111) surfaces. Especially, we focused on oxygen atom diffusion parallel and perpendicular to Ag(111) surfaces in the case of 0.5 ML coverage. In-plane oxygen atom diffusion, which corresponds to oxygen atom motion parallel to Ag(111) surfaces, makes the distance between oxygen atoms shorter, which changes the stable oxygen adsorbed structures to unstable ones. Oxygen atom diffusion into the subsurface, which correspond to oxygen atom motion perpendicular to Ag(111) surfaces, contributes to quenching the ferromagnetic states. Thus, we performed first principles investigations of oxygen atom diffusion by calculating the relevant potential energy curves. From these potential energy curves, we could obtain the diffusion barrier heights and diffusion coefficients.

2. Computational methods

We performed spin-polarized total energy calculations based on density functional theory (DFT), using the generalized gradient approximation (GGA) within the Perdew-Burke-Ernzerhof (PBE) functional for the exchange–correlation energy [16,17], which is implemented using the plane-wave and projector augmented wave method [18-21]. We applied a 600 eV cutoff to limit the plane-wave basis set without compromising computational accuracy. A 5x5x1 Monkhorst–Pack special k-point grid [22] for Brillouin zone sampling using a Gaussian smearing model of σ=0.05 eV was used. First, for the O/Ag(111) system, the Ag substrate was simulated by a slab of four fcc (2x2) Ag(111) layers. The supercell was constructed with a 1.55 nm vacuum separation between slabs. Lateral and interlayer Ag–Ag distances for this system are initially set based on the theoretically predicted equilibrium lattice constant of 0.294 nm, obtained by using a 9x9x9 k-mesh on a primitive one-atom fcc unit cell for bulk Ag. In order to determine the optimized Ag(111) surface structure, we performed calculations in which the top two layers are relaxed. The total energy convergence with this size of supercell is within 10 meV. Hence, the above parameters gave acceptable accuracy. Moreover, the magnetic moment converges to 0.01 μ_B. Because the mass of an oxygen atom is much lighter than that of an Ag atom, we expect a large difference in the time scales associated with surface relaxations and oxygen atom dynamics. Thus, as a first step, we neglected surface relaxations. In these calculations, we adopted an electric dipole correction layer in the vacuum area to compensate for the dipole interactions between the repeated slabs [23]. Lastly, electron transfer is obtained from the difference between the electron density distribution of two configurations of oxygen: isolated from and adsorbed on the substrate, using Bader charge analysis [24,25].

In Figure 1, we show 0.5 ML oxygen adsorbed structures on Ag(111) surfaces, which show ferromagnetic behaviours. In this work, we call these configurations (a) f-f and (b) h-f configurations. Two oxygen atoms in the supercell are above fcc hollow sites in the f-f configuration, and one oxygen atom is above a hcp hollow site and the other is above a fcc hollow site in the h-f configuration. In the f-f and h-f configurations, the magnetic moments of oxygen atoms remain as 0.32 μ_B and 0.49 μ_B, respectively. Also, in these configurations, the magnetic moment of the nearest Ag atom is 0.09 μ_B. For reference, we mention that the h-h configuration, which has two oxygen atoms in the supercell above hcp hollow sites, has almost the same magnetic properties as the f-f configuration. Such ferromagnetic oxygen adsorbed structures appear only for higher coverage than 0.5 ML, because the ferromagnetic superexchange interactions and RKKY interactions are appreciable between O-O only for short distances. We also estimated the Curie temperature using mean field theory. The corresponding Curie temperature is 305 K. Our previous PES studies [14,15] show that the f-f, h-h, an h-f configurations have corresponding activation barriers for O_2 dissociative adsorptions of 1.37, 1.43, and 2.17 eV, and potential energies with metastable dissociative adsorption states of 0.355, 0.543, 0.552 eV, respectively, where the sum of the energies of the Ag slab and of the isolated O_2 is chosen as the origin of the potential energy. Thus, the metastable dissociative adsorption states that we have observed have associative desorption barrier heights of 0.9~1.6 eV, which were calculated from differences between the corresponding activation barriers and the potential energies with metastable dissociative adsorption states. As expected, these desorption barriers are smaller than that for a lower coverage case (0.40 ML), whose desorption barriers are 2.68±0.17 eV based on experimental results [26]. The uptake coverage of dissociative oxygen adsorption at a surface temperature of 160 K is 0.40 ML. Thus, we can deduce that a surface temperature lower than 160 K is required to realize these 0.5 ML coverages. Nevertheless, the high coverage structures are still stable. In this work, we chose the f-f configuration as the initial structure before oxygen atom diffusion, because the f-f configuration is the most stable at a 0.5 ML coverage.

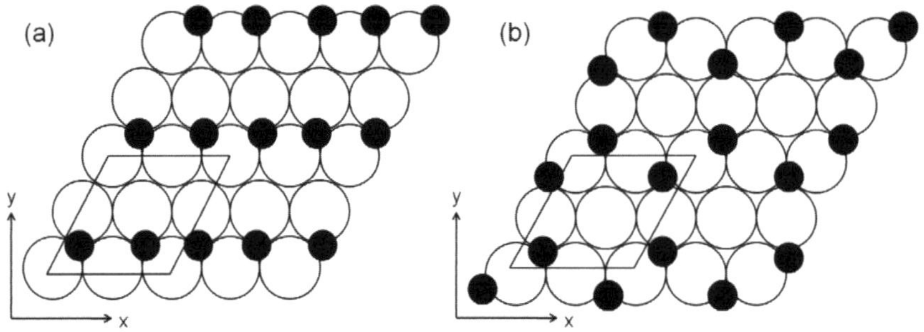

Figure 1. Oxygen adsorbed structures with (a) f-f and (b) h-f configurations. Two oxygen atoms in the supercell are above fcc hollow sites in the f-f configuration, and one oxygen atom is above a hcp hollow site and the other is above a fcc hollow site in the h-f configuration. Black and white circles show the positions of O and Ag atoms, respectively, and the parallelogram shows the supercell of fcc (2x2) Ag(111) surfaces.

3. Results and discussion

We investigated in-plane oxygen atom diffusion on Ag(111) surfaces in the case of 0.5 ML coverage. We considered that one oxygen atom moves to a hcp hollow site in the f-f configuration. In Figure 2, we show the three diffusion paths that we considered. The fixed oxygen atom (A) in Figure 2 is 0.12 nm above the first Ag(111) layer.

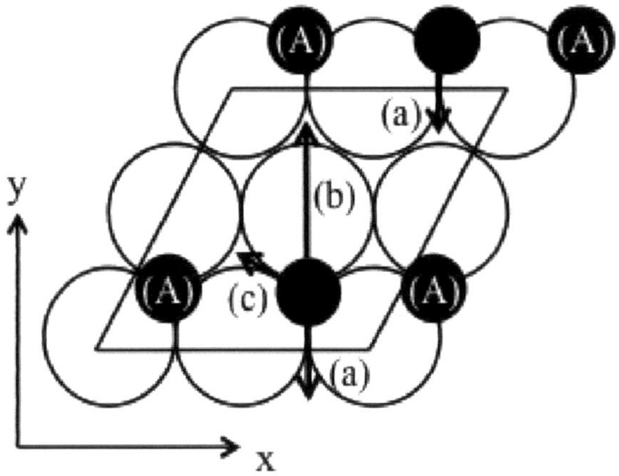

Figure 2. The three diffusion paths that we considered. Black and white circles show the positions of O and Ag atoms, respectively, and the parallelogram shows the supercell of fcc (2x2) Ag(111) surfaces. The position of oxygen atom (A) is fixed.

In Figure 3, we show potential energy curves for each diffusion path. The potential energy values are checked at 0.02 nm and 0.04 nm intervals for directions perpendicular and parallel to the Ag(111) surface, respectively. These intervals are sufficiently small so that the accuracy of the calculated potential energy curves is ensured. Here, we set the energy origin to correspond to the most stable f-f configuration. The paths along (a), (b), and (c) have diffusion barrier heights of 0.33, 1.22, and 0.80 eV, respectively. In the case of 0.11 ML coverage, which consists of one oxygen atom and fcc (3x3) Ag(111) layers, oxygen atom diffusion from a fcc hollow site to a hcp hollow site over bridge and top sites has a diffusion barrier height of 0.41 and 1.40 eV, respectively. It should be noted that the diffusion barrier heights are reduced in paths (a) and (b) and increased in path (c) because of repulsion between oxygen atoms. The diffusion along path (a) does not contribute to oxygen desorption, because this diffusion results only in a change between the f-f and h-f configurations. On the other hand, diffusion along paths (b) and (c) makes the distance between the two oxygen atoms shorter. After such diffusion, the two oxygen atoms can become desorbed as an oxygen molecule with no desorption barrier [14]. From these calculations, we found that diffusion along paths (b) and (c) prevents the realization of high coverages. We also estimated the zero point energies along these diffusion paths using a harmonic potential approximation of each potential well. These zero point energies are about 20 meV. In comparison with the diffusion barrier heights, the zero point energies do not make a significant contribution to the diffusion. Finally, we can say that the effective in-plane diffusion barrier height for breaking ferromagnetic oxygen adsorbed structures is about 0.78 eV. This value is almost the same order as the associative desorption barrier height. We also calculated the diffusion coefficient for oxygen atom diffusion Γ using the Arrhenius equation

$$\Gamma = \Gamma_0 \exp\left(-\frac{E}{k_B T}\right), \quad (1)$$

where Γ_0, E, k_B, and T are the diffusion prefactor, the effective diffusion barrier height, the Boltzmann constant, and the temperature of the system, respectively. Estimating the diffusion prefactor from the attempt frequency, which is approximately the same as the oxygen atom vibration frequency, the diffusion coefficients for in-plane diffusion are 3.5×10^{-10} nm^2/s at 160 K and 98 nm^2/s at 300 K.

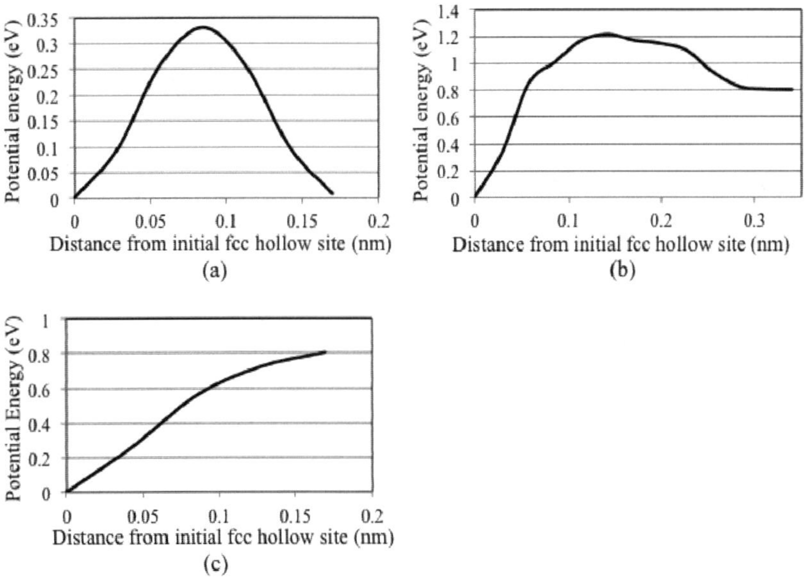

Figure 3. Potential energy curves for diffusion paths (a), (b), and (c), as indicated in Figure 2. The energy origin was set to the most stable f-f configuration.

We also focused on the penetration of one oxygen atom through the surface Ag layer from an on-surface fcc hollow site to a subsurface site in the f-f configuration. The fixed oxygen atom is 0.12 nm from the first Ag(111) layer. In Figure 4, we show the potential energy curve for oxygen atom diffusion to the subsurface. The step size perpendicular to the Ag(111) surfaces is 0.02 nm, which is sufficiently small so that the accuracy of the calculated potential energy curve is ensured. Here, we set the energy origin to the most stable f-f configuration. The diffusion barrier height for the penetration is 5.54 eV. We also performed a calculation in which the top two layers are relaxed with oxygen atoms in the topmost Ag layer. Then, the corresponding diffusion barrier height is reduced to 0.79 eV. This value is almost the same as the 0.81 eV diffusion barrier height obtained in the case of 0.11 ML coverage [27]. We also estimated the zero point energy along the diffusion path using the harmonic potential approximation. The zero point energy is about 10 meV. Thus, the effective diffusion barrier height is 0.78 eV. After the penetration, oxygen atoms gain more electrons from the Ag(111) surfaces and magnetic moments are quenched. Hence, this diffusion contributes to quenching the ferromagnetic states. It should also be noted that the corresponding diffusion coefficients for diffusion into the subsurface are 2.2×10^{-10} nm^2/s at 160 K and 62 nm^2/s at 300 K.

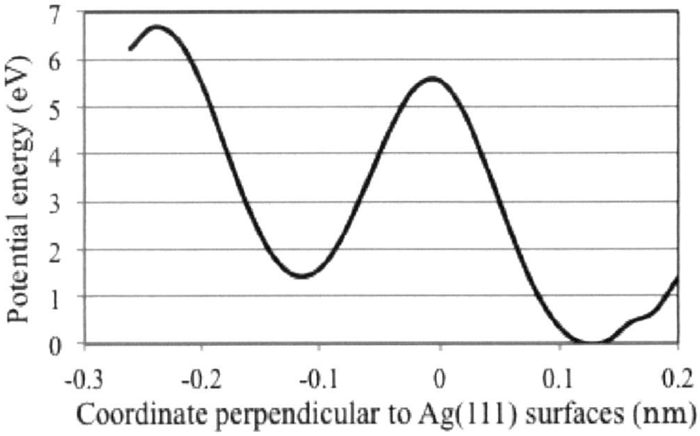

Figure 4. Potential energy curve of diffusion perpendicular to Ag(111) surfaces. The origin of the coordinates perpendicular to Ag(111) surfaces was set to the topmost Ag(111) layer. The energy origin was set to the most stable f-f configuration.

4. Summary

We performed first principles investigations on the robustness of ferromagnetic oxygen adsorbed structures on Ag(111) surfaces. We focused on oxygen atom diffusion parallel and perpendicular to Ag(111) surfaces. The in-plane diffusion can make O-O distances shorter and induce associative desorption, which reduces the oxygen coverage. On the other hand, diffusion of oxygen atoms into the subsurface can quench the magnetic moment. Consequently, such diffusion must be suppressed to realize and maintain the ferromagnetic oxygen adsorbed structures. These diffusion paths have an effective diffusion barrier of 0.78 eV. We also calculated the diffusion coefficients for oxygen atom diffusion using an Arrhenius equation. The diffusion coefficients for in-plane diffusion are 3.5×10^{-10} nm^2/s at 160 K and 98 nm^2/s at 300 K. The diffusion coefficients for diffusion into the subsurface are 2.2×10^{-10} nm^2/s at 160 K and 62 nm^2/s at 300 K. Thus, once ferromagnetic oxygen adsorbed structures form, diffusion rarely occurs in the experimentally-used low temperature region below 160 K, which is considered to be required for 0.5 ML coverage. From the results of a hyperthermal O_2 molecular beam experiment involving oxidation of Cu(111) surfaces [28], it has been reported that the oxidation process strongly depends on the O_2 translational energy, surface temperature and surface diffusion. We must optimize these parameters to realize coverages higher than 0.4 ML. Hence, we must consider these dynamics effects in the oxidation process to realize high coverages. These oxidation dynamics are ongoing studies.

Acknowledgments
This work is supported by MEXT (Ministry of Education, Culture, Sports, Science and Technology) through the G-COE (Special Coordination Funds for the Global Center of Excellence) program (H08) "Atomically Controlled Fabrication Technology", Grant-in-Aid for Scientific Research on Innovative Areas Program (2203-22104008) and Scientific Research (A) 20246011, (c) 22510107 program; by JST (Japan Science and Technology Agency) through ALCA (Advanced Low Carbon Technology Research and Development) Program " Development of Novel Metal-Air Secondary Battery Based on Fast Oxide Ion Conductor Nano Thickness Film" and Elements Science and Technology Project "New development of self-forming nano-particle catalyst without precious metals". Calculations were done using computer facilities of the Cyber Media Center (Osaka University) and the ISSP Super Computer

Center (the University of Tokyo). YK would like to extend gratitude to the Japan Society for Promotion of Science (JSPS) Young Researcher Fellowship for funding.

References

[1] Fisher B F and Klein M W 1975 *Phys. Rev. B* **11** 2025
[2] Silly F, Pivetta M, Ternes M, Patthey F, Pelz J P, and Schneider W D 2004 *Phys. Rev. Lett.* **92** 016101
[3] Negulyaev N N, Stepanyuk V S, Niebergall L, Bruno P, Pivetta M, Ternes M, Patthey F and Schneider W D 2004 *Phys. Rev. Lett.* **102** 246102
[4] Hoa N T M, Minamitani E, Diño W A, Cong B T and Kasai H 2010 *J. Phys. Soc. Jpn.* **79** 074702
[5] Minamitani E, Diño W A, Nakanishi H and Kasai H 2010 *Surf. Sci.* **604** 2139
[6] Minamitani E, Diño W A, Nakanishi H and Kasai H 2010 *Phys. Rev. B* **82** 153203
[7] Hoa N T M, Diño W A and Kasai H 2010 *J. Phys. Soc. Jpn.* **79** 113706
[8] Hoa N T M, Diño W A and Kasai H *J. Phys. Soc. Jpn.* accepted
[9] Escaño M C, Nguyen T Q, Nakanishi H and Kasai H 2009 *J. Phys.: Condens. Matt.* **21** 49221
[10] Escaño M C, Nakanishi H and Kasai H 2009 *J. Phys. Chem.* **113** 14302
[11] Tsuda M, Diño W A, Watanabe S, Nakanishi H and Kasai H 2004 *J. Phys. Condens. Matter* **16** S5721
[12] Tsuda M, Diño W A, Watanabe S, Nakanishi H and Kasai H 2004 *e-J. Surf. Sci. Nanotech.* **2** 200
[13] Zeng J, Liao S, Lee J Y, Liang Z 2010 *Int'l. Journal of Hydrogen Energy* **35** 942
[14] Kunisada Y, Nakanishi H and H. Kasai 2011 *J. Phys. Soc. Jpn.* **80** 084605
[15] Kunisada Y, Escaño M C and H. Kasai 2011 *J. Phys.: Condens. Matt.* **23** 394207
[16] Perdew J P, Burke K and Ernzerhof M 1996 Phys. Rev. Lett. **77** 3865
[17] Perdew J P, Burke K and Ernzerhof M 1997 Phys. Rev. Lett. **78** 1396
[18] Kresse G and Furthmüller J 1996 Comput. Mater. Sci. **6** 15
[19] Kresse G and Furthmüller J 1996 Phys. Rev. B **54** 11169
[20] Kresse G and Hafner J 1993 Phys. Rev. B **47** 558
[21] Kresse G and Hafner J 1994 Phys. Rev. B **49** 14251
[22] Monkhorst H J and Pack J D 1976 Phys. Rev. B **13** 5188a
[23] Neugebauer J and Scheffler M 1992 Phys. Rev. B **46** 16067
[24] Bader R F W 1990 Atoms in Molecules: A Quantum Theory (Oxford: Oxford University Press)
[25] Henkelman G, Arnaldsson A and Jonsson H 2006 Comput. Mater. Sci. **36** 354
[26] Raukema A, Butler D A, Box F M A and Kleyn A W1995 Surf. Sci. **347** 151
[27] Li W X, Stampfl C and Scheffler M 2003 Phys. Rev. B **67** 045408
[28] Moritani K, Okada M, Sato S, Goto S, Kasai T, Yoshigoe A and Teraoka Y 2004 J. Vac. Sci. Technol. A **22** 1625

Dispersion state and oxygen storage capacity properties of ceria and zirconia nanoparticles supported on alumina by the impregnation process

Masakuni Ozawa[1], Kazue Matuda[1], Yuri Miura[1], Masatomo Hattori[2], Takahiro Okouchi[3], Hiroki Yuzuriha[3], Tatunori Sugimoto[1] and Masaaki Haneda[1]

[1] Ceramics Research Laboratory, Nagoya Institute of Technology, 10-6-29 Asahigaoka, Tajimi, Gifu 507-0071, Japan
[2] Ecotopia Science Institute, Nagoya University, Furo-cho, Chikusa, Nagoya, Aichi 464-8603, Japan
[3] Graduated School of Frontier Materials Engineering, Nagoya Institute of Technology, Asahigaoka, Tajimi, Gifu 507-0071, Japan

E-mail: ozawa@nitech.ac.jp

Abstract. The characterization of the phase and oxygen storage capacity of ZrO_2, CeO_2, and mixed oxides, which were supported on Al_2O_3 by the impregnation method, was performed for the development of a better sub-catalyst in the exhaust treatment. Regarding the dispersion on the alumina support, the behavior of ZrO_2 is different from that of CeO_2; CeO_2 becomes crystalline nanoparticles on Al_2O_3, while majority of ZrO_2 do not form well-structured particles. We found that the oxygen storage capacity of cerium was improved by the addition of ZrO_2 to CeO_2 in the case of alumina-supported mixed oxides.

1. Introduction

In the field of environmental technologies, the treatment of the engine exhaust is an important issue which should be resolved to control air pollution. The three way catalyst (TWC) is used to remove carbon monoxide, hydrocarbons, and nitrogen oxides, emitted during the high- and low-temperature operation of an engine. The best performance of TWC is shown around a certain oxygen pressure corresponding to the ideal air/fuel (A/F) condition. The role of a sub-catalyst of the oxygen storage capacity (OSC), such as CeO_2 and CeO_2-ZrO_2, is very useful in order to maintain the partial pressure of oxygen with fluctuation of A/F. Therefore, Ce and Zr are the most important elements of a sub-catalyst in TWC with dispersed Pt and other noble metals on the support.[2,3] In addition, the OSC should be improved by surface and bulk properties as well as the interface in the system of CeO_2 by the addition of ZrO_2.[4] However, the dispersed state of CeO_2 and ZrO_2 themselves, when they are supported on alumina, has not been characterized even though the two metal oxides have similar fluorite structures. Since the catalyst is generally a composite powder with multi-components as well as the preparation process, the characterization of the fundamental microstructure about the nanoparticle materials is important for advanced performance and optimization of catalysts. The impregnation method using aqueous salts and support is widely used in preparing catalysts. In this work, we compare the phase and coordination state of CeO_2, ZrO_2, and CeO_2/ZrO_2, which were dispersed and supported on alumina via the impregnation process, using XRD, Raman scattering, and TEM, as well as OSC catalytic properties.

2. Experimental

2.1. Synthesis and characterization of catalyst

γ-Al_2O_3-supported $Ce_{1-x}Zr_xO_2/Al_2O_3$ (x=0, 0.2, 0.8, and 1.0) catalysts were prepared by the impregnation method, using a mixed aqueous solution of cerium nitrate $Ce(NO_3)_36H_2O$ (Wako, 99.9% purity), zirconyl nitrate $ZrO(NO_3)_26H_2O$ (Wako, 99.9% purity), and the γ-Al_2O_3 powder (Sumitomo chemical, surface area: 105 m^2/g), followed by drying and calcination at 600°C for 5 h in air. The mixture was then calcined at 800°C for 5 h in air. The mixing molar ratio of alumina:ceria and/or the zirconia catalyst was 100:30. $Pt/Ce_{1-x}Zr_xO_2/Al_2O_3$ catalysts were prepared by impregnation of a solution of $Pt(NO_2)_2(NH_3)_2$ to three supports, alumina, CeO_2/Al_2O_3, and ZrO_2/Al_2O_3, followed by drying and calcination at 600°C for 5 h in air. Catalyst samples were further heated at 1000°C for 5 h in air to examine the dispersion effect of CeO_2 and ZrO_2 and also the catalytic performance of Pt.

2.2. Characterization

XRD patterns were recorded using a Rigaku Rint X-ray diffractometer (Japan). Samples were scanned using CuKα radiation with a Ni filter over the range of 2θ from 20 to 80° at 40 kV and 20 mA. JCPDS cards were used for identification of the crystals. Clustered species in the powders were identified by laser Raman spectroscopy (JASCO NRS-3100) using a 532 nm laser. A transmission electronic microscope (TEM, Jeol JEM2000EX) was used to observe the morphology of dispersed CeO_2 and ZrO_2 in the samples.

2.3. Evaluation of catalytic properties

The OSC was measured by injecting an O_2 gas pulse of 1.0 ml at 600°C on 0.1 g of the catalyst samples preheated in flowing H_2/Ar at 800°C. The temperature programmed reduction with hydrogen (H_2-TPR) was monitored using a TCD-type detector during heat treatment of the sample in flowing H_2/Ar up to 800°C at a rate of 10°C min^{-1}. The specific surface area was measured by the nitrogen gas adsorption at 77 K after heat treatment at 200°C in flowing nitrogen for 2 h. The reaction conditions for the Pt catalyst was as following: 0.1 g of the catalyst at temperatures of 50-600°C, after heat treatment of catalysts under 10% H_2/N_2 up to 600°C, at a flow rate of 500 ml min^{-1} of mixed gases containing 3345 ppm of C_3H_6, 0.5 vol% of O_2, and a N_2 balance under a space velocity of 300,000 $lkg^{-1}h^{-1}$.

3. Results and discussion

3.1. Characterization of microstructure

Figure 1 shows the XRD patterns for the $Ce_{1-x}Zr_xO_2/Al_2O_3$ (x=0, 0.2, 0.8, and 1.0) catalysts. The XRD patterns of the samples heated at 800°C showed the formation of crystalline metal oxides and γ-Al_2O_3 in the catalysts. For the powders with x=0 and 0.2, the XRD showed the strong diffraction intensities of a cubic symmetry corresponding to fluorite-type oxides. Broad tetragonal phase patterns appeared in the composition of x=0.8 and 1.0, which were indicated by weaker diffraction intensities with broaden peaks of γ-Al_2O_3. In addition, the diffraction angle of all fluorite peaks was observed to shift to higher angles with decreasing CeO_2. The peaks of the present samples are not symmetrical because of the inhomogeneous mixed state of Ce and Zr oxides.[5] The fact that peak intensities decreased with increasing ZrO_2 suggested higher dispersion of Ce-Zr oxides on alumina when the ZrO_2 content becomes dominant.

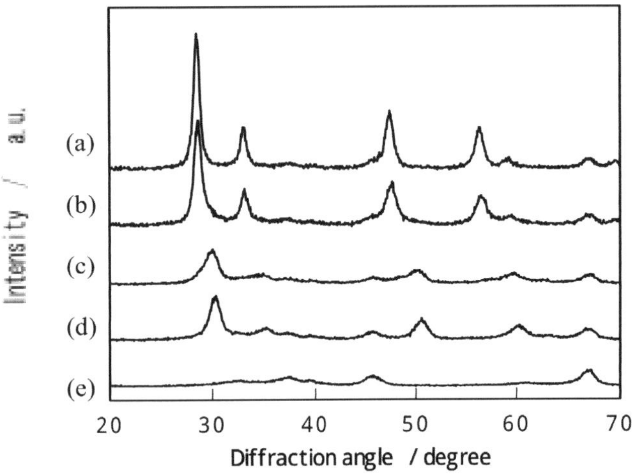

Figure 1. X-ray diffraction patterns of $Ce_{1-x}Zr_xO_2/Al_2O_3$: (a) x=0, (b) x=0.2, (c) x=0.8, (d) x=1.0, and (e) γ-Al_2O_3.

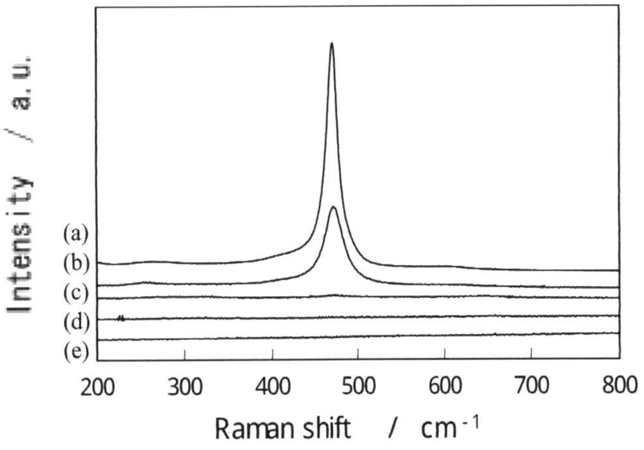

Figure 2. Raman spectra of $Ce_{1-x}Zr_xO_2/Al_2O_3$: (a) x=0, (b) x=0.2, (c) x=0.8, (d) x=1.0, and (e) γ-Al_2O_3.

Figure 2 shows the Raman spectra for $Ce_{1-x}Zr_xO_2/Al_2O_3$ (x=0, 0.2, 0.8, and 1.0). Raman scattering reveals more detailed information on the whole evolution about the symmetry of the coordinated complex between the metal and oxygen in $Ce_{1-x}Zr_xO_2$ on γ-Al_2O_3. Since the average arrangement of atoms for coherent lattice larger than ca.5 nm can be observed in the XRD, Raman scattering is preferable to detect the atomic scaled structure. The scattering intensity of the Al_2O_3 support was very weak because of the defective gamma-phase (spinel structure), as well as the very small (distorted) crystal less than 10 nm in diameter. In the present catalysts, the intensity of Raman scattering was very weak for the ZrO_2-rich samples (x=1.0, 0.8), while that of the CeO_2-rich samples showed clear scattering amplitude from highly symmetrical fluorite (CeO_2). The results suggest that the formation of the crystalline CeO_2-based solid solution occurs only in the CeO_2-rich sample, but major parts in the ZrO_2-rich sample or very small particles of the solid solution in which the coordination is disordered are in the non-crystalline state. Thus, majority of ZrO_2 exist as a highly dispersed and defective state with fluctuation of coordination between Zr and O, so that symmetry uncertainty for Zr-O interaction is assumed.

Figure 3 shows the TEM images of Al_2O_3 nanoparticles in $Ce_{1-x}Zr_xO_2/Al_2O_3$ (x=0 and 1.0). For CeO_2/Al_2O_3, the nanoparticles of CeO_2 are observed as the coverage on Al_2O_3 decreases. However for ZrO_2/Al_2O_3, there are dark grains with no particulate shape of ZrO_2 on Al_2O_3. The TEM observation is consistent with the results of Raman scattering spectra, by which a clear symmetry of Zr-O was not observed due to very low Raman intensities. The measured specific surface area (SA) was 105.1 m^2/g for a pure alumina support, 63.1 m^2/g for x=0, 63.5 m^2/g for x=0.2, 80.1 m^2/g for x=0.8, and 85.8 m^2/g for x=1.0 in $Ce_{1-x}Zr_xO_2/Al_2O_3$. The SA suggests that the coating (adsorbed) layer of ZrO_2 on the Al_2O_3 surface corresponds to a small decrease of SA. For CeO_2/Al_2O_3, the formation of CeO_2 particles on Al_2O_3, with a single nanometer size, may lead to a decrease of SA more than that of ZrO_2/Al_2O_3. The impregnation process is different from the coprecipitation of mixed oxides, because the absorption properties of each species should influence the final microstructure. In the present case, the stronger interaction (adsorption) of Zr on alumina seems to cause more dispersed ZrO_2 with a non-crystalline state which creates layer coverage on alumina.

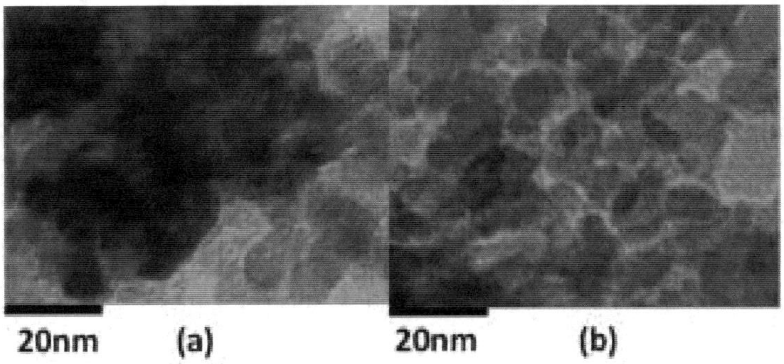

Figure 3. TEM images of (a) CeO_2/Al_2O_3 and (b) ZrO_2/Al_2O_3.

3.2. OSC and catalytic properties

Table 1 lists the total OSC for the catalysts which are measured by TPR up to 800°C, followed by oxygen pulse absorption at 600°C. The results indicate that the OSC becomes larger following the order of $Ce_{0.8}Zr_{0.2}O_2/Al_2O_3$ > CeO_2/Al_2O_3 > $Ce_{0.2}Zr_{0.8}O_2/Al_2O_3$ > ZrO_2/Al_2O_3 > Al_2O_3. The OSC should first depend on the content of CeO_2 in composite powders. On the other hand, ZrO_2 addition has improved the OSC because the best OSC is observed in $Ce_{0.8}Zr_{0.2}O_2/Al_2O_3$ in this work. Although this indicates the positive effect of ZrO_2 addition when co-impregnated to an alumina support on the OSC per Ce of $Ce_{1-x}Zr_xO_2/Al_2O_3$, more studies should be conducted by comparing the results of the bulk-state CeO_2-ZrO_2 (with no alumina). In fact, the OSC per Ce is a half of the bulk state of $Ce_{0.2}Zr_{0.8}O_2$ from the coprecipitation process [6]. It will be caused by the incomplete (noncrystalline) solid state of $Ce_{0.2}Zr_{0.8}O_2$ on Al_2O_3. The OSC is essentially the property of the bulk state to release and store oxygen in the crystal lattice when the solid solution of CeO_2-ZrO_2 is effectively used. A previous paper has proposed special crystal structures depending on the mixing of Ce and Zr and the synthesis method [7]. However, the best crystal structure has not been clear so far in the case of very small and dispersed particles on Al_2O_3. In the microstructure driven by the usual impregnation process, followed by heat treatment, all nanoparticles do not have an ideal crystal structure. Majority of the ZrO_2 species exist as a highly dispersed state on the alumina support in this work. However, with regards to the oxygen storage capacity (OSC), higher capacity properties are expected by doping ZrO_2, such as in Ce-Zr to complete the oxide solid solution. Although a more detailed examination is required for a better composite catalyst and shorter processing time when using this system in future work, the present study has clarified the state of character about the nanoparticles dispersion or symmetry of the metal and oxygen of CeO_2 or ZrO_2 on Al_2O_3 via the simple impregnation process. The effect of ZrO_2 addition to CeO_2 on the alumina support will improve the CeO_2 catalyst for realizing the dispersed state of a catalytic composition.

Table 1. Oxygen storage capacity of catalysts

X in $Ce_{1-x}Zr_xO_2$	0.0	0.2	0.8	1.0	Al_2O_3
OSC (ml/g) *1	4.0	4.7	1.7	0.1	<0.1

*1 oxygen storage capacity per gram of $Ce_{1-x}Zr_xO_2/Al_2O_3$

Finally, the performance of the Pt catalyst using three supports, Al_2O_3, CeO_2/Al_2O_3, and ZrO_2/Al_2O_3, was compared by the complete oxidation activity of propylene. The conversion efficiency to CO_2 was 91% for CeO_2/Al_2O_3, 20% for ZrO_2/Al_2O_3, 8% for Al_2O_3 at 400°C, and 100% for CeO_2/Al_2O_3, 51% for ZrO_2/Al_2O_3, and 48% for Al_2O_3 at 500°C. The result indicated that the best removal performance was obtained by the $Pt/CeO_2/Al_2O_3$ catalyst. It suggests that the strong interaction between Pt and CeO_2 improves the catalytic performance of the total oxidation of hydrocarbons. The effect of Ce, Zr, and Al oxides on oxidation over Pt is becoming an open question within the aspect of an atomic scale [8]. However, the microstructural feature of these compositions in the present work could become a useful basis to reveal the microstructure of nanometer-scale materials in complex compositions of Ce, Zr, and Al_2O_3, as well as Pt.

4. Summary

We have performed a comparative study on the phase and dispersion of CeO_2, ZrO_2, and ZrO_2-CeO_2 which are supported on alumina. Novel result on the dispersion of ZrO_2 in alumina was found using Raman scattering and TEM observation. With regards to the dispersion on the alumina support, the behavior of ZrO_2 is different from CeO_2: CeO_2 becomes crystalline nanoparticles on Al_2O_3, while majority of ZrO_2 do not form well-structured particles. The results of the catalytic performance concluded that the OSC was improved by mixing ZrO_2 to CeO_2 in the present system of $Ce_{1-x}Zr_xO_2/Al_2O_3$.

This work was supported by a project of the New Energy and Industrial Technology Development Organization, "Rare Metal Substitute Materials Development Project".

References

[1] Schnelle K B and Brown C A *Air Pollution Control Technology Handbook*; CRC Press, FL, 2002
[2] Ozawa M, Kimura M and Isogai A 1993 *J. Alloy. Compds.* **193** 72
[3] Ozawa M, 2006 *J. Alloy. Compds.***408-412** 1090
[4] Trovarelli A 1996 *Catal. Rev. Sci. Eng.* **38** 439
[5] Yashima M, Arashi H, Kakihana M and Yoshimura M 1994 *J. Amer. Ceram. Soc.* **77** 1067
[6] Ozawa M Unpublished work
[7] Varez A, Jolly J, Oliete P, Sanjuan M, Garcı´a-Gonzalez E, Jardiel T and Sanz J 2009 *Inorg. Chem.* **48** 9693
[8] Dong F, Tanabe T, Suda A, Takahashi N, Sobukawa H and Shinjoh H 2008 *Chem. Eng. Sci.* **63** 5020

Luminescence properties of $M_2Si_5N_8:Ce^{3+}$ (M = Ca, Sr, Ba) mixed nitrides prepared by metal hydrides as starting materials

Daiki Kuramoto, Hyo Sung Kim, Takashi Horikawa, Masahiro Itoh and Ken-ichi Machida

Division of Applied Chemistry, Graduate School of Engineering, Osaka University, 2-1 Yamadaoka, Suita, Osaka 565-0871, Japan

E-mail: machida@chem.eng.osaka-u.ac.jp

Abstract. Mixed metal nitrides, $M_2Si_5N_8:Ce^{3+}$ (M = Ca, Sr, Ba), were synthesized from $M_{2-y}Ce_ySi_5$ or an appropriate mixture of $MSiH_x$, Si_3N_4 and CeF_3, by a direct nitriding process in N_2 gas: 2 $MSiH_x + Si_3N_4 + CeF_3 \rightarrow M_2Si_5N_8:Ce^{3+}$. Also, charge-compensated materials, $M_2Al_zSi_{5-z}N_8:Ce^{3+}$, were prepared (from an appropriate mixture of $MSiH_x$, $MAlSiH_x$, Si_3N_4 and CeF_3) and the luminescence properties were characterized. The resultant phosphors showed green emission suitable for LED illumination by optimizing the mixing ratio of metal elements.

1. Introduction

White LEDs have come to attract attention, because they are environment-friendly in several points, such as the durability, thrifty power consumption, non-use of mercury and so on. Green emission phosphors are very important for white LEDs as a light source, *e.g.*, backlights and general solid illuminations, which need a high color rendering index since human eyes are more sensitive to green color than any other color. Conventional green phosphors such as oxides [1-4] and sulfides [5-7], however, have low thermal quenching temperature. Then sulfides are especially weak for moisture. Therefore, nitride phosphors have come to attract attention because of the wide emission spectra, the chemical stabilities and the high thermal quenching temperature due to their excellent thermal stability. The photoluminescence properties of the rare-earth ions result from lowering the gravity center of the 5d levels and decreasing the Stokes shift, because metal-nitrogen bonding has a high covalency and large crystal-field splitting effect on the luminescent centers (*e.g.*, Eu^{2+}, Ce^{3+}) caused by the surrounding N^{3-} ions [8-18]. In synthesis of phosphors, direct nitridation is easier to operate and is effective to save productions cost than the conventional solid state reaction method. That's because, almost metal nitrides, such as Ca_3N_2, EuN, CeN and so on, which are used in conventional solid state reaction method are chemically unstable and expensive [19]. Lately, we found that $Ca_2Si_5N_8:Eu^{2+}$ red phosphors prepared from CaH_2 have stronger luminescence intensity than synthesized from Ca_3N_2 powder [20]. In this research, alloy hydride powders were adopted as reactive intermediates. The synthesis conditions were optimized and the photoluminescence properties were characterized.

2. Experimental

The following alloys, MSi, MAlSi and $M_{2-y}Ce_y Al_zSi_{5-z}$ (M is an alkaline earth metal: Ca, Sr or Ba) were prepared from M, Ce, Al and Si metals by arc melting in argon atmosphere. Alloy hydrides were derived from their respective alloys by heating under high pressure (3.0 MPa) of H_2 at 723 K. Subsequently, alloy or alloy hydride powders were well mixed with Si_3N_4 and CeF_3 as Ce^{3+} ion source (the ratio was shown in Table 1) in the agate mortar and put into BN crucible. These raw material powders were adjusted to become stoichiometric amount. Total amount of Al and Si was adjusted to be 5. Then the mixtures were reacted with N_2 to synthesize $M_2Al_zSi_{5-z}N_8$:Ce^{3+} phosphors in a radio-frequency furnace filled with N_2 (0.15 MPa) using a ZrB_2 heater. The raw materials were heated in two stages: 1223 K for 3 h followed by 1723 K for 5 h. Finally, the phosphors were obtained after washing by 1N-HCl aqueous solution. The concentrations of activator Ce^{3+} ion were adjusted to 2 at% of M in all samples. The properties were evaluated by XRD, spectrophotofluorometer, NO analyzer and SEM/EDX.

Table 1. Mixing ratio for $M_2Al_zSi_{5-z}N_8$: Ce^{3+} synthesized from MSi, MAlSi, Si_3N_4 and CeF_3.

Al amount (z value)	MSi ($MSiH_x$)	MAlSi ($MAlSiH_x$)	Si_3N_4	CeF_3
z = 0	2	0	1	0.04
z = 0.5	1.5	0.5	0.83	0.04
z = 1	1	1	0.66	0.04
z = 2	0	2	0.33	0.04

3. Results and discussion

A series of XRD profiles of obtained $Ba_{2-y}Ce_yAl_zSi_{5-z}$ (z = 0 to 2) and $Ba_2Al_zSi_{5-z}N_8$:Ce^{3+} (z = 0.5, 2) phosphors are shown in Figure 1. With an increasing amount of Al, the BaAlSi ratio to $BaSi_2$ increased. Additionally, the amount of AlN impurity also increased.

Figure 2 shows the typical excitation and emission spectra of $Ba_2Al_zSi_{5-z}N_8$:Ce^{3+} and commercially used YAG:Ce^{3+} yellow phosphor for comparison. The Ce concentrations of synthesized phosphors were 2 at% of Ba. The emission spectra of these phosphors show an intense broad emission band ranging from 450 to 650 nm with a peak around 510 nm. The emission properties of the phosphors synthesized from $M_{2-y}Ce_yAl_zSi_{5-z}$ (M = Sr, Ba) were low. Also in the case of hydride powders, the emission properties were low. This is because of the low crystallinity of the synthesized phosphors due to the difficulty of direct nitridation. So, the inadequate nitridation was caused by the imbalance of the electro charge balance from Ce^{3+} ion.

Figure 3 (a) - (c) represent the typical excitation and emission spectra of $M_2Al_zSi_{5-z}N_8$:Ce^{3+} prepared from an appropriate mixture of MSi, MAlSi, Si_3N_4, CeF_3 and commercially used $Y_3Al_5O_{12}$:Ce^{3+}, YAG:Ce^{3+}, yellow phosphor for comparison. In the case of $Ca_2Al_zSi_{5-z}N_8$:Ce^{3+}, emission intensities decreased and peak top positions shifted to the longer wavelength side with increasing Al amount. On the other hand, emission properties of $Sr_2Al_zSi_{5-z}N_8$:Ce^{3+} and $Ba_2Al_zSi_{5-z}N_8$:Ce^{3+} have a different tendency from $Ca_2Al_zSi_{5-z}N_8$:Ce^{3+}. The PL properties were improved by adding Al. To add Al, Si was replaced by Al and expansion of the lattice was caused. The highest emission intensities were observed when the Al amount was 0.5. This was because of the difference in crystal structures: $Ca_2Si_5N_8$ is monoclinic, whereas $Sr_2Si_5N_8$ and $Ba_2Si_5N_8$ are orthorhombic [21, 22]. The structures of $Sr_2Si_5N_8$ and $Ba_2Si_5N_8$ had larger airspaces than that of $Ca_2Si_5N_8$, so there is little distortion of the lattice by Al addition. Figure 3 (d) shows the PL properties of $Sr_2Al_zSi_{5-z}N_8$:Ce^{3+} (z = 0, 0.5) composed from $MSiH_x$, $MAlSiH_x$, Si_3N_4 and CeF_3. The Ce concentrations were also 2 at% of Sr. The highest emission intensity (74% / YAG:Ce^{3+}) was observed when the Al amount was 0.5.

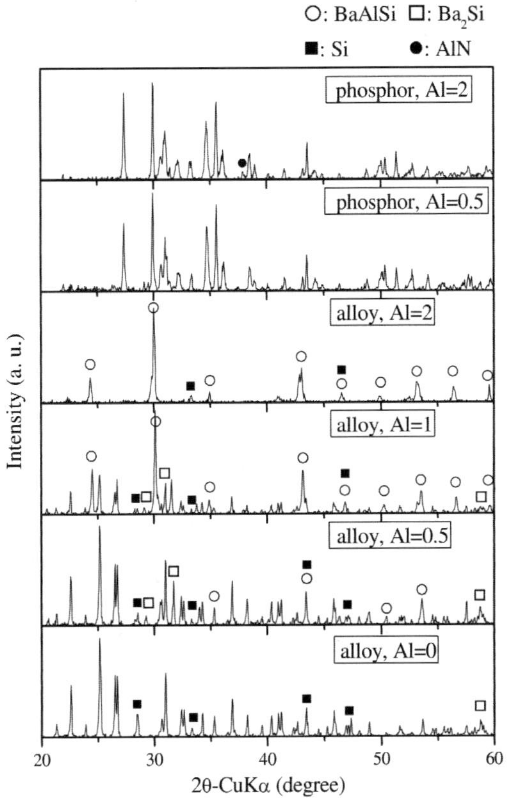

Figure 1. XRD patterns of $Ba_{1.96}Ce_{0.04}Al_zSi_{5-z}$ (z = 0 to 2) and $Ba_2Al_zSi_{5-z}N_8:Ce^{3+}$ (z = 0.5, 2). No assigned peaks are corresponding to those of $BaSi_2$ alloy and $Ba_2Si_5N_8$ phosphor, respectively.

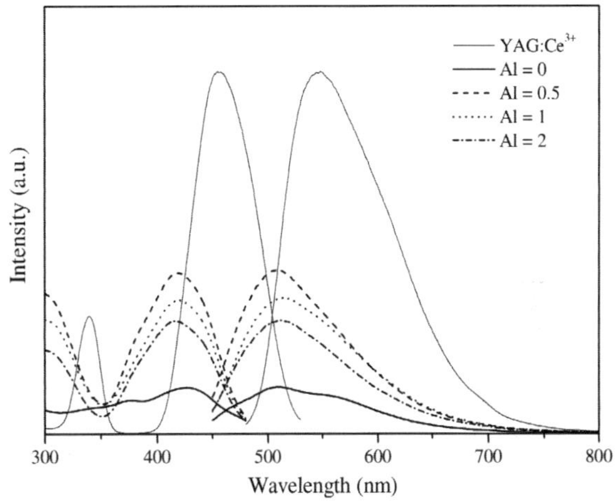

Figure 2. Photoluminescence spectra of $Ba_2Al_zSi_{5-z}N_8:Ce^{3+}$ (z = 0 to 2) from alloy powders, and commercial sample, YAG:Ce^{3+}, for comparison. The excitation and monitoring wavelength was 420 and 510 nm for obtained phosphors and 460 and 545 nm for YAG:Ce^{3+}, respectively.

These improvements of properties are because the size of powder particles became smaller by hydrogenation so that the reactivity of powders could increase. The SEM images of Sr-Al-Si alloy and Sr-Al-Si hydride powder are shown in Figure 4. Clearly, the sizes of alloys were decreased by hydridation.

From nitrogen and oxygen analysis, to adding Al caused an increase in the oxygen content of the phosphors. In addition, by increasing the Al amount, the nitrogen number changed from 8 to 7. These anion contents were lower than the stoichiometric amount. Therefore, a lattice defect existed and this also affected the photoluminescence properties.

4. Conclusions

$M_2Si_5N_8:Ce^{3+}$ (M = Ca, Sr, Ba) phosphors were synthesized from $M_{2-y}Ce_yAl_zSi_{5-z}$ or an appropriate mixture of $MSiH_x$, Si_3N_4 and CeF_3 by direct nitridation method. These phosphors can be effectively excited by violet-blue light around 400-430 nm and emit green light ranging from 450 to 650 nm with a peak around 489-528 nm. The properties of phosphor prepared from appropriate mixture of MSi, MAlSi, Si_3N_4 and CeF_3 were superior to that from $M_{2-y}Ce_yAl_zSi_{5-z}$ alloy. The activator, Ce^{3+}, condition was optimized by controlling the aluminum amount. The highest emission intensity was observed when the sample was prepared from alloy with Al = 0.5, except for $Ca_2Al_zSi_{5-z}N_8:Ce^{3+}$. In the $Ca_2Al_zSi_{5-z}N_8:Ce^{3+}$ case, the sample with Al = 0 had strong emission intensity. Then the phosphor synthesized from alloy hydride indicated strong emission intensity in $Sr_2(Al, Si)_5N_8$: Ce^{3+} with Al = 0, 0.5. From the result of photoluminescence properties, it is demonstrated that the alloy hydride powders can promote homogeneous nitridation.

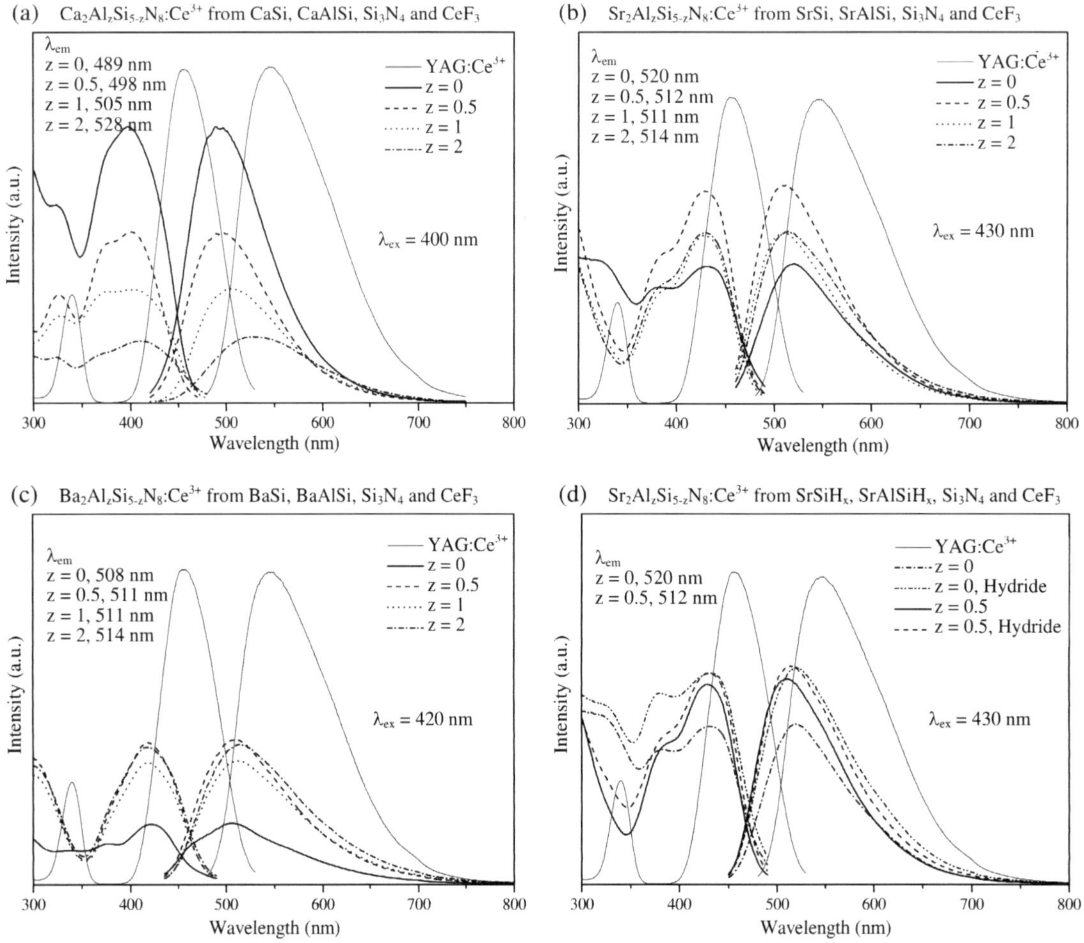

Figure 3. Photoluminescence spectra of (a) $Ca_2Al_zSi_{5-z}N_8$:Ce^{3+}, (b) $Sr_2Al_zSi_{5-z}N_8$:Ce^{3+}, (c) $Ba_2Al_zSi_{5-z}N_8$:Ce^{3+} and (d) $Sr_2Al_zSi_{5-z}N_8$:Ce^{3+} from the appropriate mixtures of raw materials and commercial sample, YAG:Ce^{3+}, for comparison. The excitation and monitoring wavelengths was 400-430 and 489-528 nm for obtained phosphors and 460 and 545 nm for YAG:Ce^{3+}, respectively.

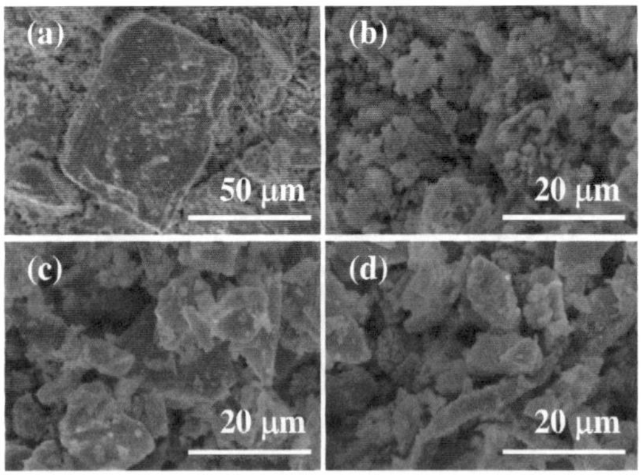

Figure 4. SEM images of (a) SrSi, (b) $SrSiH_x$, (c) SrAlSi and (d) $SrAlSiH_x$.

References

[1] Huppertz H and Schnick W 1997 *Z. Anorg. Allg. Chem.* **623** 212
[2] Hirosaki N, Xie R J, Kimoto K, Sekiguchi T, Yamamoto Y, Suehiro T and Mitomo M 2005 *Appl. Phys. Lett.* **86** 211905
[3] Xie R J, Hirosaki N, Li H L, Li Y Q and Mitomo M 2007 *J. Electrochem. Soc.* **154** J314
[4] Xie R J, Hirosaki N, Mitomo M, Uheda K, Suehiro T, Xu Y, Yamamoto Y and Sekiguchi T 2005 *J. Phys. Chem.* B **109** 9490
[5] Li Y Q, Delsing A C A, de With G and Hintzen H T 2005 *Chem. Mater.* **17** 3242
[6] Li Y Q, C M Fang, de With G and Hintzen H T 2004 *J. Solid State Chem.* **177** 4687
[7] Palilla F C, O'Reilly B E and Abbruscato V J 1970 *J. Electrochem. Soc.* **117** 87
[8] Rubio O J 1991 *J. Phys. Chem. Solids* **521** 101
[9] Lizzo S, Meijerink A and Blasse G 1994 *J. Lumin.* **59** 185
[10] Xie R J, Hirosaki N, Mitomo M, Uheda K, Suehiro T, Xu X, YamamotoY and Sekiguchi T 2005 *J. Phys. Chem.* B **109** 9490
[11] van Krevel J W H, van Rutten J W T, Mandal H, Hintzen H T and Metselaar R 2002 *J. Solid State Chem.* **165** 19
[12] Hirosaki N, Xie R J, Kimoto K, Sekiguchi T, Yamamoto Y, Suehiro T and Mitomo M 2005 *Appl. Phys. Lett.* **86** 211905
[13] Song X, He H, Fu R, Wang D, Zhao X and Pan Z 2009 *J. Phys. D: Appl. Phys.* **42** 065409
[14] Li Y Q, Delsing A C, de With G and Hintzen H T 2005 *Chem. Mater.* **17** 3242
[15] Uheda K, Hirosaki N, Yamamoto Y, Naito A, Nakajima T and Yamamoto H 2006 *Electrochem. Solid State* **9** 22
[16] Watanabe H and Kijima N 2009 *J. Alloys Compd.* **475** 434
[17] Höppe H A, Lutz H, Morys P, Schnick W and Seilmeier A 2000 *J. Phys. Chem. Solids* **61** 2001
[18] Li Y Q, van Steen J E J, van Krevel J W H, Botty G, Delsing A C A, DiSalvo F J, de With G and Hintzen H T 2006 *J. Alloys Compd.* **417** 273
[19] Piao X, Machida K, Horikawa T and Hanzawa H 2007 *Appl. Phys. Lett.* **91** 041908
[20] Lei B, Machida K, Horikawa T and Hanzawa H 2009 *J. Leng, K. A. Asundi and W. Ecke Proc. SPIE* **7493**
[21] Schlieper T and Schnick W 1995 *Z. Anorg. Allg. Chem.* **621** 1037
[22] Schlieper T, Milius W and Schnick W 1995 *Z. Anorg. Allg. Chem.* **621** 1380

Luminescence properties of $CaAlSiN_3$:Eu^{2+} mixed nitrides prepared by carbothermal process

Hyo Sung Kim[1], Takashi Horikawa[1], Hiromasa Hanzawa[2] and Ken-ichi Machida[1]
[1]Division of Applied Chemistry, Graduate School of Engineering, Osaka University, 2-1 Yamadaoka, Suita, Osaka 565-0871, Japan
[2]Graduate School of Engineering Science, Osaka University, Toyonaka, Osaka 560-8531, Japan

E-mail: machida@chem.eng.osaka-u.ac.jp

Abstract. Mixed metal nitride ($CaAlSiN_3$:Eu^{2+}) phosphor was synthesized from an appropriate mixture of $CaCN_2$, $CaCO_3$, AlN, Si_3N_4, Eu_2O_3 and C by the carbothermal nitriding method in a N_2 gas flow: 1/2 $CaCN_2$ + 1/2(1-2x)$CaCO_3$ + AlN + 1/3Si_3N_4 + $x$$Eu_2O_3$ + C → $(Ca_{1-x}Eu_x)AlSiN_3$ + CO_2↑. The resultant phosphor gave an intense red emission and good high-temperature durability for LED solid illuminations by optimizing some synthesis conditions.

1. Introduction

Recently, red phosphors have attracted much attention as one of the materials to improve the color rendering of white LEDs used as backlights for LCD panels and general illumination lamps, etc. As important properties of such red phosphors, being excitable by blue or near UV light and having a lower thermal quenching characteristic are required. However, conventional red phosphors such as oxides phosphor [1-4] are generally not excitable by such visible or near UV light. In addition, their emissions are considerably quenched at the temperature of 150-200 °C and sulfide phosphors [5-7] are very sensitive to moisture and have a low thermal quenching temperature. Among Eu^{2+}-doped nitride phosphors for LEDs used in solid illuminations, $CaAlSiN_3$:Eu^{2+} was found to show excellent optical properties as the red color phosphors because of the adequate thermal stability and lack of environmental hazard, non-toxicity, high emission efficiency and high quenching temperature [8]. Generally, critical preparation conditions including high temperature up to 1800 °C and high pressure more than 0.9 MPa is required to obtain high-quality $CaAlSiN_3$:Eu^{2+} phosphors. Due to the chemical and thermal inertness of the starting materials, AlN and Si_3N_4, for example, are widely used. Additionally, Ca_3N_2 used as a Ca component for $CaAlSiN_3$:Eu^{2+} in the conventional synthesis method is expensive and deliquescent. In this work, therefore, a carbothermal reduction and nitriding (CRN) method using raw material mixtures of $CaCN_2/Ca(CH_3COO)_2$, $CaCN_2/CaCO_3$ and $CaCN_2/Ca(HCOO)_2$ for the Ca component was applied to synthesize the $CaAlSiN_3$:Eu^{2+} phosphor and the luminescence properties were characterized.

2. Experimental

Stoichiometric amounts of $CaCN_2$ (95%), $Ca(CH_3COO)_2$ (95%), $CaCO_3$ (95%) and $Ca(HCOO)_2$ (95%) were used to prepare mixtures of $CaCN_2/Ca(CH_3COO)_2$, $CaCN_2/CaCO_3$ and $CaCN_2/Ca(HCOO)_2$. Each mixture was then mixed thoroughly with AlN (95%), α-Si_3N_4 (99.5%), Eu_2O_3 (> 99.99%) and C (graphite) and put into a BN crucible in a radio frequency furnace using a ZrB_2 heater. Thereafter, the powder mixture was heated at 1550°C for 6 h under the normal pressure of nitrogen gas (0.1 MPa) to obtain $CaAlSiN_3:Eu^{2+}$ and then the products were washed by an alkaline solution. The phase identification of resultant samples was made by powder X-ray diffraction (XRD) analysis with Cu $K\alpha$ radiation (RINT2200, Rigaku). Photoluminescence (PL) spectra of the obtained phosphor powders were measured on a fluorescence spectrophotometer (F-4500, Hitachi) at room temperature.

3. Results and discussion

The $CaAlSiN_3:Eu^{2+}$ phosphor was synthesized by the CRN method without using the deliquescent Ca_3N_2 raw material. Figure 1 shows the XRD patterns of the products prepared with various raw materials. The resultant samples consisted of $CaAlSiN_3:Eu^{2+}$ and AlN, of which the latter was non-reacted residue of raw material mixtures. However, molar ratio of $CaCN_2:CaCO_3$ (0:1), sample case, formed low crystallization of the $CaAlSiN_3$ phase. There is a clearly shown difference with other samples, which is an effect of the $CaCN_2$ raw material. After washing with HCl or NaOH solution, the impurity phase still remained and the body color of obtained phosphors was yellowish-orange.

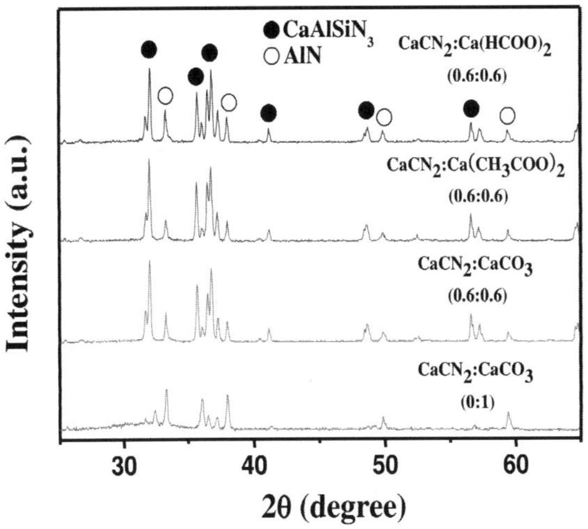

Figure 1. XRD patterns of $CaAlSiN_3:Eu^{2+}$ (2 at%) prepared with various raw material mixtures.

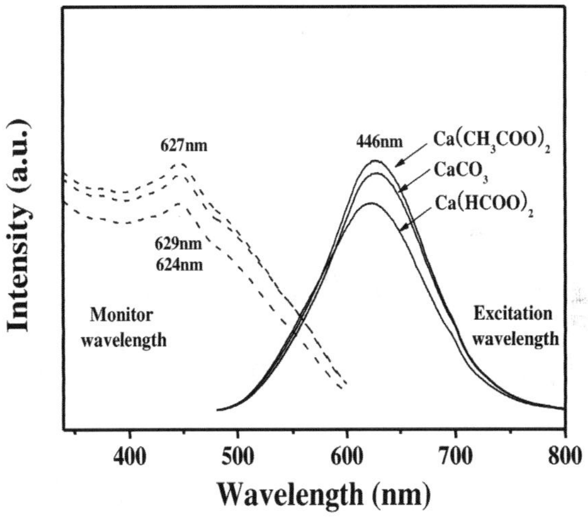

Figure 2. PL spectra of $CaAlSiN_3:Eu^{2+}$ (2 at%) prepared with various raw material mixtures.

The PL spectra of $CaAlSiN_3:Eu^{2+}$ phosphors prepared from the various raw material mixtures are shown in Figure 2. The raw materials composed of some combinations of $CaCN_2$ and $Ca(HCOO)_2$, $CaCO_3$ or $Ca(CH_3COO)_2$ powders with an optimized molar ratio of 0.6:0.6 (to AlN and 1/3 Si_3N_4) without carbon powder. The emission spectra measured were peaking at 627, 629 and 624 nm and the relative emission intensity values (*vs.* commercially available $YAG:Ce^{3+}$ (P46-Y3)) were about 40, 38 and 35% for phosphors prepared with $Ca(HCOO)_2$, $CaCO_3$ and $Ca(CH_3COO)_2$, respectively. Due to the AlN impurity and high oxygen content as explained later, the emission wavelength observed on present phosphors shifted lower than that of commercially available $CaAlSiN_3:Eu^{2+}$ ($CASN:Eu^{2+}$) red phosphor.

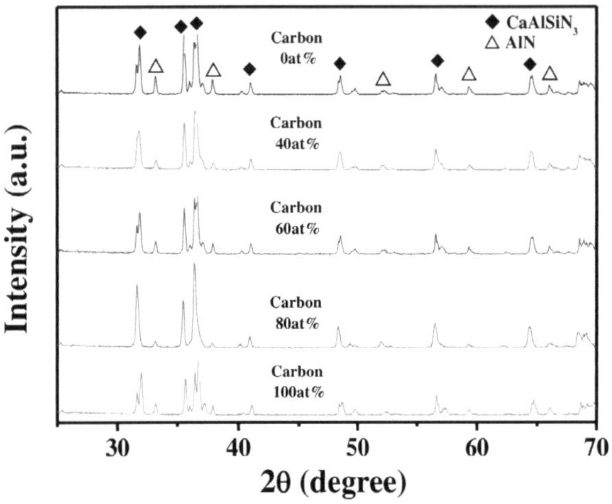

Figure 3. XRD patterns of the $CaAlSiN_3:Eu^{2+}$ (2 at%) prepared from $CaCN_2$-$CaCO_3$ based mixtures with various amounts of C powder.

Table 1. Oxygen and nitrogen contents in the resultant $CaAlSiN_3:Eu^{2+}$ phosphors.

Preparation condition $CaCN_2$: CO_3 (0.6 : 0.6)	O / N analysis (mass%)	Emission wavelength (nm)	Excitation wavelength (nm)
Additive Carbon 0at%	1.08 / 13.51	643	445
Additive Carbon 40at%	0.77 / 15.57	648	446
Additive Carbon 60at%	1.21 / 18.97	643	446
Additive Carbon 80at%	0.68 / 18.22	648	448
Additive Carbon 100at%	2.12 / 39.68	638	445

Figure 3 shows XRD patterns of the $CaAlSiN_3:Eu^{2+}$ (2 at%) phosphors obtained by using the equimolar $CaCN_2$-$CaCO_3$ based mixtures with various amounts of C powder. In all $CaAlSiN_3:Eu^{2+}$ samples in Figure 3, an AlN impurity phase remained, but the intensity of XRD peaks was systematically decreased with increasing amount of C powder. By washing them with the HCl solution, the $CaAlSiN_3:Eu^{2+}$ phosphor was obtained in a high-quality form and the body color was deep red in contrast to the samples prepared without C powder. This suggests that the carbon additive is effective for promoting the solid-state reaction among the raw materials, so that the high-quality $CaAlSiN_3:Eu^{2+}$ phosphor could be produced with good reproducibility.

The oxygen and nitrogen contents in the resultant $CaAlSiN_3:Eu^{2+}$ (2 at%) phosphors are summarized in Table 1, where the samples were prepared from the $CaCN_2$-$CaCO_3$ based mixtures of AlN, Si_3N_4 and Eu_2O_3 with various amounts of C powder. As shown in Table 1, the N/O ratio is increased with increasing amounts of C powder and the N content of the sample prepared with a 100 at% amount of C powder was much closer than the theoretical one (30.64 %).

The PL spectra of the $CaAlSiN_3:Eu^{2+}$ phosphor prepared from the $CaCN_2$-$CaCO_3$ based mixtures with various amounts of C powder are shown in Figure 4. The emission spectra were peaking at about 648 nm and the relative emission intensity was maximized (about 85% vs. $YAG:Ce^{3+}$) on the sample prepared with 80% of C powder, although N content (18.22 mass%) of the sample was considerably low compared with the theoretical value of 30.64 mass%. For the sample using 100 at% of C powder, the excess amount of nitrogen (39.68 – 30.64 = 9.04 mass%) should suggest the formation of Ca_3N_2 other than $CaAlSiN_3:Eu^{2+}$ and such alkaline earth nitride reacts with H_2O to convert to CaO or $Ca(OH)_2$. So that the emission intensity could be lower than that of the $CaAlSiN_3:Eu^{2+}$ phosphor prepared from the mixture containing the 80 at% of C powder.

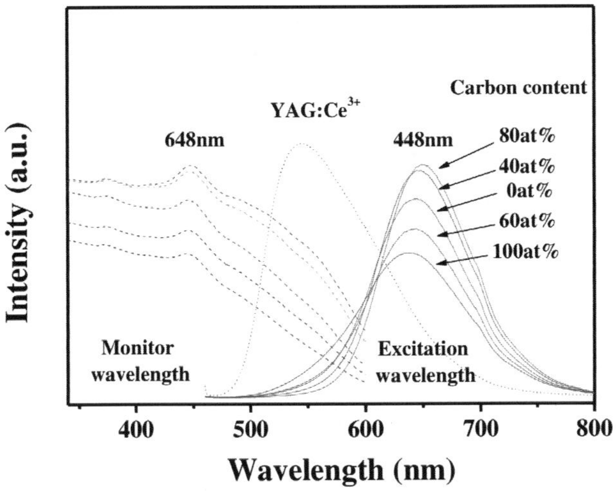

Figure 4. PL spectra of the $CaAlSiN_3:Eu^{2+}$ (2 at%) prepared from $CaCN_2$-$CaCO_3$ based mixtures with various amounts of C powder.

4. Conclusions

The excellent red phosphor of $CaAlSiN_3:Eu^{2+}$ is synthesized using the raw material mixture of $CaCN_2$, $CaCO_3$, AlN and Si_3N_4 with the molar ratio of 0.6, 0.6, 1.0 and 1.0 for the host lattice and the additives of Eu_2O_3 as the dopant (about 2 at% vs. Ca^{2+} ion) and C powder as the reductant (about 80 at%) by the CRN method. The optimum concentration of Eu^{2+} ion is about 2 at% and the relative emission intensity vs. YAG:Ce^{3+} is about 85%. The phosphor is efficiently excited by 460 nm blue light of InGaN LED chips and gives an emission band peaking at 648 nm.

References

[1] Huppertz H, Schnick W and Anorg Z 1997 *Allg. Chem.* **623** 212
[2] Hirosaki N, Xie R J, Kimoto K, Sekiguchi T, Yamamoto Y, Suehiro T and Mitomo M 2005 *Appl. Phys. Lett.* **86** 211905
[3] Xie R J, Hirosaki N, Li H L, Li Y Q and Mitomo M 2007 *J. Electrochem. Soc.* **154** J314
[4] Xie R J, Hirosaki N, Mitomo M, Uheda K, Suehiro T, Xu Y, Yamamoto Y and Sekiguchi T 2005 *J. Phys. Chem.* B **109** 9490
[5] Li Y Q, Delsing A C A, de With G and Hintzen H T 2005 *Chem. Mater.* **17** 3242
[6] Li Y Q, C M Fang, de With G and Hintzen H T 2004 *J. Solid State Chem.***177** 4687
[7] Palilla F C, O'Reilly B E and Abbruscato V J 1970 *J. Electrochem. Soc.* **117** 87
[8] Piao X, Machida K, Horikawa T, Hanzawa H, Shimomura Y and Kijima N *Chem. Mater.* **19** 4592

Discrimination of wireless electromagnetic signals by electro-optic modulators using an array of patch antennas embedded with orthogonal gaps

Yusuf Nur Wijayanto, Hiroshi Murata and Yasuyuki Okamura
Graduate School of Engineering Science, Osaka University
1-3 Machikaneyama, Toyonaka, Osaka 560-8531, Japan

E-mail: wijayanto@ec.ee.es.osaka-u.ac.jp

Abstract. We proposed a new wireless electromagnetic signal discrimination device using electro-optic modulators and an array of planar patch antennas embedded with orthogonal gaps. Wireless electromagnetic signals can be detected and converted directly to lightwave signals through optical modulation. The magnitude, phase, and polarization of the wireless signal can be measured precisely and identified. By using meandering gap structures, the directivity of the wireless signal conversion can be tuned more precisely. Furthermore, the directivity of the wireless signal conversion in two-dimensional space can also be obtained by comparing two orthogonal modulated lightwave signals from two orthogonal optical waveguides. The design and analysis of the device and experiments at an operational frequency of 26 GHz are presented.

1. Introduction

Wireless-fibre technologies with low power consumption have been developed to establish a sustainable society. Wireless technologies provide convenience, mobility, and flexibility for users [1]. However, the wireless microwave signals are easily affected by environmental noise and high transmission loss. Their allocated bandwidth is relatively narrow. On the other hand, optical fibre technologies provide large bandwidth, low transmission loss, no inductance, and no interference [2]. The combination of wireless microwave and optical fibre technologies has been developed in order to exploit their advantages. The combined technologies can be used for communication, remote sensing, electromagnetic compatibility testing, identification, and so on.

Conversion between wireless microwave and lightwave signals is important in wireless-fiber systems, which can be obtained using optical modulators and optical detectors with microwave antennas. Several converters composed of discrete optical modulators and discrete antennas have been reported [3-4]. The attenuation of the received signal through the cable becomes a problem in high-frequency operation. Electro-optic (EO) converters using antenna-coupled optical modulation electrodes were also proposed for high-frequency operation with low distortion [5-8]. They are composed of antennas, resonant electrodes, and connection lines. Good impedance matching conditions and precise frequency tuning between the planar structures are required in order to obtain effective optical modulation. However, complete matching and precise tuning are rather difficult.

We have proposed EO wireless-optical converters using patch antennas embedded with a narrow gap [9,10]. The devices are composed of only patch antennas and there are no other planar elements on the substrate. The proposed devices have simple compact structures and are operated with no precise tuning and no external power supply. Optical modulation with extremely low distortion was demonstrated. We have also proposed EO wireless-optical converters using patch antennas embedded with orthogonal gaps for dual-polarized wireless signal detection [11]. However, the modulation index was low owing to the limitation by the lightwave transit-time effect [12, 13].

In the present paper, a new EO modulator using planar patch antennas embedded with orthogonal gaps is proposed for wireless signal discrimination. The modulation index of the proposed devices can be enhanced by approximately three times by adding a number of the patch antennas in array form using a new design for the compensation of the transit-time effects. The magnitude and phase of the wireless signal is obtainable at the same time. The wireless directivity in the one-dimensional (1-D) plane and two-dimensional (2-D) space can also be tuned by meandering gap structures for phase-reversal. Therefore, the wireless electromagnetic signals can be identified and discriminated using the proposed device.

In the following sections, we will discuss the proposed device structure and analyze its characteristics in detail. The preliminary experimental results and advanced measurement setup for identifying wireless signals are also presented.

2. Device structures

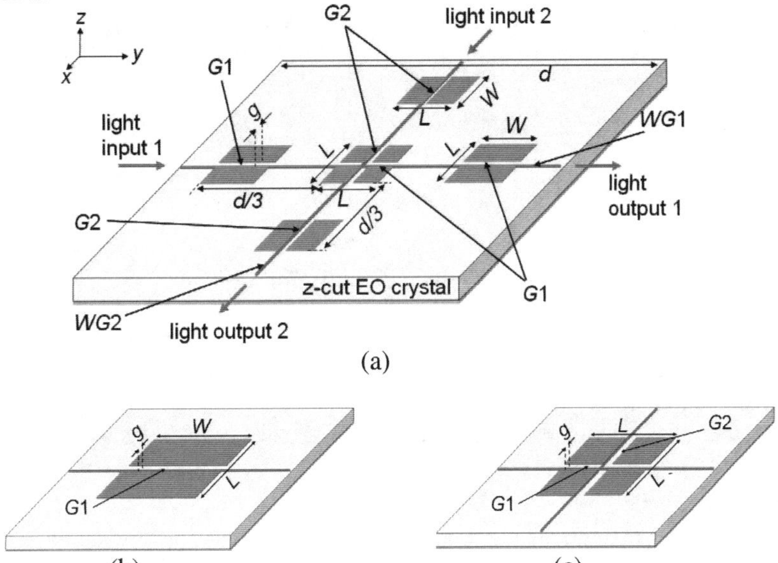

Figure 1. (a) Basic structure of the wireless signal discrimination device using a combination of patch antennas with gaps: the entire device structure, (b) the patch antenna with a gap, and (c) the patch antenna with orthogonal gaps.

Figure 1 shows the basic structure of the proposed device, which is fabricated on an EO crystal substrate. The proposed device is composed of two orthogonal optical waveguides, one patch antenna with orthogonal gaps, and four patch antennas with a single gap. The length, L, of each patch antenna is set as a half wavelength for the designed wireless electromagnetic signal. The patch antenna with orthogonal gaps is set at the centre of the device, where the size is set to LxL. The patch antennas with a single gap are set surrounding the patch antenna with orthogonal gaps, where the size is set to LxW. The width, W, of the patch antennas with a single gap can be set to be longer than a half wavelength but slightly shorter than one wavelength of the microwave signal in order to avoid unwanted higher-order mode effects. The width of the gaps is of micrometer order. The orthogonal optical waveguides are located under the edge of the gaps. A buffer layer is inserted between the substrate and the antennas. The reverse side of the substrate is covered with a ground electrode.

When a wireless signal at the design frequency with an arbitrary polarization is irradiated to the proposed device, a resonant standing-wave microwave current is induced on the patch antenna surface. The induced current can be separated into two components along the *x*- and *y*-axes. In the proposed device, displacement currents are induced across the gap, G_1, normal to the *x*-axis and the gap, G_2, normal to the *y*-axis, due to the continuity of the current flow. As a result, strong electric fields are also induced across the gaps. The strong electric field across the gaps can be used for optical modulation. Therefore, the wireless signal can be converted to the lightwave signal through optical modulation by the induced electric field across the gaps. The modulated lightwaves are obtained from two orthogonal optical waveguides. The magnitude and phase of the wireless signal can be obtained by detecting the modulated lightwaves. The polarization of the wireless signal can also be identified by comparing the two orthogonal modulated lightwaves. Furthermore, the directivity of the wireless signal conversion is precisely controllable using meandering gap structures to compensate for the degradation of the optical modulation owing to the transit-time effect. Therefore, the wireless signals with different irradiation angles can be discriminated through optical modulation using the proposed device.

3. Patch antennas

3.1. Patch antenna with a single gap

The patch antenna with a gap shown in Figure 1(b) is considered. This antenna has a gap, G_1, parallel to the *y*-axis. When a wireless electromagnetic signal with a polarization in the *x*-axis is irradiated to the patch antenna with a normal irradiation angle, a standing-wave electromagnetic surface current is induced on the patch antenna along the *x*-axis [14-17]. The induced surface current along the *x*-axis, K_x, can be expressed

$$K_x(x,t) = K_{xp}\cos(\omega_m t)\cos\left(p\frac{2\pi}{\Lambda_m}x\right), \tag{1}$$

where K_{xp} is the amplitude of the surface current along the *x*-axis, ω_m, is the wireless microwave signal angular frequency, and Λ_m is the microwave wavelength. The patch antenna length, L, is set to the half wavelength of the wireless electromagnetic signal, when the fundamental mode is induced at the resonant frequency, i.e., $p = 1$. In a standard patch antenna with no gap, the surface current for the fundamental mode becomes maximum at $x = 0$ along the *x*-axis.

In the patch antenna with a narrow gap, the displacement current must be induced across the gap for the current continuity requirement [18-20]. The induced electric field, E_x, across the gap is obtained by time integration of the displacement currents as

$$E_x(0,t) = \frac{1}{\varepsilon}D_x(0,t) \propto \int K_x(0,t)dt = K_{x1}\sin(\omega_m t), \tag{2}$$

where D is the electric flux density ($D = \varepsilon E$), and ε is the permittivity. The strong electric field across the gap can be used for optical modulation.

The analysis of the patch antenna with the gap, G_2, parallel to the *x*-axis can be obtained using a similar procedure. When a wireless electromagnetic signal with polarization along the *x*-axis is irradiated to the patch antenna, a standing-wave electromagnetic surface current is induced on the patch antenna along the *y*-axis. The standing-wave electromagnetic surface current, K_y, along *y*-axis and the induced electric field, E_y, across the gap for the half-wavelength patch antennas are expressed as

$$K_y(y,t) = K_{y1}\cos(\omega_m t + \varphi)\cos\left(\frac{2\pi}{\Lambda_m}y\right) \tag{3}$$

$$E_y(0,t) = \frac{1}{\varepsilon}D_y(0,t) \propto \int K_y(0,t)dt = K_{y1}\sin(\omega_m t + \varphi), \tag{4}$$

where K_{y1} is the amplitude of the surface current along the y-axis, and φ is the mutual phase shift between the x- and y-current components.

An arbitrary polarization of the wireless signal can be also analyzed. The standing-wave current surface can be separated into two components along the x- and y-axes. Therefore, strong electric fields are induced across the two orthogonal gaps in both patch antennas.

3.2. Patch antenna with orthogonal gaps

A patch antenna with two orthogonal gaps is shown in Figure 1(c), the size of which was set to be *LxL* for receiving dual polarization wireless signals. When a wireless electromagnetic signal with an arbitrary polarization is irradiated to the patch antenna with a normal irradiation angle, a standing-wave electromagnetic surface current is induced on the patch antenna along the x- and y-axes. The standing-wave electromagnetic surface currents, K_x and K_y, along the x- and y-axes can be expressed by Equations (1) and (3), respectively. Then, displacement currents must be induced across both gaps in order to satisfy the current continuity requirement. Therefore, the induced electric fields across the gap, E_x and E_y, are obtained. These electric fields can be also expressed by Equations (2) and (4), respectively. The strong electric field across the two gaps can be also used for optical modulation.

By measuring the modulation indices and phases, the polarization of the wireless signal can be identified completely.

3.3. Numerical analysis

The field distributions on the surfaces of the patch antennas were analyzed using HFSS 3-D electromagnetic field analysis software. Figure 2 shows the calculated electric field distributions on the substrate surface when microwave signals with different polarization conditions are irradiated to the device.

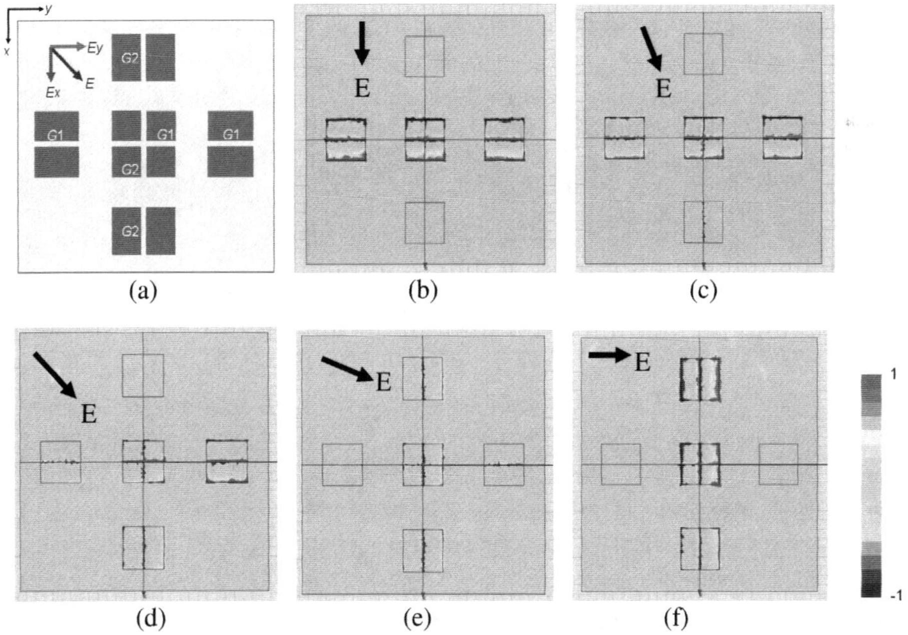

Figure 2. (a) Top view of the patch antenna array. Field distributions in the z-component on the substrate surface with the polarization at (b) 0 degrees, (c) 30 degrees, (d) 45 degrees, (e) 60 degrees, and (f) 90 degrees.

First, when the wireless signal with a linear polarization along the x-axis (0 degrees) is irradiated to the patch antenna array, as shown in Figure 1(a), a strong electric field is induced across the gap, G_1,

normal to the x-axis and a negligible electric field is induced across the other gap, G_2, parallel to the x-axis, as shown in Figure 2(b). When a linear polarized wireless signal with an angle of 30 degrees to the x-axis is irradiated to the device, strong electric fields are induced across both orthogonal gaps, as shown in Figure 2(c). However, the magnitudes of the electric fields across the gaps are different. Then, when a linear polarized wireless signal with an angle of 45 degrees to the x-axis is irradiated to the device, strong electric fields are induced across both orthogonal gaps with almost the same magnitude, as shown in Figure 2(d). Figure 2(e) shows the electric field distribution when the wireless signal with the 60-degree polarization condition is irradiated to the device. The electric field magnitudes along the gaps are the inverse of those for the 30-degree polarization condition. Finally, when the polarization condition of the radiated wireless signal is rotated 90 degrees (along the y-axis), a strong electric field is induced across the gap, G_2, normal to the y-axis and a negligible electric field is induced across the gap, G_1, parallel to the y-axis, as shown in Figure 2(f). Using these characteristics, the linear polarization condition of the wireless signals can be identified.
Furthermore, when a microwave signal with circular polarization states is irradiated to the device, the induced electric fields across all gaps have the same magnitude but their phases differ by 90 degrees between the x- and y-axes. Therefore, the circular polarization of the wireless signals can be measured using the proposed device.

4. Optical modulation

4.1. Transit-time effect

In order to calculate modulation indices through the EO effect, the microwave electric field that would be observed by the lightwave propagating in the optical waveguides should be considered while taking the transit-time effect into account.
The transformation for considering the transit-time effect for a lightwave propagating in the optical waveguide, WG_1, along the y-axis can be expressed by $y' = y - v_{g1}t$, where y' denotes the point of the lightwave in the coordinate system moving with the lightwave, and v_g is the group velocity of the lightwave. Therefore, the microwave electric fields would be observed by the lightwave propagating along the waveguide WG_1 at the h-th patch antenna in the array structure become

$$\begin{aligned} E_x^{opt}(y) &= E_x\left(t = \frac{y - y'}{v_{g1}}\right) \\ &\propto K_{x0} \sin\left(\omega_m \frac{y - y'}{v_{g1}} + d(h-1)n_0 k_{my}\, y \sin\theta_1\right) \\ &= K_{x0} \sin(k_m n_{g1}\, y + d(h-1)n_0 k_{my}\, y \sin\theta_1 + \zeta_1)\,, \end{aligned} \tag{5}$$

where n_{g1} is the group index of the lightwave propagating in the waveguide WG_1 ($n_{g1} = c/v_{g1}$), ζ_1 is the initial phase of the lightwave in waveguide WG_1 ($\zeta_1 = k_m n_{g1} y'$), and d is the distance of each unit cell. Using a similar method, the microwave electric fields that would be observed by the lightwave propagating, E^{opt}, along waveguide WG_2 in the unit cell device is expressed as

$$E_y^{opt}(x) = K_{y0} \sin(k_m n_{g2}\, x + d(h-1)n_0 k_{mx}\, x \sin\theta_2 + \zeta_2 + \varphi)\,, \tag{6}$$

where n_{g2} is the group index of the lightwave propagating in WG_2 ($n_{g2} = c/v_{g2}$), and ζ_2 is the initial phase of the lightwave in WG_2 ($\zeta_2 = k_m n_{g2} x'$).

4.2. Modulation index

The lightwaves propagating in the orthogonal optical waveguides, WG_1 and WG_2, are modulated by the induced electric field across gaps G_1 and G_2 through optical modulation. The phase-modulated lightwaves by the wireless signal are obtained. The modulation indices, $\Delta\phi_1$ and $\Delta\phi_2$, of the phase-modulated lightwaves can be determined taking into account the overlapping between of the induced electric fields across the gaps and the lightwave in the cross section, which are expressed as

$$WG_1 : \Delta\phi_1(\theta) = \frac{\pi r_{33} n_e^3}{\lambda}\Gamma\left[\int_{-\frac{d}{3}-\frac{W}{2}}^{-\frac{d}{3}+\frac{W}{2}} E_x^{opt}(y)dy + \int_{-\frac{L}{2}}^{\frac{L}{2}} E_x^{opt}(y)dy + \int_{\frac{d}{3}-\frac{W}{2}}^{\frac{d}{3}+\frac{W}{2}} E_x^{opt}(y)dy\right] \quad (7)$$

$$WG_2 : \Delta\phi_2(\theta) = \frac{\pi r_{33} n_e^3}{\lambda}\Gamma\left[\int_{-\frac{d}{3}-\frac{W}{2}}^{-\frac{d}{3}+\frac{W}{2}} E_y^{opt}(x)dx + \int_{-\frac{L}{2}}^{\frac{L}{2}} E_y^{opt}(x)dx + \int_{\frac{d}{3}-\frac{W}{2}}^{\frac{d}{3}+\frac{W}{2}} E_y^{opt}(x)dx\right], \quad (8)$$

where λ is the wavelength of lightwave propagating in the optical waveguides, r_{33} is the EO coefficient, n_e is the extraordinary refractive index of the substrate, Γ is a factor expressing the overlapping between the induced microwave electric field and the lightwave, and L and W are the lengths of the patch antenna.

The modulation indices, $\Delta\phi_1$ and $\Delta\phi_2$, are function of the irradiation angle of the wireless signal, as shown in Equations (7) and (8), respectively. These indices contain information of the irradiation angle of the wireless signal along the *yz*- and *xz*-planes independently. The 1-D directivities of the wireless electromagnetic signal using the proposed device can be obtained. The directivity can be tuned using meandering gap structures for phase-reversal.

4.3. Phase-reversal

The degradation of the optical modulation can be compensated by phase-reversal by considering the transit-time effect. As a result of the electric field distribution, the direction of the *z*-component of the electric field under the gap edge is reversed between the edges. The characteristics of the *z*-component can be used for phase-reversed structures. We have proposed an EO wireless-optical converter using quasi-phase-matching array of patch antennas with a gap [13]. The degradation of the optical modulation can be compensated for by shifting the position of the patch antennas with a gap along the optical waveguide. The device has been worked properly to increase the modulation index.

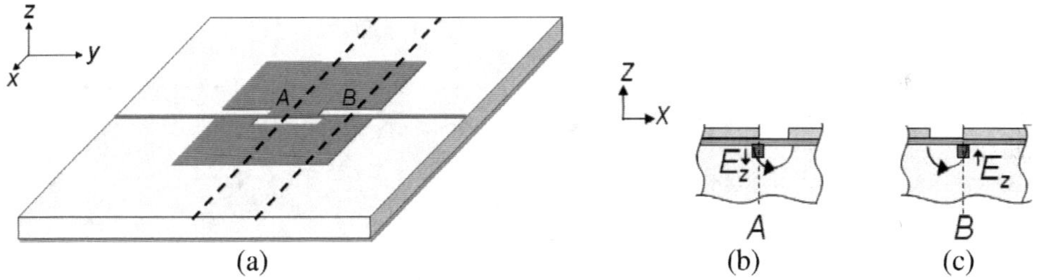

Figure 3. (a) Patch antenna with a meandering gap, (b) cross-sectional view along line-A, and (c) cross-sectional view along line-B.

Here, phase reversal by meandering gap structures is proposed. The degradation of the optical modulation with tuning for several wireless irradiation angles can be obtained using meandering gap structures. The typical structure of patch antennas with a meandering gap is shown in Figure 3. The cross-sectional views for the structures *A* and *B* are also shown in Figures 3(b) and 3(c), respectively. The directions of the electric fields across the optical waveguide are reversed along the *z*-component. Therefore, control of the modulation polarity along the optical waveguide is obtainable. These characteristics are very useful for the design of the directivity in the array structures.

5. Experiments

As a preliminary experiment, a prototype device composed of only patch antennas embedded with orthogonal gaps was designed and fabricated. The prototype device was designed at a 26-GHz microwave operational frequency using a 0.4-mm-thick *z*-cut $LiTaO_3$ crystal. The patch antennas were 0.8-mm squares having 5-μm-wide gaps at the centre. In the device, 4x4 patch antennas were aligned with a separation of 5.5 mm.

In the device fabrication, 4x4 channel optical waveguides for single-mode operation at a wavelength of 1.55 μm was fabricated in 12 hours using the annealed proton-exchange method with benzoic acid at 513 K [21,22]. A thin SiO_2 buffer layer was deposited on the surface of the substrate after optical waveguide fabrication. Then, patch antennas with gaps were fabricated using a 1-μm-thick aluminium film on the buffer layer by thermal vapour deposition, a standard photolithography technique, and a lift-off process. The edges of the gaps were placed on the optical waveguides for efficient conversion. Thermal annealing at 623 K was also performed for 1 hour in order to increase the optical waveguide performance. Finally, a 1-μm-thick aluminium film was deposited on the reverse side of the substrate as a ground electrode.

In the device measurement, two lightwaves of 1.55 μm in wavelength from a distributed-feedback laser were coupled to the fabricated device. A wireless microwave signal of 200 mW from a microwave signal generator was irradiated to the fabricated device using a horn antenna. The light output spectra were observed and monitored using an optical spectrum analyzer.

The measured lightwave spectra are shown in Figures 4(a) and 4(b), when the polarization of the microwave signal was set to 0 degrees (x-polarization). In this case, optical sidebands were clearly observed in the output lightwave from the optical waveguide WG_1, and no optical sideband was observed from the waveguide WG_2, as shown in Figures 4(a) and 4(b). Furthermore, Figure 5 shows the measured optical modulation for the wireless irradiation angle dependence. The experimental results approximately coincided with the calculation results.

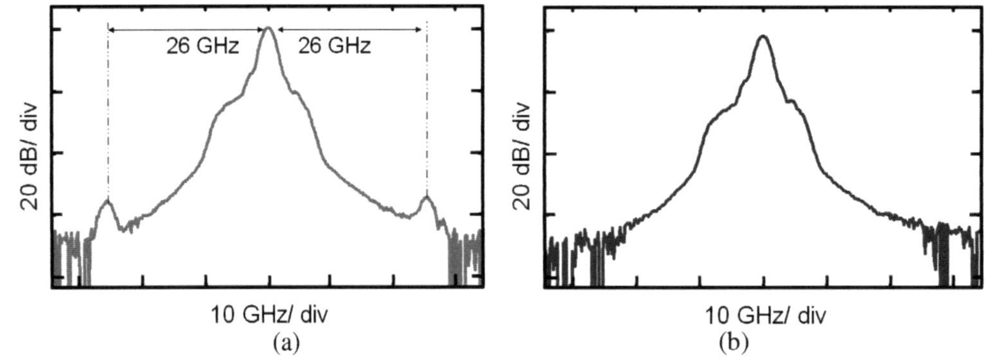

Figure 4. Measured lightwave output spectra from optical waveguides (a) WG_1 and (b) WG_2 when the wireless electromagnetic signal polarization was set to be 0 degrees.

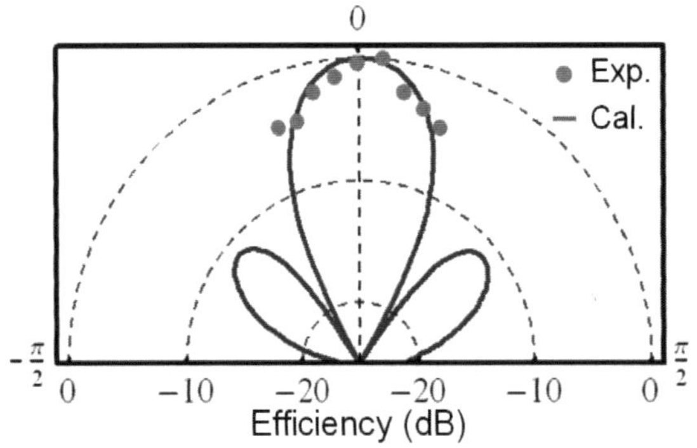

Figure 5. Measured optical modulation for the wireless irradiation angle dependence.

Based on the measurement results, the basic operations of the device were successfully confirmed. The polarization of the wireless electromagnetic signal can be identified. The magnitude of the wireless electromagnetic signal can be also measured. The 1-D directivity of the wireless signal conversion was also measured and confirmed.

6. Advanced device for 2-D directivity

6.1. Design

In order to realize more advanced functionality with 2-D directivity for discriminating space-division-multiplexed (SDM) or multiple-input-multiple-output (MIMO) wireless signals, a new device using an *M*x*N* array of the proposed device is also proposed, as shown in Figure 6. The device is composed of *M*x*N* orthogonal optical waveguides and an *M*x*N* array of patch antennas embedded with gaps, as shown in Figure 1(a) (dashed-line area in Figure 6). The optical waveguide is placed under the edge of the gaps for effective modulation.

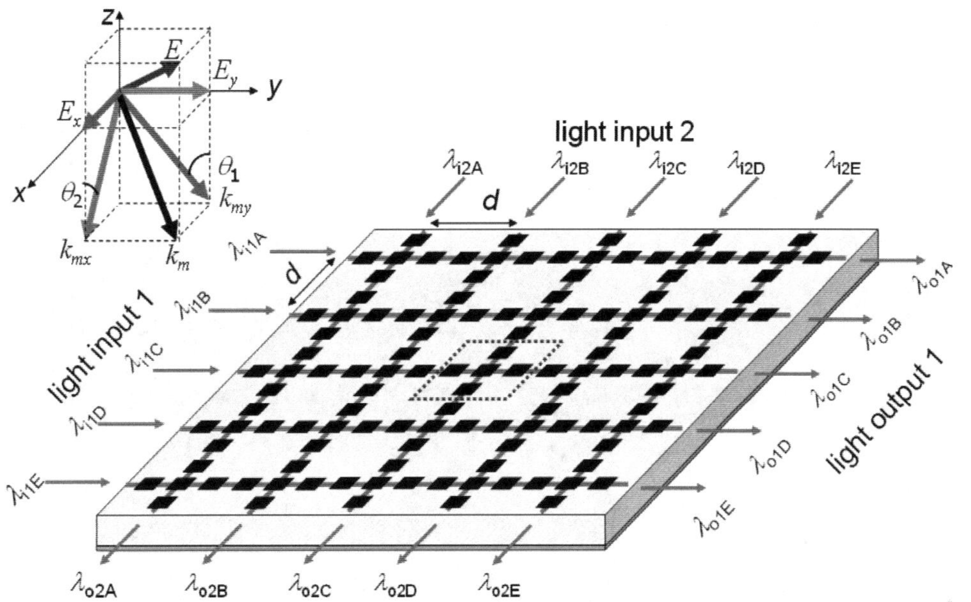

Figure 6. Wireless signal discrimination device using *M*x*N* orthogonal optical waveguides and patch antennas with gaps, where the dashed-line area is as shown in Figure 1(a).

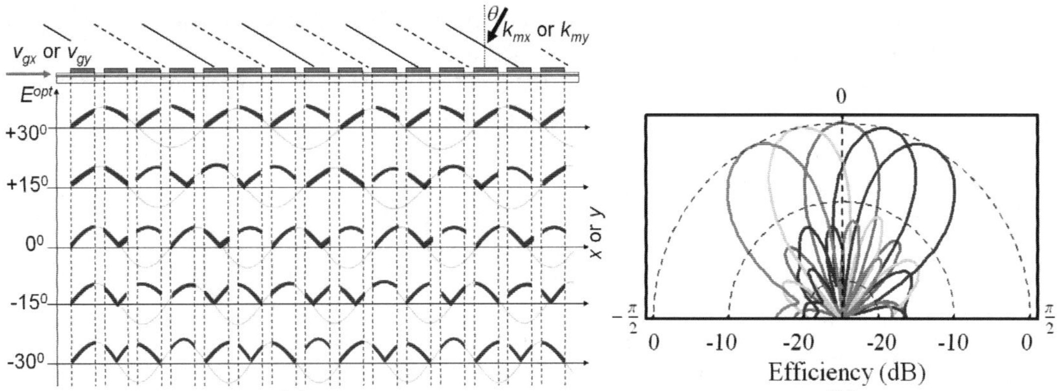

Figure 7. (a) Several patterns of patch antenna arrays with meandering gap structures, where the thick line is E^{opt} with phase-reversed structures and (b) the calculation results for the proposed device according to the directivity in 1-D using the meandering gap patterns.

When the polarization of the wireless signal at 45 degrees is irradiated to the device with an arbitrary wireless irradiation angle, strong electric fields are induced across all gaps with same magnitude strength. The induced electric fields can be converted to lightwaves through optical modulation. Using meandering structures, the degradation of the optical modulation can be compensated for by tuning the directivity. The effective optical modulation function of the irradiation angle of the wireless signal can be obtained by modifying Equations (7) and (8) for several array structures. The electric field that would be observed by the lightwave should be considered.

In the device design, the parameters were set as follows: f_m = 26 GHz, n_o = 1, n_g = 2.1, d = 5.5 mm, L = 0.8 mm, and $M = N = 5$. First, considering the 1-D directivity of the wireless signal conversion, the patch antennas with meandering gap structures for several wireless signal angles are shown in Figure 7(a). The calculation results with the meandering gap structures are shown in Figure 7(b) for 1-D directivity analysis. The 1-D directivity was analyzed independently along the *yz*-plane (θ_2) and the *xy*-plane (θ_1).

6.2. Analysis

Based on the designed 1-D directivity of the wireless signal conversion, the 2-D directivity can be obtained by considering the two orthogonal modulated lightwaves. The 2-D directivity can be measured by multiplying or comparing the modulated lightwaves.

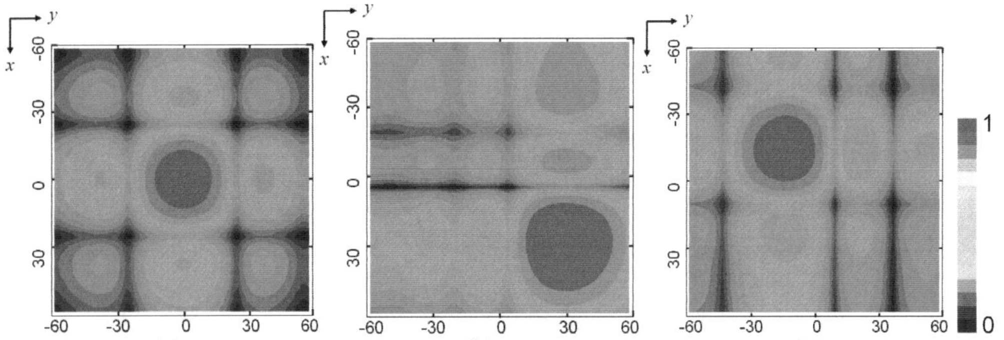

(a) (b) (c)

Figure 8. Calculated 2-D directivity of the wireless signal conversion for wireless irradiation angles of (a) $\theta_1 = \theta_2 = 0$ degrees, (b) $\theta_1 = \theta_2 = 30$ degrees, and (c) $\theta_1 = \theta_2 = -15$ degrees.

Figure 8 shows the typical calculated 2-D directivity of the wireless signal conversion in the *xy*-plane for several wireless irradiation angles. The first combination of the two orthogonal modulated lights was set to be effective for the normal wireless irradiation angle, as shown in Figure 8(a). The other combinations of the orthogonal modulated lights for wireless irradiation angle in the *xz*-plane (θ_2) and the *yz*-plane (θ_1) are shown in Figure 8(b) for $\theta_1 = \theta_2 = 30$ degrees and Figure 8(c) for $\theta_1 = \theta_2 = -15$ degrees. The proposed device has *M*x*N* 2-D directivity combinations of the wireless signal conversion, according to the number of orthogonal modulated lightwaves. Therefore, the proposed device can be used for discriminating the wireless electromagnetic signals.

7. Conclusion

The discrimination of the wireless electromagnetic signal by EO modulators using planar patch antennas embedded with orthogonal gaps was proposed. The wireless signal can be received, separated, and converted to a lightwave directly and simultaneously using the device. The directivity of the wireless signal conversion can be tuned by controlling the phase reversal with meandering gap structures. The typical basic operations of the device were demonstrated successfully. The magnitude, phase and polarization of the wireless signals can be measured and identified through optical modulation. The discrimination directivity of the wireless signal in two-dimensional space can be obtained by considering the two orthogonal modulated lightwaves. The proposed device can be applied to SDM/ MIMO wireless signals and electromagnetic compatibility test systems.

At present, we are attempting to demonstrate and measure the performance of the device in wireless-over-fibre links.

Acknowledgments
The authors would like to thank Drs. H. Shiomi and K. Kitatani for their valuable advice and comments regarding the analysis and discussion. The present study was supported in part by a Grant-in-Aid for Scientific Research and a grant from the Osaka University Global COE Program, "Center for Electronics Device Innovation", both from the Ministry of Education, Culture, Sports, Science and Technology (MEXT) of Japan.

References
[1] Ramirez-Iniguez R, Idrus S M, Sun Z 2008 *Optical Wireless Communications: IR for Wireless Connectivity* (Florida: CRC Press.)
[2] Iezekiel S 2009 *Microwave Photonics: Device and Applications* (Chichester: John Wiley & Sons)
[3] Kawanishi T, Oikawa S, Higuma K, Sasaki M, Izutsu M 2002 *IEICE Trans. Electron.* **E85-C** 1
[4] Murphy T E and Pagan V R Phase Modulated Radio-Over-Fiber Link with Electro-Optic Down Conversion Using a Delay-Line Interferometer *Proc. SPIE* (San Francisco, USA, 2011) 7936 79360L doi:10.1117/12.875695
[5] Sheehy F T, Bridges W B, and Schaffner J H 1993 *IEEE Photon. Tech. Lett.* **5** 3
[6] Bridges W B, Burrows L J, Cummings U V, Schaffner J H, Scherer A, and Sheehy F T Antenna-Coupled Millimeter-Wave Electro-Optic Modulators For 20 to 100GHz Proc. International Topical Meeting on Microwave Photonics (Kyoto, Japan, Dec. 1996) pp. 21-24
[7] Shinada S, Kawanishi T, and Izutsu M 2007 *IEEE Trans. Electron.* **E90-C** 5
[8] Murata H, Miyanaka R, and Okamura Y Millimeter-Wave Wireless Signal Discrimination Device Using Electro-Optic Modulator with Antenna-Coupled Electrodes and Polarization-Eeversed Structures of Ferroelectric Optical Crystal *Proc. 41st European Microwave Conference* (Manchester, UK, October 2011) pp EuMC/EuMIC11-1
[9] Wijayanto Y N, Murata H, and Okamura Y 2011 Novel Electro-Optic Microwave-Lightwave Converters Utilizing a Patch Antenna Embedded with a Narrow Gap *IEICE Electron. Express* **8** 7
[10] Wijayanto Y N, Murata H, and Okamura Y 2012 *Electron. Lett.* **48** 1
[11] Wijayanto Y N, Murata H, and Okamura Y 2011 Passive Electromagnetic Field Sensors Using Electro-Optic Crystals with Metal Planar Antennas and Narrow Gaps *Proc. Int. Sym. Materials Science and Innovation for Sustainable Society: ECO-MATES* (Osaka, Japan Nov. 2011) pp PT 7-13
[12] Murata H, Matsunaga S, Enokihara A and Okamura Y 2005 *Electron. Lett.* **41** 8
[13] Yariv A 1989 *Quantum electronics* (New York: Wiley)
[14] Balanis C A 1997 *Antenna Theory: Analysis and Design* (New York: John Wiley & Sons, Inc.)
[15] Garg R, Bartia P, Bahl I, and Ittipiboon A 2001 *Microstrip Antenna Design Handbook* (Norwood: Artech House, Inc.)
[16] Bahl I J and Bhartia P 1980 *Microstrip Antennas* (Dedham: Artech House)
[17] Gupta V R and Gupta N 2004 *Microwave and Optical Tech. Lett.* **40** 2 pp 158-160
[18] Gupta K C, Garg R, Bahl I, and Bhartia P 1996 *Microstrip Lines and Slotlines* (Norwood: Artech House)
[19] Zhang K, and Li D 2008 *Electromagnetic Theory for Microwaves and Optoelectronics* (Berlin Heidelberg: Springer-Verlag)
[20] Fitzpatrick R 2008 *Maxwell's Equation and the Principles of Electromagnetism* (Hingham: Infinity Science Press)
[21] Savatinova I, Tonchev S, Todorov R, Armenise M N, Passaro V M N, and Ziling C C 1996 *Journal of Lightwave Tech.* **14** 3 pp 403-409
[22] Tada K, Murai T, Nakabayashi T, Iwashima T, and Ishikawa T 1987 *Japanese Journal of Applied Physics* **26** 1 3 pp 503–504

Phase analysis and optical evaluation of ceria-zirconia-terbia prepared by coprecipitation method

Masaya Amimoto[1], Masaaki Haneda[1] and Masakuni Ozawa[1]
[1] Ceramic Research Laboratory, Nagoya Institute of Technology, 10 Asahigaoka, Tajimi, Gifu 507-0071, Japan,

E-mail: amimoto.m@crl.nitech.ac.jp

Abstract. A series of ceria-zirconia containing terbia samples were prepared by coprecipitation, and were characterized by XRD, Raman spectroscopy and UV-vis diffuse reflectance measurements. A single-phase solid solution with a cubic fluorite-type structure or tetragonal distorted fluorite structure was obtained for the samples heated at 900°C. Diffuse reflection spectra of the powders showed a broad band at 320-340 nm, which is assigned to interband transitions of CeO_2 in the UV region. The effect of ZrO_2 on the interband and absorption band of CeO_2 was investigated in a series of solid solutions.

1. Introduction

Mixed metal oxides in the system of ZrO_2-CeO_2 are useful for various applications such as ceramics, catalysts, and pigments [1,2]. The electronic band structures, which govern the transport and optical properties, depend on the mixed state of cations in the same fluorite-type structure. In the system of ZrO_2-CeO_2, one cubic, one monoclinic and three tetragonal forms exist, depending on the Zr concentration. These are assessed by UV-vis absorption and fluorescence spectroscopy. Also, based on these optical properties, new pigments containing CeO_2 (yellow) or Pr^{3+} or Tb^{3+} doped CeO_2 (brown or red) have recently been proposed by several researchers.[3-5] Since commercial yellow or red pigments contain toxic elements such as cadmium and lead that are hazardous to health and the environment, it is necessary to develop environmentally-friendly pigments. In the present work, we synthesized solid solution compounds of $Zr_xCe_{0.95-x}Tb_{0.05}O_{2-\delta}$ by the coprecipitation method, and examined their UV-Vis diffuse reflectance characteristics.

2. Experimental

2.1. Synthesis of raw materials

The $Zr_xCe_{0.95-x}Tb_{0.05}O_{2-\delta}$ powder samples, where x=0, 0.20, 0.40, 0.60, 0.80 and 0.95, were synthesized by coprecipitation using $ZrO(NO_3)_2 2H_2O$, $Ce(NH_4)_2(NO_3)_6$, $Tb(NO_3)_3 6H_2O$ and aqueous ammonia as starting materials. Aqueous solutions containing metal salts, corresponding to the composition $Zr_xCe_{0.95-x}Tb_{0.05}O_{2-\delta}$ (x=0-0.95), were prepared. Aqueous ammonia was added to the solution, and the mixed solution was agitated sufficiently. The obtained coprecipitates from mixed solution of salts and ammonia were filtered, and dried at 110°C. Powders were calcined at 600°C for 3 h, and then heated at 900°C for 3 h in air.

2.2. Evaluation of synthesized powder

2.2.1. Powder X-ray diffraction and Raman spectroscopy

Powder X-ray diffraction (XRD) patterns were recorded in order to determine the crystal phases in the products. The patterns were recorded using a Rigaku Rint X-ray Diffractometer (Japan) with Ni filtered CuKα radiation (λ=1.5418 Å) over the range of 2θ from 20 to 80° at 40 kV and 20 mA to study the products. The JCPDS cards were used to identify the resultant phases. Raman spectroscopy of the samples was carried out using a laser with a wavelength of 532 nm.

2.2.2. Colour characterization

The CIE L*a*b* system was used for evaluation of the chromatic properties. The parameters were measured using a spectrophotometer (Konica Minolta CM-2600d, Japan). In this system, L* is the colour lightness (L* is 0 for black and L* is 100 for white), a* is the green (-)/red (+) axis, b* is the blue (-) /yellow(+) axis.

2.2.3. UV-vis diffuse reflectance spectroscopy

UV-vis diffuse reflection spectra were measured from 240 to 800 nm using a Hitachi U-3000 Spectrophotometer (Japan) with a $BaSO_4$ coated integrating sphere. The inner diameter of the integrating sphere is 60 mm, and the aperture ratio is 7.8%.

Kubelka-Munk functions for the samples were calculated using the formula

$$F(R) = \frac{(1-R)^2}{2R} = \frac{K}{S}, \tag{1}$$

where R is the reflectance, K is the absorption coefficient, and S is the scattering coefficient. If it is assumed that the ratio of the scattered to incident light is constant, the spectra of the K-M function can be regarded as relative absorption spectra.

3. Results and discussion

3.1. Powder X-ray diffraction and Raman Spectroscopy

The XRD patterns for $Zr_xCe_{0.95-x}Tb_{0.05}O_{2-\delta}$ (x=0, 0.2, 0.4, 0.6, 0.8, and 0.95) samples heated at 900°C are shown in Figure 1. The patterns revealed that a single phase solid solution with a cubic fluorite-type structure was obtained for $Zr_xCe_{0.95-x}Tb_{0.05}O_{2-\delta}$ in the range of x=0 to 0.80. For x=0.20 to 0.80, the XRD results indicated lower crystallinity with broader peaks than that for x=0. The diffraction peaks shifted toward higher angle with increasing x (Zr content) in the region from x=0 to x=0.80. The Raman spectra for $Zr_xCe_{0.95-x}Tb_{0.05}O_{2-\delta}$ (x=0, 0.2, 0.4, 0.6, 0.8 and 0.95) samples heated at 900°C are shown in Figure 2. For x=0, a sharp band around 455 cm^{-1} and a broad band at 582 cm^{-1} were observed. For x=0.2 to 0.6, weak bands around 300 and 600 cm^{-1} became prominent with increasing Zr. Based on a published study on the CeO_2-ZrO_2 system [6], these were attributed to displacement of oxygen atoms from their ideal fluorite lattice positions and to oxygen vacancies, respectively. For x=0.8 and 0.95 several bands associated with the tetragonal phase, which has six Raman-active modes, were detected. In the CeO_2-ZrO_2 system one cubic, one monoclinic and three tetragonal (two metastable phases and a stable phase) forms exist, depending on the Zr content [7]. The XRD results suggest that a metastable tetragonal phase with a distorted fluorite-type structure exists in the range from x=0.2 to 0.6, a stable tetragonal phase exists for x=0.8, and a stable tetragonal phase and a minor monoclinic phase exist for 0.95. For x=0, a band at 583 cm^{-1} appeared due to the presence of Tb. This is probably associated with a cubic phase with a fluorite-type structure including oxygen vacancies produced by the presence of Tb^{3+}.

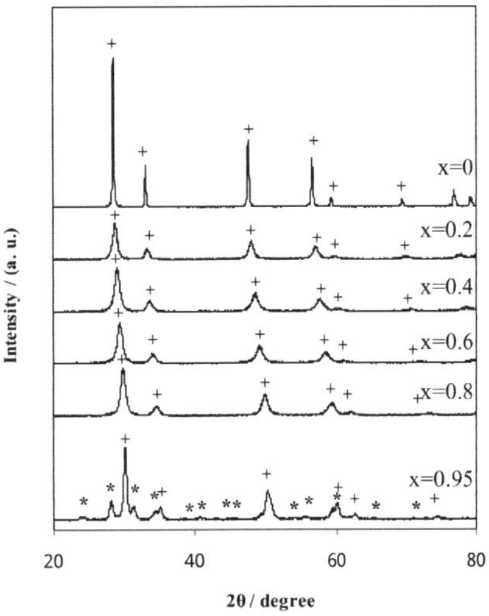

Figure 1. X-ray diffraction patterns of $Zr_xCe_{0.95-x}Tb_{0.05}O_{2-\delta}$ (x=0, 0.20, 0.40, 0.60, 0.80, and 0.95) samples heated at 900°C (+, cubic or tetragonal; *, monoclinic).

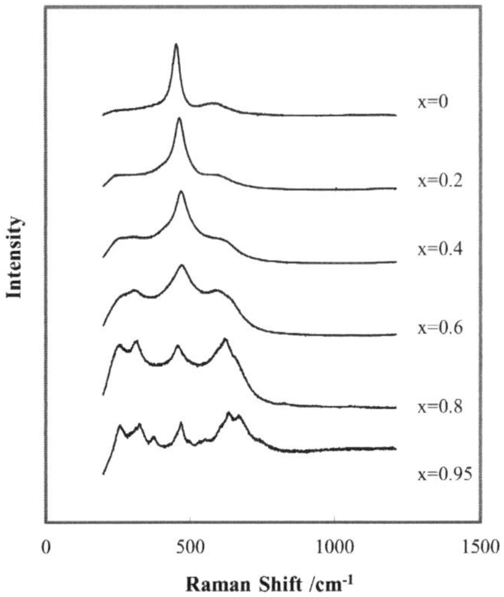

Figure 2. Raman spectra of $Zr_xCe_{0.95-x}Tb_{0.05}O_{2-\delta}$ (x=0, 0.20, 0.40, 0.60, 0.80, and 0.95) samples heated at 900°C.

3.2. Colour characterization

The L*a*b* parameters measured for a series of the samples of $Zr_xCe_{0.95-x}Tb_{0.05}O_{2-\delta}$ heated at 900°C are listed in Table 1. The lowest L* value of 35, the lowest a* value of 7.1, and the lowest b* value of 8.1 were observed for x=0. The L* value increased with increasing x in the range x=0 to 0.6 and decreased with increasing x in the range x=0.6 to 0.95. The highest L* value was 73.2 for x=0.6. The highest a* value was 18.4 for x=0.95. The a* value was higher than 10 for x=0.2 to 0.95, compared to a value of around 0 for the CeO_2-ZrO_2 system, and the $Zr_xCe_{0.95-x}Tb_{0.05}O_{2-\delta}$ sample powders appeared slightly more reddish. The b* value increased with increasing x, and the highest b* value was 45.7 for x=0.95. The colour of the sample powder was dark brown for x=0, yellow-brown from x=0.2 to 0.8 and vivid orange for x=0.95.

3.3. UV-Vis diffuse reflectance spectroscopy

Figure 3 shows the UV-Vis diffuse reflectance spectra of $Zr_xCe_{0.95-x}Tb_{0.05}O_{2-\delta}$ (x=0, 0.20, 0.40, 0.60, 0.80 and 0.95) powders. A strong absorption maximum was found in the range 290-300 nm for all the samples. In addition, another peak appeared in the range 320-340 nm for x=0 to 0.60. Although the former absorption at around 300 nm is common, its intensity changed with the Ce content. This was identified as being due to $Ce^{4+}\rightarrow O^{2-}$ (at 297 nm) charge transfer transitions for Ce-O in each crystal phase with different Zr content. The latter absorption is strongly affected by the Zr content and is assigned to interband transitions in CeO_2 and its solid solution. In CeO_2 doped with ZrO_2, the absorption due to interband transitions occurs at longer wavelengths than 400 nm, so that the band gap falls in the blue wavelength region. The complementary light yellow colour is observed because of the blue absorption associated with the modified interband structure in the crystalline solid solution. We found that Tb-doped CeO_2 and Tb-doped ZrO_2 and their solid solutions absorbed light in a broad range of wavelengths and showed new colours. Other peaks appeared around 500 nm for Tb-doped CeO_2 (x=0) and around 390 nm for Tb-doped ZrO_2 (x=0.95). These peaks were very broad and are not observed in CeO_2-ZrO_2 samples. Therefore, they are considered to be the effect of Tb doping of the samples. These peaks did not appear clearly for Tb doping with x=0.20 to 0.80. However, the absorption in the wavelength region below 600 nm was slightly higher than the case without Tb, and hence the sample appeared yellow-brown.

Table 1. Colour coordinate values for $Zr_xCe_{0.95-x}Tb_{0.05}O_{2-\delta}$ (x=0, 0.20, 0.40, 0.60, 0.80, and 0.95) samples heated at 900°C in CIE L*a*b*.

Zr content / mol	L*	a*	b*
x=0	35.0	7.1	8.1
x=0.2	57.5	11.6	23.5
x=0.4	67.2	12.8	35.2
x=0.6	73.2	10.5	37.0
x=0.8	68.6	15.0	38.2
x=0.95	69.4	18.4	45.7

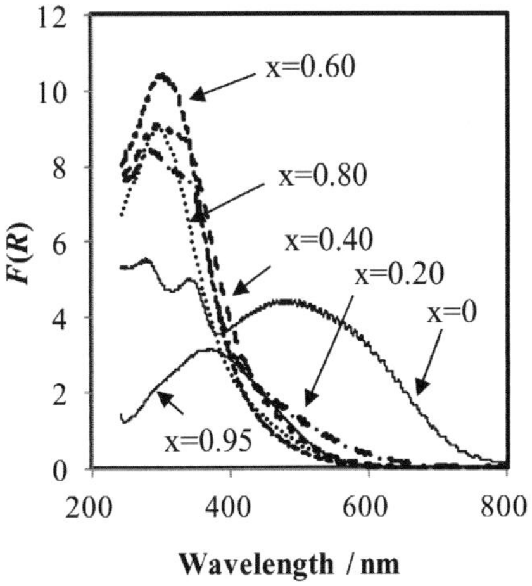

Figure 3. UV-Vis diffuse reflectance spectra of $Zr_xCe_{0.95-x}Tb_{0.05}O_{2-\delta}$ (x=0, 0.20, 0.40, 0.60, 0.80, and 0.95) samples heated at 900°C.

4. Summary

$Zr_xCe_{0.95-x}Tb_{0.05}O_{2-\delta}$ powders were synthesized by the coprecipitation method, and their UV-Vis absorption properties were characterized. From the XRD measurements, a mixed oxide with a fluorite-type cubic structure or a tetragonal distorted fluorite-type structure was formed in the single phase for x=0 to 0.8. By doping CeO_2 with ZrO_2, the band gap of the crystal fell in the blue region of the visible spectrum, so that the mixed oxide powders had a yellowish colour. Thus, the addition of Tb to CeO_2-ZrO_2 will allow the possibility of controlling the optical properties.

Acknowledgments
Work supported by Ministry of Education, Culture, Sports, Science and Technology, of Japanese Government (Grant-in-Aids No.22655069), and the New Energy and Industrial Technology Development Organization (NEDO, Rare Metal Substitute Materials Development Project).

References

[1] Ozawa M, Kimura M and Isogai A 1993 *J. Alloy. Compds.* **193** 72
[2] Trovarelli A 1996 Catal. Rev. Sci. Eng. **38** 439
[3] Sreeram K J, Aby C P and Nair B U 2008 *Chemistry Letters* **37** 902
[4] García A Llusar M Calbo J Tena M A and Monrós G 2001 Green Chemistry **3** 238
[5] Kumari L S, Rao P P and Koshy P 2010 *J. Am. Ceram. Soc.* **93** 1402
[6] Reddy B M, Bharali P, Saikia P, Park S E, van den Berg M W E, Muhler M, and Gruenert W 2008 *J. Phys. Chem. C* **112** 11729
[7] Yashima M, Arashi H, Kakihana M and Yoshimura M 1994 *J. Amer. Ceram. Soc.* **77** 1067

Journal of Physics: Conference Series **379** (2012) 012019 doi:10.1088/1742-6596/379/1/012019

Clock field in arrayed magnetic logic gates with a magnetic force microscope tip

H Nomura, Y Imanaga, Y Hiratsuka and R Nakatani

Department of Materials Science and Engineering, Graduate School of Engineering, Osaka University, 2-1 Yamada-oka, Suita, Osaka 565-0871, Japan

Email: nomura@mat.eng.osaka-u.ac.jp

Abstract. A magnetic logic gate (MLG), which is based on magnetic quantum dot cellular automata (MQCA), is capable of NAND/NOR logic operations. By arranging MLGs in a two-dimensional periodic array, a highly functional circuit can be created. However, NAND/NOR gates are difficult to form into a two-dimensional periodic arrangement. Here, we propose NOT/AND/ORs gate based on MLGs, which can be arranged in a two-dimensional periodic array. To demonstrate logic operations, we performed numerical simulations based on the macro-spin model. To execute logic operation in the arrayed structure, we used the stray field from a magnetic force microscope tip.

1. Introduction

A magnetic logic gate (MLG)[1], which is based on magnetic quantum dot cellular automata (MQCA) [2], can perform NAND/NOR logic operations via magnetostatic interactions between adjacent magnetic dots. The MLG has the advantages of high integration density, information non-volatility, high resistance to ionized radiation, low dissipation energy and stability at room temperature. In general, the MLG is composed of elliptical dots. The dots are small enough to take a single-domain state. Because of their shape anisotropy, the magnetization directions of the dots at remnant states are bistable. Therefore, digital information of "0"/"1" can be stored as the direction of dot magnetization. Logic operations are executed by applying an external magnetic field (clock-field). In previous studies, a NAND/NOR gate [3] and a similar gate [4] were experimentally demonstrated.

By arranging the gates in a two-dimensional periodic array, we can create highly functional circuits. However, NAND/NOR gates are difficult to form into a two-dimensional periodic arrangement. In the NAND/NOR gate, input dots are placed on the easy/hard axis of the output dot. Therefore, the number of adjacent dots, which defines the behavior of the gate, changes when the gates are arranged in a two-dimensional periodic array.

In this study, we propose NOT/AND/OR gates that can be arranged in a two-dimensional periodic array. To demonstrate local logic operations of the NOT/AND/OR gate, we performed numerical simulations using a macro spin model. To apply a local clock field to the gate, we use the stray field from a magnetic force microscope (MFM) tip.

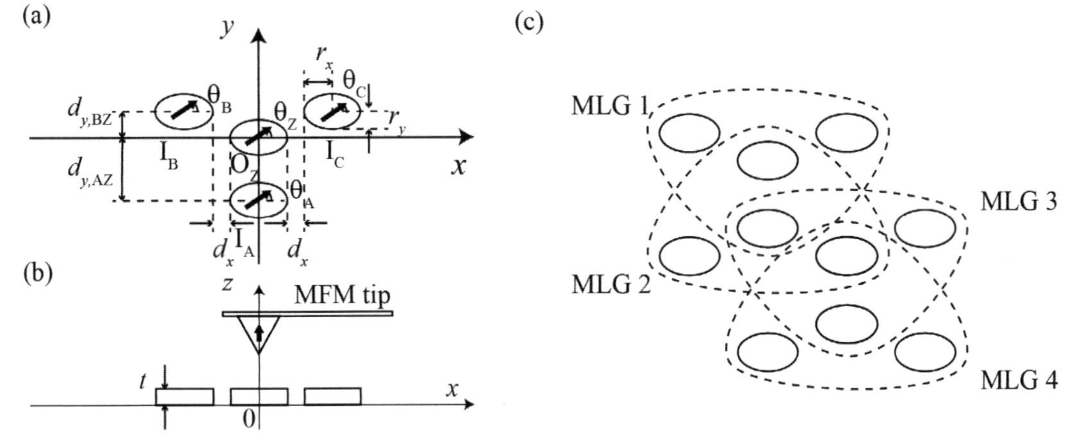

Figure 1. Schematic diagram of NOT/AND/OR gate. (a) Top view and (b) front view and (c) NOT/AND/OR gate in two-dimensional periodic array.

2. NOT/AND/OR gate

Figures 1(a) and (b) show schematic diagrams of the NOT/AND/OR gate. The gate is composed of three input dots (I_A, I_B and I_C) and an output dot (O_Z). The binary states of 0 and 1 correspond to negative and positive polarity of the *x* component of the dot magnetization, respectively. I_A and O_Z are anti-ferromagnetically coupled, whereas I_B and O_Z or I_C and O_Z are ferromagnetically coupled. The magnetization vector of the O_Z points parallel to the direction of the majority stray field from I_A, I_B and I_C. Therefore, the gate works as a NOT/AND/OR gate [($I_C = I_B$; O_Z = NOT I_A), ($I_A = 0$; $O_Z = I_B$ OR I_C), ($I_A = 1$; $O_Z = I_B$ AND I_C)]. By sharing the dots with adjacent gates, the NOT/AND/OR gates can be arranged in a two-dimensional array [Figure 1(c)].

3. Simulation methods

In our numerical simulation, we made four assumptions. (1) Each dot occupies a single-domain state. This is a good assumption when the dot size becomes smaller than the 100-nm semi-major axis. (2) The MFM tip is assumed to be a magnetic dipole. (3) The magnetization of the dots is in the *x*-*y* plane, based on the demagnetization of the dots by thin films. (4) The major interaction between the dots and the tip is a magnetostatic interaction. For Permalloy dots with a gap wider than 10 nm, other factors such as the exchange interaction between the dots and the intrinsic anisotropy were negligible. Similar approaches were taken in studies of coupled nanomagnets [5]. The total energy of the gate is

$$E = -\sum_i \mu_0 \boldsymbol{m}_i \cdot \left(\boldsymbol{H} + \frac{1}{2} \sum_j \mathrm{N}_{i,j} \boldsymbol{M}_j \right), \tag{1}$$

where $\boldsymbol{m}_i$, $\boldsymbol{M}_j$, $\boldsymbol{H}$ and μ_0 are the magnetic moment of dot $i = I_A, I_B, I_C, O_Z$, the magnetization of dot *j*, the clock field and the permeability of a vacuum, respectively. $\mathrm{N}_{i,j}$ is the demagnetization tensor. When $i = j$, $\mathrm{N}_{i,j}$ is calculated from the demagnetizing factor of the ellipsoid [6], and $\mathrm{N}_{i,j}$ for $i \neq j$ is calculated by averaging the stray field [7] from dot *j* at each point in dot *i* in 10-nm steps. After $\boldsymbol{H}$ is changed, we calculate the metastable/stable state of the system by oscillating the magnetization of the dots at an angle of 0.1°. The parameters used here are as follows: $|\boldsymbol{m}|$= 800 kA/m, r_x = 50 nm, r_y = 30 nm, d_x = 30 nm, $d_{y,ZA}$ = 108 nm, $d_{y,ZB}$ = 44 nm, t = 20 nm. The magnetization and volume of the MFM tip are 800 kA/m and 1×10^{-21} m^3, respectively.

4. Results and discussion

Figures 2(a-d) show the threshold tip position for magnetization reversal of O_Z. The *y* positions of the tip are fixed at 0 nm. Based on the symmetrical shape of the gate, binary states of I_A are fixed at 1. The initial binary states of the input dots in Figure 2(a-d) are (I_A, I_B, I_C) = (1, 1, 1), (1, 0, 1), (1, 1, 0) and

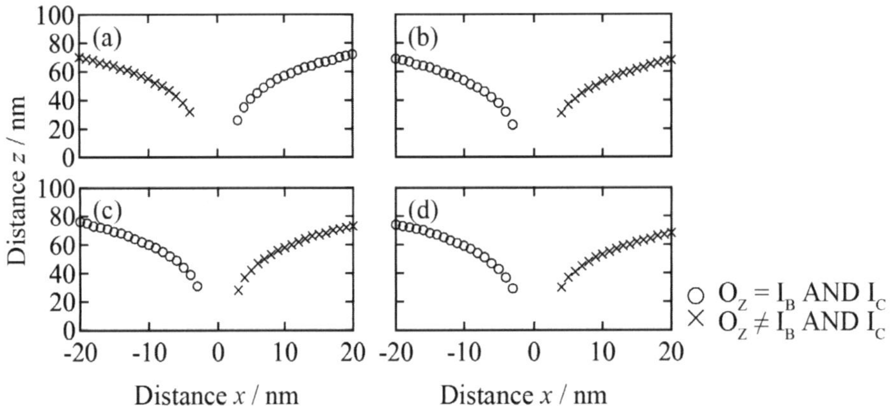

Figure 2. Threshold tip position for magnetization reversal of O_Z. Initial binary states of input dots are (a) (I_A, I_B, I_C) = (1, 1, 1), (b) (1, 0, 1), (c) (1, 1, 0) and (d) (1, 0, 0).

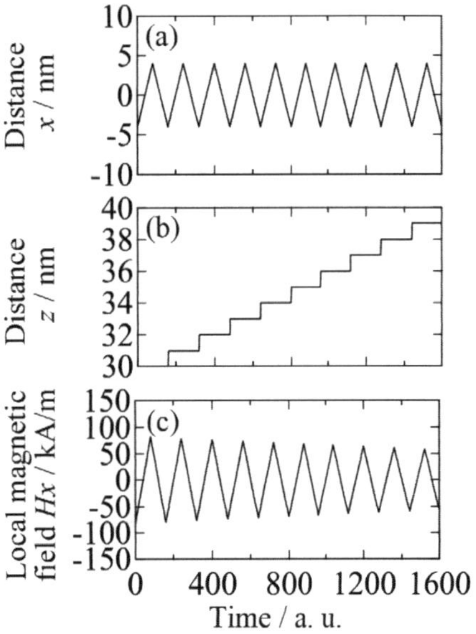

Figure 3. (a), (b) MFM tip motion for logic operation with NOT/AND/OR gate and (c) x component of stray field from MFM tip at the center of O_Z.

(1, 0, 0), respectively. The circle and cross symbols represent the tip positions where the magnetization of O_Z shows the correct logic operation results (O_Z = I_B OR I_C) and erroneous results ($O_Z \neq I_B$ OR I_C), respectively. To set the binary state of 0 or 1 to O_Z, the MFM tip should be oscillated along the x axis. When we fix the center of the oscillation at $x = 0$, the threshold heights corresponding to the correct operation results are higher than those for the erroneous operation results. Therefore, by retracting the tip from the sample surface, we can execute the logic operation using the stray field from the MFM tip.

Figures 3(a) and (b) show MFM tip motion for the logic operation. The threshold heights corresponding to correct logic operation at $|x|$ = 4 nm are 35 nm, 32 nm, 39 nm and 37 nm, respectively. On the other hand, those corresponding to the erroneous operation results are 32 nm, 31 nm, 37 nm and 30 nm, respectively. Therefore, to execute the logical operation, the tip is oscillated along the x-axis with an amplitude of 4 nm and lifted from 30 nm to 39 nm. Figure 3(c) shows the x component of the stray field from the MFM tip at the center of O_Z.. The stray field is oscillated as a chopping wave and the amplitude is decreased from 82.0 kA/m to 57.7 kA/m.

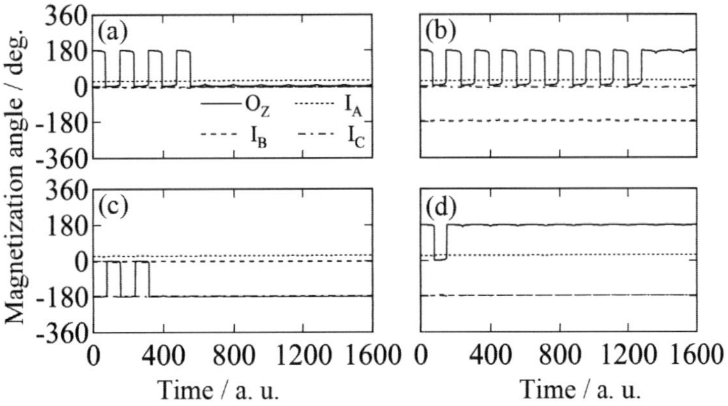

Figure 4. Change of the magnetization angles of the dots with MFM tip motion. Initial binary states of input dots are (a) (I_A, I_B, I_C) = (1, 1, 1), (b) (1, 0, 1), (c) (1, 1, 0) and (d) (1, 0, 0).

Figures 4(a-d) show the change of the magnetization angles of the dots with MFM tip motion. The initial binary states of the input dots are (I_A, I_B, I_C) = (1, 1, 1), (1, 0, 1), (1, 1, 0) and (1, 0, 0) respectively. When the tip-sample distance is less than the threshold heights, the magnetizations of O_Z are oscillated due to the MFM tip motion. When the tip reaches the threshold height, the binary state of O_Z is fixed to the NOT/AND/OR logic operation results. The threshold heights are 33 nm, 32 nm, 38 nm and 31 nm, respectively. It is noteworthy that disturbances to the magnetization angle of the input dots by the MFM tip are negligible. Therefore, we can locally execute logic operations of arrayed NOT/AND/OR gates with an MFM tip.

5. Summary

We demonstrate NOT/AND/OR gates, which can be arranged in a two-dimensional periodic array, with an MFM tip. The clock field from the MFM tip can be considered as a localized oscillating magnetic field. By using a bit line and a word line of a general MRAM circuit, a similar clock field can be applied to the arrayed MLG structures without mechanical moving parts such as the MFM tip. With arrayed MLG structures, highly functional circuits such as adder circuits will be realized in the near future.

Acknowledgements

This work was supported by Priority Assistance for the Formation of Worldwide Renowned Centers of Research from the Global COE Program (Project: Center of Excellence for Advanced Structural and Functional Materials Design) from the Ministry of Education, Culture, Sports, Science and Technology (MEXT), Japan, and was also partly supported by KAKENHI Grant-in-Aid for Young Scientist B (22760503) from Japan Society for the Promotion of Science (JSPS).

References

[1] Haque S A, Yamamoto M, Nakatani R and Endo Y 2004 *J. Magn. Magn. Mater.* **282** 380
[2] Cowburn R P and Welland M E 2000 *Science* **287** 1466
[3] Nomura H and Nakatani R 2011 *Appl. Phys. Express.* **4** 013004
[4] Imre A, Csaba G, Li L, Orlov A, Bernstein G H and Porod W 2006 *Science* **311** 205
[5] Worledge D C, 2004 *Appl. Phys. Lett.* **84** 2847
[6] Osborn J A, 1945 *Phys. Rev.* **67** 351
[7] Chang H 1961 *Br. J. Appl. Phys.* **12** 160

Effect of interfacial reaction on electrical conduction across the interface between n-type gallium nitride and contact materials

Masakatsu Maeda[1], Takao Yamasaki[2] and Yasuo Takahashi[1]

[1] Joining and Welding Research Institute, Osaka University, 11-1 Mihogaoka, Ibaraki, Osaka 567-0047, Japan
[2] Graduate School of Engineering, Osaka University, Suita, Osaka 565-0871, Japan

E-mail: maeda@casi.osaka-u.ac.jp

Abstract. The present paper reports electrical properties at the interface between n-type GaN and Ti-based contact layers formed by radio-frequency magnetron sputter deposition under various conditions. TiN contacts deposited using N_2 gas are non-ohmic, *i.e.*, formation of TiN adjacent to GaN is not a necessary condition for developing ohmic properties. On the other hand, Ti contacts deposited using Ar gas are ohmic in the as-deposited state, even though a layer of Ti_2N is formed between GaN and Ti during the deposition. It is also shown that Ti deposition on undoped GaN produces ohmic contacts. The nitrogen vacancies increased in the sub-interface of GaN are essential for developing ohmic properties. The interfacial reaction between Ti and GaN to form nitrogen vacancies is affected by the partial pressure of N_2 during the deposition.

1. Introduction

Gallium nitride (GaN) is a direct transition wide band-gap III-V compound semiconductor commonly used in light-emitting and laser diodes that exhibit high luminous efficacy and long service life [1, 2]. In addition, its high electron mobility, breakdown voltage and saturated electron drift velocity make GaN suitable for high-frequency operated power electronic devices [3, 4]. Power electronic devices are widely installed in electrical and electronic facilities to manage power flow and to improve energy efficiency. Owing to these functions of the devices, the facilities achieve reduced energy consumption without deteriorating their performance. Recently, the devices are considered as one of the key for establishing dispersed renewable power systems [5] and, as a further goal, a sustainable society. The emerging applications demand high power capability, high energy efficiency and long service life. Furthermore, all power electronic devices have to exhibit extremely high reliability, since only one failure can halt the entire system. Replacing conventional Si-based power electronic devices with GaN-based ones will improve the functions of power electronic systems, *i.e.*, the demands can be satisfied by putting GaN-based power electronic devices into practical use. However, several problems have to be overcome to realize GaN-based power electronic devices.

One of the major problems is to form low-resistance ohmic contacts at the interfaces between GaN and metallic outer circuits. The difference in electronic structure of the materials can form a Schottky barrier. The barrier interferes with charge transportation across the interface, similar to a resistor or a capacitor, and it generates Joule heat. The heat generation rate becomes high under high power and frequency operation, which corresponds to the field to which GaN-based power electronic devices are going to be applied. Repetitive thermal expansion and shrinkage induced by the heat deteriorate the reliability of the interface. By suppressing the heat generation, the reliability and the energy efficiency of the devices can be improved. Therefore, forming low-resistance ohmic contacts is important.

To form low-resistance ohmic contacts on GaN, it is necessary to reduce the height and width of the Schottky barrier. An appropriate contact material has to be formed adjacent to GaN to lower the barrier. One of the requirements for the appropriate contact material is its work function. In the case of n-type GaN, it has to be shallower than the conduction-band edge of GaN, which is 4.11 eV below the vacuum level. TiN is one of those materials with a sufficiently shallow work function of 3.74 eV [6, 7]. On the other hand, to make the Schottky barrier thin, it is effective to increase the carrier density in the GaN just under the contact material. The density can be increased by implantation of dopants. Formation of TiN by interfacial reaction between GaN and Ti generates nitrogen vacancies in GaN, which work as donors at an energy level close to the conduction band edge of GaN. Therefore, TiN is formed adjacent to GaN generally by interfacial reaction between GaN and multilayered metallic film containing Ti layer [7-10]. However, this method has a problem. Interlayer reactions during heat treatment form several intermetallic compounds in the film [10]. The compounds have high electric resistances and brittle nature. Thus, it is preferred to form TiN retaining the rest of the contact film metallic. However, it is difficult due to the high miscibility and diffusivity of nitrogen in Ti [11-13]. Recently, Maeda *et al.* have developed a method to form Ti(C,N), an alternative contact material to TiN, by which ohmic conduction properties were obtained preventing the formation of intermetallic compounds [13]. The results raise a new question regarding which factor dominates the ohmic conduction properties: the formation of TiN adjacent to GaN or the increase in donor density by the interfacial reaction.

Luther *et al.* have investigated the evolution of Ti, TiN and Ti/TiN contact resistances by annealing in Ar or N_2 [6]. They concluded that the presence of TiN adjacent to GaN is a necessary condition for ohmic contact formation and that the interfacial reaction assists in lowering the contact resistance. However, their results indicate that TiN contacts formed directly on n-type GaN fail to be ohmic in the as-deposited state. The contact became ohmic after annealing at 673 K for 60 s in Ar. On the other hand, the Ga-Ti-N ternary phase diagram indicates that TiN can equilibrate with GaN [14], *i.e.*, no interfacial reaction takes place during annealing. Therefore, the appearance of ohmic properties by annealing cannot be correlated with the interfacial reaction between GaN and TiN. On the contrary, Lin *et al.* have argued that nitrogen vacancies produced in GaN under the Ti contact play the main role in development of ohmic conduction and that the ohmic behavior cannot be attributed to the presence of TiN, based on their XPS analysis of Ti/GaN interfaces [15]. However, they failed to form N–Ti bonds in the as-deposited state even though the GaN surfaces were free of oxide films. Thus, the factor dominating the development of ohmic conduction is still unclear.

The present study has been implemented to clarify the dominating factor for ohmic conduction of contact interfaces on n-type GaN: whether it is the formation of TiN itself or the increased nitrogen vacancy in the GaN sub-interface. Various deposition conditions were employed to form TiN or Ti contact films on GaN. Electrical conduction profiles and interfacial structures were analyzed to investigate the effects of interfacial reactions between GaN and the films during the deposition on the electrical conduction properties. It is also demonstrated that ohmic contacts can be formed on n-type GaN without annealing by employing deposition conditions that enhance the mechanism dominating ohmic conduction.

2. Experimental procedure

Two types of GaN substrates were used in the present study. One was a monolithic single crystal wafer of n-type GaN of which the thickness and carrier density were 350 μm and 4.8×10^{18} cm^{-3}, respectively. The other was a 6.5-μm-thick undoped GaN single crystal layer formed epitaxially on the (0001) plane of sapphire. The surface orientation of the undoped GaN was the (0001) Ga-face.

Table 1. Samples and their preparation conditions.

Sample No.	Substrate	Target	Atmosphere
1	n-type GaN	TiN	N_2, 8.0 Pa
2	n-type GaN	Ti	Ar, 8.0 Pa
3	undoped GaN	Ti	Ar, 8.0 Pa
4	n-type GaN	Ti	N_2, 8.0 Pa
5	n-type GaN	TiN	Ar, 8.0 Pa

Therefore, only the (0001) Ga-face was used also with the n-type GaN. The substrates were cut to 4.0 mm square and were degreased by acetone applying ultrasonic vibration.

The substrates were set on a water-cooled copper holder to suppress the temperature increase and the reactions between the contact film and the substrate during the deposition. 1.0-mm-wide aluminum ribbons were employed to set the substrates on the holder of a radio-frequency (RF) magnetron sputter deposition apparatus. The ribbons also work as deposition masks. The distance between the target and the substrates was 50 mm. The deposition chamber was first evacuated to 4.0×10^{-5} Pa. High-purity Ar or N_2 gas was then introduced into the chamber up to 8.0 Pa. Immediately after careful sputter-cleaning of the surfaces of targets and substrates at the RF power of 200 W and the sputtering time of 600 and 300 s, respectively, deposition was implemented using Ti or TiN targets. The samples used in the present study are listed in Table 1. The thickness of the deposited films was varied in the range of 30–90 nm to fit the analysis to be conducted with the samples by controlling the RF power and the deposition time. The preparation condition for Sample 1 was employed to form TiN directly on GaN suppressing the interfacial reaction between GaN and the contact film. Therefore, the donor density in GaN is hardly expected to increase under this deposition condition. On the contrary, the condition for Sample 2 was employed to suppress the formation of TiN by the reaction between Ti and the atmosphere. Nitrogen atoms dissolved into and/or reacted with the Ti film will be supplied only from the GaN substrate. Thus, the donor density in GaN sub-interface is expected to increase by these interfacial interactions. The condition for Sample 3 was employed to suppress the initial donor density in the GaN substrate. Since the condition, except for the substrate, is the same as that for Sample 2, the electrical properties of Sample 3 will reveal the interfacial interactions during the deposition. The conditions for Samples 4 and 5 were employed to compare the effect of the target material and the atmosphere on the development of ohmic conduction.

The microstructures of the deposited films and the interfaces were analyzed by X-ray diffraction (XRD) and transmission electron microscopy (TEM). The TEM samples were prepared by focused ion beam machining. Electrical properties were estimated by direct-current conduction test at 273 K.

3. Results and discussion

The deposited films of Sample 1 appear in gold color, implying that TiN is successfully formed in the films. To confirm the formation of TiN, the sample was analyzed by XRD. Figure 1 shows the XRD pattern of Sample 1. In the pattern, only one peak corresponding to TiN 200 [16] appears with strong peaks of GaN [17] and a weak peak of αTi [18]. Therefore, the contact film consists of TiN and αTi. Formation of αTi will occur due to the atomic detachment of Ti atoms from TiN target by sputtering. Missing of all other peaks than 200 of TiN indicates that TiN in the film has been formed retaining a crystallographic relation with the GaN substrate being

$$100_{\mathrm{TiN}} \,//\, 0001_{\mathrm{GaN}} \,. \tag{1}$$

Although the relation is different from that observed in previous reports [10, 13], the existence of such a relation implies the formation of TiN adjacent to GaN.

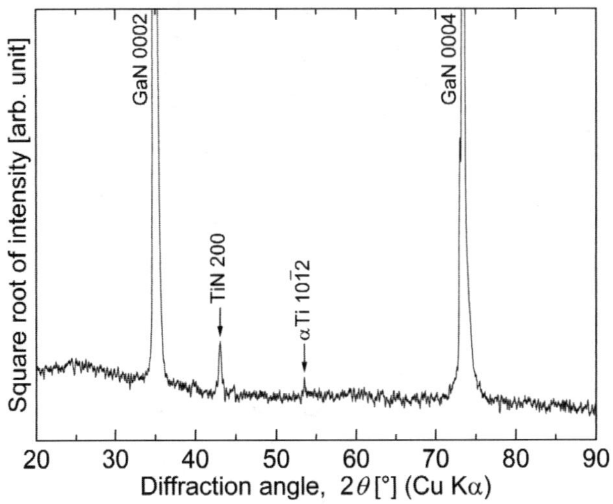

Figure 1. XRD pattern of Sample 1 prepared using TiN target and N_2 gas. The deposition was performed at an RF power of 300 W for 1800 s, which resulted in a film thickness of 90 nm.

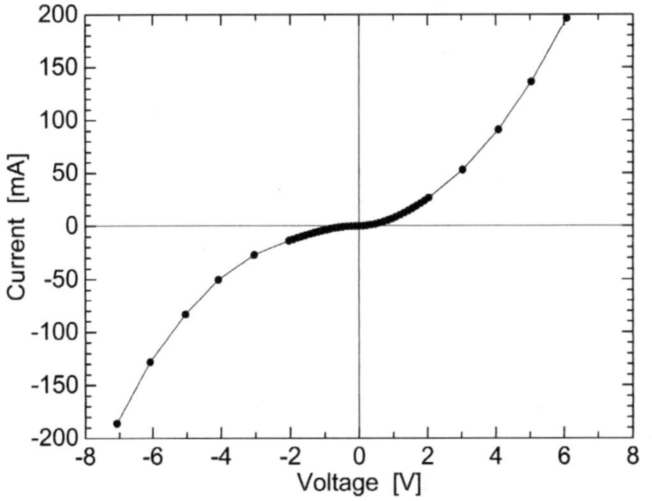

Figure 2. Current-voltage profile taken from Sample 1, on which two 1-mm-distant TiN contacts are formed.

The electric conduction profile between two 1.0-mm-distant contact films on Sample 1 is shown in Figure 2. The profile shows a non-linear current-voltage relation. Especially, the current drops to a nearly negligible level at voltages below 0.5 V, *i.e.*, the contacts are non-ohmic. The Schottky barrier height derived from the conduction profile is 0.46 eV [19, 20]. This barrier height agrees well with that of the TiN contact formed by annealing of Ti on n-type GaN reported by Lin *et al.* [15]. The result clearly shows that an ohmic contact cannot be developed only by forming TiN adjacent to GaN. Therefore, the interfacial reaction between Ti and GaN to form nitrogen vacancies in GaN under the contact will be rather necessary for developing ohmic conduction at TiN / n-type GaN contact interface than the TiN formation itself.

Sample 2 was prepared to enhance the formation of nitrogen vacancies in GaN by suppressing the interaction between Ti and the atmosphere. Figure 3 shows the cross-sectional structure of the contact interface formed on Sample 2. In the bright-field image (BFI) shown in Figure 3(a), a 12-nm-thick layer is observed adjacent to GaN. The electron diffraction pattern (EDP) taken from this area shown in Figure 3(b) consists of net patterns of GaN and Ti_2N. In the dark-field image (DFI) of the same area using the Ti_2N 020 diffraction shown in Figure 3(c), only the layer adjacent to GaN appears bright. Therefore, it is confirmed that a thin layer of Ti_2N is formed adjacent to GaN during the deposition of

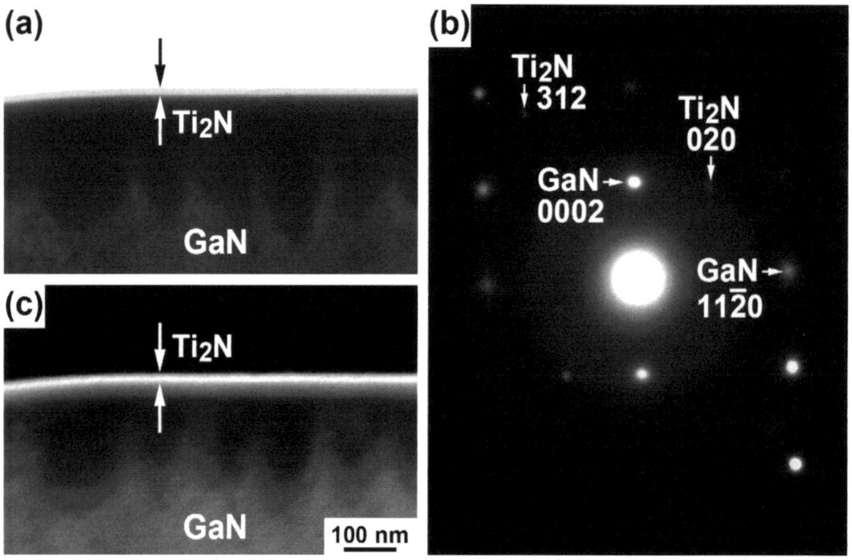

Figure 3. Interfacial structure of the contact formed on Sample 2. (a) BFI; (b) EDP corresponding to the area shown in (a); (c) DFI of the same area as (a) using Ti_2N 020.

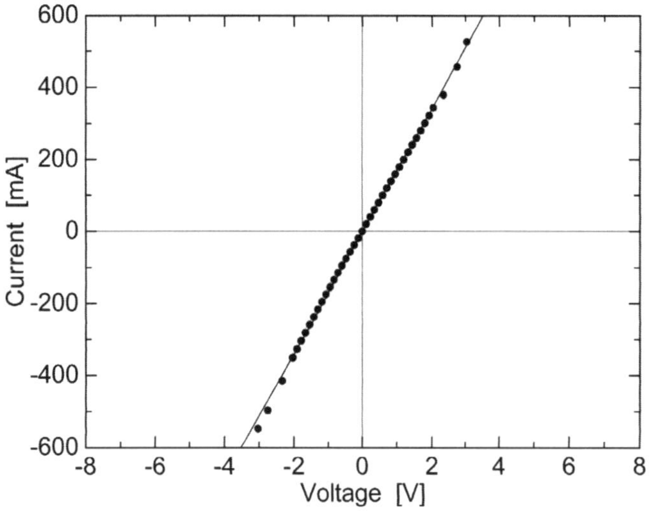

Figure 4. Current-voltage profile taken from Sample 2. Ohmic conduction is obtained even though Ti_2N is formed adjacent to n-type GaN.

Sample 2. Since the nitrogen constituting Ti_2N originates only from the GaN substrate, a considerable amount of nitrogen vacancies must be formed in the GaN just under the contact.

Ti_2N is not suitable for ohmic contact formation with n-type GaN. The Fermi level of Ti_2N is approximately 2 eV deeper than that of TiN [21], *i.e.*, Ti_2N forms a Schottky barrier 2 eV higher than TiN. As shown in Figure 2, the contact fails to be ohmic due to the formation of 0.46-eV-high Schottky barrier. Therefore, a significant interference to electrical conduction by the barrier will be expected. However, the electrical conduction profile of Sample 2 depicted in Figure 4 clearly shows a proportional relation between the current and voltage, *i.e.*, the contacts are ohmic even though a layer of Ti_2N is formed. Ohmic conduction in the presence of a high Schottky barrier can occur by tunneling mechanism. Since tunneling requires thinning of the barrier by introducing a high carrier density in the GaN sub-interface, the achievement of ohmic conduction with Sample 2 indicates that the interfacial reaction to form Ti_2N has introduced a sufficient amount of nitrogen vacancies in GaN just under the contact.

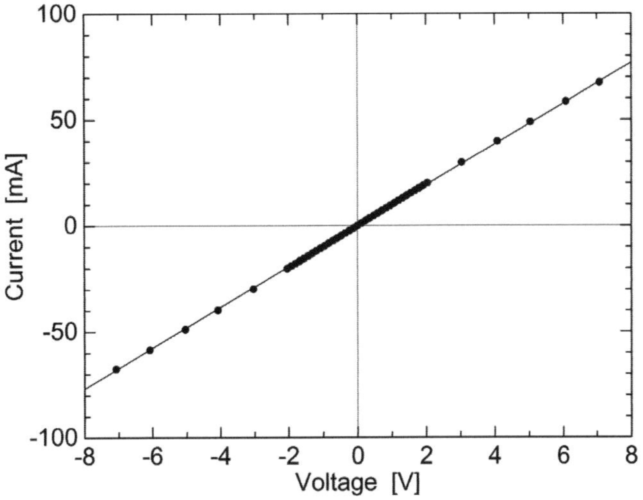

Figure 5. Current-voltage profile taken from Sample 3. Ohmic conduction is obtained with an as-deposited contact on undoped GaN.

To confirm the formation of a high carrier density zone in GaN by the interfacial reaction during the deposition, Sample 3 was prepared with undoped GaN under the same deposition condition as Sample 2. The initial carrier density in undoped GaN must be very low at 273 K, since it is a wide-bandgap intrinsic semiconductor in which only a small amount of thermally activated carriers exist. Therefore, the increase in carriers is all attributed to the increase in nitrogen vacancies due to the interfacial reaction. The electrical conduction profile of Sample 3 is shown in Figure 5. Although the electrical conductivity of Sample 3 appears approximately one-eighteenth that of Sample 2, the relation between the current and voltage is proportional, *i.e.*, the contacts are ohmic. This result indicates that a high carrier density sufficient for developing ohmic conduction has been achieved in the GaN sub-interface. The interfacial reaction during the deposition has introduced a high density of nitrogen vacancies.

Interfacial reaction between the deposited film and GaN plays an essential role in the formation of ohmic contacts on n-type GaN. The reaction has to be controlled to achieve the optimum performance of the contacts. Luther *et al.* [6] and Lin *et al.* [15] reported that Ti contacts require annealing to be ohmic. In the present study, however, Samples 2 and 3 successfully achieve ohmic conduction in the as-deposited state. The difference in sample preparation between the previous reports and the present study is the surface treatment before deposition. The previous studies used chemical etching, *i.e.*, the surfaces are exposed to ambient air after drying. The native oxides on the surfaces of GaN can be removed by aqua regia [15]. However, exposure to air after the treatment reforms the oxides; otherwise the surfaces of GaN are covered by radicals which terminate the surface dangling bonds of GaN and protect from oxidation. These surface layers will prevent the reaction until the sample is annealed, *i.e.*, the role of annealing in development of ohmic conduction is not only to initiate the interfacial reaction, but also to remove these layers. On the other hand, the samples prepared in the present study were cleaned by substrate sputtering and the deposition was conducted immediately without exposing the cleaned surfaces to air. Therefore, the present study has shown that the surface of GaN free of the surface layers reacts with Ti even during the deposition.

In the present study, the effects of the target and the atmosphere selection on the interfacial reaction are investigated using the reactive surface of GaN. Figure 6 shows the electrical conduction profiles of Samples 4 and 5. A distinct difference in electrical conduction properties is observed: the contacts on Sample 4 are non-ohmic, whereas those on Sample 5 are ohmic. The difference between the samples is the partial pressure of N_2. The partial pressure of N_2 is kept at 0.8 Pa for Sample 4. Ti atoms sputtered out of the target fly through the gas. The distance between the target and the substrate was 50 mm, which is approximately one order of magnitude longer than the mean free path of the atoms. Therefore, Ti atoms collide several times with N_2 molecules and react to form TiN during their flight to the

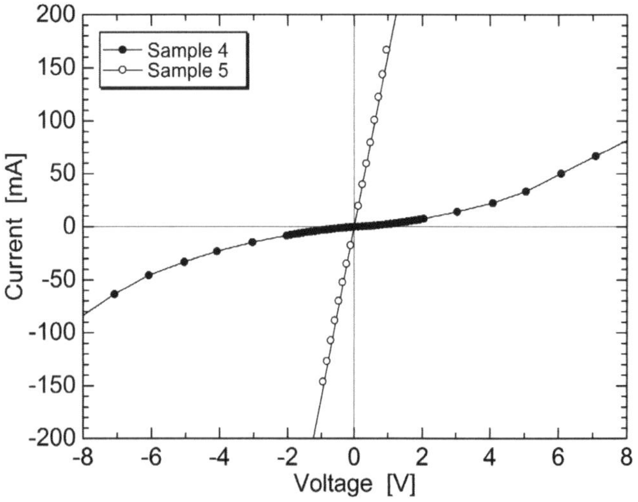

Figure 6. Current-voltage profiles of Samples 4 and 5.

substrate, *i.e.*, most of the Ti atoms have become TiN when they reach the substrate. TiN will not react with GaN. Nitrogen vacancies are not formed in the GaN sub-interface and the contacts are non-ohmic. On the other hand, the pressure for Sample 5 is considered to be very low. Although Ti atoms and N_2 gas are emitted by sputtering of the TiN target, the number of N_2 molecules will be insufficient to increase its partial pressure. Thus, Ti atoms seldom react with N_2 gas, *i.e.*, Ti atoms will reach the substrate as Ti. Ti deposited on GaN reacts immediately with GaN and forms nitrogen vacancies in the GaN sub-interface to attain ohmic conduction. The results shown in Figure 6 indicate that whether the target is Ti or TiN has little effect on the development of ohmic contacts, whereas the partial pressure of N_2 in the sputtering gas has a critical effect on it.

Samples 2, 3 and 5 have ohmic contacts in their as-deposited state, *i.e.*, without annealing. Generally, formation of ohmic contacts requires high-temperature annealing after deposition of metallic layers. However, such a high-temperature process can deteriorate the performance of the devices and other contacts formed before the ohmic contact. Consequently, to form ohmic contacts without annealing is preferred. The key point of the contact formation process employed in the present study will be the deposition of Ti or TiN immediately after the sputter-cleaning of the surfaces of the substrates. The surface of GaN becomes reactive by sputter-cleaning enough to form Ti_2N during the deposition of Ti, as shown in Figure 3. The sputter-cleaning removes oxide films on the surfaces of GaN and introduces defects in the GaN subsurface. The former point enhances the interfacial reaction to generate nitrogen vacancies in GaN, since the surface of GaN free of oxide films can react directly with the deposited Ti. With regard to the latter point, the amount of defects formed by sputter-cleaning at the RF power of 200 W for 300 s is considered to be negligible since Sample 1 which was also sputter-cleaned under the same condition fails to be ohmic even though TiN is better than Ti_2N as an ohmic contact material for n-type GaN. Therefore, ohmic contacts can be formed to n-type GaN without annealing by preparing a clean and reactive surface of GaN, which enhances the interfacial reaction during the deposition of Ti and generates nitrogen vacancies in GaN.

4. Conclusions

Ti-based contact films are formed on GaN under various conditions to investigate the dominant factor to develop ohmic conduction. The following points are clarified.

1) TiN formed directly on n-type GaN shows non-ohmic conduction indicating that the formation of TiN is not essential for development of an ohmic contact.
2) Ti deposited on n-type GaN shows ohmic conduction, even though Ti_2N is formed at the interface during the deposition process by the reaction between Ti and GaN. Nitrogen vacancies are formed and the carrier (donor) density is increased by the reaction.

3) Ti deposited on undoped GaN shows ohmic conduction, indicating that a significant increase in carrier density is achieved due to the interfacial reaction. This result supports the idea that the ohmic contact is developed by the formation of nitrogen vacancies.
4) The partial pressure of N_2 during RF magnetron sputter deposition affects the conduction properties of the contact interface. A low partial pressure of N_2 is needed to form low resistance ohmic contacts.
5) Ohmic contacts for n-type GaN can be formed without annealing by preparing a clean and reactive surface of GaN, which enhances the interfacial reaction and generates nitrogen vacancies in GaN.

Acknowledgements
The authors express their gratitude to Prof. H. Mori and Mr. E. Taguchi for their kind permission and assistance to use facilities in the Research Center for Ultra-High Voltage Electron Microscopy, Osaka University. The present study is supported by the Japan Society for Promotion of Science through Grant-in-Aid for Scientific Research (No. 21560746).

References
[1] Nakamura S, Mukai T and Senoh M 1991 *Jpn. J Appl. Phys.* **30** L1998
[2] Akasaki I 2007 *J. Cryst. Growth* **300** 2
[3] Trew R J, Shin M W and Gatto V 1997 *Solid-Stat. Electron.* **41** 1561
[4] Zhang A P, Ren F, Anderson T J, Abernathy C R, Singh R K, Holloway P H, Pearton S J, Palmer D and McGuire G E 2011 *Critic. Rev. Solid Stat. Mater. Sci.* **27** 1
[5] Blaabjerg F, Chen Z and Kjaer S B 2004 *IEEE Trans. Power Electron.* **19** 1184
[6] Luther B P, Mohney S E and Jackson T N 1998 *Semicond. Sci. Technol.* **13** 1322
[7] Mohammad S N 2004 *J. Appl. Phys.* **95** 7940
[8] Fan Z, Mohammad S N, Kim W, Aktas Ö, Botchkarev A E and Morkoç H 1996 *Appl. Phys. Lett.* **68** 1672
[9] Smith L L, Davis R F, Liu R-J, Kim M J and Carpenter R W 1999 *J. Mater. Res.* **14** 1032
[10] Maeda M, Matsumoto N, Hatakawa H and Takahashi Y 2009 *Quart. J. Jpn. Weld. Soc.* **27** 204s
[11] 1996 *Binary Alloy Phase Diagrams* 2nd edition plus updates on CD-ROM ed Massalski T B, Okamoto H, Subramanian P R and Kacprzak L (Materials Park, OH: ASM International)
[12] Metin E and Inal O T 1989 *Metall. Trans. A* **20A** 1819
[13] Maeda M, Matsumoto N and Takahashi Y 2010 *Materials Science and Technology (MS&T'10) Conf. Proc. CD-ROM (Houston, TX, 17-21 October 2010)* (Warrendale, PA: The Minerals, Metals & Materials Society) pp 2732–2742
[14] Mohney S E, Luther B P and Jackson T N 1996 *Mater. Res. Soc. Symp. Proc.* **395** 843
[15] Lin Y-J, Chen Y-M, Cheng T-J and Ker Q 2004 *J. Appl. Phys.* **95** 571
[16] *Powder Diffraction Files* (Newtown Square, PA: ICDD) 38-1420
[17] *Powder Diffraction Files* (Newtown Square, PA: ICDD) 50-0792
[18] *Powder Diffraction Files* (Newtown Square, PA: ICDD) 44-1294
[19] Koyama Y, Hashizume T and Hasegawa H 1999 *Solid-Stat. Electron.* **43** 1483
[20] Jang J-S, Park S-J and Seong T-Y 2002 *Phys. Stat. Sol. A* **194** 576
[21] Eibler R 1993 *J. Phys.: Condens. Matter* **5** 5261

Journal of Physics: Conference Series **379** (2012) 012021 doi:10.1088/1742-6596/379/1/012021

Electrical properties and structure of contact interface between Ti_3SiC_2 and p-type GaN

Aiman bin Mohd Halil[1], Masakatsu Maeda[2] and Yasuo Takahashi[2]

[1] Graduate School of Engineering, Osaka University, 2-1 Yamadaoka, Suita, Osaka 565-0871, Japan

[2] Joining and Welding Research Institute, Osaka University, 11-1 Mihogaoka, Ibaraki, Osaka 567-0047, Japan

Email: aiman@casi.osaka-u.ac.jp

Abstract. The purpose of the present study is to investigate the electrical properties and structure of the contact interface between Ti_3SiC_2 and p-type GaN. To form Ti_3SiC_2 on GaN, a Ti-Si-C ternary film was deposited by radio-frequency magnetron sputtering and then annealed at 1073 K for a very short period. The electric conduction properties of the contact interface were examined. The as-deposited amorphous Ti-Si-C film on p-type GaN shows non-ohmic conduction with a Schottky barrier height (SBH) of 0.89 eV. The SBH is reduced to 0.70 eV by annealing to form Ti_3SiC_2. On the other hand, a prolonged Ar ion bombardment of the GaN surface during sputter cleaning reduces the SBH of the as-deposited Ti-Si-C film to 0.76 eV. However, the effect of ion bombardment is not retained after annealing.

1. Introduction

The new era of sustainable society will be established based on renewable energy sources, being independent from exhaustible energy sources such as fossil and nuclear power generation. The power generation method will shift from concentrated power plants to dispersed cells [1] and windmills combined with smart grids [2, 3]. Each cell requires high efficiency power electronic systems at the interface with the grid for conditioning the electricity [4, 5]. The power electronic devices in such applications have to be highly energy-efficient, and capable of handling high power with a long service life.

Today, most of power electronic devices are still made of silicon. However, due to limitations of the physical properties of silicon, it is important to replace the devices with better alternative materials for use in the future power electronics. Two of the most promising candidates for next-generation power electronic devices are gallium nitride (GaN) and silicon carbide (SiC). Both of these wide-bandgap semiconductors allow operation at higher temperatures, power densities, frequencies and voltages with lower leak currents than silicon-based devices [6].

To form an electric circuit, GaN and SiC have to be connected with metallic materials. The ideal contact between these semiconductors and metals is one that allows smooth transportation across the interface. Such an interface shows a constant resistance indicated by proportional current-voltage (*I–V*) characteristics obeying the Ohm's law. Contact interfaces between semiconductors and metals with such properties are referred to as ohmic contacts. For p-type semiconductors, an ohmic contact can be formed by connecting with a metallic material of which work function (ϕ_m) is the same or higher than the electron affinity (χ) and the bandgap (E_g) of the semiconductor, *i.e.*,

$$e\phi_m \geq \chi + E_g \ , \tag{1}$$

where e is the elementary charge. However, the energy levels of the valence band edges ($\chi + E_g$) for p-type GaN and SiC are very deep: 7.50 and 6.46 eV, respectively. There is no pure metal element which satisfies condition (1). In the case that the work function of the connected metal is lower than the total of the electron affinity and the bandgap of the semiconductor, the electrons in the metal flow into the semiconductor until the Fermi levels of these two materials become equal, forming a depleted zone of carriers (holes) in the semiconductor under the contact. The zone interferes with carrier transportation across the interface. This interference is known as Schottky barrier. Carriers moving across such a contact interface require energy to get over the barrier. Thus, Joule heat is generated at the interface, which deteriorates the energy efficiency and the reliability of the device. Therefore, a technology is demanded to establish low-resistance ohmic contacts by lowering the Schottky barrier height (SBH) and/or by thinning the carrier depleted zone.

This problem has been researched actively worldwide and some important solutions have been reported. It was found that Ti_3SiC_2 formed adjacent to SiC provides a good ohmic contact for p-type SiC [7-12]. Ti_3SiC_2 is considered to work as a narrow-bandgap intermediate semiconductor layer which reduces the SBH [8, 9]. On the other hand, no appropriate material for low-resistance ohmic contacts to p-type GaN has been found so far. Jang *et al.* reported that low-resistance and thermally stable ohmic characteristic is achieved with a Au/Ni/Ru/Ag/Ni multilayered contact on p-type GaN [13]. In another paper, Chen *et al.* suggested that a low-resistance ohmic contact is achieved with the formation of NiO and a specific microstructure by annealing Au/Ni/p-type-GaN in an oxidative ambient [14]. However, most of the solutions are usually difficult to be reproduced by other researchers, even if the reported procedures are followed. This implies that the reported processes do not correspond only to the formation of the preferred structure at the contact interface. Therefore, a technology to control the interfacial structure has to be established based on knowledge of the formation mechanism of low-resistance ohmic interfaces between metals and GaN or SiC. This is a difficult task to tackle, since the suggested contacts are formed by complex interfacial reaction between the substrate and multilayered films [13].

GaN shows some degree of similarity with SiC in terms of lattice parameters, coefficient of thermal expansion and bandgap as shown in Table 1 [15-18]. These parameters are three of the most important factors which affect directly the formed contact structure and electrical properties. Therefore, there is a high possibility that Ti_3SiC_2 would also achieve an ohmic contact with p-type GaN. The present study demonstrates this idea by forming Ti_3SiC_2 on p-type GaN. To form a Ti_3SiC_2 layer on GaN, a Ti-Si-C ternary film with a composition stoichiometrically equivalent to Ti_3SiC_2 is deposited on GaN and then annealed at 1073 K.

Table 1. Properties of GaN and SiC.

Properties	GaN	4H-SiC
Crystal system	Hexagonal	Hexagonal
Lattice parameter, a / nm	0.3189	0.3073
c / nm	0.5186	1.0053
Coefficient of thermal expansion, α / K^{-1}	5.59×10^{-6}	4.47×10^{-6}
Bandgap, E_g / eV	3.39	3.26
Electron affinity, χ / eV	4.11	3.20

The SBH ($e\phi_{SB}$) at the interface between Ti_3SiC_2 and p-type SiC is estimated as 1.39 eV by the Schottky-Mott model [19] described by

$$e\phi_{SB} = \chi + E_g - e\phi_{Ti_3SiC_2} \,, \tag{2}$$

where the work function of Ti_3SiC_2 ($\phi_{Ti3SiC2}$) is taken as 5.07 eV [17]. This prediction shows that an ohmic contact cannot be formed by simply connecting Ti_3SiC_2 and p-type SiC, but an additional mechanism is needed to compensate the SBH. The interfacial reaction between SiC and the deposition film, which consists of Ti and Al, to form Ti_3SiC_2 is expected to bare the role: the reaction does not form only Ti_3SiC_2, but also defects in SiC which lower the SBH by increasing the carrier density [8, 9].

On the other hand, the SBH at the interface between Ti_3SiC_2 and p-type GaN is estimated as 2.43 eV. In contrast with the contact between Ti_3SiC_2 and p-type SiC, the interfacial reaction between the deposited Ti-Si-C ternary film and GaN cannot be used for Ti_3SiC_2 formation and defect introduction in GaN under the contact. In the present study, changing the carrier density is attempted by Ar ion bombardment during sputter-cleaning of the GaN surface just before the deposition. By performing a prolonged bombardment, it is expected that severe radiation damage is induced in the subsurface of the substrate. The damage corresponds to vacancy defect formation. Vacancies in GaN are known to act as dopants [19]. Thus, the increase in vacancy defect concentration results in an increase in carrier density. The effects of carrier density on the electrical properties of the contact have been discussed in several reports [19, 20].

The objective of the present study is to clarify the electrical properties and structure of the contact interface between Ti_3SiC_2 and p-type SiC, including the effect of vacancy defects introduced by Ar ion bombardment.

2. Experimental procedure

A 2.0-μm-thick p-type GaN epitaxially grown on a 330-μm-thick sapphire (0001) wafer with a 2.3-μm-thick GaN buffer layer was used in the present study. The carrier density and the surface orientation were 3×10^{17} cm^{-3} and the (0001) Ga-face, respectively. The wafer was cut into 4.0-mm-square substrates. The substrates were cleaned with acetone applying ultrasonic vibration just before the deposition. The substrates were fixed in a radio-frequency (RF) magnetron sputter deposition apparatus with 1.0-mm-wide Al masking ribbons. Before the deposition, careful sputter cleaning of the surfaces of the Ti-Si-C alloy target and substrates was performed. The bombardment of the substrates was performed during this process. Some of the substrates were subjected to 300 s of sputter cleaning to remove native oxide layer on the surfaces, whereas the other substrates were subjected to prolonged bombardment for 1800 s to remove the oxide layer and to introduce a high density of vacancy defects in the substrates. After the sputter cleaning, Ti-Si-C ternary films were deposited on the substrates. Both sputter cleaning and deposition were performed under 0.8 Pa of 99.9999% high-purity Ar and an RF power of 200 W. The target was a Ti-Si-C ternary alloy sintered disk with the composition of $Ti_{65}Si_{22}C_{13}$. By performing deposition for 600 s with this target, 20-nm-thick films with the composition of $Ti_{49}Si_{18}C_{33}$ were successfully deposited on all GaN substrates. The composition of the Ti-Si-C ternary films was close to the stoichiometric composition of Ti_3SiC_2. After the deposition, some of the specimens were subjected to annealing at 1073 K in a vacuum of 1.3×10^{-3} Pa. The specimens were cooled down immediately after reaching the temperature of 1073 K, i.e., the specimens were hold at 1073 K only for a very short time.

The structures and electrical properties of the as-deposited and annealed specimens were analyzed by X-ray diffraction (XRD) and direct current (DC) conduction test at room temperature.

3. Results and discussion

Figure 1 shows the structural change in the specimens caused by annealing at 1073 K. In the XRD pattern of the as-deposited specimen, only strong peaks of GaN and sapphire are identified, i.e., no peak related to the Ti-Si-C phases in the deposition film is found. It is likely that the deposition film is amorphous. The XRD pattern of the annealed specimen appears different from that of the as-deposited specimen. Peaks corresponding to various crystallographic planes of Ti_3SiC_2 appear, indicating that the Ti_3SiC_2 has a randomly oriented polycrystalline structure. Thus, it is proven that the Ti_3SiC_2 layer can be formed on GaN by deposition of the Ti-Si-C ternary film stoichiometrically close to Ti_3SiC_2 and subsequent annealing at 1073 K.

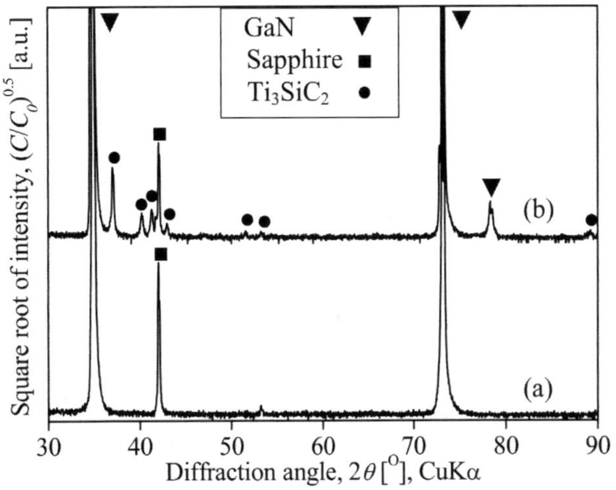

Figure 1. Structural change for specimens sputter-cleaned for 300 s (a) before and (b) after the annealing at 1073 K.

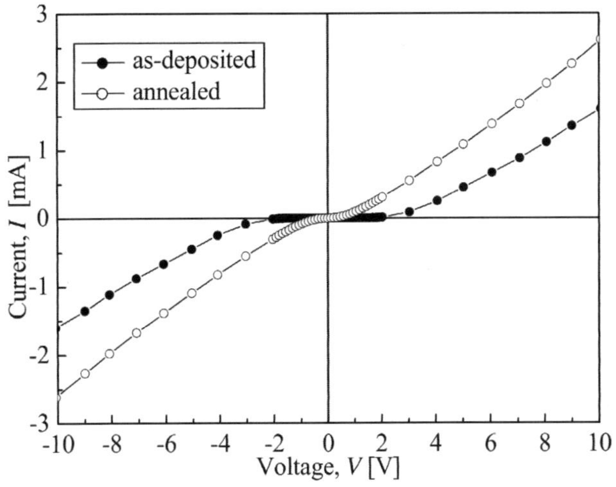

Figure 2. Electric conduction profiles of specimens sputter-cleaned for 300 s, in as-deposited and annealed state.

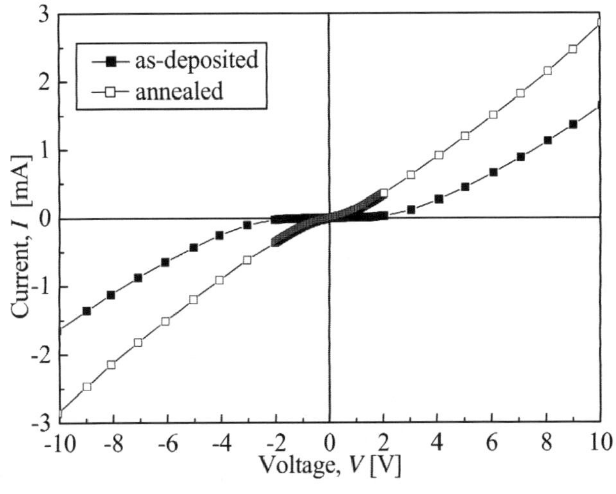

Figure 3. Electric conduction profiles of specimens sputter-cleaned for 1800 s, in as-deposited and annealed state.

Figure 2 shows the change in the electrical conduction profile by annealing of the specimens sputter-cleaned for 300 s. The curves of both the as-deposited and the annealed specimens show a non-linear relation between voltage and current, indicating that the contacts of both specimens are not ohmic. However, the difference in the rising voltage of the current indicates that the SBH is reduced by some amount by annealing to form Ti_3SiC_2. To evaluate the reduction of the SBH, the conduction profiles are analyzed by the thermionic emission model [21] given by

$$I_0 = AA^*T^2 \exp\left(-\frac{e\phi_{SB}}{kT}\right), \tag{3}$$

where I_0, A, A^*, T and k are the saturated current, the contact area, the effective Richardson constant of 1.038×10^6 $Am^{-2}K^{-2}$ for p-type GaN [21], the absolute temperature of the specimen and the Boltzmann constant, respectively. The SBHs of the as-deposited and the annealed specimens are 0.89 and 0.70 eV, respectively. Thus, the reduction in SBH for 0.19 eV is attributed to the structural change in the contact film from amorphous to polycrystalline Ti_3SiC_2. Furthermore, the SBH of the interface between Ti_3SiC_2 and p-type GaN (the annealed specimen) is 1.73 eV lower than the value predicted by the Schottky-Mott model. Such lowering of the SBH is discussed in detail by Mohammad [19].

Figure 3 shows the change in the electrical conduction profile by annealing of the specimens sputter-cleaned for 1800 s. The profile and its change by annealing show a similar tendency with those shown in Figure 2: both the as-deposited and the annealed specimens are not ohmic. The SBHs of the specimens are 0.76 and 0.69 eV, respectively.

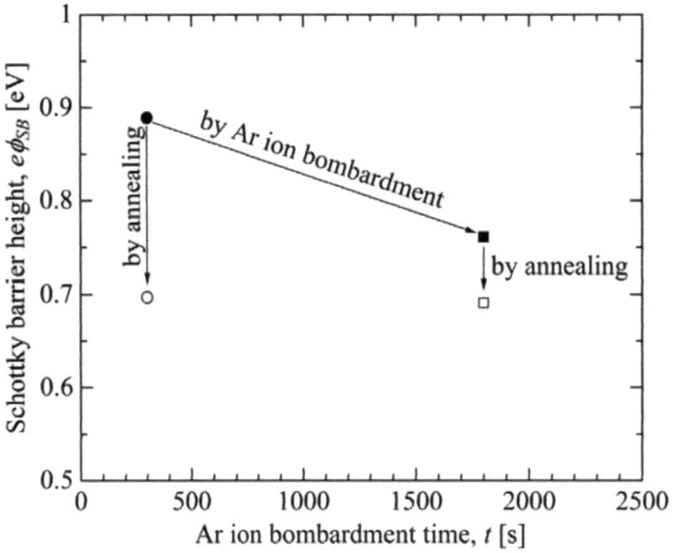

Figure 4. Effects of Ar ion bombardment and annealing on SBH reduction.

Figure 4 summarizes the effects of Ar ion bombardment and annealing on the reduction of the SBH. The extensive Ar ion bombardment for 1800 s prior to deposition reduces the SBH by 0.13 eV compared to the specimens subjected to conventional sputter-cleaning for 300 s. The reduction is considered to have been brought about by defects in the GaN sub-surface introduced by the bombardment. However, the difference between the SBH of the annealed specimens is negligible. It seems that the defects are recovered almost completely by annealing to form Ti_3SiC_2 and the effect of the defects on reduction of the SBH is weakened.

4. Conclusions

A Ti_3SiC_2 layer can be formed adjacent to p-type GaN by deposition of an amorphous Ti-Si-C ternary film stoichiometrically close to Ti_3SiC_2 and subsequent annealing at 1073 K for a short time. The SBH of the Ti_3SiC_2 contact on p-type GaN is 0.70 eV, which is 1.73 eV lower than the theoretically predicted value. A prolonged Ar ion bombardment of the GaN surface reduces the SBH of the as-deposited Ti-Si-C film. However, the effect of ion bombardment is weakened by annealing to form Ti_3SiC_2.

References

[1] Blaabjerg F, Chen Z and Kjaer S B 2004 *IEEE Trans. Power Electron.* **19** 1184
[2] Abe R, Taoka H and McQuilkin D 2011 *IEEE Trans. Smart Grid* **2** 399
[3] Farhangi H 2010 *IEEE Power and Energy Mag.* **8** 18
[4] Chu T-Y, Tsang S-W, Zhou J, Verly P G, Lu J, Beaupré S, Leclerc M and Tao Y 2012 *Solar Energy Mater. Solar Cell.* **96** 155
[5] Peng B and Chen J 2009 *Coordinat. Chem. Rev.* **253** 2805
[6] Madsen L D 2001 *J. Electron. Mater.* **30** 1353
[7] Crofton J, Beyer L, Williams J R, Luckowski E D, Mohney S E and Delucca J M 1997 *Solid State Electron.* **41** 1725
[8] Johnson B J and Capano M A 2004 *J. Appl. Phys.* **95** 5616
[9] Tsukimoto S, Nitta K, Sakai T, Moriyama M and Murakami M 2004 *J. Electron. Mater.* **33** 460
[10] Veisz B and Pécz B 2004 *Appl. Surface Sci.* **233** 360
[11] Tsukimoto S, Ito K, Wang Z, Saito M, Ikuhara Y and Murakami M 2009 *Mater. Trans.* **50** 1071
[12] Maeda M, Nonomura K, Takahashi Y and Tenyama K 2007 *Proc. 8th Int. Conf. on Brazing, High Temperature Brazing and Diffusion Welding (LÖT 2007) (Aachen, Germany, June 2007)* (Düsseldorf, Germany: DVS Verlag) pp 84-87
[13] Jang H W and Lee J L 2004 *Appl. Phys. Lett.* **85** 4421
[14] Chen L-C, Ho K-K, Jong C-H, Chiu C C, Shih K-K, Chen F-R, Kai J J and Chang L 2000 *Appl. Phys. Lett.* **76** 3703
[15] Davydov S Y 2007 *Semicond.* **41** 696
[16] Sumakeris J, Sitar Z, Ailey-Trent K S, More K L and Davis R F 1993 *Thin Solid Films* **225** 244
[17] Gao M, Tsukimoto S, Goss S H, Tumakha S P, Onishi T, Murakami M and Brillson L J 2007 *J. Electron. Mater.* **36** 277
[18] Okumura H, Ohta K, Ando K, Rühle W W, Nagatomo T and Yoshida S 1997 *Solid State Electron.* **41** 201
[19] Mohammad S N 2004 *Philos. Mag.* **84** 2559
[20] Kim J K, Kim K-J, Kim B, Kim J N, Kwak J S, Park Y and Lee J-L 2001 *J. Electron. Mater.* **30** 129
[21] Cinar K, Yildirim N, Coskun C and Turut A 2009 *J. Appl. Phys.* **106** 073717

Optimization of control parameters of a boost converter for energy harvesting

M Tanaka, P Hemthavy and K Takahashi

Department of International Development Engineering, Tokyo Institute of Technology, I4-11, 2-12-1 O-okayama, Meguro-ku, Tokyo 152-8552, Japan

E-mail: pasomphone@ide.titech.ac.jp

Abstract. In order to maximize the energy efficiency for energy harvesting, a simplified circuit model using a boost converter and a capacitor is proposed. The circuit model allows the analysis of the whole system's theoretical energy efficiency and determination of the optimum control parameters. The optimum duty ratio is determined analytically and the optimum switching period numerically. This is important information for designing a switching controller that achieves both low power consumption and accurate maximum power point tracking. Moreover, the proposed theoretical energy efficiency is helpful for component selection when designing boost converters.

1. Introduction

Energy harvesting with small solar cells or thermoelectric generators requires an efficient storage method. Compared to large-scale generation, power consumption in an energy harvester is a fatal energy loss for achieving high efficiency. The main source of power consumption is the switching controller of the DC-DC converter that operates the generating device at its maximum power point (MPP). Two methods of maximum power point tracking (MPPT) for energy harvesting have typically been used: fractional open-circuit voltage (FOCV) [1]-[3] and perturb and observe (P&O)[4]. The FOCV method achieves low power consumption for only one comparator and a few discrete components but has low tracking accuracy. On the other hand, the P&O method exploits very accurate MPP calculations but requires a high power consumption for complex circuits such as microcontrollers or digital signal processors. Therefore, an ideal switching controller can improve energy efficiency by achieving both low power consumption and accurate MPPT. The optimum control parameters, which are output by the ideal switching controller, should be clarified theoretically before the design stage. In the present study, in order to achieve high efficiency, a capacitor is used as a storage device with a low equivalent series resistance (ESR). Additionally, a boost converter is used for MPPT because of its excellent transmission efficiency for electric power [3]. The objective is to introduce an equivalent energy harvester model and obtain the optimum control parameters for maximizing the energy efficiency.

2. Modeling an energy harvester

In order to analyze the electric behavior of an energy harvester for different switching periods, the equivalent circuit model shown in Figure 1 is used. The power source should be operated around its MPP. Therefore, the i-v characteristics can be assumed to be a tangential line at the MPP. The i-v characteristics can be expressed as

$$v = 2V_{\mathrm{MP}} - R_{\mathrm{MP}}\, i, \tag{1}$$

where V_{MP} and I_{MP} are the power source output voltage and current at the MPP, and R_{MP} is defined as $\frac{V_{\mathrm{MP}}}{I_{\mathrm{MP}}}$.

For the same reason, the diode forward voltage v_{f} for a current i is assumed to be

$$v_{\mathrm{f}} = V_{\mathrm{D}} + R_{\mathrm{D}}\, i. \tag{2}$$

Additionally, the storage capacitor has a large capacitance so that its voltage can be assumed to be a constant V_{C} during one switching period.

Moreover, the components for the boost converter and capacitor should have low resistance compared to R_{MP} in order to improve energy efficiency. Thus, the total resistance of the current path in the on-state can be assumed to be nearly equal to that in the off-state. This can be expressed as

$$\begin{aligned} R &= R_{\mathrm{MP}} + R_{\mathrm{L}} + R_{\mathrm{DS}} \\ &\approx R_{\mathrm{MP}} + R_{\mathrm{L}} + R_{\mathrm{D}} + R_{\mathrm{C}}. \end{aligned} \tag{3}$$

Kirchhoff's voltage law for the on- and off-states can be written as

$$L\frac{di_{\mathrm{on}}}{dt} + Ri_{\mathrm{on}} = 2V_{\mathrm{MP}}, \tag{4}$$

when the switch is on and

$$L\frac{di_{\mathrm{off}}}{dt} + Ri_{\mathrm{off}} = 2V_{\mathrm{MP}} - V_{\mathrm{C}} - V_{\mathrm{D}}, \tag{5}$$

when the switch is off.

In order to analyze the current flow, a constant switch-on time of T_{on} and a switch-off time of T_{off} are considered. If the number of switching operations is defined as n, the current in the on-state $i_{\mathrm{on}(2n-1)}$ is expressed as

$$i_{\mathrm{on}(2n-1)} = (I_{2n-2} - \frac{2V_{\mathrm{MP}}}{R})\,\mathrm{e}^{-\frac{Rt}{L}} + \frac{2V_{\mathrm{MP}}}{R}, \tag{6}$$

where I_{2n-2} is a boundary condition when switching from off to on. Likewise, the current in the off-state $i_{\mathrm{off}(2n)}$ is expressed as

$$i_{\mathrm{off}(2n)} = (I_{2n-1} - \frac{2V_{\mathrm{MP}} - V_{\mathrm{C,D}}}{R})\,\mathrm{e}^{-\frac{Rt}{L}} + \frac{2V_{\mathrm{MP}} - V_{\mathrm{C,D}}}{R}, \tag{7}$$

where I_{2n-1} is a boundary condition when switching from on to off and $V_{\mathrm{C,D}} = V_{\mathrm{C}} + V_{\mathrm{D}}$. The boundary conditions I_{2n-2} , I_{2n-1} and I_{2n} are related by the recurrence equations

$$I_{2n-1} = I_{2n-2}\mathrm{e}^{-\frac{RT_{\mathrm{on}}}{L}} + \frac{2V_{\mathrm{MP}}}{R}(1 - \mathrm{e}^{-\frac{RT_{\mathrm{on}}}{L}}), \tag{8}$$

and

$$I_{2n} = I_{2n-1}\mathrm{e}^{-\frac{RT_{\mathrm{off}}}{L}} + \frac{2V_{\mathrm{MP}} - V_{\mathrm{C,D}}}{R}(1 - \mathrm{e}^{-\frac{RT_{\mathrm{off}}}{L}}). \tag{9}$$

When the initial condition I_0 is 0, the recurrence equations can be expressed as

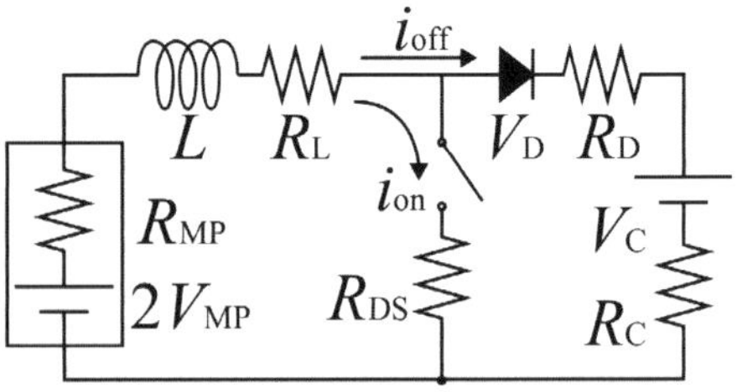

Figure 1. Equivalent circuit model of an energy harvester.

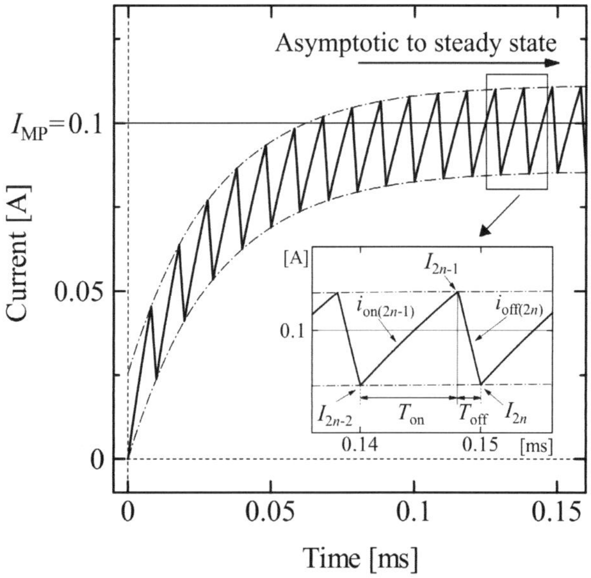

Figure 2. Current oscillation obtained from equations (6) and (7) using boundary conditions set by equations (10) and (11) with available specifications.

$$I_{2n-2} = \left(\frac{2V_{\mathrm{MP}}}{R} - \frac{V_{\mathrm{C,D}}(1-\mathrm{e}^{-\frac{RT_{\mathrm{off}}}{L}})}{R(1-\mathrm{e}^{-\frac{RT}{L}})}\right)\left(1-\mathrm{e}^{-\frac{(n-1)RT}{L}}\right), \tag{10}$$

and

$$\begin{aligned} I_{2n-1} =& \left(\frac{2V_{\mathrm{MP}}}{R} - \frac{V_{\mathrm{C,D}}\mathrm{e}^{-\frac{RT_{\mathrm{on}}}{L}}(1-\mathrm{e}^{-\frac{RT_{\mathrm{off}}}{L}})}{R(1-\mathrm{e}^{-\frac{RT}{L}})}\right)\left(1-\mathrm{e}^{-\frac{(n-1)RT}{L}}\right) \\ &+ \frac{2V_{\mathrm{MP}}}{R}\left(1-\mathrm{e}^{-\frac{RT_{\mathrm{on}}}{L}}\right)\mathrm{e}^{-\frac{(n-1)RT}{L}}, \end{aligned} \tag{11}$$

where $T = T_{\mathrm{on}} + T_{\mathrm{off}}$.

The resulting current oscillation is shown in Figure 2 based on equations (6),(7),(10) and (11). Figure 2 is based on the experimental specifications shown in Table 1, with T_{on} =8.09 [μs], T_{off} =1.91 [μs] and V_{C}=7.5 [V]. Figure 2 indicates that the top and bottom boundary conditions converge immediately from the initial conditions. Thus, the steady state current can be expressed using the boundary conditions of equations (10) and (11) at $n \to \infty$ as

$$i_{\mathrm{on}} = -\frac{V_{\mathrm{C,D}}\left(1-\mathrm{e}^{-\frac{RT_{\mathrm{off}}}{L}}\right)}{R\left(1-\mathrm{e}^{-\frac{RT}{L}}\right)}\mathrm{e}^{-\frac{Rt}{L}} + \frac{2V_{\mathrm{MP}}}{R}, \tag{12}$$

and

$$i_{\mathrm{off}} = \frac{V_{\mathrm{C,D}}\left(1-\mathrm{e}^{-\frac{RT_{\mathrm{on}}}{L}}\right)}{R\left(1-\mathrm{e}^{-\frac{RT}{L}}\right)}\mathrm{e}^{-\frac{Rt}{L}} + \frac{2V_{\mathrm{MP}} - V_{\mathrm{C,D}}}{R}. \tag{13}$$

The energy stored in the capacitor during one switching period can be calculated using equations (12) and (13). The energy E_{ct} is expressed as

$$E_{ct} = V_C \int_0^{T_{off}} i_{off} dt. \tag{14}$$

However, the field effect transistor (FET) operates as an ideal switch without R_{DS} in the model. The switching loss E_{sw} and gate drive loss E_{gd} at the FET should be considered for energy harvesting. The switching loss E_{sw} is approximated as

$$E_{sw} = \frac{I_{on} V_{off} T_{r,f}}{6}, \tag{15}$$

where I_{on} is the current when the switch is changing from on to off, V_{off} is the voltage across the diode and the capacitor, and $T_{r,f}$ is the total current rise or fall time required for switching. Additionally, the gate drive loss E_{gd} can be expressed as

$$E_{gd} = Q_g V_{gs}, \tag{16}$$

where Q_g is the gate charge and V_{gs} is the gate-source voltage.

On the other hand, an ideal boost converter for energy harvesting must operate the power source at its MPP and store the energy in the capacitor without loss. Thus, the energy efficiency of the energy harvester model during one switching period can be defined as

$$\eta = \frac{E_{ct} - E_{sw} - E_{gd}}{E_{MP}}, \tag{17}$$

where $E_{MP} = I_{MP} V_{MP} T$.

Using equations (12),(13),(14),(15) and (16), equation (17) can be expressed as

$$\begin{aligned}\eta =& \frac{R_{MP} V_C}{R V_{C,D}} \left(\frac{\tilde{g}_{C,D}^2 (1 - e^{-(1-d)\tilde{T}})(1 - e^{-d\tilde{T}})}{\tilde{T}(1 - e^{-\tilde{T}})} - \tilde{g}_{C,D}(1-d)(\tilde{g}_{C,D} - 2) \right) \\ &- \frac{I_{on} V_{off} T_{r,f} + 6 Q_g V_{gs}}{6 I_{MP} V_{MP} T},\end{aligned} \tag{18}$$

where $\tilde{g}_{C,D} = \frac{V_{C,D}}{V_{MP}}$, $\tilde{T} = \frac{RT}{L}$ and d is the duty ratio expressed as $\frac{T_{on}}{T}$. The duty ratio can be optimized analytically by differentiating equation (18) with respect to d. The optimized duty ratio d_{op} is then written as

$$d_{op} = 1 + \frac{1}{\tilde{T}} \log \left(\frac{(\tilde{g}_{C,D} - 2)(1 - e^{-\tilde{T}})}{2\tilde{g}_{C,D}} + \sqrt{\frac{(\tilde{g}_{C,D} - 2)^2 (1 - e^{-\tilde{T}})^2}{4\tilde{g}_{C,D}^2} + e^{-\tilde{T}}} \right). \tag{19}$$

Using equation (18) with d set to d_{op}, as given by equation (19), the relationship between the switching period and energy efficiency is plotted in Figure 3 for the experimental specifications shown in Table 1 and V_C=7.5 [V]. The calculation for Figure 3 used a power source MPP of 150 [mW], which is similar to the output often used in solar harvesting. The optimum switching period for achieving both a higher power source output and a lower FET energy loss is 9.29 [μs] from Figure 3, which gives an energy efficiency of 92.7%.

3. Experimental results

To validate the theoretical energy efficiency, the experimental energy efficiency is calculated by measuring the time Δt for charging the capacitor from V_L =7.45 [V] to V_H =7.55 [V]. Figure 4 shows the experimental circuit for measuring the capacitor voltage as a function of time, which is designed using the components shown in Table 1.

Table 1. Specifications for the experiment.

Components	Manufacturer	Model number	Symbols	Values
Power source	MATSUSADA KOA	PL-36-2.2 MOS2CT52A15R0F	V_{MP} R_{MP}	1.5 [V] 15 [Ω]
Inductor	Panasonic	ELC18B471L	L R_L	470 [μH] 210 [mΩ]
FET	TOSHIBA	TPC6006-H	R_{DS} $T_{r,f}$ Q_g	78 [mΩ] 7.0 [ns] 2.4 [nC]
Diode	TOSHIBA	CMS16	V_D R_D	0.32 [V] 200 [mΩ]
Capacitor	BHC	ALS3-224NP040	C R_C	225 [mF] 6.0 [mΩ]

The stored energy E_{ce} in the capacitor during one switching period is experimentally obtained to be

$$E_{ce} = \frac{1}{2}C(V_H^2 - V_L^2)\frac{T}{\Delta t}. \tag{20}$$

Since E_{ce} is influenced by the FET switching loss, equation (17) can be equivalently expressed as

$$\eta = \frac{E_{ce} - E_{gd}}{E_{MP}}, \tag{21}$$

where E_{gd} is calculated using equation (16), with Q_g =2.4 [nC] and V_{gs} =5.0 [V]. Table 2 shows the values of d_{op} calculated using equation (19) for different switching periods.

Table 2. Calculated values of d_{op} using equation (19) for different T in the experiment.

T [μs] Frequency	0.316 3.16 [MHz]	1.00 1.00 [MHz]	3.16 316 [kHz]	10.0 100 [kHz]	31.6 31.6 [kHz]	100 10.0 [kHz]	316 3.16 [kHz]
d_{op}	0.808	0.808	0.808	0.809	0.816	0.869	0.953

The experimental results are plotted in Figure 3. Approximation errors for the FET switching loss occur for low switching periods, but this is not important for determining the optimum

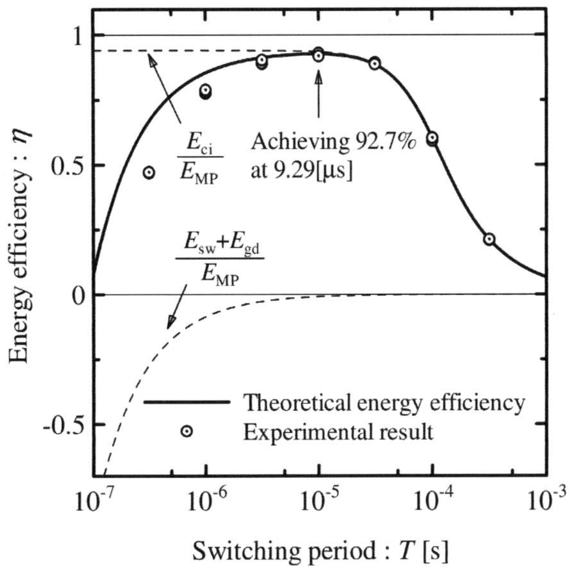

Figure 3. Relationship between switching period and energy efficiency.

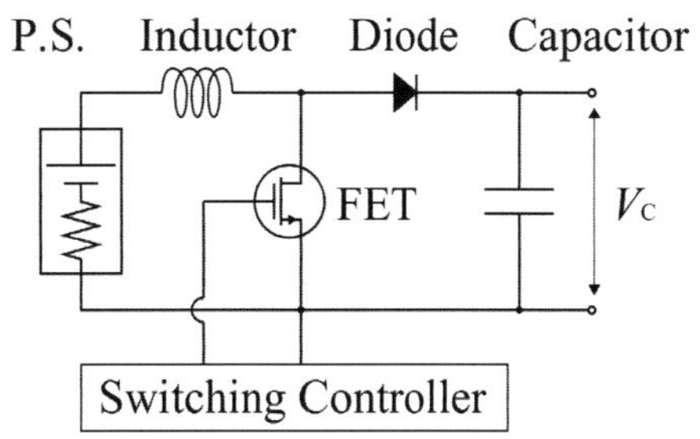

Figure 4. Experimental circuit for measuring the capacitor voltage.

switching period. The theoretical energy harvesting model and energy efficiency are validated experimentally in the high efficiency region.

4. Conclusion

This paper focuses on the optimum control parameters for a boost converter for energy harvesting. The optimum duty ratio is determined analytically and the optimum switching period numerically. The results provide important information for the design of an ideal switching controller for energy harvesting. Additionally, the proposed energy harvesting model and theoretical energy efficiency are helpful for component selection when designing boost converters.

References

[1] Dondi D, Bertacchini A, Brunelli D, Larcher L and Benini L 2008 *IEEE Trans. Ind. Electron.* **55**, 2759
[2] Simjee F I and Chou P H 2008 *IEEE Trans. Power Electron.* **23** 1526
[3] Chini A and Soci F 2010 *Electron. Lett* **46** 296
[4] Femia N, Petrone G, Spagnuolo G and Vitelli M 2005 *IEEE Trans. Power Electron.* **20** 963

Optimum condition of boost switching regulator for charging tiny electric energy to capacitor

Y Honda, M Tanaka, Y Sekiguchi, P Hemthavy and K Takahashi

Department of International Development Engineering, Tokyo Institute of Technology, I4-11, 2-12-1 Ookayama, Meguro-ku, Tokyo 152-8552, Japan

E-mail: honda.y.ag@m.titech.ac.jp

Abstract. Some energy sources are left unused because they can only generate a small amount of energy. In order to utilize the small amount of electric energy generated from these energy sources, an electric power storage system is investigated. The storage system consists of a capacitor and a switching regulator. In order to obtain an analytic solution of the storage system, the system is approximated by an equivalent circuit, i.e., the energy source is assumed to have a linear I-V characteristic, and a linear approximation is introduced for the diode, the FET, and so on. The exact analytic solution for the circuit is obtained by normalized parameters. The optimum duty ratio is obtained as a function of the parameters such as stored voltage and power source voltage.

1. Introduction

Various kinds of energy sources exist, but some of the sources have not been used, because they can generate only a small amount of electric energy. As a result, the energy generated from these sources is wasted. On the other hand, energy sources such as fossil fuel and nuclear power can generate larger amounts of electric energy, but they cause environmental problems such as air pollution, radioactive pollution, and global warming. Usage of the tiny electric energy mentioned above can be a solution to the problem. In addition, the tiny electric energy can contribute to the power supply for rural areas far from power plants.

In order to utilize the tiny electric energy, accumulation of the energy is necessary because it is unsteady for usage. Thus, a storage device and a boost converter are used to accumulate the energy [1]. A capacitor or a rechargeable battery is used commonly as the storage device. A switching regulator is used as the boost converter for small amounts of electric power to large amounts of electric power [2-4]. Since the electric energy generated is small, the system must be designed and controlled to maximize the charging efficiency of stored energy per input energy.

Though the numerical solution of the boost converter gives a precise design of an efficient system [5], the design is discrete and can be adapted to only limited situations. On the other hand, the analytic solution of the boost converter gives a tolerant design of the system which can be adapted to all situations. As for the control of the system, tracking of the power source voltage [6, 7] may give good efficiency, if the system is designed well. On the other hand, an analytic solution can be given to the design of an efficient control which can be adapted to all circuits. Thus, the analytic solution is significant for both the design and control of the system. It also gives a large range of view on the design of the system from a small scale to a large scale, meanwhile numerical simulation gives us a limited range of view.

The purpose of this paper is to obtain an analytic solution of the system. Firstly, we assumed a system consisting of a switching regulator as the boost converter and a capacitor as the storage device. Secondly, we modelled an equivalent circuit of the system, i.e., the energy source is assumed to have a linear I-V characteristic, and a linear approximation is introduced for the diode, the FET, and so on. Thirdly, we obtained the analytic solution of the system. Finally, we obtained the optimum control parameters expressed as a function of circuit parameters from the analytic solution, and discussed the adaptability of the model.

2. Boost switching regulator and its equivalent circuit

Figure 1 shows a typical circuit of the boost switching regulator. The circuit can superimpose the energy charged in the inductor by inputting pulse waveform to the FET. Thus, the circuit can boost the voltage. In order to model a linear equivalent circuit of the boost switching regulator, the resistance component of the inductance, diode, and FET is assumed as linear. As the simplest power source, the power source in Figure 1 is assumed to have I-V characteristics as shown in Figure 2. For the purpose of accumulating tiny energy, the capacitance of the capacitor must be large enough. Therefore, the capacitor can be assumed as a battery with a constant voltage. An equivalent circuit of the boost switching regulator is shown in Figure 3.

L, R_L, R_D, R_S, and R_P are the inductance, the resistance of the inductor, the forward resistance of the diode, the ON resistance of the FET, and the internal resistance of the power source, respectively. The circuit parameter is defined as a collective term of resistance and inductance. V_{in} is the open circuit voltage of the power source and V_{out} is the voltage of the capacitor. The voltage gain is defined as $\tilde{g} = V_{out}/V_{in}$. While the FET function as ON, the resistance of the FET is small enough compare to the resistance of the diode and the capacitor. Thus, the current flowed into the capacitor side is negligible.

When the switch is on 1, the current flows through the inductor and FET. The energy is stored in the inductor. When the switch is on 2, the energy is released for accumulation in the capacitor. T_1 and T_2 are the time interval when the switch is on 1 and 2, respectively. $T = T_1 + T_2$ is the switching period. The duty ratio of the switching regulator is defined as $\tilde{d} = T_2/T$. The control parameter is defined as a collective term of the switching period and the duty ratio. The state of current in one switching cycle depends on the number of switching. After a sufficient number of switching, the state of the current reaches a steady state in Figure 4. Applying the Kirchhoff's law of the voltage to Figure 3 gives

$$V_{in} = L\frac{di}{dt} + R_1 i \qquad \text{(switch on 1)} \quad (1)$$

$$V_{in} - V_{out} = L\frac{di}{dt} + R_2 i, \qquad \text{(switch on 2)} \quad (2)$$

where $R_1 = R_L + R_S + R_P$ and $R_2 = R_L + R_S + R_D$. Equation (1) and Equation (2) are solved for i under the condition of a continuous current. Subsequently, the current can be expressed as

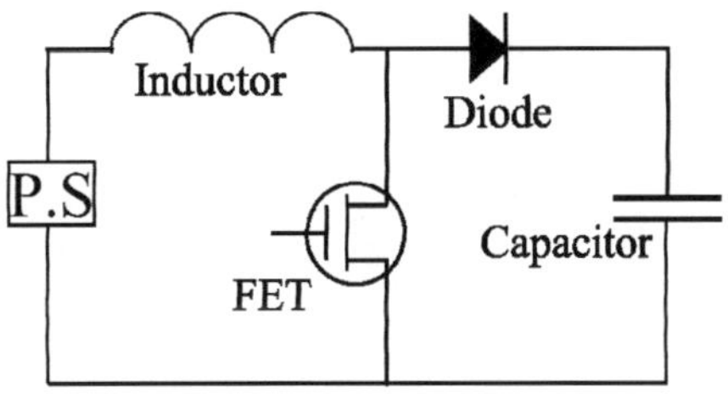

Figure 1. Boost switching regulator.

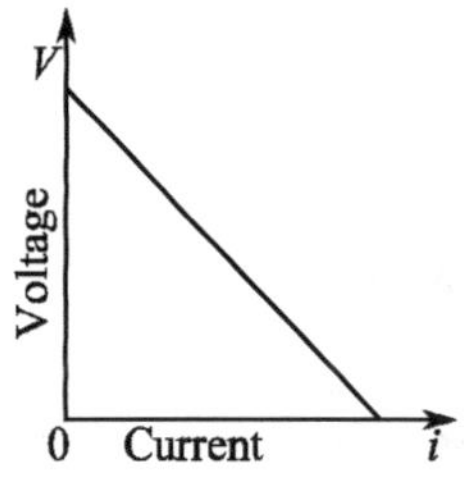

Figure 2. I-V characteristic of the power source.

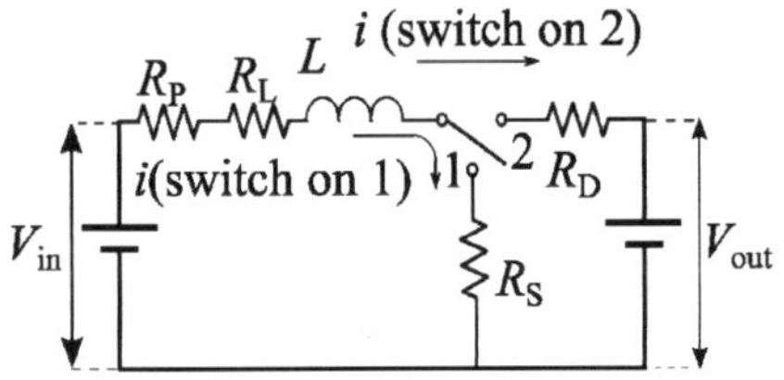

Figure 3. Equivalent circuit.

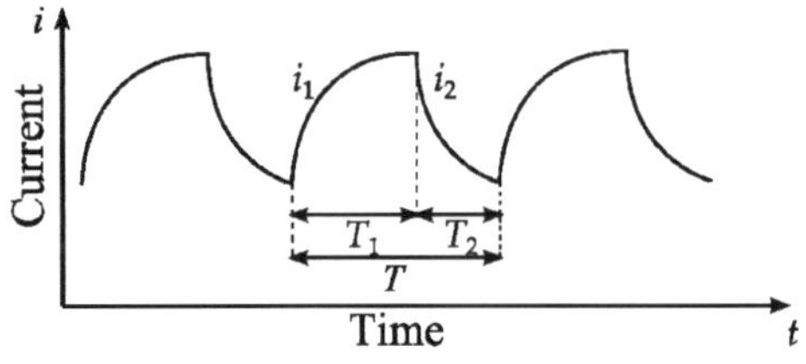

Figure 4. Current and switching period.

$$i=\begin{cases}\dfrac{V_{\rm in}}{R_1}\dfrac{\left(\dfrac{1-\tilde{g}}{\tilde{r}}-1\right)\left(1-e^{-\tilde{r}\tilde{d}\tilde{T}}\right)}{1-e^{-\left(1-\tilde{d}+\tilde{r}\tilde{d}\right)\tilde{T}}}e^{-\frac{R_1}{L}t}+\dfrac{V_{\rm in}}{R_1} & \text{(switch on 1)} \quad (3)\\[2ex] \dfrac{V_{\rm in}}{R_1}\dfrac{\left(\dfrac{1-\tilde{g}}{\tilde{r}}-1\right)\left(1-e^{-\left(1-\tilde{d}\right)\tilde{T}}\right)}{e^{-\tilde{r}\tilde{T}-\left(1-\tilde{d}\right)\tilde{T}}-e^{-\tilde{r}\left(1-\tilde{d}\right)\tilde{T}}}e^{-\frac{R_2}{L}t}+\dfrac{V_{\rm in}-V_{\rm out}}{R_2}, & \text{(switch on 2)} \quad (4)\end{cases}$$

where $\tilde{r}=R_2/R_1$, $\tilde{T}_1=R_1T_1/L$, $\tilde{T}_1=R_1T_2/L$, and $\tilde{T}=\tilde{T}_1+\tilde{T}_2$.

The open circuit voltage of the power source $V_{\rm in}$ and the circuit parameters are determined when the circuit is designed. The voltage of capacitor $V_{\rm out}$ increases with increasing charge in the capacitor during the operation. Therefore, the voltage gain is a given parameter. The optimum condition is defined as the condition which maximizes the electric charge stored in the capacitor per unit time. In order to obtain the optimum condition, the control parameters should be optimized according to the circuit parameters and the voltage gain.

3. Derivation of optimum condition

The charge accumulated in the capacitor in one switching cycle is

$$q=\int_{T_1}^{T} i\,dt\ . \quad (5)$$

The optimum control parameters can be obtained by maximizing

$$\frac{q}{T}=\frac{V_{\rm in}}{R_2}\left\{\frac{1}{\tilde{T}}\frac{\left(1-e^{-\left(1-\tilde{d}\right)\tilde{T}}\right)\left(1-e^{-\tilde{r}\tilde{d}\tilde{T}}\right)}{1-e^{-\left(1-\tilde{d}+\tilde{r}\tilde{d}\right)\tilde{T}}}\left(1+\frac{\tilde{g}-1}{\tilde{r}}\right)-\left(\tilde{g}-1\right)\tilde{d}\right\}\ . \quad (6)$$

The optimum duty ratio $\tilde{d}$ should be chosen so that

$$\frac{\partial\left(q/T\right)}{\partial\tilde{d}}=0 \quad (7)$$

is satisfied. Equation (7) leads to

$$\tilde{g}=1-\tilde{r}+\frac{\tilde{r}^2}{\tilde{r}-\dfrac{-e^{-(1-\tilde{d})\tilde{T}}+\tilde{r}e^{-\tilde{r}\tilde{d}\tilde{T}}+2\left(1-\tilde{r}\right)e^{-\left(1-\tilde{d}+\tilde{r}\tilde{d}\right)\tilde{T}}+\tilde{r}e^{-(2-2\tilde{d}+\tilde{r}\tilde{d})\tilde{T}}-e^{-(1-\tilde{d}+2\tilde{r}\tilde{d})\tilde{T}}}{\left(1-e^{-\left(1-\tilde{d}+\tilde{r}\tilde{d}\right)\tilde{T}}\right)^2}}\ . \quad (8)$$

Since $\partial^2\left(q/T\right)/\partial\tilde{d}^2$ is always negative, equation (8) is the optimum condition. As shown in equation (8), the optimum duty ratio can be obtained as a function of the switching period, the circuit parameters, and the gain. In order to monitor the relation between the gain and the optimum duty ratio, equation (8) is plotted in Figure 5 with given parameters of $\tilde{r}=1$ and $\tilde{T}=1, 5, 10$. We can plot Figure 5 independent of the specific value of each circuit parameter because the circuit parameter is normalized. The gain $\tilde{g}$ increases with increasing voltage of the capacitor. The optimum duty ratio is obtained according to the voltage of the capacitor.

Using the optimum duty ratio, equation (6) is a function of $\tilde{T}$ and $\tilde{r}$. In equation (4), $\tilde{T}$ appears in the term

$$f\left(\tilde{T},\tilde{r}\right)=\frac{1}{\tilde{T}}\frac{\left(1-e^{-\left(1-\tilde{d}\right)\tilde{T}}\right)\left(1-e^{-\tilde{r}\tilde{d}\tilde{T}}\right)}{1-e^{-\left(1-\tilde{d}+\tilde{r}\tilde{d}\right)\tilde{T}}}. \tag{9}$$

Figure 6 shows the relation between $\tilde{T}$ and $f\left(\tilde{T},\tilde{r}\right)$. In Figure 6, $f\left(\tilde{T},\tilde{r}\right)$ monotonically decreases with $\tilde{T}$. In order to maximize $f\left(\tilde{T},\tilde{r}\right)$, $\tilde{T}$ should be determined as small as possible. Minimizing the $\tilde{T}$ by increasing the frequency of the FET is not efficient because energy loss of the switching of the FET increases. Thus, $\tilde{T}$ should be minimized using circuit elements with small resistance and an inductor with large inductance. The optimum period $\tilde{T}$ should be determined when the circuit is designed in order to make an efficient circuit.

4. Discussion on optimum duty ratio

In the equivalent circuit, a reverse current of diode might exist because the diode is replaced with the resistor. When the reverse current flows, the assumption of the equivalent circuit breaks down. When $i>0$ is satisfied, the reverse current does not flow. Thus, the condition which the reverse current does not flow is obtained as

$$\tilde{g}<1+\frac{\tilde{r}e^{-\tilde{r}\tilde{d}\tilde{T}}\left(1-e^{-\left(1-\tilde{d}\right)\tilde{T}}\right)}{1-e^{-\tilde{r}\tilde{d}\tilde{T}}}. \tag{10}$$

The optimum duty ratio is needed to satisfy equation (10). Equation 8 shows the relation of the voltage gain and the optimum duty ratio. Equation 10 shows the applicable condition of the equivalent circuit. From the difference g' which is obtained by subtracting the right side of equation (8) from that of equation (10) is

$$g'=e^{-\left(1-\tilde{d}\right)\tilde{T}}\left(1-e^{-\tilde{r}\tilde{d}\tilde{T}}\right)^2+\tilde{r}e^{-\left(1-\tilde{d}+\tilde{r}\tilde{d}\right)\tilde{T}}\left(1-e^{-\tilde{r}\tilde{d}\tilde{T}}\right)\left(1-e^{-\left(1-\tilde{d}\right)\tilde{T}}\right). \tag{11}$$

Equation (11) is always positive. Thus, the equivalent circuit is valid as the model of the system when the optimum duty ratio is used.

From equation (8), a relation between the gain and optimum duty ratio is obtained. Switching the FET is not necessary for charging electricity when the voltage gain is smaller than $1/1+\tilde{r}$. When the voltage gain is larger than $1/1+\tilde{r}$, the switch is controlled with the optimum duty ratio according to the voltage gain. A value as small as possible for the normalized switching period $\tilde{T}$ is determined.

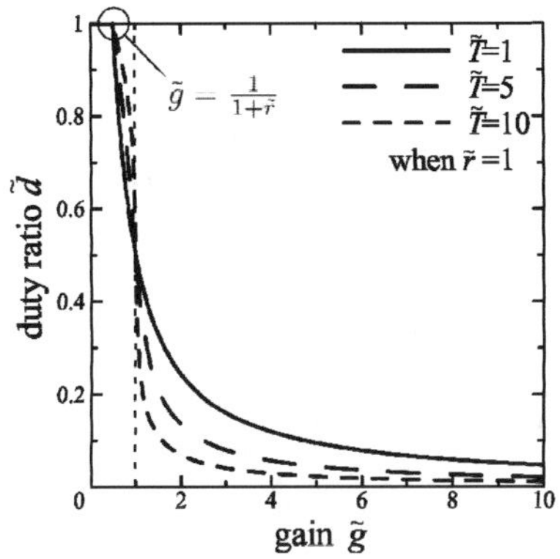

Figure 5. Relation between $\tilde{g}$ and $\tilde{d}$.

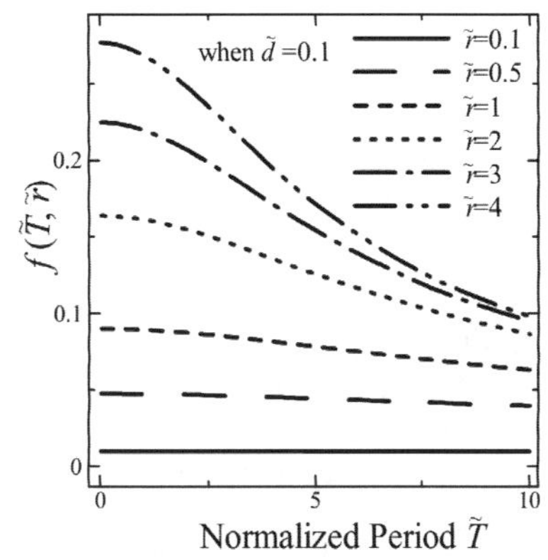

Figure 6. Relation between $\tilde{T}$ and $f\left(\tilde{T},\tilde{r}\right)$.

In Figure 5, the optimum duty ratio exists when the gain is smaller than 1. This result suggests that the usage of the switching regulator is more efficient than the direct connection between the power source and also the capacitor, in the case that the capacitor voltage is less than V_{in}. The area A in Figure 7 shows the charge stored into the capacitor during one cycle in the case of direct connection, i.e., the charge is obtained as

$$q_{\text{A}} = \frac{V_{\text{in}} - V_{\text{out}}}{R_2} T \ . \tag{12}$$

The area B in Figure 7 shows the stored energy during one cycle in the case of using the switching regulator, i.e., the charge is obtained as

$$q_{\text{B}} = \frac{V_{\text{in}}}{R_1} \frac{L}{R_2} \frac{\left(1 - e^{-\tilde{T}}\right)\left(1 - e^{-\tilde{r}\tilde{d}\tilde{T}}\right)}{1 - e^{-\left(1-\tilde{d}+\tilde{r}\tilde{d}\right)\tilde{T}}} \left(1 + \frac{\tilde{g} - 1}{\tilde{r}}\right) + \frac{V_{\text{in}} - V_{\text{out}}}{R_2} T_2 \ . \tag{13}$$

The term of $q_{\text{B}} - q_{\text{A}}$ is positive, if the duty ratio is controlled according to equation (8). Therefore, the condition of equation (8) is optimum when the gain is $1/1 + \tilde{r} < \tilde{g} < 1$.

5. Conclusion

In this paper, the optimum condition which maximizes the stored energy into the capacitor per unit time is obtained from the analytic solution of an equivalent circuit of the system. The optimum duty ratio is controlled by the voltage of the capacitor. An optimum switching period as small as possible should be determined in the design of the system. The equivalent circuit is valid as the model of the system when the optimum duty ratio is used. The optimum condition can be applied even if the gain is smaller than 1. In making the equivalent circuit, we assumed the diode as an ideal diode and the FET as a switch and ON resistance. For the purpose of obtaining the analytic solution, the non-linear element is neglected. Thus, in future studies, we should take into account the neglected factor such as the effect of non-linear elements when we design and control the actual system.

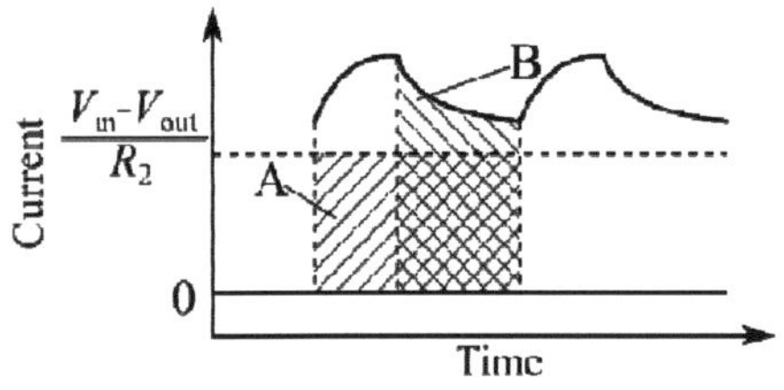

Figure 7. Current and stored energy.

References

[1] Wang W, Wang N and Hayes M 2011 MEIME Applied Mechanics and Materials **63-64** 978-982
[2] Cao X, Chiang W J and King Y C 2007 *IEEE Trans. Ind.* **22** 679-685
[3] Amorndechaphon D, Premrudeepreechacharn S and Higuchi K 2010 *TENCON IEEE Region 10 Conference* 1192-1197
[4] Kimball J W, Flowers T L and Chapman P L 2004 *IEEE POWER ELECTRONICS LETTERS* **2 NO. 3** 96-99
[5] Howe R M 2008 *ICSC* 1001-1006
[6] Tan Y K and Panda S K 2011 *IEEE TRANSACTIONS ON POWER ELECTRONICS* **26 NO.1**, 38-50
[7] Wang W, Wang N and Hayes M 2011 *Applied Mechanics and Materials* **63-64** 978-982

Effect of PVP on the low temperature bonding process using polyol prepared Ag nanoparticle paste for electronic packaging application

Jianfeng Yan[1], Guisheng Zou[1], Aiping Wu[1], Jialie Ren[1], Jiuchun Yan[2], Anming Hu[3], Lei Liu[3], Y.Norman Zhou[1,3]

[1.] Department of Mechanical Engineering, Tsinghua University, Beijing 100084, China
[2.] State Key Laboratory of Advanced Welding & Joining, Harbin Institute of Technology, Harbin 150001, China
[3.] Department of Mechanical and Mechatronics Engineering, University of Waterloo, 200 University Avenue West, Waterloo, ON, N2L 3G1, Canada

E-mail: yanjf09@mails.tsinghua.edu.cn

Abstract. There has been an increasing interest in developing low temperature interconnection process using metal nanoparticles. In this study the Ag nanoparticles (NPs) for this low temperature bonding process applications were prepared based on the polyol method using polyvinylpyrrolidone (PVP) as the protecting agent. The effect of PVP on the Ag nanoparticle size, nanoparticle solution viscosity and the bondability of the Ag nanoparticle paste were studied. The silver nanoparticle with diameter of 90 nm, 40 nm and 20 nm were synthesized by adjusting PVP concentration. The bonding processes using different kind of Ag nanoparticle paste were conducted. The joint with shear strength of 50 MPa were formed using Ag nanoparticle (NP) paste prepared with appropriate PVP concentration.

1. Introduction

Soldering technology based on tin/lead (Sn/Pb) solders has been playing an important role in electronic packaging. However, Pb and its compounds result in serious health threat to humans when deposing the Pb containing electronic devices in the environment. Emerging regulations have targeted the elimination of Pb usage in electronic assemblies in many countries [1, 2]. Particularly, the high temperature solders are still widely used due to no suitable alternative. Therefore, there is a strong drive to find Pb-free alternatives especially for high temperature applications.

As Pb-free electronic interconnection materials, the low temperature bonding technology using metal nanoparticles (Ag, Cu, etc) have attracted people's attention in the recent years [3-12]. As the size of metal nanoparticles is reduced, their surface energy increases, which lead to some special properties such as increase of diffusion coefficient and decrease of sintering temperature [13-16]. Due to the size effect the metal nanoparticles can be used as the low temperature bonding materials. After the bonding is completed, the nanoparticles are changed into microscaled or bulk materials. Consequently, the apparent melting point of bonding materials increases to that of the bulk materials.

Therefore, the metal nanoparticle paste, which can be sintered at low temperature, has the potential to be a lead-free, high-performance interconnect material[17]. Based on this principle, the bonding technique using metal nanoparticles (Ag, Cu, etc) has been studied by several research groups [3-12]. Usually the metal nanoparticle paste was prepared by mixing the organic components to the commercial nanoparticles in order to prevent the self-cohesion. However, the added organic components will bring detrimental effect to the sintering and the bonding process. Recently, our group has developed a new method to prepare metal nanoparticle paste, which contains low organic component, based on the polyol method. This new Ag nanoparticle paste was prepared through the polyol synthesis of Ag naoparticles and the subsequent concentration without a further addition of organic component. Thus the organic content in this new nanoparticle paste is significantly lower than that in the conventional Ag nanoparticle paste [7, 9-12]. During our Ag nanoparticle preparation, polyvinylpyrrolidone (PVP) was used as the protecting agent to prevent the self-cohesion of the chemical reduction created Ag nanoparticles. In this study, the effect of the added polyvinylpyrrolidone (PVP) concentration on the Ag nanoparticle size, the Ag solution viscosity and the bondability of Ag nanoparticle paste were studied.

2. Experimental procedure

Different kinds of silver nanoparticle solutions were synthesized by the modified polyol process in this study[18]. First, silver nitrite was dissolved into ethylene glycol (EG) to form a solution with an $AgNO_3$ concentration of 0.15 M. On the other hand, the EG solution of polyvinylpyrrolidone (PVP-K30, with viscosity about 2.5mPa.s) with a given PVP concentration such as 0.045 M (mol/L), 0.45 M, 1.35 M, were prepared. After the two kinds of mentioned solutions were heated to the chemical-reaction temperature of 160 ^{o}C, at a constant heating rate of 20 ^{o}C/min. The PVP solution was injected into the silver nitrite solution at a constant rate. The color of the solution changed from the colorless to gray, indicating the precipitation of Ag NPs. When the reaction was completed, the solution was rapidly cooled by adding chilled deionized (DI) water. Three kinds of Ag NP pastes (A, B and C) were prepared by directly condensing the chemically reduced Ag NP solutions using different PVP concentration (0.045 M, 0.45 M, 1.35 M) without a further addition of organic component.

The bondability of Ag nanoparticle pastes (A, B and C) prepared with different PVP concentrations were investigated by bonding the silver plated Cu discs, which simulate the metal-coated semiconductor chip and substrate respectively, in air. The shape of Cu specimens used for the bonding experiment is show in Figure 1. The Ag nanoparticle pastes were applied on the faying surfaces of Cu discs. Then the samples were dried at 80 ^{o}C on the hotplate to further remove the low boiling point components in the paste. Afterwards, put the smaller Cu disc on the bigger Cu disc stage. These prepared bonding specimens were heated up to the bonding temperature under assistant pressure of 5 MPa for 30 min in air atmosphere. The bonding strength was evaluated as shear strength measured using the Thermal-Mechanical Simulator Gleeble 1500D with a displacement speed of 5 mm /min under room temperature. Three samples were tested for the same bonding parameters.

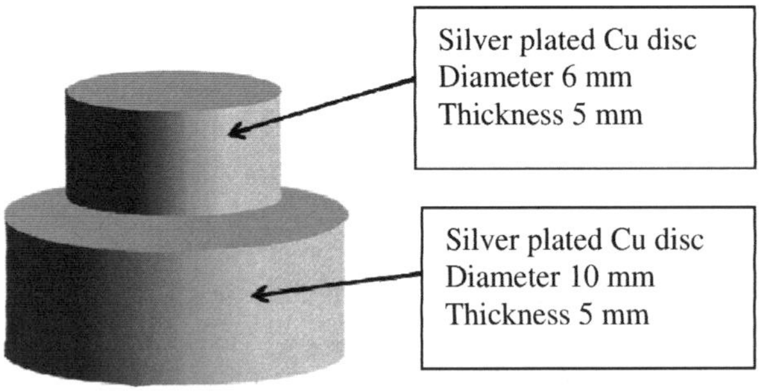

Figure 1. Schematic illustration of the bonding specimens.

3. Results and discussion

3.1. Effect of PVP concentration on the characteristics of Ag nanoparticle paste and the joint strength

The PVP concentration had a significant influence on the particle size distribution of the synthesized Ag NPs. The added PVP has a function to prevent the chemically generated Ag nanoparticles from aggregation. Figure 2 shows SEM images of Ag nanoparticles obtained when using different PVP concentration. As the PVP concentration was low (0.045 M), the PVP was not sufficient to completely cover the surface of Ag NPs. The initially generated Ag nanoparticles tended to merge into larger particles. Therefore, the generated Ag nanoparticles had a larger diameter size of about 90 nm. With increasing PVP concentration (0. 45 M), more of the generated Ag nanoparticles were covered by the PVP opportunely. The diameter of Ag nanoparticles is of about 40 nm. When the PVP concentration was increased to (1.35 M), the corresponding Ag nanoparticles had a smaller diameter of 20 nm.

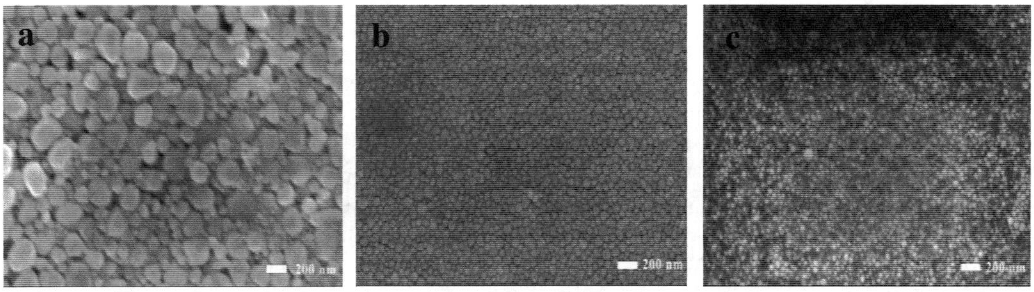

Figure 2. SEM images of Ag nanoparticles synthesized with different PVP concentration (a) 0.045 M, (b) 0.45 M, (c) 1.35M.

The viscosities of the Ag nanoparticle solutions before the condensing, which were detected using a Rheometer, increased with PVP concentration increasing. Ag nanoparticle solution synthesized with the PVP concentration of 0.045 M has a viscosity of 4.2 cP (mPa.s). As the concentration of PVP increased to 0.45 M, the corresponding viscosity of Ag nanoparticle solution slightly increased to 4.4 cP. When further increasing PVP concentration to 1.35 M, the viscosity increased to 6.0 cP. It seems that the viscosity of the solvent is strongly depended on PVP at high concentrations. The viscosity of solution is higher, when the PVP concentration increases. Since the PVP concentration will affect the viscosity of the Ag nanoparticle solutions, it is reasonable to deduce that the PVP concentration has a significant effect on the viscosity of the concentrated Ag nanoparticle paste.

The bond characteristics for Ag nanoparticle pastes (A, B and C) were evaluated at bonding pressure of 5 MPa in air for 30 min. Three bonded samples were made for each level. The PVP has a great effect on the bondability of the Ag nanoparticle paste. As shown in Table 1, the shear strengths of the joints using Ag nanoparticle pastes (A, B and C) are 26 MPa, 50 MPa and 5.0 MPa, respectively. Based on the above analysis, it can be seen that the PVP concentration affects the generated Ag nanoparticle sizes, solution viscosities and the bondability of the Ag nanoparticle paste, as shown in Table 1.

Table 1. Properties of Ag nanoparticle pastes prepared with different PVP concentrations.

Ag nanoparticle paste	PVP concentration (M)	Size diameter (nm)	Viscosity (cP)	Average shear Strength (MPa)
A	0.045	90	4.2	26
B	0.45	40	4.4	50
C	1.35	20	6.0	5.0

3.2. Microstructures of the joints using different Ag nanoparticle paste

Figure 3 displays the fracture appearance photographs of copper disc joints using Ag nanoparticle pastes (A, B and C). Figures 4 display the corresponding SEM images of the fracture appearances of the joints using Ag nanoparticle pastes (A, B and C). For the joint using Ag nanoparticle paste A (Figure 4a), the fracture appearance is coarse and many pit-cracks in the sintered Ag layers are observed. It is attributed to the low viscosity of the Ag nanoparticle paste prepared with low PVP concentration. Due to the existence of cracking in the sintered Ag layer, the average joint strength using Ag nanaparticle paste A is about 26 MPa. As shown in Figure 3b and Figure 4b, smooth and dense fracture surfaces were observed when using the Ag nanoparticle paste B and C as the bonding materials. The dimple structures are obvious which indicate the good joints were formed under these conditions. The PVP in the Ag nanoparticle paste has a function to adjust viscosity and to prevent the cracking of the sintered Ag layer. The joint strength using Ag nanoparticle paste B is significantly increased to 50 MPa since the cracking of the Ag paste was effectively prevented. This demonstrated that the Ag nanoparticle paste prepared with the PVP concentration of 0.45 M has a better bondability. The Figure 3c and Figure 4c display the smooth and dense fracture surfaces of the joints using Ag nanoparticle paste C. Although obvious fracture traces are observed in Figure 4c, the joints strength using Ag nanoparticle paste is as low as only 5 MPa. The results of qualitative analyses by EDX of three areas of the joints using different Ag nanoparticle pastes are given in Figure 5. As displayed in Figure 5a and Figure 5b, only pure Ag elements were detected, confirming that the bonding layer was consisted of only silver and no organic materials remained in those areas. However, in Figure 5c, a large carbon peak and a smaller oxygen peak were detected in addition to the silver peaks. This result indicates that organic materials remained in this area. The organic materials bring negative effects on the joint formation, because they prevent surface diffusion, lattice diffusion from the surface and grain boundary diffusion of nanoparticles during the sintering process, leading to low strength bonding between the sintered layer and Ag coating. In this case, the corresponding joint strength decreased significantly. This result demonstrated that, besides preventing the aggregation of the generated Ag nanoparticles, the PVP has the additional function of preventing cracking of the dried Ag paste during bonding process. When the PVP concentration is low, the cracking of the Ag nanoparticle paste is likely to happen. As the PVP concentration in the Ag nanoparticle paste is high, the remained organic components may reduce the joint quality. The optimization of PVP concentration is critical to the low temperature bonding process quality using polyol synthesized nanoparticle Ag paste.

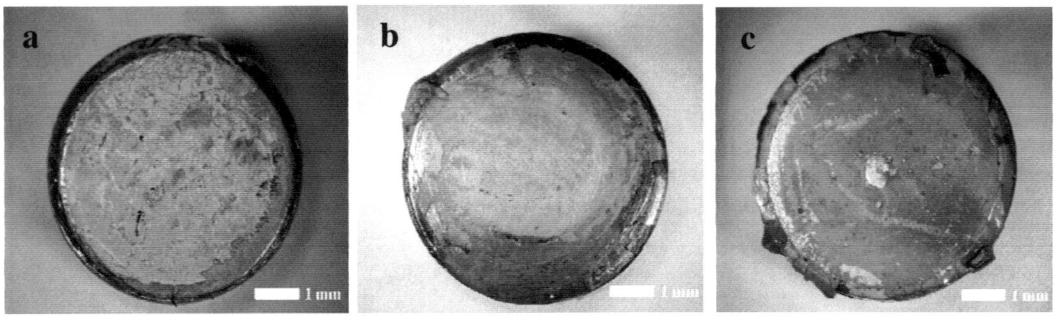

Figure 3. Photographs of the fractured joints bonded using different Ag nanoparticle paste (a) Ag nanoparticle paste A; (b) Ag nanoparticle paste B; (c) Ag nanoparticle paste C.

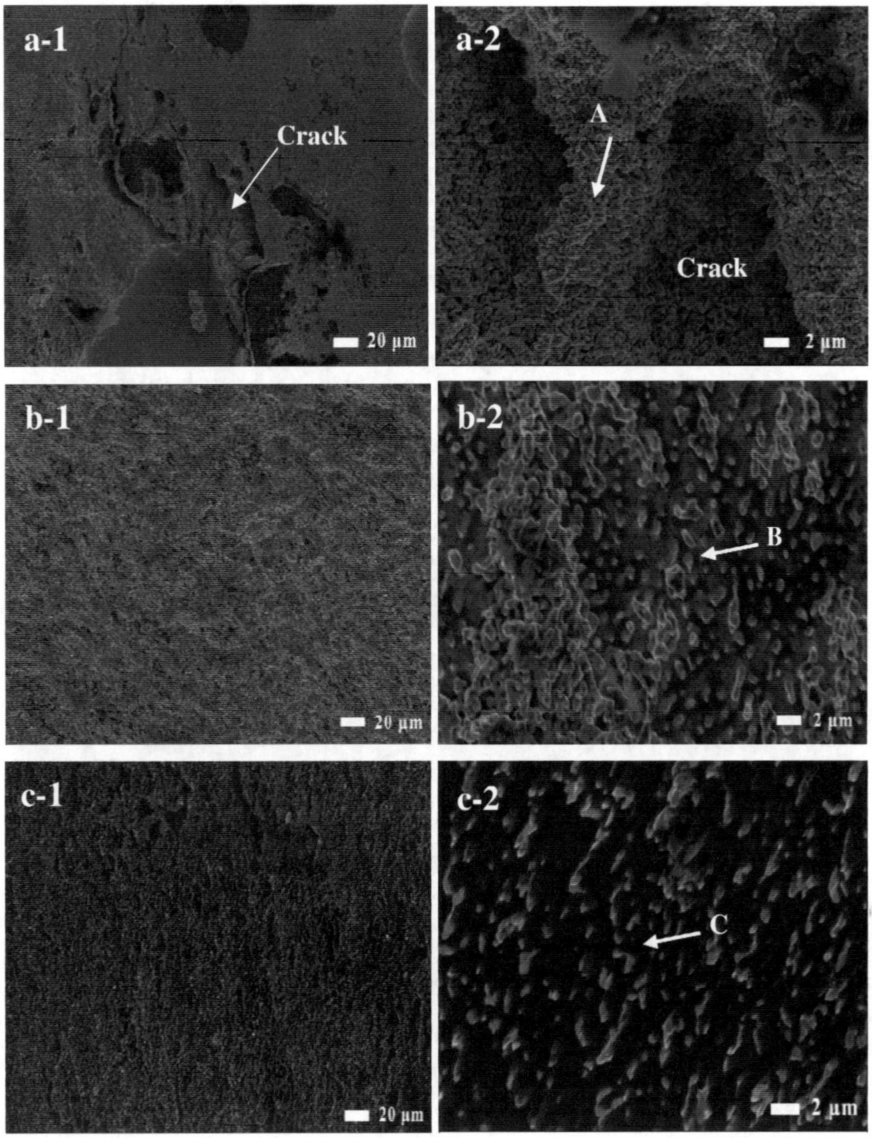

Figure 4. SEM images of microstructures of the joints using different Ag nanoparticle paste (a) Ag nanoparticle paste A, (b) Ag nanoparticle paste B, (c) Ag nanoparticle paste C.

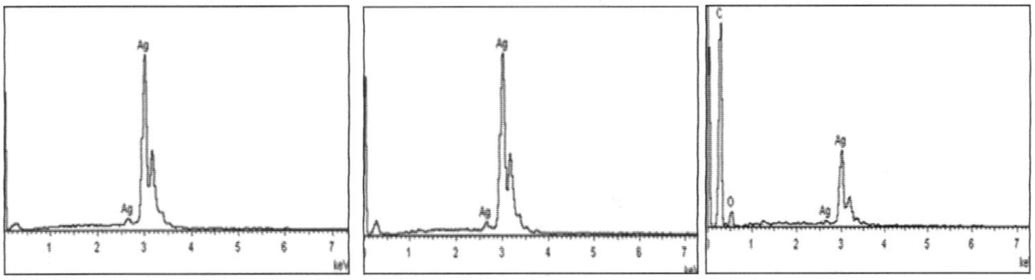

Figure 5. EDX spectra of different areas in Figure .4 (a) area A, (b) area B, (c) area C.

4. Conclusions

The low temperature bonding process using Ag nanoparticle pastes prepared through polyol method with different PVP concentration was studied. The PVP concentration has a significant influence on the size and solution viscosity of Ag nanoparticles and the bondability of Ag nanoparticle paste. The low PVP concentration (0.045M) would lead to cracking of the sintered Ag layer. As the PVP concentration in the Ag nanoparticle paste is much high, the remained organic components would reduce the joint quality. The strong joint with shear strength of 50 MPa was formed using Ag NP paste prepared with appropriate PVP concentration under the sintering conditions of 250 °C bonding for 30 min at 5 MPa.

Acknowledgements

This research was supported by National Natural Science Foundation of China (Grant No. 51075232), by Tsinghua University Initiative Scientific Research Program (Grant No.2010THZ 02-1) and by State Key Lab of Advanced Welding & Joining, Harbin Institute of Technology (Grant No. AWPT-Z12-04).

References

[1] Li Y, Moon K, Wong C P 2005 Electronics without lead. *Science.* **308** 1419-20
[2] Abtew M, Selvaduray G 2000 Lead-free Solders in Microelectronics. *Mat. Sci. Eng.R.* **27** 95-141
[3] Yan J, Zou G, Hu A, Zhou Y 2011 Preparation of PVP coated Cu NPs and its applications for low-temperature bonding. *J Mater Chem.* 21 15981-6
[4] Alarifi H, Hu A, Yavuz M, Zhou Y 2011 Silver nanoparticle paste for low-temperature bonding of copper. *J Electron Mater.* **40** 1394-402
[5] Lei T G, Calata J N, Lu G Q, Chen X, Luo Shufang 2010 Low-temperature sintering of nanoscale silver paste for attaching large-area (>100mm^2) chips. *IEEE T Compon Pack T.* **33** 98-104
[6] Hu A, Guo J Y, Alarifi H, Patane G, Zhou Y, Compagnini G, Xu C X. 2010 Low temperature sintering of Ag nanoparticles for flexible electronics packaging. *Appl Phys Lett.* **97** 153117
[7] Ide E, Angata S, Hirose A, Kobayashi K F 2005 Metal-metal bonding process using Ag metallo-organic nanoparticles. *Acta Mater.* **53** 2385-93
[8] Bai J G, Calata J N, Lu G Q 2007 Processing and characterization of nanosilver pastes for die-attaching SiC devices. *IEEE. T. Electron Pack.* 30 241-5
[9] Maruyama M, Matsubayashi R, Iwakuro H, Isoda S, Komatsu T 2008 Silver nanosintering: a lead-free alternative to soldering. *Appl Phys A.* **93** 467-70
[10] Morita T, Ide E, Yasuda Y, Hirose A, Kobayashi K 2008 Study of bonding technology using silver nanoparticles. *JPN J Appl Phys.***47** 6615–22.
[11] Bai J G, Lu G Q 2006 Thermomechanical reliability of low-temperature sintered silver die attached SiC power device assembly. *IEEE. T. Device. Mat. Re.* **6** 436-41
[12] Akada Y, Tatsumi H, Yamaguchi T, Hirose A, Morita T, Ide E 2008 Interfacial bonding mechanism using silver metallo-organic nanoparticles to bulk metals and observation of sintering behavior. *Mater.Trans.* **49** 1537-45
[13] Jiang Q, Zhang S H, Li J C 2004 Grain size-dependent diffusion activation energy in nanomaterials. *Solid State Commun.* **130** 581-4
[14] Buffat P, Borel J P 1976 Size effect on the melting temperature of gold particles. *Phys Rev A.* **13** 2287-98
[15] Moon K-S, Dong H, Maric R, Pothukuchi S, Hunt A, Li Y, Wong C P 2005 Thermal behavior of silver nanoparticles for low-temperature interconnect applications. *J Electron Mater.* **34** 168-75
[16] Zhang M, Efremov M Y, Schiettekatte F, Olson E A, Kwan A T, Lai S L, Wisleder T, Greene J E, Allen L H 2000 Size-dependent melting point depression of nanostructures: Nanocalorimetric measurements. *Phys Rev B.* **62** 10548-57
[17] Bai J G, Zhang Z Z, Calata J N, Lu G Q 2006 Low-Temperature Sintered Nanoscale Silver as a Novel Semiconductor Device-Metallized Substrate Interconnect Material. *IEEE. T. Electron Pack.* **29** 589-93
[18] Sun Y, Xia Y 2002 Shape-Controlled Synthesis of Gold and Silver Nanoparticles. *Science.* **298** 2176-9

Study on erosion resistance characteristics of Fe-MWCNT composite plating with respect to lead-free solder

J Watanabe[1, 3], N Sekimori[2], K Hatsuzawa[1], T Uetani[2] and I Shohji[3]

[1] Packaging Development Team, Module Business Promotion Department, Nagano OKI Electric Co., Ltd., 965 Mimitori, Komoro, Nagano 384-0084, Japan

[2] Research Section, Research and Development Division, Hakko Corporation, 4-5 Shiokusa 2-chome, Naniwa, Osaka 556-0024, Japan

[3] Graduate School of Engineering, Gunma University, 1-5-1 Tenjin-cho, Kiryu, Gunma 376-8515, Japan

E-mail: watanabe626@oki.com

Abstract. Sn-3mass%Ag-0.5mass%Cu lead-free solder causes strong erosion of Fe plating, which is used in hand soldering. In order to improve the durability of the soldering iron tip, we tried Fe-MWCNT (Multi-walled Carbon Nanotube) composite plating on the tip. Microstructures and the erosion resistance of the Fe-MWCNT composite plating under heating were examined. On the basis of the SEM/EDS analysis, it was confirmed that there were MWCNTs in the Fe-MWCNT composite plating. Even when heated to 673 K, MWCNTs were found to be present in the plating film under a stable condition. It was shown that the performance of erosion resistance for lead-free solders under high temperature (673 K) is improved due to the MWCNT addition to the Fe plating.

1. Introduction

In recent years, the use of chemical substances for electronic materials devices has been reconsidered. Since 2006, the EU has enforced the RoHS directive, which is a set of strict regulations for using harmful chemical substances in electronic materials. The use of lead has been prohibited by RoHS. Therefore, we have to use lead-free solder when we mount electronic parts on printed circuit boards, and Sn-3mass%Ag-0.5mass%Cu lead-free solder (SAC305) is currently being used. One of the problems caused by using SAC305 is the erosion of the soldering machine. In the case of soldering with SAC305, a solder bath for flow soldering was damaged in a short amount of time. It is assumed that SAC305 has a high content of Sn and a high melting point of about 30 K compared with Sn-37mass%Pb (Sn37Pb), which has been used thus far. The erosion mechanism of the flow soldering machine by molten SAC305 solder was elucidated by previous research [1, 2]. In order to prevent erosion of the flow soldering machine, nitriding treatment is performed on the surface of the materials of the flow soldering machine. On the other hand, although a similar erosion phenomenon to that affecting the flow soldering machine has been also found in the hand soldering iron tip, a practical solution has not been suggested (Figures 1 and 2).

Different from the materials of the flow soldering machine, the hand soldering iron tip requires moderate wettability of SAC305 solder. Then, Fe plating is usually conducted on the surface of the tip. However, the Fe-plated soldering iron tip erodes faster by SAC305 solder. This is caused by increase in the Sn content, which reacts with Fe more than Sn37Pb solder. Moreover, the reaction between the Sn and Fe is also related to the temperature of the hand soldering iron tip: the hotter the tip, the faster the reaction. The temperature is a contributing factor to the reduced life of the hand soldering iron tip. In the case of the automatic soldering robot, the soldering iron tip was eroded by SAC305 solder more than twice as fast as when Sn37Pb was used. In addition to the difference in solder type, the hardness of solder is also affected by accumulative damage to the tip, because the hardness of SAC305 solder is higher than that of the Sn37Pb solder. Since a hard SAC305 solder wire touches the same point of the tip repeatedly in the soldering robot, physical damage by wear occurs easily.

Reducing erosion of the soldering iron tip should be done with a view toward reducing waste and being more friendly to the environment. To prevent erosion of the soldering iron tip, we focused on using multi-walled carbon nanotube (MWCNT) composite plating [3-5], which has high strength and high thermal conductivity. We coated the soldering iron tip with easily available MWCNTs, and tested the erosion resistance. It was reported that the Ni-MWCNT composite plating decreased the maximum erosion depth by approximately 50% compared to standard Ni plating [6]. In this study, we focus on MWCNT composite plating for the purpose of reducing erosion of the tip caused by SAC305. The microstructure, wettability, and erosion resistances properties of the MWCNT composite plating were evaluated.

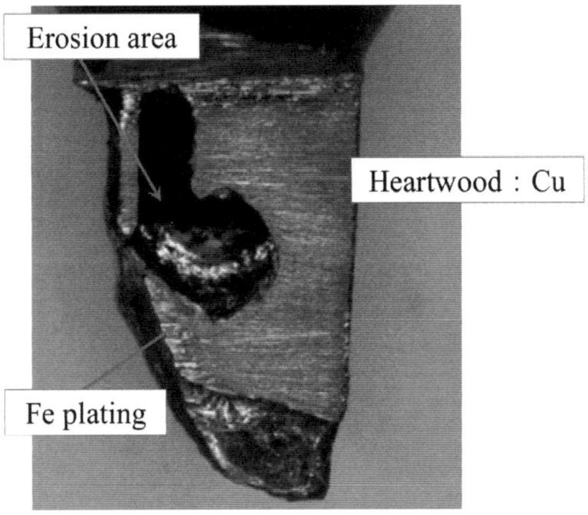

Figure 1. Erosion of soldering iron tip.

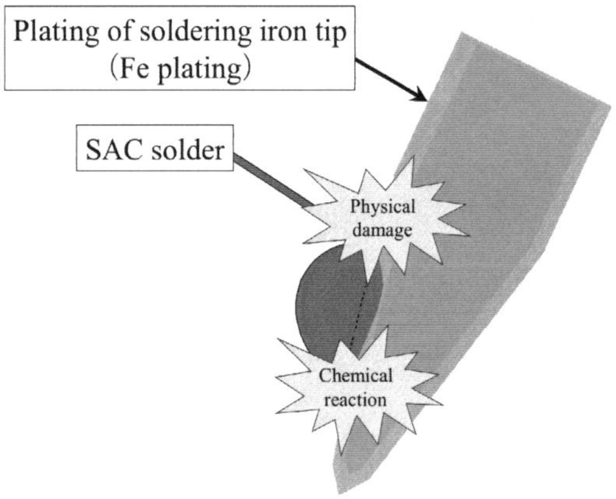

Figure 2. Erosion mechanism of soldering iron tip.

2. Experimental

2.1. Preparation of Fe-MWCNT composite plating

Table 1 shows plating conditions of the Fe and Fe-MWCNT composite plating. A sample preparation was made by performing electrolytic Fe plating on the Cu surfaces. We used a commercial MWCNT called VGCF-H manufactured by Showa Denko Co. Ltd. An $FeSO_4$ solution that was adjusted to about pH 3 including MWCNT was used for the plating bath. Current density was set to 1.0 A/dm^2. When we adjusted the plating solution, sodium lauryl sulphate was used as a dispersant and to achieve an anti-pit effect. It was stirred continuously to disperse the MWCNT in the plating solution.

Table 1. Plating conditions.

	Characteristic plating Wettability test	Erosion test
Base metal	Cu	Cu
Form of base	Flat plate	Column
Plating metal	Fe (Electrolytic)	Fe (Electrolytic)
Composite material	MWCNT (VGCF-H)	MWCNT (VGCF-H)
Addition agent	Sodium lauryl sulphate	Sodium lauryl sulphate
Plating thickness	100 μm	100 μm
Current density	1.0 A/dm^2	1.0 A/dm^2
Amount of MWCNT in plating solution	1 g/L	0.5, 3 g/L

2.2. Characteristics of plating sample

The prepared Fe-MWCNT composite plating was evaluated using a scanning electron microscope (SEM), energy dispersive X-ray spectroscopy analysis (EDS), and electron back scattering diffraction pattern (EBSP) to investigate the microstructure of the composite plating and the effects of heating on the plating. SEM and EDS analyses were performed with acceleration voltages of 1 kV and 5 kV. EBSP analysis was performed using an acceleration voltage of 15 kV.

2.3. Solder wettability test of SAC305

The solder wettability test was carried out at temperatures of 523, 573, and 623 K, comparing SAC305 resin flux core solder wire with Fe-MWCNT composite plating to see if wettability changed or not, when composite MWCNT was used. The test method is shown in Figure 3. The prepared Fe plating and Fe-MWCNT composite plating samples included 1 g/L of MWCNT in the plating solution. Wettability properties were measured according to JIS-Z 3198-3. The surfaces of the samples were cleaned using 2-propanole and were dried well. 0.3 g of solder was placed on the plating samples and heated to the test temperature. After the solder melted, the temperature was held for 30 s. Thereafter, the plating samples were cooled to room temperature in order to allow crystallization, and the height of solder measured. The wetting rate was calculated from equations (1) and (2).

$$S_R = \frac{D-H}{D} \times 100 \tag{1}$$

$$D = 1.2 \times V^{\frac{1}{3}}, \tag{2}$$

where S_R is the wetting rate, D is the diameter of the theoretical solder sphere, and H is height of the measured solder. V is the solder volume used in the test. In this test, V was 39.7 mm^3, as calculated by measuring the solder height after heating. The samples were heated on a hot plate.

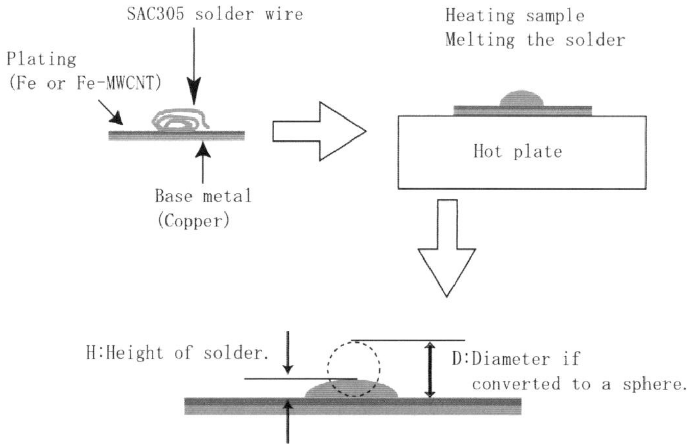

Figure 3. Wettability test method.

2.4. Erosion test

To evaluate the erosion characteristic of the Fe and Fe-MWCNT composite plating, the erosion test shown in Figure 4 [7] was conducted using SAC305 at 673 K assuming the actual temperature of the soldering iron tip. Fe and Fe-MWCNT composite plating were coated on a Cu column (φ 5.4 mm) to a thickness of 100 μm. To investigate the effects of MWCNT composite, two different amounts of MWCNT were used in the plating solution 0.5 and 3 g/L. The soldering iron tip temperature was held at 673 K. The flux core solder wire (φ 1.0 mm) and the soldering iron tip came into contact horizontally every 3 seconds for a total of 3000 times. The total test time was 9000 seconds. Each time the solder wire comes into contact with the soldering iron tip, 5 mm of lead wire melts before the solder wire moves back, breaking the contact. After this test was completed, a cross sectional cut was made through the soldering iron tip, and the maximum erosion depth was measured. Chrome plating was performed on the surface, and the plating was used as a marker to measure the maximum erosion depth. Fe plating was performed as a reference for standard conditions.

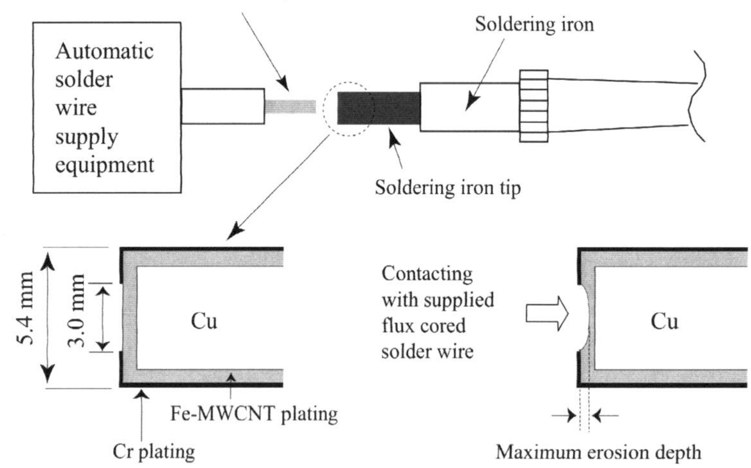

Figure 4. Erosion test method [7].

3. Results and discussion

3.1. Microstructure analysis

Figure 5 shows the appearance of plating samples before and after heating. Before heating, the surface of the Fe plating was glossy, and the surface of the Fe-MWCNT composite plating was rough. After heating to 673 K, there was not much change in the Fe plating, but the color of the Fe-MWCNT composite plating was changed to deep blue. The Fe-MWCNT composite plating has an additional oxidation mechanism compared with the Fe plating. Figure 6 shows SEM images of the surfaces of the Fe and Fe-MWCNT composite plating. The appearance of needle-like MWCNT (White) in the Fe plating is obvious in the composite plating.

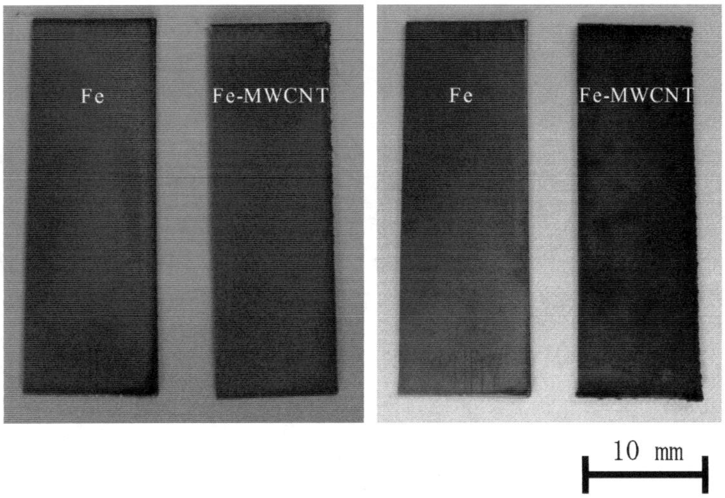

Figure 5. Fe plating and Fe-MWCNT composite plating.
Left: Non-heating. Right: After heating at 673 K for 1 h.

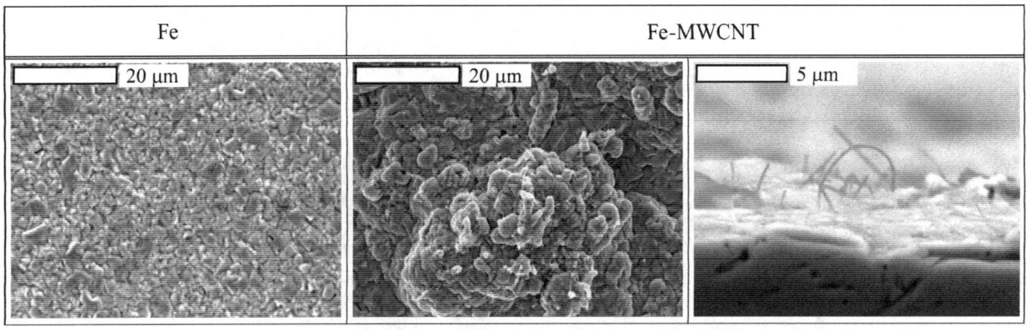

Figure 6. SEM images of plating sample surface.

Figure 7 shows a cross-sectional SEM image of the Fe-MWCNT composite plating after heating at 673 K for 1 h. The white area is Fe, and the black spots indicate MWCNT in the figure. The formation of a composite with MWCNT was confirmed not only on the plating surface but also in the plating film on the basis of the cross-sectional analysis. The results of EDS mapping analysis of the composite plating after heating at 673 K for 1 h are shown in Figure 8. The MWCNTs in the Fe plating did not change their size even after heating. The results showed that most of the MWCNTs remained in the composite plating even after it was heated to 673 K. Figure 9 shows the EBSP analysis results for the composite plating. Note that a trace amount of Fe_3C were precipitated at the grain boundaries of Fe. The MWCNT has a stable chemical structure, and therefore Fe_3C particles formed at grain boundaries due to heating. It has been reported that Fe_3C particles suppress the diffusion of Fe into SAC305 solder [8]. Therefore, we can improve the erosion resistance of the composite plating by the effective dispersion of Fe_3C particles.

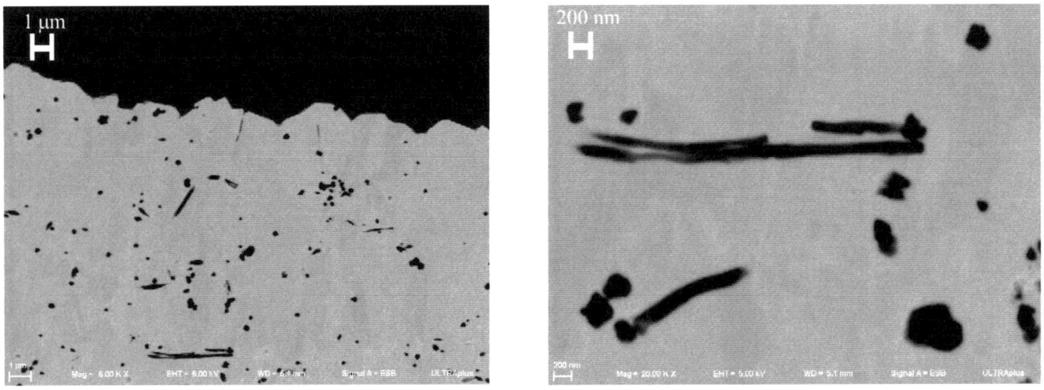

Figure 7. Cross-sectional SEM images of Fe-MWCNT composite plating after heating at 673 K for 1 h.

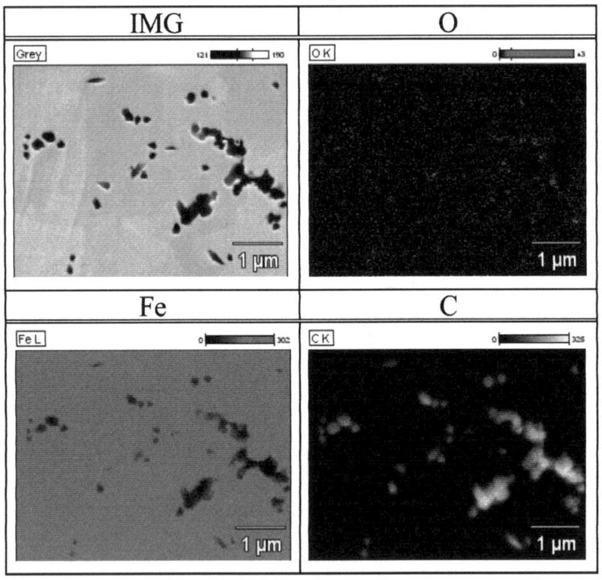

Figure 8. EDS mapping of Fe-MWCNT composite plating after heating at 673 K for 1 h.

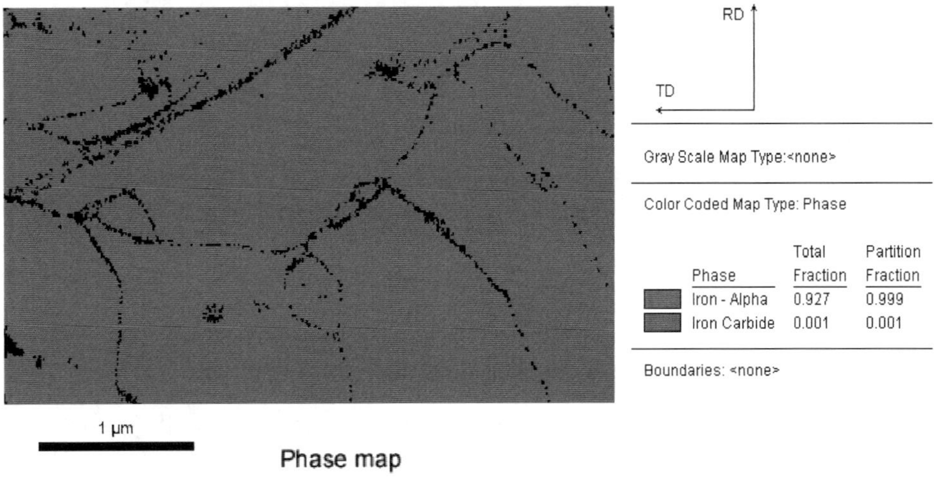

Figure 9. EBSP mapping of Fe-MWCNT composite plating after heating at 673 K for 1 h.

3.2. Solder wettability test of SAC305

Figure 10 shows the results of the solder wettability test. It was found that Fe plating has good wetting properties at all temperatures. Fe-MWCNT composite plating showed negative trends in the temperature range investigated compared to Fe plating. This means MWCNT affects the wetting properties of the solder spread. Figure 11 shows the results of the cross-sectional analysis using SEM/EDS after the wettability test. It was found that Fe-Sn compounds formed at the interface between solder and plating in both platings. The Fe-MWCNT composite plating had a worse wetting rate than the Fe plating. The worst result was at 523 K; however; SAC305 was not repelled and it has 50% wetting rate, which means that the Fe-MWCNT composite plating does not repel SAC305 solder and solder wets the surface of the plating. Moreover, Fe-Sn compounds formed at the wetting interface in the same manner as the Fe plating. Thus, Fe-MWCNT composite plating is expected to be applicable to the soldering iron tip.

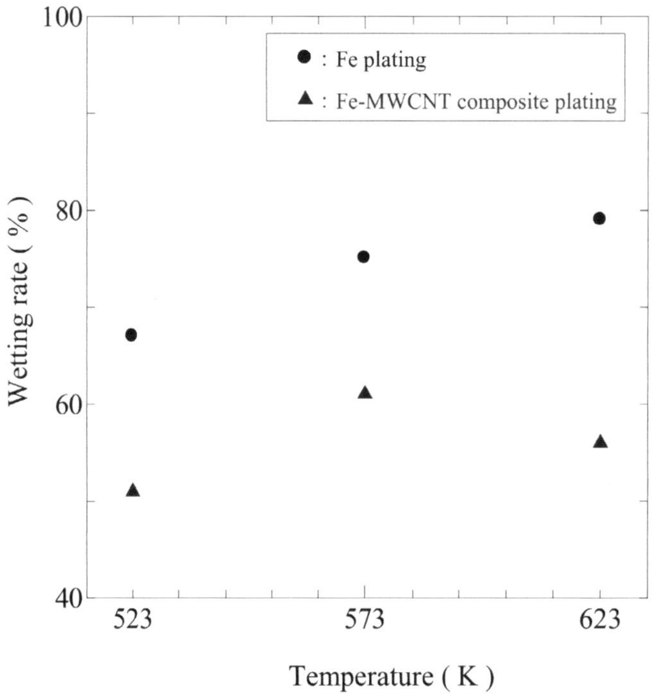

Figure 10. Temperature dependence on wetting rate for Fe plating and Fe-MWCNT composite plating.

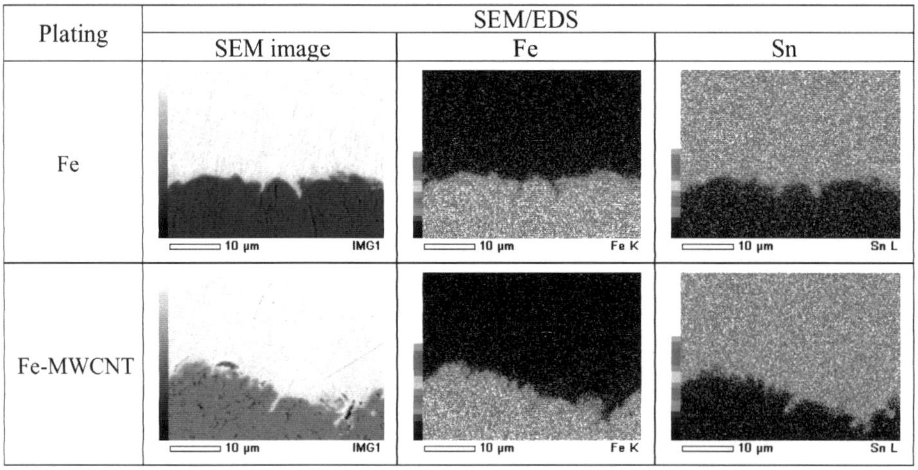

Figure 11. Results of SEM/EDS mapping analysis after the wettability test.

3.3. Erosion test of SAC305

Erosion test results are shown in Figure 12. The maximum erosion depth decreased approximately 23% in the Fe-MWCNT composite plating compared to the standard Fe plating. The effect of the MWCNT concentration in the plating solution on the maximum erosion depth is negligible. It seems that the amount of MWCNT in the plating solution does not affect the amount of MWCNT taken into the plating. It is assumed that the improvement of erosion resistance is due to the mitigation of physical damage by the action of MWCNT as a solid lubricant [9], because MWCNT has strong heat resistance and excellent abrasion characteristics. In addition, traces of Fe_3C that formed at the grain boundaries could suppress the diffusion of Fe into the SAC305. The inhibition effect would be one of the factors that improve erosion resistance.

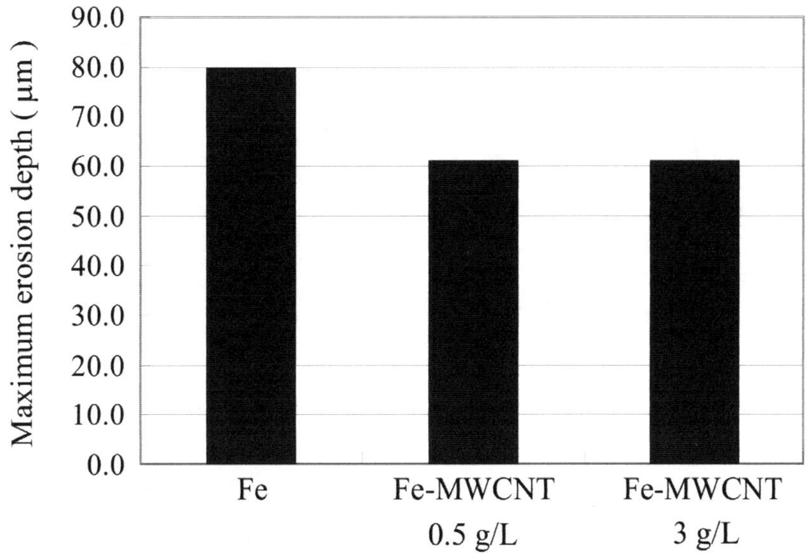

Figure 12. Results of an erosion test at 673 K.

4. Conclusions

The Fe-MWCNT composite plating was performed using an electrolytic sulphate plating bath. MWCNTs form a composite not only on the Fe plating surface but also in the plating film. In the Fe-MWCNT composite plating, MWCNTs were stable in Fe plating under heating conditions at 673 K. It is expected that MWCNT has excellent high-temperature wear properties when MWCNT contacts molten solder. A trace amount of Fe_3C was observed at the Fe grain boundaries in the composite plating by EBSP measurement. Fe_3C has the effect of inhibiting the diffusion of Fe into SAC305. This is expected to improve erosion resistance by increasing the number of Fe_3C particles. Although the Fe-MWCNT composite plating decreases the wettability of SAC305, SAC305 is not repelled on the surface of the composite plating and it has 50% wetting rate. Thus, SAC305 can be applied to the soldering iron tip. It was shown that the Fe-MWCNT composite plating was effective in reducing erosion of the plating layer by SAC305 compared to the standard Fe plating. This could be due to the reduction in physical damage because of the excellent high-temperature wear characteristics of MWCNT and the prevention of the diffusion of Fe into the SAC305 by the generation of Fe_3C.

References

[1] Serizawa K, Takemoto T and Yamamoto K 2009 *Proc. 15th Symposium on Microjoining and Assembly Technology in Electronics* **15** 379-382
[2] Takemoto T, Nishikawa H, Serizawa K and Yamamoto K 2009 *Proc. 15th Symposium on Microjoining and Assembly Technology in Electronics* **15** 383-386
[3] Oberlin A, Endo M and Koyama T 1976 *J. Cryst. Growth* **32** 335
[4] Iijima S 1991 *Nature* **354** 56-58
[5] Iijima S and Ichihashi T 1993 *Nature* **363** 603-605
[6] Miyazaki M, Watanabe J, Hatsuzawa K and Arai S 2010 *Proc. 16th Symposium on Microjoining and Assembly Technology in Electronics* **16** 235-238
[7] Takemoto T, Takemoto M, Matsunawa A, Uetani T and Yamazaki M 2002 *Proc. 8th Symposium on Microjoining and Assembly Technology in Electronics* **8** 179-182
[8] Kawamoto T, Yamauchi A, Kawakubo S, Irisawa A, Kurokawa K and Tanaka J, 2011 *Proc. Japan Institute of Electronics Packaging* **25** 165-166
[9] Arai S, Fujimori A, Murai M and Endo M 2008 *Materials Letters* **62** 3545-3548

Bonding of copper to silicon chips using vapor-deposited tin film

T Fujimoto[1], S Fukumoto[1], T Miyazaki[1], Y Kashiba[2], K Shiotani[1] and K Fujimoto[1]

[1]Graduate School of Engineering, Osaka University, 2-1 Yamadaoka, Suita, Osaka 565-0871, Japan
[2]Mitsubishi Electric Corporation, 8-1-1 Tsukaguchihonmachi, Amagasaki, Hyogo 661-8661, Japan

E-mail: takashi@mapse.eng.osaka-u.ac.jp

Abstract. A bonding process between a Cu plate and a Si chip using a vapor-deposited tin film has been developed. The microstructures of the bond interface and the reliability during thermal cyclic tests were investigated. Intermetallic compounds (IMCs) of Cu3Sn and/or Au-Cu-Sn were formed at the bond interface. The stress concentration initiated fracture in the IMC layer at the corners of the bond area and the initial cracks in the IMC layer propagated through the Al layer in the metallized layer on the Si chip. Joints using a deposited tin film showed much longer lifetime than solder joints.

1. Introduction

Nowadays, power devices are widely used as key technologies for energy saving. It has been expected that Al lead wires on Si chips will be replaced by Cu to deal with high current, which makes power devices operate at high temperatures of nearly 473 K [1–3]. The Cu lead is usually soldered to Si chips. However, it is difficult to secure the reliability of the joint when a solder is used at near its melting point. In order to obtain a reliable high-temperature joint, it could be favorable to form an intermetallic compound (IMC) layer at the bond interface. Since the electrical resistivity of an IMC such as Cu_6Sn_5 or Cu_3Sn is smaller than that of Pb-5Sn solder, which is a typical high-temperature solder [4, 5], joints whose interlayer consists of an IMC could also improve the energy efficiency of the power electronic system. The solid-liquid interdiffusion (SLID) process has been applied for mechanical and electrical interconnects [6–8].

The SLID bonding process is illustrated in Figure 1. The process can be divided into four stages: (i) A low melting point metal is inserted between the base metals. (ii) Heat treatment under appropriate pressure is carried out, and the insert metal becomes liquid. (iii) Solid-liquid interdiffusion occurs between the base metals and the liquid phase. (iv) Further interdiffusion results in complete vanishing of the liquid phase to leave an IMC layer at the bond interface. By using the SLID process, high melting point compounds are formed. For example, in the case of using Sn films to bond Cu, the bonding temperature is approximately 573 K, which is higher than the melting temperature of Sn (505 K). The melting temperature of the obtained bond layer of Cu_6Sn_5 (688 K) or Cu_3Sn (949 K) was higher than that of Sn [9, 10].

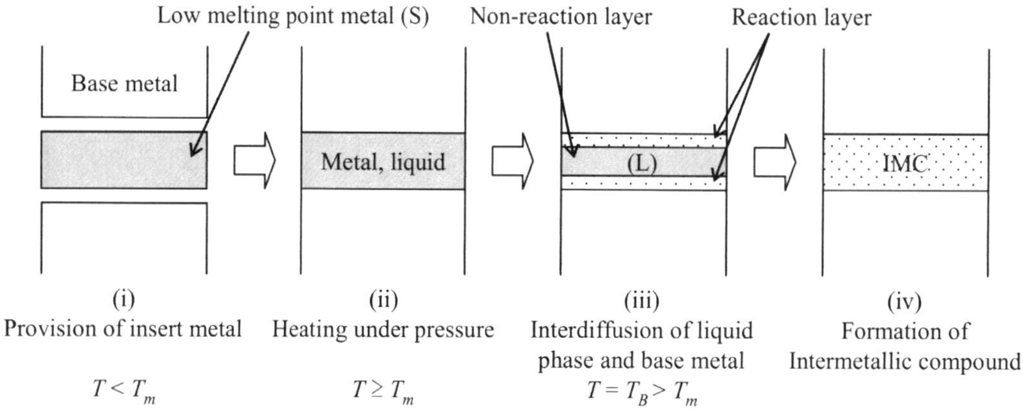

Figure 1. Schematic showing the flow of the SLID process. (T_m = melting temperature, T_B = bonding temperature)

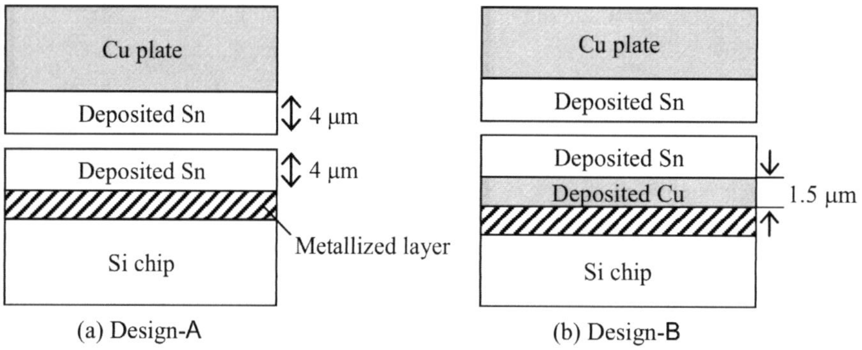

Figure 2. Designs of faying surfaces. The metallized layer on Si chip was composed of Al/Ni/Au.

Generally, an IMC is known to be hard and have a small coefficient of thermal expansion. In the case of an IMC formed between a Cu lead and Si chip, high thermal stress due to the different coefficients of thermal expansion of the materials will be induced between the Si chip and Cu. However, there are few studies of the reliability of such joints. The aim of this study is to investigate the microstructures of the bond interface and the reliability during thermal cyclic tests.

2. Experimental procedures

The materials used were oxygen-free Cu plates (4×4×0.2 mm^3) and Si chips (14.8×7.7×0.255 mm^3). The Cu plates were polished with emery paper of #4000. They were then pickled in a 5 % HCl solution, and rinsed in ethanol. One of the surfaces of the Si chip had a metallization layer. The layer consisted of Al/Ni/Au in order from the surface of Si. Figure 2 shows the designs of the faying surfaces. The faying surface of the Cu plate was coated with Sn by vapor deposition. The thickness of Sn film was 4 μm. Either Sn alone or Sn and Cu were coated on the metallization layer on the Si chip. The Cu plates were bonded to the Si chips in a nitrogen atmosphere at 573 K under a pressure of 10 MPa. The bonding times were varied from 300 to 3000 s. Some bare Cu plates were soldered to the Si chips with Sn-5Sb to obtain the reference data. The microstructures at the bond interface were observed by scanning electron microscopy (SEM) with energy dispersive X-ray spectroscopy (EDX). X-ray diffraction (XRD) analysis was carried out to identify the IMC layer. The hardness of each layer

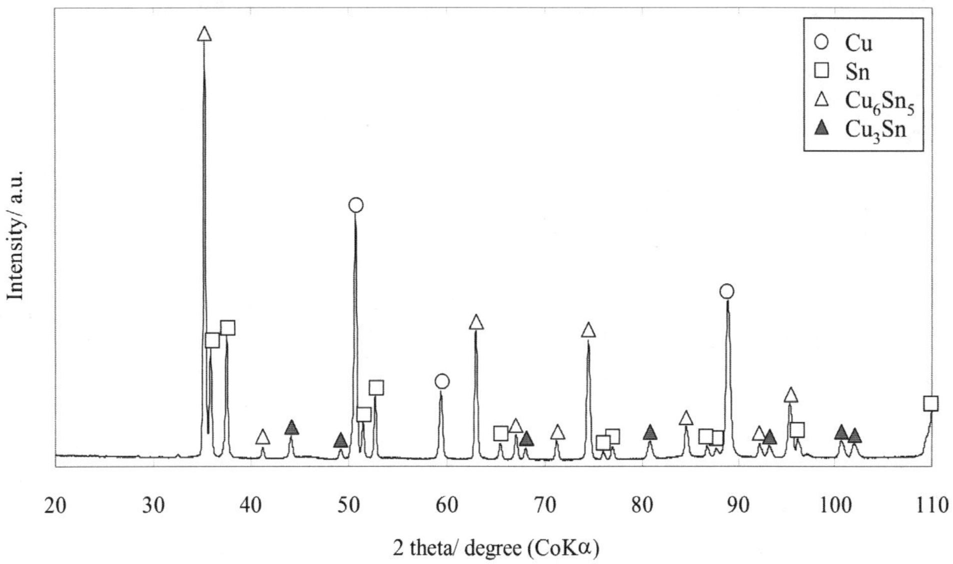

Figure 3. Micro-focus XRD pattern on the faying surface of Cu plate with a deposited Sn layer. The sample was heated to 573 K.

in the bond interface was measured by nanoindentation tests (Nano Test nano-indentation system, Micro Materials Ltd).
The loading/unloading rates, peak load, and holding time were 0.25 mN/s, 5 mN, and 60 s, respectively. The reliability of the joints was evaluated by thermal cyclic tests. The lower and higher temperatures were 318 K and 448 K, respectively. The bond interfaces were observed by scanning acoustic tomography (SAT).

3. Results and discussion

3.1. Microstructures of bond interface
It has been reported that Sn immediately reacts with Cu to form a Cu_6Sn_5 compound on deposition of a Sn-Cu composite [11]. The copper plate with the Sn film was heat treated at 573 K for 0 s to confirm the existence of Sn. Figure 3 shows the XRD pattern taken from the heat-treated Cu plate with the deposited Sn layer. In the case of heating to 573 K, Cu, Sn, Cu_6Sn_5, and Cu_3Sn phases were identified. Therefore, it is considered that liquid Sn remains at the bond interface at the beginning of bonding. Figure 4 shows typical cross-sectional SEM images of bond interfaces in the case of using the design-A faying interface. In this case, the liquid Sn reacted with the Cu plate and with the Au layer in the metallized layer on the Si chip, and two interlayers were formed. By EDX and XRD analysis, Cu_3Sn and Au-Cu-Sn IMC were detected. The Cu_3Sn was thinner than the Au-Cu-Sn IMC for a bond time of 300 s. It is believed that the Au-Cu-Sn IMC layer includes phases such as Cu_6Sn_5 and $AuSn_2$. The boundary between the Cu_3Sn layer and Au-Cu-Sn IMC layer is ragged. With increasing bonding time, the Cu_3Sn phase increased while the Au-Cu-Sn intermetallic phase decreased. Since the initial Sn layer had already been consumed at a bonding time of 300 s, Cu diffused through the Cu_3Sn layer from the Cu plate and reacted with the Cu_6Sn_5 phase in the Au-Cu-Sn IMC layer to grow Cu_3Sn.

Figure 5 shows typical cross-sectional SEM images of bond interfaces in the case of using the design-B. In this case, an intermediate layer of Cu_3Sn was formed between the Cu plate and the deposited Cu layer, and there was no Au-Cu-Sn IMC found as was the case when using design-A. Since the deposited Sn layer lay next to the deposited Cu layer, the molten Sn probably reacted with

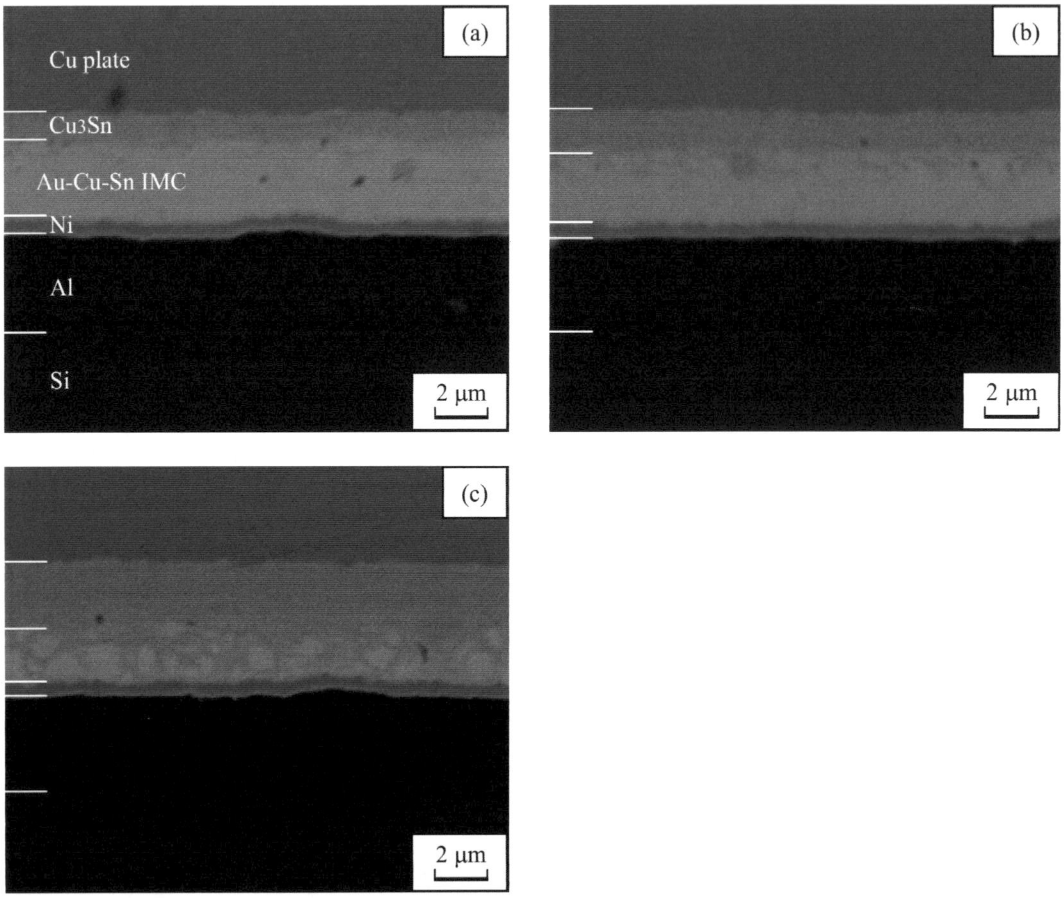

Figure 4. Cross-sectional SEM images of bond interface using design-A. Bonding time was: (a) 300 s, (b) 1200 s, and (c) 3000 s.

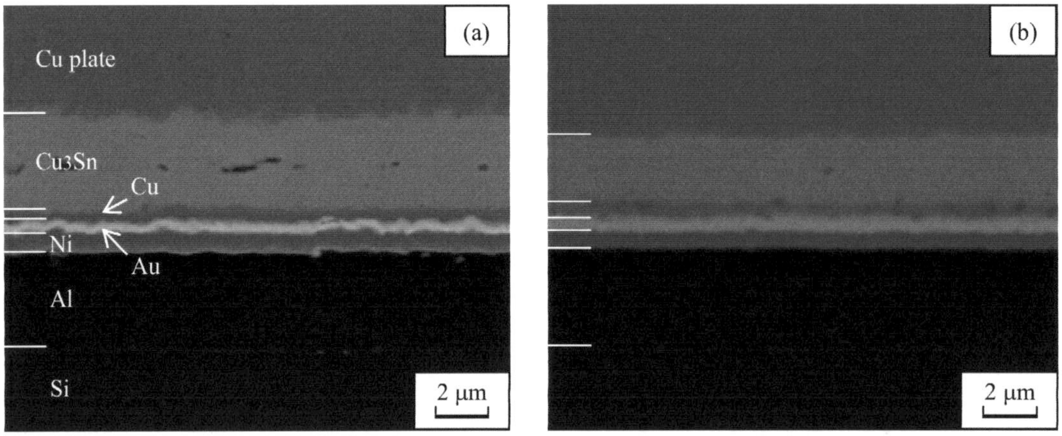

Figure 5. Cross-sectional SEM images of bond interface using design-B. Bonding time was: (a) 300 s, (b) 3000 s.

Table 1. Hardness of each layer in bond interface.

Layer	Hardness (GPa)
Cu	1.94
Cu_3Sn	6.54
Au-Cu-Sn IMC	8.71
Al	1.56
Si	14.3

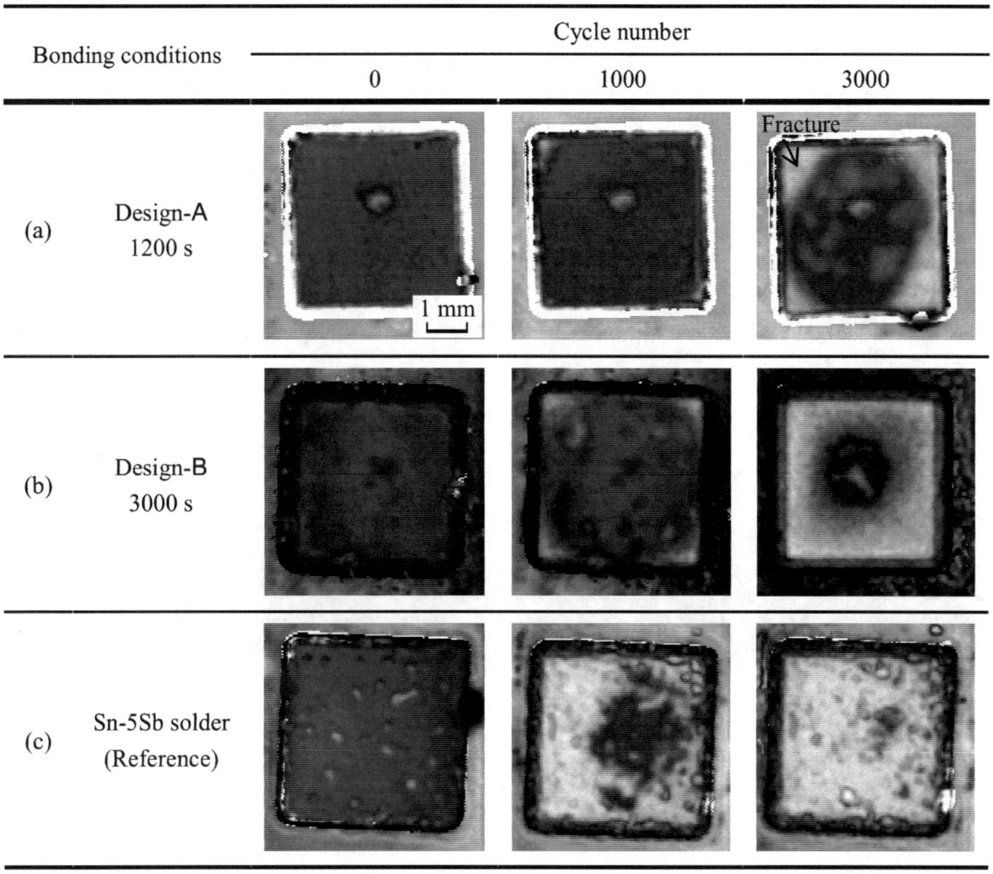

Figure 6. SAT images at the bond interface between Cu plate and Si chip.

that layer first and foremost but very little with the Au layer. Even if the bonding time increased, there was no appreciable change of phase in the bond layer.

The hardness of each layer in the bond interface is shown in Table 1. The hardness was the highest for the Au-Cu-Sn IMC layer and the lowest for the Al layer. It is generally known that Cu_3Sn and Cu_6Sn_5 show similar hardnesses [12, 13]. As mentioned above, the Au-Cu-Sn IMC layer includes Cu_6Sn_5 and other unidentified phases. Therefore, the unidentified phases are probably harder than Cu_6Sn_5.

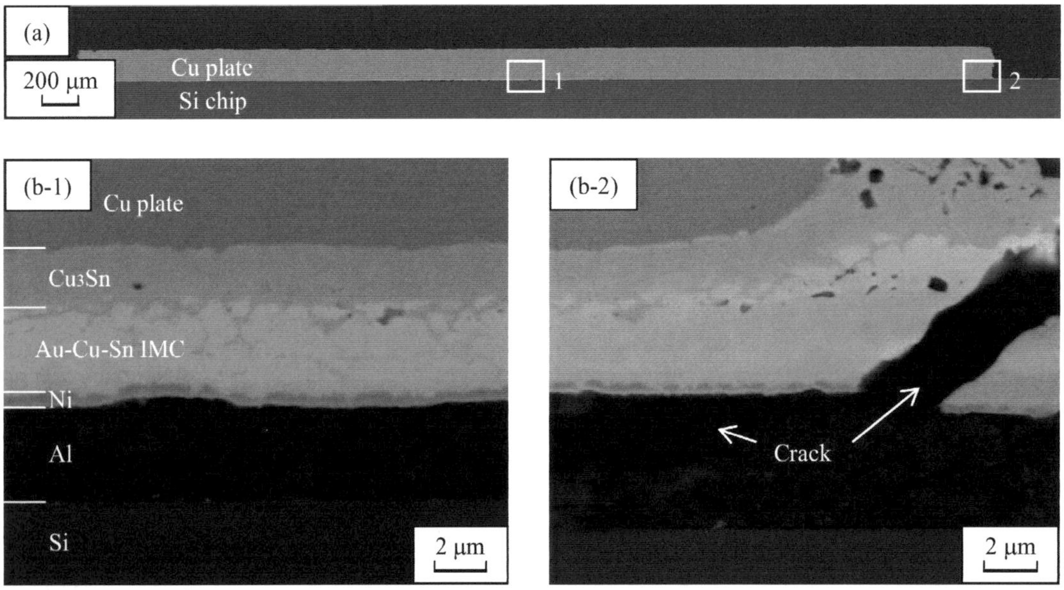

(b) Highlighted areas in (a).

Figure 7. Crack propagation path at the bond interface between Cu plate and Si chip by temperature cyclic test. Cycle number is 3000. Bonding was carried out with design-A faying surface for 3000 s.

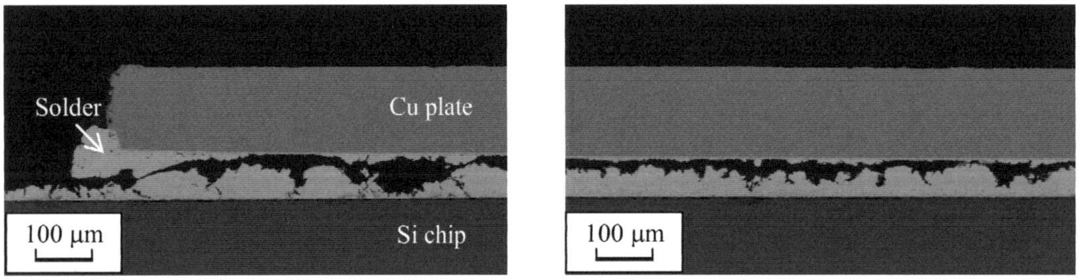

(a) At the corner (b) At the center

Figure 8. Crack propagation path at the bond interface between Cu plate and Si chip by temperature cyclic test. Cycle number is 3000. Bonding was carried out with Sn-5Sb solder paste for 300 s.

3.2. Reliability during thermal cyclic tests

The SAT images at the bond interface between Cu plate and Si chip before (0 cycle) and after the thermal cyclic test (1000, 3000 cycles) are shown in Figure 6. The samples in Figure 6(a) and 6(b) correspond to those in Figure 4(b) and 5(b), respectively. The bright region presents the fracture part at the bond area. Initiation of cracks occurred at the corners of the bond area where stress is concentrated for all samples. In the case of the solder joints, almost the whole interface fractured even after 1000 cycles (Figure 6(c)). On the other hand, the joints whose interlayer consists of an IMC showed longer lifetimes than the solder joints (Figures 6(a) and 6(b)), but there was no obvious differences in reliability between using the design-A and using the design-B.

Figures 7 and 8 show the crack propagation path at the bond interface between the Cu plate and Si chip during temperature cyclic tests. In the case of joints whose interlayer consists of Cu_3Sn and Au-Cu-Sn IMC, the crack passed through not only through the IMC layer but also the Al layer. The cracks in the IMC layer were observed only at the corner of the bond area and the Al layer was almost fractured completely. Therefore, initiation of cracks occurred in the IMC layer and the cracks propagated through the Al layer. Similar fractures were found in the case of joints with design-B whose interlayer consists of Cu_3Sn and Cu. On the other hand, in the case of solder joints, an overall fracture occurred in the solder layer. As mentioned in the last section, the bond layer consists of a hard IMC layer and a soft metallized layer. Hence, it is considered that initiation of cracks first occurred at the corner of the hard and brittle IMC layer where stress is concentrated, and the cracks propagated through the soft Al layer.

4. Conclusions

A bonding process of a Cu plate and Si chip using a vapor-deposited tin film has been developed. Cu_3Sn and/or Au-Cu-Sn IMC were formed at the bond interface. The Au-Cu-Sn IMC layer consists of several phases. Inserting a Cu layer between the vapor-deposited Sn layer and the Au layer of the Si chip prevents formation of Au-Cu-Sn IMC. In the thermal cyclic test, the joints using the deposited tin film showed much longer lifetime than solder joints. In the case of joints whose interlayer consists of IMCs, cracks occurred in the IMC layer of the joint and propagated through the Al layer in the metallized layer of the Si chip.

References

[1] Nakajima D, Sudo S, Tada K, Shikano T, Sasaki T and Kashiba Y 2005 *11th Symposium on "Microjoining and Assembly Technology in Electronics"* **11** 433
[2] Hiramitsu S and Ohta H. 2009 *15th Symposium on "Microjoining and Assembly Technology in Electronics"* **15** 105
[3] Kashiba Y 2011 *Journal of the Japan Welding Society* **80** 285
[4] Çadirli E, Böyük U, Kaya H, Marasli N, Aksöz S and Ocak Y 2010 *J. Electron. Mater.* **40** 195
[5] Fields R J, Low III S R and Lucey, Jr. G K 1991 *Proc. TMS Symposium* (Cincinnati) 165
[6] Tanaka A, Miyazaki T, Nishioka S, Matsushima M, Fukumoto S, and Fujimoto K 2010 *16th Symposium on "Microjoining and Assembly Technology in Electronics"* **16** 165
[7] Guth K, Siepe D, Görlich J, Torwesten H, Roth R, Hille F, and Umbach F 2010 *Proc. Power Conversion, Intelligent Motion (PCIM)* (Nuremberg) 232
[8] Klumpp A, Ramm P, and Wieland R 2010 *Proc. Design, Automation & Test in Europe Conference & Exhibition (DATE)* (Dresden) 1678
[9] Bader S, Gust W, and Hieber H 1995 *Acta metall. mater.* **35** 329
[10] Li J F, Agyakwa P A, and Johnson C M 2011 *Acta Materialia* **59** 1198
[11] Lee C C and Chen Y C 1996 *Thin Solid Film* **286** 213
[12] Chromik R R, Vinci R P, Allen S L, and Notis M R 2003 *J. Mater. Res.* **18** 2251
[13] Yang P F, Lai Y S, Jian S R, Chen J, and Chen R S 2008 *Materials Science and Engineering A* **485** 305

Microstructural development at weld interface between Zr-based glassy alloy and stainless steel by resistance microwelding

S Fukumoto[1], M Minami[1], A Soeda[1], M Matsushima[1], M Takahashi[2], Y Yokoyama[3] and K Fujimoto[1]

[1] Graduate School of Engineering, Osaka University, 2-1 Yamadaoka, Suita, Osaka 565-0871, Japan.

[2] Joining and Welding Research Institute, Osaka University, Ibaraki, Osaka, Japan

[3] Institute for Materials Research, Tohoku University, Sendai, Miyagi, Japan

E-mail: fukumoto@mapse.eng.osaka-u.ac.jp

Abstract. Zr-based bulk metallic glasses are expected to be welded to conventional structural alloys. Dissimilar welding of metallic glasses to stainless steel was carried out by resistance microwelding. The metallurgical analysis of the weld interface revealed the welding mechanism. A thin reaction layer was formed between the two liquid materials. The melting of stainless steel should be limited to obtain sound joints.

1. Introduction

Zr-based bulk metallic glasses (BMGs), which are kinds of amorphous alloys, not only have superior mechanical properties such as high strength and high elastic limit but also a low Young's modulus. It is expected that high-performance devices could be developed using BMGs for energy saving. On the other hand, it is well known that BMGs are often crystallized due to heat hysteresis during welding processes, which causes embrittlement in a weld and a heat affected zone (HAZ). A fast heating and cooling rate are necessary to avoid crystallization in the fusion welding of BMGs. It has been reported that electron beam welding[1-2] and laser welding [3-5] are appropriate fusion processes to weld BMGs successfully. The authors have reported that resistance micro-spot welding is also a suitable fusion welding process to weld BMG without crystallization even in an air atmosphere.[6-7] Recently, BMGs are also expected to be welded to conventional structural alloys such as stainless steels to produce functional small devices such as pressure sensors. However, as is well known, it would be unfavorable to form brittle intermetallic compounds (IMCs) in a weld in the case of dissimilar welding. The reaction between dissimilar metals must be limited to avoid brittlement in the weld.

In the present study, the microstructural development of a weld interface between Zr-based BMG and austenitic stainless steel formed by resistance microwelding was investigated to clarify the welding process. The relation between interfacial strength and the microstructure was also investigated.

2. Experimental procedures

The materials used are $Zr_{50}Cu_{30}Ni_{10}Al_{10}$ bulk metallic glass and type 304 austenitic stainless steel (SS) sheets of 0.2 mm in thickness. The geometry of the sheet was 20×6 mm. The faying surfaces of the

BMG sheet were polished by #2000 emery paper and cleaned with acetone. The physical properties are shown in Table 1. The BMG was welded to the stainless steel by resistance micro-spot and micro-seam welding. The experimental setups are shown in Figure 1. The welding current was varied from 300 to 700 A, and the weld time and the electrode force were 9 ms and 49 N, respectively, in the case of resistance micro-spot welding, and Cu-Cr electrodes with an R-type tip-face were used. The welding current, impulse time including on-time and off-time, welding speed, and electrode force are the parameters in resistance micro-seam welding. In the present study, only the welding current was varied from 600 to 950 A, and the other parameters, that is to say, impulse time of 30 ms, welding speed of 10 mm/s, and electrode force of 49 N, were kept constant. On-time is 33% of impulse time, while off-time is 66% of impulse time. The basic welding mechanism of both processes are similar.[8] Resistance seam welding is like resistance "multi-spot" welding. Therefore, in the present work, both processes are treated as the same process to explain microstructural development.

Cross sections were observed by optical microscopy and scanning electron microscopy (SEM). The fracture surfaces were also observed by SEM and energy dispersive X-ray analysis (EDX). The weld interface was observed by transmission electron microscopy (TEM) (JFM-2100). The TEM sample was prepared by the focused ion beam system. The joint strengths of the seam-welded joints were evaluated by a tensile shear test with a crosshead speed of 0.017 mm/s. The fracture surfaces of the spot-welded joints were obtained by the peel test. The experimental setups of the mechanical tests are shown in Figure 2.

Table 1. Physical properties of materials.

	$Zr_{50}Cu_{30}Ni_{10}Al_{10}$	SUS304
Electrical resistivity (10^{-8} Ωm)	205	72
Thermal conductivity (W/m·K)	~10	16.7
Melting point (K)	1133	1673~1723
Glass transition temperature (K)	706*	-

* *The values depend on the cooling rate.*

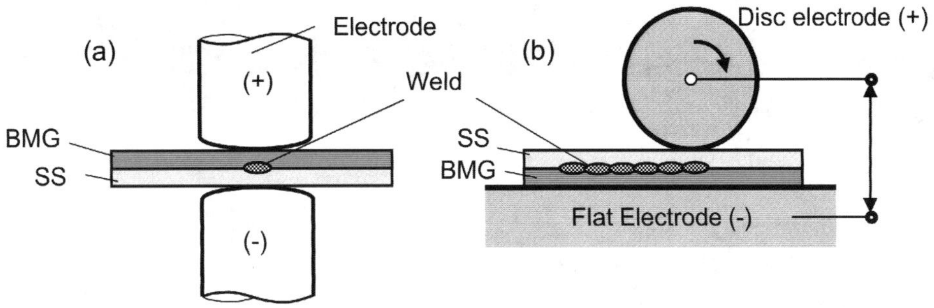

Figure 1. Experimental setups of resistance micro-spot , (a); and micro-seam, (b), welding.

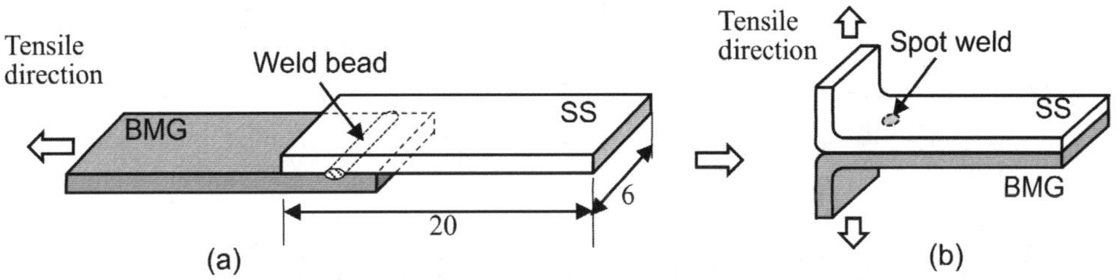

Figure 2. Schematic illustration of (a) tensile shear and (b) peel tests. (unit: mm)

3. Results

3.1. Cross sections

Figure 3 shows cross sections of resistance micro-spot welded dissimilar joints for various welding currents. At a welding current of 300 A, no fusion nuggets were observed. As the welding current increased, the stainless steel started to melt and penetrate into the BMG sheet. At a welding current of 700 A, an oval fusion nugget was formed and many cracks were introduced due to the formation of brittle IMCs.

The fusion of both alloy sheets is obvious when the welding current is higher than 500 A as shown in Figure 3b. On the other hand, a weld nugget was not observed in a weld made at 300 A. It has been reported that a weld nugget is invisible as long as the BMG maintains an amorphous structure, even if there is a fusion zone.[7] The electric resistivity of Zr-based BMG is three times larger than that of stainless steel, and the thermal conductivity of BMG is less than that of stainless steels (Table 1). Moreover, the melting point of BMG is less than that of stainless steel. Therefore, even at a small welding current, the BMG could have melted, even though a fusion zone was not observed. In this study, the BMG was squeezed out in a liquid state or a supercooled liquid state at the edges of the weld interface to form an expulsion. Once the stainless steel melts, both the liquid phases mix together, resulting in the formation of a brittle IMC nugget. Therefore, the amount of the liquid phase of stainless steel should be limited at the weld interface as shown in Figure 3a.

Figure 4 shows a detailed weld interface, in which only limited fusion occurs in the stainless steel sheet. At the outmost region, squeezed out BMG is observed. There are principally three kinds of interfaces. The contact between sheets was insufficient and there are some small gaps, hereafter called interface-A. At the center is the weld interface, where fusion of the stainless steel occurred and where it mixed with the liquid BMG in a limited area, which is called interface-C. Between interface-A and interface-C is interface-B, where both materials show good contact without fusion of the stainless steel.

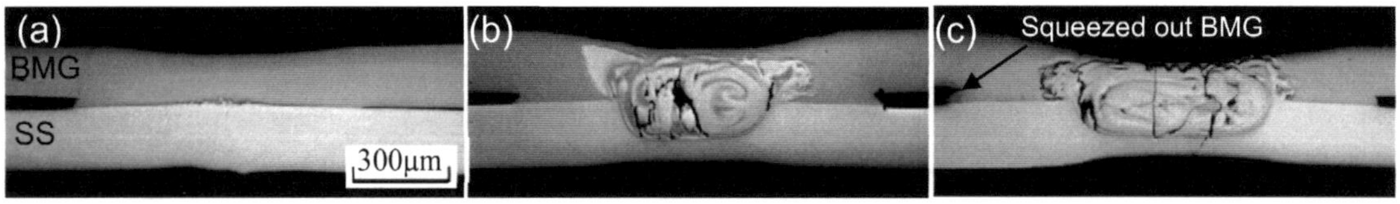

Figure 3. Cross sections of resistance micro-spot welds at various welding currents. (a) 300 A, (b) 500 A, (c) 700 A.

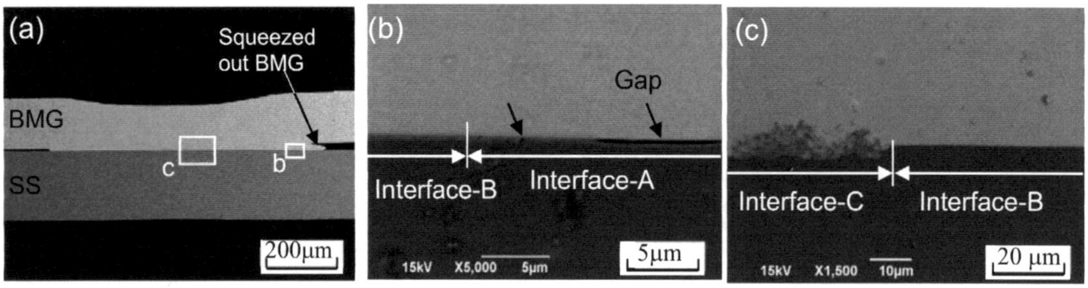

Figure 4. Three kinds of weld interface between BMG and stainless steel. Images (b) and (c) correspond to the highlighted areas in (a).

3.2. Fracture surfaces

Figure 5 shows fracture surfaces on a BMG sheet with a micro-spot weld with a welding current of 300 A, before the formation of an IMC nugget. Squeezed out BMG, like an expulsion, was observed. Three kinds of regions—in addition to the squeezed out BMG—were observed on the fracture surface.

At the center, there is the brightest region, which contains not only Zr, Al, Cu and Ni but also Fe and Cr, and where no flat-surface morphology was observed (Figure 5b). It is believed that the region corresponds to interface-C, where a limited reaction occurred between the liquid BMG and stainless steel. A small IMC reaction layer formed at this region. Next to this region, there is another region where the surface morphology was relatively flat, while small amounts of elements of stainless steel were detected (Figure 5c). The region most likely corresponds to interface-B. At the outmost region, there are many straight lines in one direction (Figure 5d). These line patterns were transferred from the surface of the stainless steel, and no Cr or Fe was detected there. It is believed that the stainless steel in a solid state did not react but contacted with the BMG in a liquid or supercooled liquid state at this region. Therefore, this region probably corresponds to interface-A.

As described above, the welding mechanism of resistance seam welding is basically similar to that of resistance spot welding. Figure 6 shows the fracture surface of a micro-seam weld. A brittle fracture surface morphology, including elements of stainless steel, was observed at the center area of the weld bead and there was also dimple morphology in places. The transfer printing of the surface of the stainless steel sheet were also observed at the outmost region. Between those two regions, a slightly bright region was observed in comparison with the BMG base alloy, which means the area includes a small amount of elements of stainless steel. The regions are almost similar to those observed on the fracture surfaces of the resistance micro-spot welds. Each region increased with an increase in the welding current.

3.3. Interfacial strength

It is difficult to estimate the interfacial strength of resistance micro-spot welds accurately since the area is relatively small. Therefore, the joint strength of resistance micro-seam welds was evaluated since the weld area is larger than that of micro-spot welding. Figure 7 shows the shear breaking force of the seam welds and the shear stress against each region for various welding currents. Each region was measured on the fractured surfaces of BMGs. Most of the joints were fractured at the weld interface. When the welding current was 950 A, a fracture occurred in the BMG base alloy.

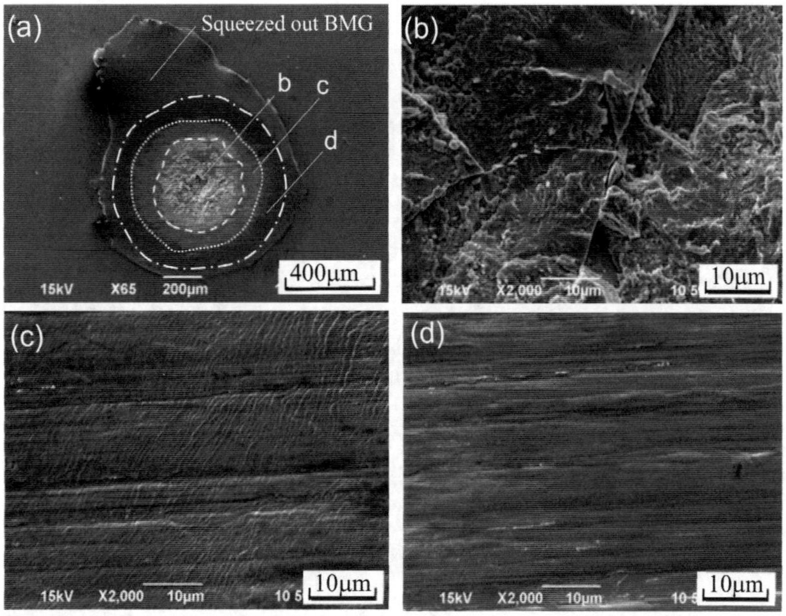

Figure 5. Fracture surface of BMG sheet of resistance micro-spot welding at a current of 300 A. (a) Macroscopic view. Images (b) to (d) correspond to the higher magnifications of each region indicated in (a).

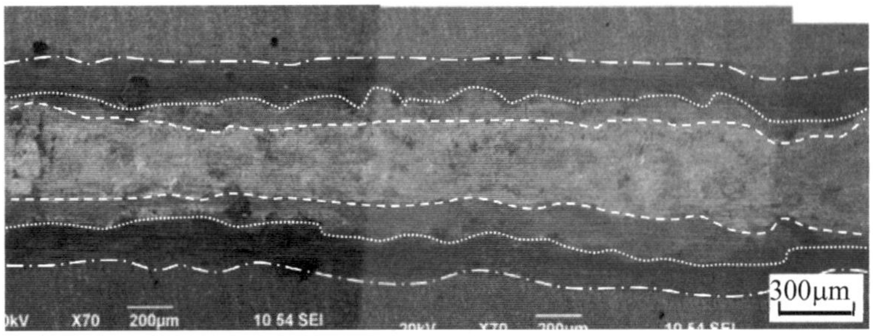

Figure 6. Typical fracture surface of BMG sheet of resistance micro-seam welding. Each kind of white dotted line corresponds to that shown in Figure 5a.

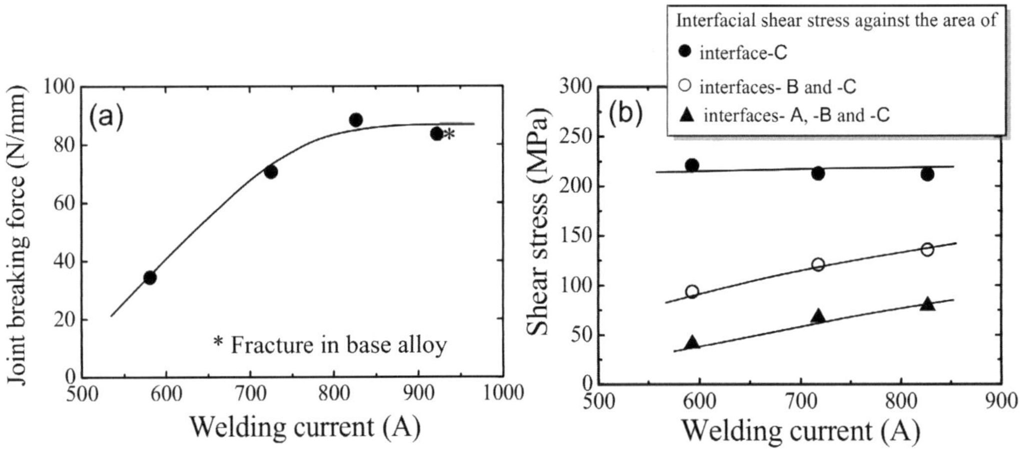

Figure 7. Joint shear breaking force of resistance micro-seam welds, (a); and interfacial shear stress for various weld interfaces, (b).

The interfacial shear breaking force of interfacial fractured joints increases as the welding current increases because the weld area increases. The weld area includes three kinds of regions as described in Section 3.2. The shear stress against the area including all kinds of interfaces, that is to say interfaces-A, -B and -C, increased with an increase in the welding current (Figure 7b). The shear stress against the area including interfaces-B and -C also increased with an increase in welding current. On the other hand, the shear stress of the joint against interface-C, which is indicated as solid circles in Figure 7b, shows almost a constant value of approximately 220 MPa regardless of the welding current. It is believed that interface-C, where a limited fusion of stainless steel occurs, is most likely dominant in terms of joint strength.

4. Discussion

A weld interface has three kinds of interfaces as shown in Figure 4. Among them, interface-C is dominant for joint strength. Detailed microstructures of interfaces-B and -C are shown in Figure 8. A reaction layer and a plane interface were observed at the weld interface. They correspond to interface-C and interface-B, respectively. An amorphous structure was maintained in the BMG sheet except for the area close to interface-B. A reaction layer was formed at weld interface-C. A eutectic structure was observed in the reaction layer, and the selected area diffraction pattern reveals that one of the phases was $Fe_{23}Zr_6$ (Figures 9a and 9b). Therefore, it is believed that interface-C was produced by a reaction between molten stainless steel and molten BMGs.

A crystalline phase of the stainless steel was observed in the amorphous phase close to both interface-B and the reaction layer, that is interface-C. The area is shown by white dotted lines in Figure 8a, and there is not only a crystalline phase but also an amorphous phase (Figures 9c and 9d). It is believed that the molten stainless steel flows into the liquid phase of BMG from the reaction layer to form such an area. Since the stainless steel was melted at interface-C, the temperature at interface-C must be higher than 1673 K. As shown in Table 1, the melting point of BMG is much lower than that of stainless steel. Hence, the BMG would be the liquid phase near interface-B. On the other hand, interface-B was a plane, which suggests the stainless steel was in a solid state at interface-B during the welding. Figure 10 shows the distribution of Zr in the stainless steel near interface-B. Zr was detected in the range of 100 to 150 nm from interface-B, and the concentration decreases with an increase in the distance. Although the weld time was just 9 ms, the diffusion should not be ignored near the weld interface. The welding mechanism at the weld interface-B is based on the interdiffusion between solid stainless steel and liquid BMG. So interface-B contributes to the joint strength to some degree, although it is probably smaller than that of interface-C. Further study is necessary to clarify the relation between interfacial strength and the microstructure.

5. Conclusions

Zr-based BMG was welded to stainless steel by resistance microwelding. The weld interface consisted of three kinds of interfaces. There was a reaction layer composed of a eutectic structure at the center of the weld interface. Next to the reaction layer, there was an interface formed by the liquid/solid interdiffusion between BMG and stainless steel. There are some gaps in the outermost region where interfacial strength cannot be expected. An interface with a limited reaction layer might be dominant for the joint strength.

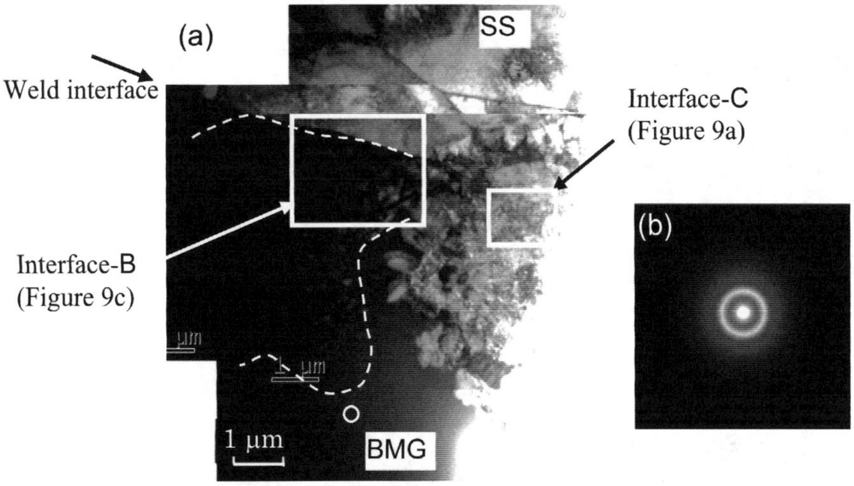

Figure 8. Microstructures of weld interfaces-B and -C of a resistance micro-spot weld at a welding current of 300 A, (a); and the selected area diffraction pattern of BMG sheet, (b). The selected area is shown by an open circle.

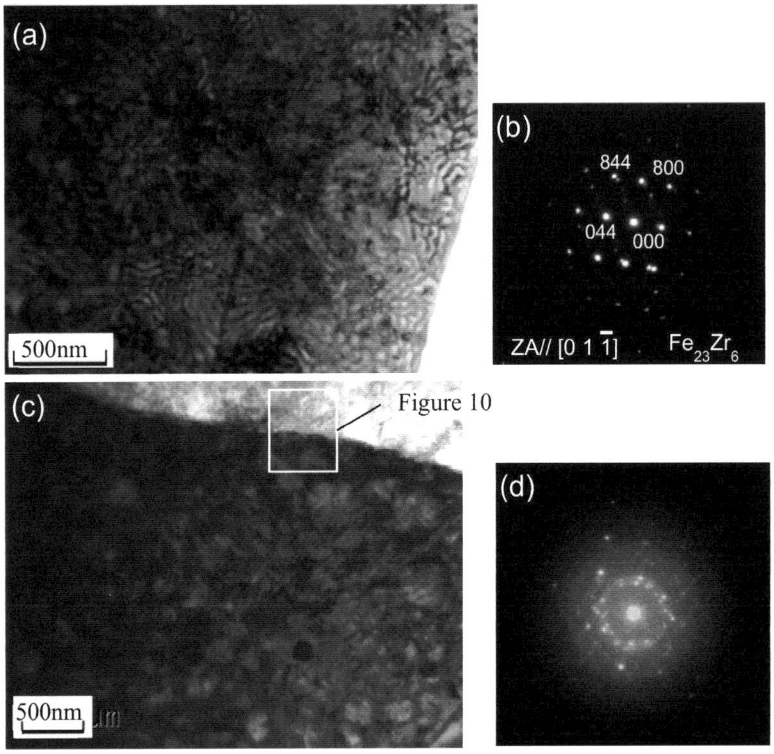

Figure 9. Detailed microstructures of highlighted areas shown in Figure 8a. (a)-(b) Microstructure of reaction layer and selected area diffraction patterns. (c)-(d) Microstructure of BMG sheet near interface-B and selected area diffraction patterns.

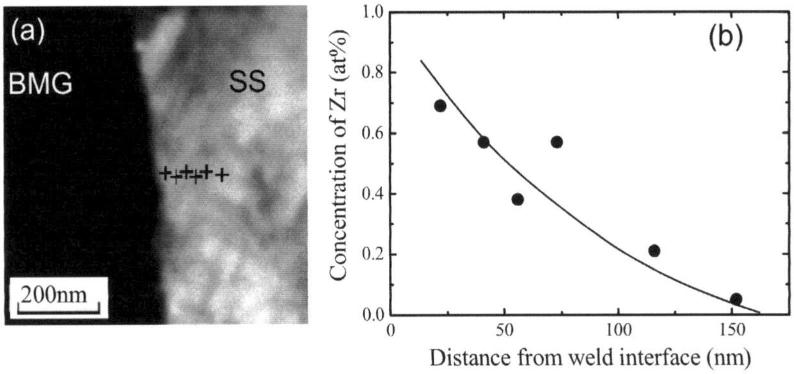

Figure 10. Distribution of Zr in stainless steel near weld interface-B, which is the highlighted area in Figure 9c.
(a) Bright-field image, (b) Concentration of Zr near weld interface.
Crosses indicate analysis points. The diameter of electron beam is 1 nm.

Acknowledgements
This work was performed under the inter-university cooperative Research Program of the Institute for Materials Research, Tohoku University and the Cooperative Research Program of Joining and

Welding Research Institute, Osaka University. This work was partially supported by the Ministry of Education, Science, Sports and Culture, Grant-in-Aid for Scientific Research (C), 21560750, 2011.

References

[1] Yokoyama Y *et al* 2002 *Mater. Trans. JIM* **43** 2509
[2] Kawamura Y and Ohno Y 2001 *Mater. Trans. JIM* **42** 2476
[3] Li B et al 2006 *J. Alloy and Compound* **1-2** 118
[4] Kim J H *et al* 2007 *Mater. Sci. Eng. A* **449-451** 872
[5] Tsumura T and K Nakata 2011 *Weld. Inter.* **27** 491
[6] Fukumoto S *et al* 2007 *Mater. Sci. Forum* **562-565** 1307
[7] Fujiwara K, Fukumoto S et al 2008 *Mater. Sci. Eng. A* **498** 302
[8] Karagoulis M J 2000 Resisatace seam welding *ASM Handbook Vol.6 Welding, Brazing, and Soldering* ed Olson D L *et al* (Materials Park, OH) pp238-246

Numerical analysis of interfacial deformation and temperature rise during ultrasonic Al ribbon bonding

Yasuo Takahashi[1], Shinji Suzuki[2], Yusuke Ohyama[2] and Masakatsu Maeda[1]

[1] Joining and Welding Research Institute, Osaka University, 11-1 Mihogaoka, Ibaraki, Osaka 567-0047, Japan
[2] Graduate School of Engineering, Osaka University, Suita, Osaka, Japan

E-mail: takasy@jwri.osaka-u.ac.jp

Abstract. The plastic deformation, thermal behaviour and interfacial state during ultrasonic bonding of Al ribbon with substrate were analyzed based on numerical simulations. The numerical simulations were carried out by finite difference and element methods. The change of bond interface temperature with the bonding time t was simulated. It is suggested that the interfacial frictional behaviour between ribbon and substrate does not only affect the temperature rising behaviour but also the stress situation and plastic deformation behaviour, i.e., the ribbon deformation and the frictional slip behaviours influence each other. It is also found that partial constraint (frictional slip) at the bond-interface increases interfacial shear stress drastically and controls the frictional heat input and temperature rise of the Al ribbon.

1. Introduction

In recent years, thick Al ribbon bonding has been applied to the packaging of power devices such as IGBT (Insulated Gate Bipolar Transistor) modules to secure high power density and to improve power controlling [1, 2]. These are very important for green innovations in the field of car electronic technologies, and wind and solar power controlling. The ultrasonic Al ribbon bonding is affected by temperature rise of the bond area and ribbon deformation [2]. The temperature rise is related to frictional slip behavior at the bond-interface. There have been few numerical studies of Al ribbon bonding, although there exist a few studies of experimental investigation. Basic knowledge on the bonding mechanism is essential for further development of ultrasonic bonding. However, it is difficult to analyze the mechanism experimentally due to the fine-scale phenomenon and short-time process of the bonding. Even now, the phenomenon occurring at the interface is not perfectly simulated, although many numerical studies of wire bonding have been reported [3-8]. In the present study, the effect of ultrasonic vibration on plastic deformation and temperature rise is, therefore, simulated in order to understand the bond interfacial state and the softening effect induced by the ultrasonic vibration.

2. Numerical modelling and simulation procedure

Figure 1 illustrates the ultrasonic ribbon bonding. Al ribbon is used in the power electronics packaging. Ultrasonic vibration is applied in the longitudinal direction of Al ribbon as illustrated in Figure 1. The area enclosed by dotted lines was considered in the present study, i.e., a two-dimensional model was adopted [2, 3]. The heat flow and deformation in the z direction were ignored because the length of

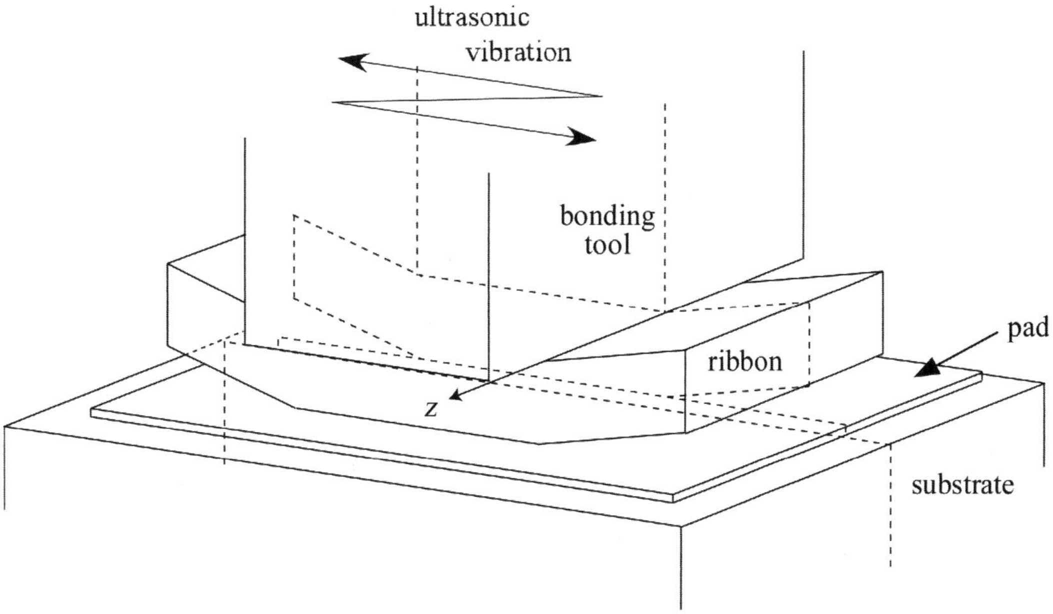

Figure 1. Schematic illustration of Al ribbon bonding with flat-tool.

Table 1. Thermal conductivity and specific heat of Al, SiO_2 and WC.

Name	Symbol	Al	SiO_2	WC	Unit
Thermal conductivity	h	2.37×10^{2}	1.38	8.00×10^{1}	$W\ m^{-1}K^{-1}$
Density	ρ	2.70×10^{3}	2.19×10^{3}	2.70×10^{3}	$kg\ m^{-3}$
Absolute temperature	T				K
Specific heat	$C_p = a_1 + a_2T + a_3/T^2 + a_4T^2$				$J\ kg^{-1}K^{-1}$
where	a_1	31.38	43.93	43.39	
	a_2	-1.64×10^{2}	3.88×10^{-2}	8.62×10^{-3}	
	a_3	-3.60×10^{5}	-9.69×10^{5}	-9.33×10^{5}	
	a_4	2.08×10^{-6}		-1.03×10^{-6}	

substrate in the z direction is much greater than the thickness of substrate. It was assumed that the bottom of the substrate was kept at 300 K (in contact with a large heat sink) [2]. The top of the bonding tool was also kept at 300 K. The bonding tool was made of tungsten carbide (WC). It was assumed that the heat was not conducted to the ambient air. A silica plate was adopted as a substrate because Al ribbon could be connected to silica easily under the constant applied pressure together with ultrasonic vibration, i.e., the ultrasonic vibration (vibration behavior of bonding tool) for silica was more stable during bonding than that of Si-chip. Also, the thermal conductivity of silica is much lower than that of Si-chip. The temperature rising process of Al ribbon adhered to silica can, therefore, be simulated more clearly. The material constants of thermal behavior are listed in Table 1.

Figure 2 shows the mesh pattern of the two-dimensional model. The mesh division is enough to solve the visco-plastic deformation of Al ribbon, because the eight-node isoparametric elements were adopted [3]. The time increment had to be less than $\Delta t = 1\text{x}10^{-10}$ s for calculating heat conduction if the mesh division was finer than that of Figure 2 [2]. The mesh pattern in Figure 2 is adequate for simulating interfacial deformation and heat conduction. It is assumed that the bonding tool and silica substrate are rigid bodies but the heat conduction occurs through them. Al thin electric pad on the

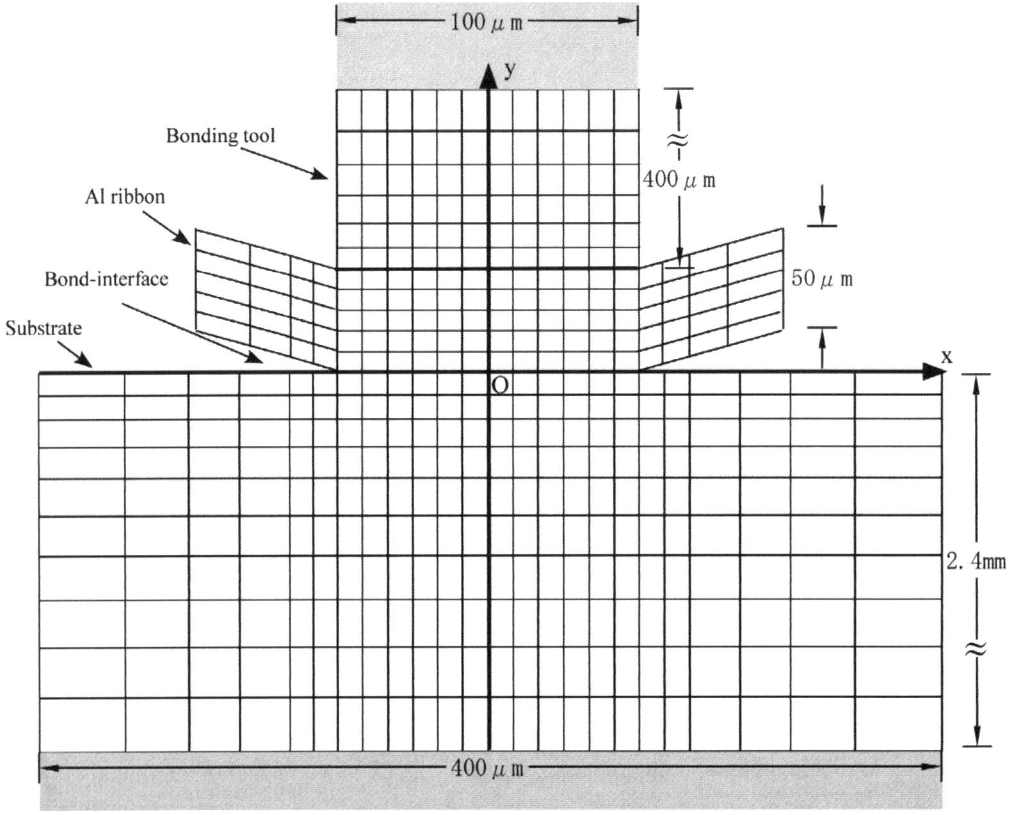

Figure 2. Mesh pattern in cross section of tool, Al ribbon and substrate, corresponding to the dotted area in Figure 1. The electric pad is ignored.

substrate is ignored so that the slipping mode can be simplified. Only the Al ribbon can be deformed during the bonding period when ultrasonic vibration is applied under the constant bonding pressure P_B ($= F/S$), where F is the bonding load and S is the head area of the bonding tool. The bonding tool is too hard to deform plastically although it can vibrate. The ultrasonic vibration is parallel to the x direction. It is assumed that the interface (tool-interface) between tool and ribbon is always fixed, because of simplification, i.e., the heat input is not produced at the tool interface. The amplitude A_o changes with time t, as solid adhesion occurs at the bond-interface between ribbon and substrate along the x-axis. The average amplitude A_b of Al ribbon vibration at the bond-interface differs from A_o as the adhesion occurs at the bond-interface. This is due to the frictional slip. The top amplitude A_o and the vibration ratio $R_d = A_b/A_o$ decrease step by step if the adhesion occurs gradually. In other words, if the adhesion progress is not steady, then A_o and R_d fluctuate. The change in A_o and R_d should be related to the frictional slip phenomena at the bond-interface, but it is very difficult to explain the relationship between the frictional slip and the solid adhesion quantitatively. Therefore, in the present study, the change of A_o and R_d with time t was given as a boundary condition but the friction coefficient μ was not given for the bond-interface because the relation between adhesion and friction coefficient was unknown.

Figure 3 shows the three conditions (a), (b) and (c) for change of A_o (red color) and R_d (blue color). The dotted lines just suggest the actual changes in A_o and R_d. The values shown by solid lines were given to A_o and R_d in calculation. The steady state adhesion progress is supposed in the present study. The ultrasonic vibration amplitude does not change up and down. The steady state (stable) bonding process is often observed under proper bonding conditions unless the ramp force is given and the gross slip motion can be produced even under the constant load [9, 10]. The ratio R_d was decreased linearly from 1.000 to 0.999 in the first stage. For example, the first stage is t = 0 - 40 ms for Figure 3 (a).

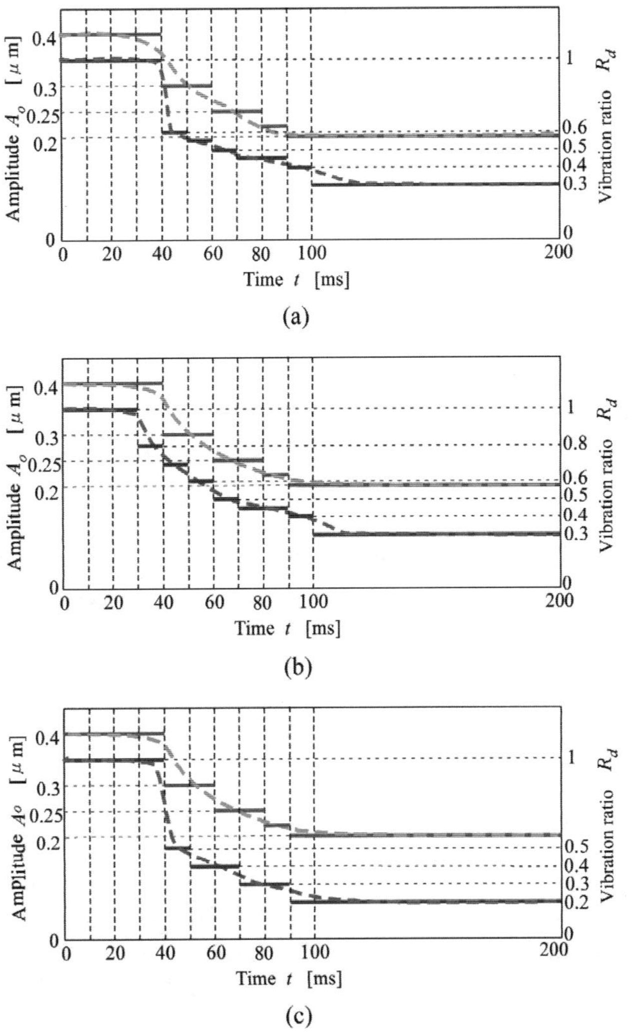

Figure 3. Change in vibration amplitude A_o and ratio R_d as a boundary condition.

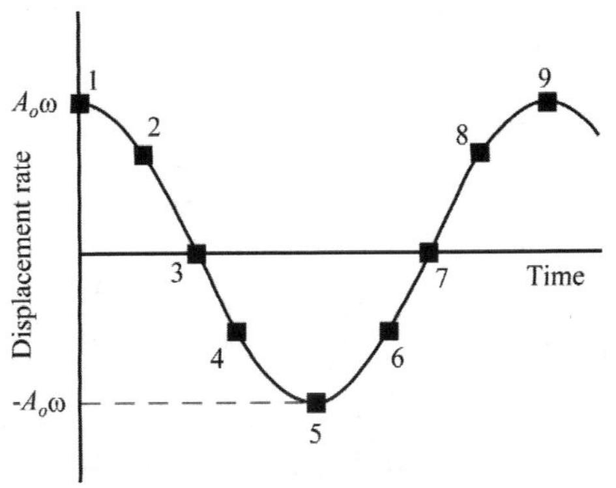

Figure 4. Schematic illustration of ultrasonic vibration wave pattern (displacement rate at the interface between tool and ribbon).

If the displacement (ultrasonic vibration) u is given by

$$u = A_o \sin \omega t, \tag{1}$$

where ω is the circular frequency (= $2\pi f_v$), the ultrasonic power $P_w(t)$ is given by

$$P_w(t) = F_x(t) \cdot A_o \omega \cdot \cos \omega t, \tag{2}$$

where F_x is the force in the x direction. The average P_w is obtained by $4F_x \cdot A_o \cdot f_v$ [11]. The displacement rate at the interface between tool and ribbon can be given by

$$(du/dt)|_{\text{at tool}} = A_o \omega \cos \omega t. \tag{3}$$

Therefore, the change in the displacement rate at the tool interface is given by Figure 4. The time increment Δt for calculating Al ribbon deformation was set by $T_v/8$ (=$1/8f_v$ = 2.08x10^{-6} s), where T_v is the period of ultrasonic vibration and f_v is the frequency (= 60 kHz). Al ribbon deformation in unit cycle was calculated in 8 steps, as shown in Figure 4. It is necessary to solve F_x at finite elements of the bond-interface for obtaining the real power (heat) input to the bond-interface. The shear stress at the bond-interface can be obtained by calculating the deformation of Al ribbon. In the present study, it was assumed that the heat input was obtained by the frictional slip at the bond-interface. When the plastic deformation of Al ribbon was calculated, it was assumed that the temperature T (of each element) was kept constant in the period T_v unit cycle of ultrasonic vibration. This assumption makes it easy to give the heat inputs to the elements mated at the bond-interface. The shear stress τ_{xy} at the bond-interface was solved for each finite element in each step Δt. The heat input Q_{in} to each finite element at the bond-interface was given by

$$Q_{in} = \tau_{xy} \cdot S_b \cdot (du/dt)_x|_{\text{at bond-interface}}, \tag{4}$$

where S_b is the area of the finite element at the bond-interface and $(du/dt)_x|_{\text{at bond-interface}}$ is the displacement rate of the nodal point of the Al ribbon side at the bond-interface. This corresponds to the local frictional slip rate and it can change along the position x due to the plastic deformation of Al ribbon but it was constant for each step Δt (=$T_v/8$). It was assumed that the average of $(du/dt)_x|_{\text{at bond-interface}}$ at time t was equal to $A_o \omega \cdot R_d$ at time t. The heat input Q_{in} was assigned only to the elements (Al and silica) mated at the bond-interface in proportional to the volume of each element. Al ribbon was assumed to be a rate-sensitive material with material constants listed in Table 2 for the low temperature region of Al. The equivalent strain rate $\dot{\bar{\varepsilon}}$ is given by

$$\dot{\bar{\varepsilon}} = A_O \left[1 + B\exp\left(-C\bar{\varepsilon}\right)\right]\left(\frac{D_o bG}{kT}\right)\left(\frac{\bar{\sigma}}{G}\right)^n \exp\left(-\frac{Q_C}{RT}\right), \tag{5}$$

where A_O, B and C are a non-dimensional constant, $\bar{\varepsilon}$ the equivalent strain, D_o a frequency factor, b the Burgers vector, G the shear modulus, k the Bolzmann's constant, T the absolute temperature of the specimen, $\bar{\sigma}$ the equivalent stress, n the stress exponent, Q the activation energy for the visco-plastic flow, and R the gas constant [3, 6]. The finite element method (eight-node isoparametric elements) was used [3]. As the equivalent stress increases, n can be high enough to prevent stress enhancement but it is assumed in the present study that n is constant.

On the other hand, the heat conduction was simulated by the time step Δt = 1x10^{-10} s, assuming the heat input was constant for a few cycles ($m \cdot T_v$, where m is an integer). The cycle number m was increased as the bonding progressed from early stage to latter stage, because the task would be very hard if a unit cycle was adopted for solving ribbon-deformation, i.e., the distribution of temperature T (the heat flow) was solved under the assumption that the heat input due to frictional slip was constant for several cycles. The heat flow was calculated by a finite difference method [2]. The assumption on heat input can reduce the hard task (numerous simulations) of the heat flow phenomenon by the ultrasonic vibration power.

Table 2. Material constants for aluminum deformation at low temperatures.

Name	Symbol	Al	Unit
Equivalent strain rate	$\dot{\bar{\varepsilon}}$		s^{-1}
Equivalent strain	$\bar{\varepsilon}$		
Equivalent stress	σ		Pa
Absolute temperature	T		K
Boltzman's constant	k	1.3806504×10^{-23}	$J\,K^{-1}$
Gas constant	R	8.314	$J\,mol^{-1}K^{-1}$
Frequency factor	D_o	0.1×10^{-4}	m^2s^{-1}
Constant	A_o	2.14×10^{5}	
Constant	B	5.60×10^{2}	
Constant	C	1.00×10^{2}	
Burgers vector	b	2.86×10^{-10}	m
Stress exponent	n	6.4	
Activation energy	Q	75.4	$kJ\,mol^{-1}$
Melting temperature	T_m	933	K
Shear modulus	$G(T) = a_G(T/T_m)^2 + b_G(T/T_m) + c_G$		$N\,m^{-2}$
where	a_G	-9.4×10^{9}	
	b_G	-5.3×10^{9}	
	c_G	2.9×10^{10}	

3. Results and discussion

Figure 5 shows the temperature change at the center of the bond-interface. Strictly speaking, it is the temperature of two elements of ribbon-side closed to the origin O $(x, y) = (0, 0)$. Figure 5 (a), (b) and (c) are the calculated results for the boundary conditions of Figure 3 (a), (b) and (c), respectively. Because the amplitude A_o and the vibration ratio R_d are given in stairs-like patterns as shown in Figure 3, the temperature changes discontinuously. The conditions (a) and (c) correspond to the case when the partial adhesion occurs suddenly at t = 40 ms. On the other hand, the condition (b) is for the case when the adhesion more gradually occurs. The temperature rising of Figure 5 (a) or (c) is very different from that of Figure 5 (b). This suggests that if the adhesion is gradually produced, then the temperature rise can be facilitated. The temperature at $t \approx 50$ ms of Figure 5 (a) is larger than that of Figure 5 (c). The vibration ratio of condition (c) is less than that of (a) at t = 40 - 50 ms. The slight difference in R_d can also control the temperature rising, resulting in the difference of temperature rising between (a) and (c) at t = 40 - 60 ms. Figures 5 (a) and (b) imply that if the frictional slip continues by the latter stage of bonding, then temperature cannot fall easily. Figure 5 (c) suggests that the further adhesion occurrence at the latter stage of bonding reduces temperature rising of the specimen, i.e., the specimen is cooled down rapidly if the interface is fixed without friction. The early sudden sticking at the bond-interface reduces the temperature rising and makes the bonding process unstable. As seen in Figure 5, the temperature is roughly kept constant after t = 110 ms because the frictional slip is assumed to be produced after full adhesion. In rotational frictional pressure welding [12], the specimen can keep rotating at high bonding temperature after full adhesion is produced, till the specimen is cooled down to the room temperature. The adhered (bonded) interface is not necessarily fixed at high temperatures, i.e., the frictional slip occurs even after bonding so that temperature rise can occur.

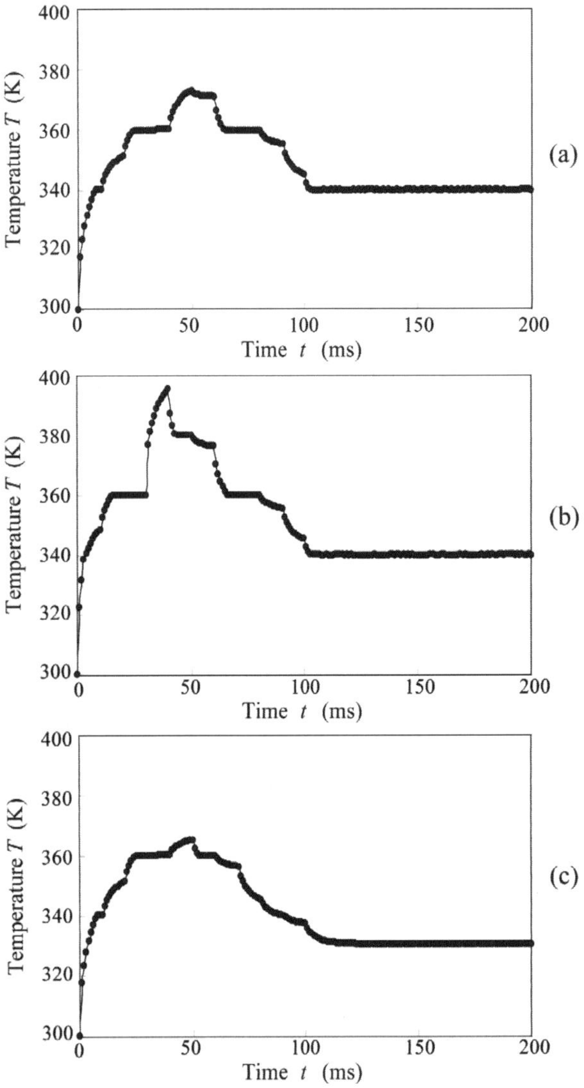

Figure 5. Temperature change of Al ribbon (bottom side) with time. (a), (b) and (c) are the calculated results for conditions of Figure 3 (a), (b) and (c), respectively.

Figure 6 shows the change in the temperature distribution with time under the condition of Figure 3 (a). Because the substrate is made of silica, the heat is not easily conducted to the bottom. Al ribbon is heated up dominantly and the temperature is roughly uniform. This makes it easy to analyze the deformation behavior of Al ribbon during bonding. The temperature rise of Al ribbon due to frictional slip is about 80 - 100 K. The bonding tool is also heated up, as seen in Figure 6 (b). After that, the temperature of Al ribbon and tool decreases and is nearly equal to that of substrate, as shown in Figure 6 (d), and the whole is cooled down.

The ratio R_d is close to unity in the early stage of bonding and decreases with increasing t. This is an assumption in the present study but possible and right in general because the bonding (adhesion) is produced. If R_d is perfectly equal to unity (R_d = 1.00), then the interface between Al ribbon and substrate can slide freely. Therefore, the deformation of Al ribbon for R_d = 1.00 is not facilitated by the ultrasonic vibration. In other words, it is just the deformation given by applying the bonding pressure P_B without ultrasonic vibration (A_o = 0.00 μm). The maximum equivalent stress of Al ribbon does not increase, compared with the value of P_B. However, if R_d decreases a little, the equivalent stress increases very much as an effect of the ultrasonic vibration [2].

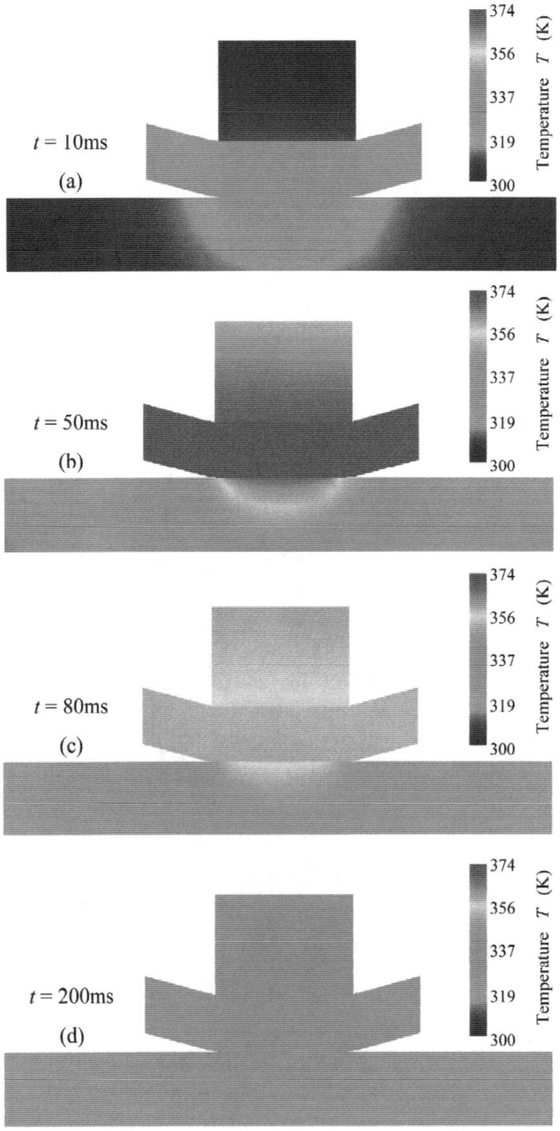

Figure 6. Temperature distribution of Al ribbon, tool and substrate. (a) t = 10 ms, (b) t = 50 ms, (c) t = 80 ms and (d) t = 200 ms. The calculation was carried out under the condition of Figure 3 (a). Because of shear deformation, the compressive deformation can be neglected in drawing the temperature distribution.

Figure 7 shows the calculated results of the equivalent stress distribution under the condition of Figure 3 (a) (P_B = 28 MPa). Even if R_d = 0.9999, the equivalent stress of Al ribbon goes up to 190 - 250 MPa (seven to ten times the P_B value) immediately after ultrasonic vibration is turned on, as seen in Figure 7 (a) at T = 300 K, $t \approx 0$ ms. This is derived from the shear stress enhanced by the frictional slip. The value of the shear stress at the central area was obtained by $\tau_{xy} \approx 125$ MPa from the calculated results, being much grater than $\sigma_y \approx P_B$. If R_d is constant, the stress decreases as shown in Figure 7 (b) at t = 8 ms. Al becomes soft and the shear stress decreases. The frictional work also decreases. As the deformation can be enhanced, the adhesion progresses. The deformation-restriction and hardening occurs and then the frictional work is again produced, resulting in temperature rise. If temperature increases further, Al ribbon becomes soft again and deforms to harden, repeatedly. The adhesion goes on further, and R_d decreases and the interface constraint still increases. The equivalent stress increases even if temperature increases ($T \approx 360$ K) as seen in Figure 7(c) at $t \approx 40$ ms. The

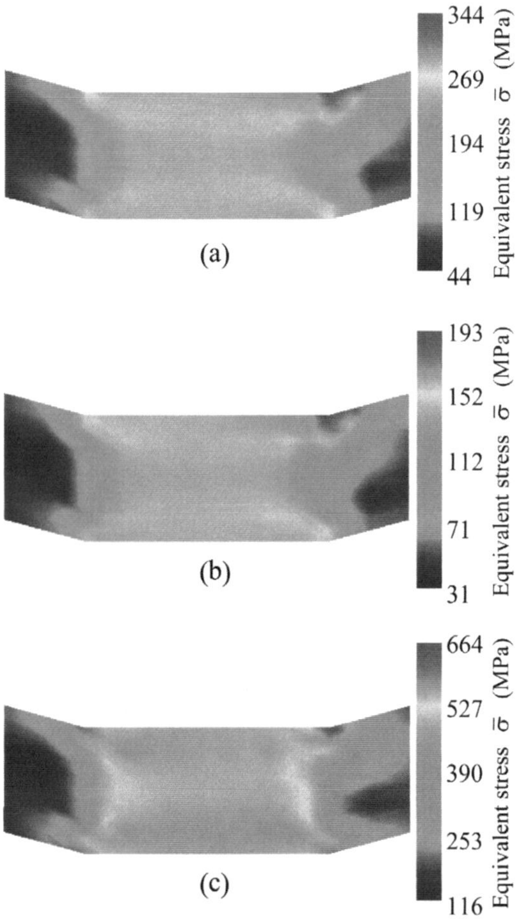

Figure 7. Equivalent stress distribution in Al ribbon. The calculating condition is Figure 3 (a) and P_B = 28 MPa. The ultrasonic vibration (displacement rate) is $A_o\omega$ in the right-hand direction. (a) at $t \approx 0$ ms (T = 300 K, A_o = 0.4 μm, R_d = 0.9999), (b) at $t \approx 8$ ms ($T \approx 340$ K, A_o = 0.4 μm, R_d = 0.9999) and (c) at $t \approx 40$ ms ($T \approx 360$ K, A_o = 0.3 μm, R_d = 0.6).

shear stress decreases as temperature further increases and then the frictional work finally decreases, i.e., the heat input decreases. The specimen begins to be cooled down. To sum up, frictional slip, heating, softening and hardening are produced repeatedly by applying the ultrasonic vibration. The shear stress often fluctuates. Actually, the adhesion may be broken down discontinuously, if the shear stress increases very much. As reported elsewhere [13], if temperature rises to T = 400 K, then the equivalent stress goes down clearly.

Figure 8 shows the stress distributions along the bond-interface (Al ribbon side) at t = 40 ms of Figure 7 (c). The displacement rate due to ultrasonic vibration is applied to the right-hand side (step 1 in Figure 4). The abscissa is the number of nodal point along the bond-interface as shown in Figure 2. No. 13 corresponds to the origin O, and No. 1 and 25 are the edge points of the bond-interface. The shear stress τ_{xy} is roughly constant and about 500 MPa, which is close to the equivalent stress as shown in Figure 7 (c). The shear deformation is dominant. The stress σ_y in the y direction is compressive in the right-hand edge of the bond-interface and tensile in the left-hand edge. The equivalent stress becomes larger in both edges. Because the friction is assumed to be gross slip, the friction coefficient is not constant at the position x, and the friction coefficient in the central area (No. 5 to No. 21 of nodal points) is greater than ten. The adhesion must already be produced but the frictional slip continues after adhesion. If this is not so, the temperature does not rise after t > 40 ms. Because the temperature rises even after t > 40 ms as seen in Figure 5, it is suggested that adhesion and frictional slip occur simultaneously.

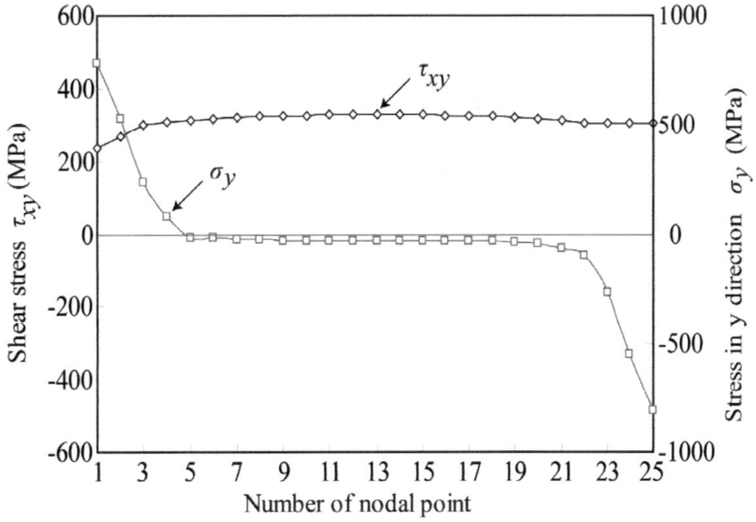

Figure 8. Stress distribution at t = 40 ms under the condition of Figure 3 (a) ($T \approx$ 360 K, P_B = 28 MPa) in the case when the ultrasonic vibration is applied in the right-hand direction.

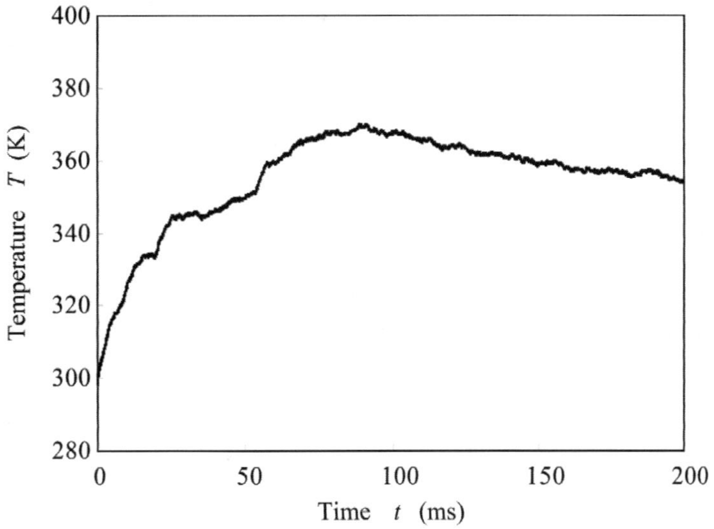

Figure 9. Experimental result of temperature change with time. The bonding time t = 0 ms was defined when the ultrasonic power was begun to be inputted. The bonding condition was F = 7.0 N, P_w = 3.8 W, f_v = 60 kHz.

Figure 9 shows an experimental result of temperature change of bonding interface with bonding time. The temperature was measured by a micro thermo-couple embedded in the bonding interface. The experimental condition was F = 7.0 N, P_w = 3.8 W, f_v = 60 kHz. The width of Al ribbon was 1.0 mm. The tool width was 0.25 mm (2.5 times the calculating condition, 0.1 mm, because it was difficult to embed the micro thermo-couple). As the bonding area was about 0.25 mm^2, $P_B \approx$ 28 MPa. As can be seen in Figure 9, temperature keeps rising until t = 80 ms, fluctuates and then decreases for t > 100 ms. The maximum temperature is about $T \approx$ 370 K at t = 70 - 90 ms. The maximum temperature is nearly equal to that of Figure 5 (a) or (c). In calculating, the bonding area was assumed to be 0.1 x

10^{-6} m^2 (100 μm x 1 mm). The average shear stress at the bond-interface is 500 MPa from Figure 8. The displacement rate due to frictional slip is $(du/dt) = A_o \omega R_d \approx 6.8 \times 10^{-2}$ m/s. The heat input is, therefore, about 3.4 W as obtained by Eq. (4). This is nearly equal to that of the experimental condition. There is no difference between calculated and experimental results with respect to ultrasonic power input and applied force.

On the other hand, the temperature decreases more slowly than that of the calculated results in Figure 5. The heat input from the top of the Al ribbon must occur at the latter stage of bonding due to the frictional slip at the tool interface. The heat input at the tool interface was neglected in calculated results, although the frictional slip at the bond-interface is taken into account ($R_d > 0.2$). It is suggested that it is necessary to take into consideration the frictional slip at the tool interface in the latter stage of bonding. It is found from calculated and experimental results that the ribbon temperature rises rapidly due to the frictional slip in the early stage of bonding, i.e., the temperature goes up to 360 - 400 K in 50 ms. The frictional work decreases after the ribbon temperature increases. The three slip conditions are roughly classified into 1) free slip, 2) frictional slip and 3) fixed bond-interface. It is, however, suggested that the bond-interface is not fixed perfectly in the period when the ultrasonic vibration is inputted under high temperature conditions.

4. Conclusions

It is very important to understand the thick Al ribbon bonding applied to power electronics packaging. The deformation and thermal phenomena of Al ribbon due to ultrasonic vibration were simulated. The main results are shown as follows.

1) The equivalent stress of Al ribbon is strikingly enhanced by the ultrasonic vibration. This is due to the frictional slip at the bond-interface. If the amplitude is kept constant, as the frictional constraint increases, the stress becomes higher.

2) If the frictional condition is the same, as temperature increases, the equivalent stress decreases. However, if the frictional constraint increases due to the partial adhesion, the stress increases and deformation is facilitated. The deformation of aluminum with softening and hardening is repeated by the ultrasonic vibration and temperature increases. This is related to the adhesion phenomena.

3) The ribbon temperature is heated up to about 360 - 400 K rapidly in 40 - 50 ms when frictional heat due to the ultrasonic power is generated at the bond-interface.

4) The frictional slip can occur even after adhesion at high temperatures. Adhesion and frictional slip occur simultaneously.

References

[1] Takahashi Y and M Maeda 2006 *Smart Processing Technology High Temp. Soc. Jpn.* **1** 163
[2] Takahashi Y, Maeda M, Suzuki S and Ohyama Y 2011 *Quart. J. Jpn. Weld. Soc.* **29** 138s
[3] Takahashi Y and Inoue M 2002 *J. Electronic Packaging* ASME **124** 27
[4] Ding Y, Kim J K, Tong P 2006 *Mechanics of Materials* **38** 11
[5] Ding Y and Kim J K 2008 *Microelectronics Reliability* **48** 149
[6] He J, Guo Y, Lin Z 2008 *J. Phys. D Appl. Phys.* **41** 1
[7] Liu Y and Luk T 2008 *IEEE Trans. Electron. Packag. Manufac.* **31** No.1 61
[8] Takahashi Y 2009 *IEICE Trans. Electron. Jpn.* (in Japanese) **J92-C** 581
[9] Ute G, Martin S R and Herbert R 2009 *IEEE Trans. Comp. Packag. Technol.* **32** 794
[10] Lum I, Mayer M and Zhou Y 2006 *J. Electro. Mater.* **35** 433
[11] Shah A, Mayer M, Qin I, Huynh C, Zhou Y and Meyer M 2010 *J. Phys D* **43** 1
[12] Hasegawa M, Asada I, Ozawa H and Ogura K 2009 *Proc. 81th Meeting of Interface bonding Committee* (Tokyo /Jpn Weld. Soc.) Doc. IJ-06-09
[13] Suzuki S, Takahashi Y, Ohyama Y and Maeda M 2011 *Proc. ECO-MATES* 2011 (Osaka Jpn/ JWRI, Osaka Univ.) **2** 225.

A crack initiation and propagation simulation and the fatigue characteristics of solder joints considering the material property changes

M Matsushima, Y Shishihara, H Matsunami, S Fukumoto and K Fujimoto

Graduate School of Engineering, Osaka University, 2-1 Yamadaoka, Suita, Osaka 565-0871, Japan

E-mail: matsushima@mapse.eng.osaka-u.ac.jp

Abstract. The re-working of the manufacturing process from the reliability evaluation after the production process is a significant cost and loss of energy. In-vehicle electronic devices are exposed to multiple environmental loads such as thermal and vibrational loads. The effects of the material property changes in the thermal cycle load on the fatigue life of solder joints were estimated with our fatigue simulation for the purpose of constructing a design method considering the fatigue characteristic changes in the thermal cycle. The fatigue lives were estimated with and without considering the creep property changes measured by indentation tests. The fatigue ductility index and the coefficients increased toward the reported values by considering the creep property changes.

1. Introduction

In recent manufacturing, sustainability for reducing the environmental load is necessary, as is the cost reduction against the worldwide economic depression. The re-working of the manufacturing process from the reliability evaluation after the prototyping or the production process is a significant cost and loss of energy. Estimating the fatigue life at the design step is a huge contribution toward reduction of the environmental load.

It is experimentally known that the low cycle fatigue life of solder joints in electronic devices is fit by Coffin-Manson's law [1][2]. The fatigue life of solder joints under thermal cycle loading estimated by Coffin-Manson's law has been reported [3][4]. In the condition of high acceleration such as resonance frequency, the fatigue life can also be estimated by Coffin-Manson's law [5]. For these estimations, the inelastic strains are calculated with the finite element method (FEM) by applying the initial material properties and structure. However, the inelastic strain changes according to the structural changes by crack initiation. In vibration tests after thermal cycle loading, the degradation of the fatigue life is impossible to estimate by a simple cumulative damage rule [5]. Furthermore, the material properties of the solder under thermal cycle loading changes according to the microstructural changes [6]. Analyses considering the structural changes with crack initiation and propagation have been executed [7]; however, the property changes described above were not considered.

In this paper, we perform a crack simulation considering the material property changes according to the microstructural change with strain and thermal loading as well as the structural change with crack initiation and propagation in addition to the initial stress-strain distribution during the construction of electronic devices. This simulation clarifies the effects of material property changes on fatigue life and is used to construct a fatigue life estimation system against multiple environmental stresses.

2. Sample and experimental method

2.1. Sample
Sn-3Ag-0.5Cu solder joints of a ball grid array (BGA) package with 272 terminals in 1.27 [mm] pitch assembled at the center of a glass epoxy board (FR-4 100 [mm] x 100 [mm] x 1.6 [mm]) are the target of this research (Figure 1). The BGA packages were assembled with the temperature profile in which the peak temperature is 240 [°C], which is 20 [°C] over the melting point of Sn-3Ag-0.5Cu solder.

2.2. Thermal cycle loading
A thermal shock test chamber (Espec TSV-40S) is used for the thermal cycle test. The temperature ranges are -40/80 [°C], -40/125 [°C], and -40/150 [°C]. The cycle time of 30 [min] consists of the 4 [min] ramp times and 11 [min] hold times. The failure criteria is the threshold of voltage (20 [mV]) of the daisy chain circuit connecting every solder bump in series during the passage of a 1.0 [mA] current.

2.3. Dynamic indentation test
The dynamic indentation test was executed to estimate the material property changes under thermal cycle stress. A bonding tester (RHESCA PTR-1100) with a Vickers' indentator was used for the test, in which a constant load of 0.15 [N] was held for 15, 50, and 100 [s].

3. FEM model and analysis

3.1. FEM model
The FEM analysis is executed to estimate the stress and strain distribution generated in the solder joint under thermal loading. Figure 2 shows the 1/8 model for the analysis. The corner solder bump, which is first loaded with the maximum stress and fractures in the experiments, is meshed in detail with 50 [μm] elements. The crack of approximately 10 [μm] is also modelled in the corner bump for the crack propagation analysis. The mesh size around the crack tip is 5 [μm]. Table 1 shows the material properties used in the simulation. The solver is ANSYS 11.0 SP1.

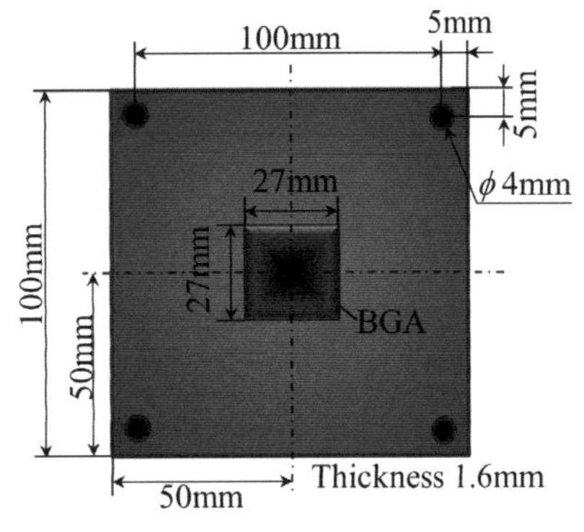

Figure 1. Size and position of BGA package on the substrate.

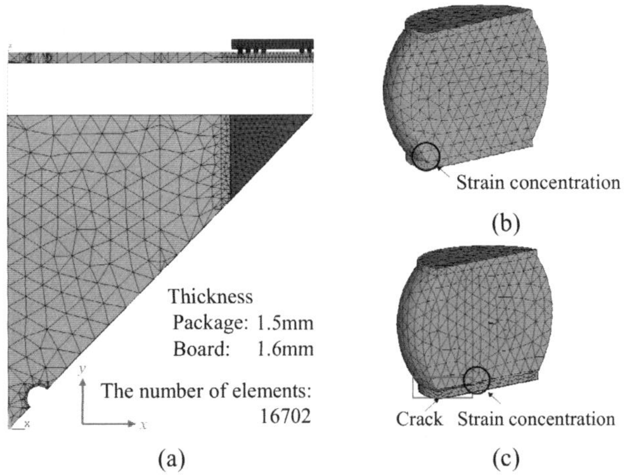

Figure 2. Finite element model of the BGA assembled board. (a) 1/8 model overview, (b) the corner solder bump in the crack initiation phase and (c) the corner solder bump in the crack propagation phase.

Table 1. Initial material properties used in the simulation.

	FR-4	Package	Sn-3Ag-0.5Cu	Cu
Young's Modulus	19	15	(See Table 2)	129.8
Poisson's ratio	0.18	0.24	0.38	0.3
Mass density [kg/m^3]	1930	2300	2100	8800
Linear expansion coefficient [ppm]	15	27	21.7	16.5

Table 2. Temperature dependent mechanical properties of Sn-3Ag-0.5Cu.

Temperature[°C]	Young's modulus [GPa]	Yield stress [MPa]	Tangent modulus [MPa]
-50	41.3	48	1100
-15	40.8	36	900
25	38.2	30	700
75	35.5	23	450
125	33.8	17	250

3.2. Thermal cycle analysis

The temperature profiles in the thermal cycle analysis are generated with the same conditions as in the experiments. The temperature ranges are -40/80 [°C], -40/125 [°C], and -40/150 [°C]. The temperature is held for 11 [min] at the maximum and minimum temperature and is changed in the 4 [min] ramp time.

3.3. Material properties

Table 1 shows the initial material properties used in the simulation. The elasto-plastic creep analysis is executed with the solder properties shown in Table 2. The bilinear kinematic hardening rule with the parameters in Table 2 and the Norton Law expressed by the following equations are applied in the elasto-plastic creep analysis [8]:

$$d\varepsilon/dt = A\sigma^n , \tag{1}$$

$$A = \exp(-0.008T^2 + 0.3313T - 58.923) , \tag{2}$$

$$n = -0.0481T + 13.575 , \tag{3}$$

where T expresses the temperature ([°C]).

4. Crack initiation and propagation simulation

4.1. Simulation and assumptions

The crack initiation and propagation simulation method is developed to investigate the effect of material property changes on the fatigue characteristics. The process leading to the failure of solder joints is divided into two phases: the crack initiation phase and the crack propagation phase. The crack simulation method is suggested in accordance with hypotheses about the relationship between the inelastic strain range under fatigue stress and the crack initiation life N_c or the thermal fatigue life N_f, described later. The thermal fatigue life is estimated based on the strain changing with crack initiation in the simulation. The effects of material property changes are considered by the changed properties in the calculation of the strain range.

Figure 3 is the algorithm of the crack initiation and propagation simulation. The circles in Figure 2(b) and (c) are the strain concentrated points at the solder bump in each phase, respectively, where $\Delta\varepsilon_{in}$(N), $\Delta\varepsilon_{in\_crack}(N)$ express the inelastic strain range at the Nth cycle in each phase. The changing

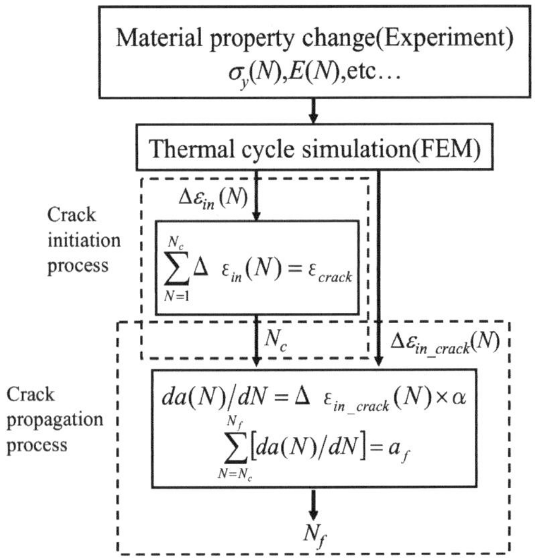

Figure 3. Algorithm of thermal fatigue life prediction from material property changes.

material properties in the thermal cycle such as yield stress σ_y, and Young's modulus, are applied to the FEM analysis to calculate the inelastic strain range $\Delta\varepsilon_{in}$(N) and $\Delta\varepsilon_{in\_crack}$(N) at the concentrated point. The crack initiation life N_c and the thermal fatigue life N_f are based on these inelastic strain ranges. The hypotheses in the simulation are as follows.

[Hypothesis 1] Crack initiation phase

It is reported that the ductile crack of structural steel is generated when the cumulative strain reaches a certain value [9]. In the crack initiation phase in our simulation, it is hypothesized that the crack initiates when the cumulative strain ranges reach the constant value of the critical cumulative inelastic strain ε_{crack}, as in the following equation:

$$\sum_{N=1}^{Nc}\{2\times\Delta\varepsilon_{in}(N)\}=\varepsilon_{crack}\,. \tag{4}$$

[Hypothesis 2] Crack propagation phase

It is also reported that the crack simulation shows good correspondence between the crack tip opening displacement (CTOD) and the crack propagation rate Δa in the simulation and the experiment with the hypothesis that the crack propagates when the inelastic strain at the crack tip reaches a certain value [10]. In our simulation, the crack propagation rate is decided from the inelastic strain at the tip of the crack, shown in the following equations:

$$da(N)/dN=2\times\Delta\varepsilon_{in\_crack}(N)\times\alpha\,, \tag{5}$$

$$\sum_{N=N_c}^{N_f}a(N)=a_f\,, \tag{6}$$

where $da(N)/dN$ is the crack propagation rate at the Nth cycle, α is the crack propagation coefficient, and a_f is the critical crack length to failure.

4.2. Inelastic strain range in crack propagation

Elasto-plastic creep analyses were executed using the corner bump model with 10, 100 and 200 [μm] crack lengths under thermal cycle loading. The inelastic strain range at the tip of the crack changed only a few percent with the different lengths of the cracks. Therefore, the inelastic strain range $\Delta\varepsilon_{in\_crack}(N)$ is calculated using the model with the 10 [μm] crack in the corner bump in our simulation.

4.3. *Crack initiation and propagation condition*

The critical crack initiation inelastic strain ε_{crack} and the crack propagation coefficient α are defined based on observation of the intersection of the BGA assembled board after the thermal cycle with the temperature range of -40/150 [°C] is loaded. The 10 [μm] crack is observed after 300 [cycle]. Therefore, ε_{crack} is defined as the following equation using the inelastic strain range at the first cycle $\Delta\varepsilon_{in}(N=1)$ of the same temperature condition:

$$\varepsilon_{crack} = 2 \times \Delta\varepsilon_{in}(N=1) \times 300 . \tag{7}$$

The crack propagation rate is calculated to be approximately 0.46 [μm/cycle]. The α is hypothesized in the following equation using the inelastic strain range at the first cycle ($\Delta\varepsilon_{in\_crack}(N=1)$) of the same load condition:

$$\alpha = 0.46 / \{2 \times \Delta\varepsilon_{in\_crack}(N=1)\} . \tag{8}$$

The criterion of failure in the simulation is that the crack beginning at one end of the solder bump reaches the other end. Therefore, the failure crack length a_f is the diameter of the solder bump, 600 [μm].

If every change of the inelastic strain range in the thermal cycle is calculated by applying the changed material properties, the calculation cost would be tremendous. Therefore, the crack initiation process was divided into several steps and the inelastic strain ranges are calculated in the first cycle of each step using the material properties at that point. The inelastic strain ranges during each step are calculated with linear interpolation.

5. Material property changes under thermal cycle stress

5.1. *The method of investigating the fatigue characteristics according to the material property changes*

The fatigue lives under the three temperature ranges of the thermal cycle loadings, -40/80 [°C], -40/125 [°C], and -40/150 [°C], are estimated using the suggested crack initiation propagation

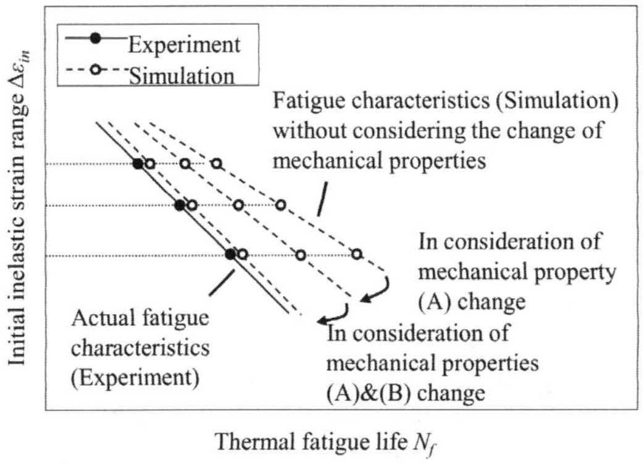

Figure 4. Investigating the effects of material property changes on the simulated fatigue characteristics.

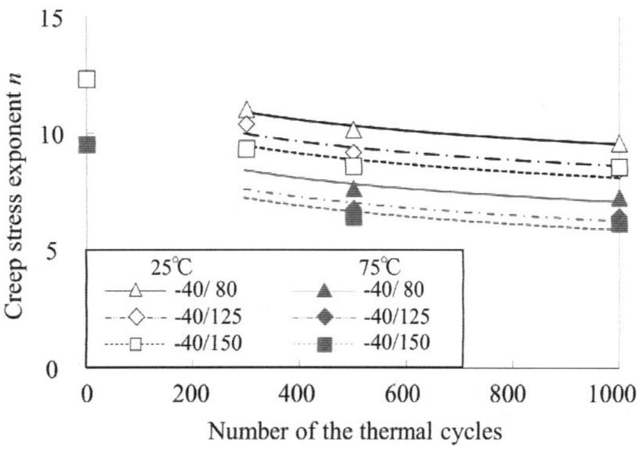

Figure 5. Creep stress exponent changes under thermal cycle stress.

simulation. Figure 4 shows the concept of the method to investigate the effects of material property changes on the fatigue characteristics. The figure shows the actual fatigue characteristics and the ones estimated in the simulation with/without considering material property changes. The inelastic strain range $\Delta\varepsilon_{in}$ on the vertical axis and the thermal fatigue life N_f on the horizontal axis are both logarithmic. The bold line shows the actual fatigue characteristics for the relationship between the initial $\Delta\varepsilon_{in}$ calculated in the FEM analysis and the fatigue life obtained by the thermal cycle test. The broken line shows the simulated fatigue characteristics with the fatigue life calculated in this simulation based on the inelastic strain ranges $\Delta\varepsilon_{in}$ and $\Delta\varepsilon_{in\_crack}$.
The actual fatigue characteristics are expected to be obtained as the result of material property changes in the thermal cycle stress. Therefore, the effects of material property changes on the fatigue characteristics are investigated by comparing the fatigue characteristics simulated with and without material property changes in the simulation.

5.2. The material property effects on the fatigue characteristics
Each material property, such as yield stress, Young's modulus, tangent modulus, and creep exponent, is changed. Then, the crack initiation and propagation simulations with the range -40/75 [°C] as the in-use environment for in-vehicle electronic devices [11] are executed to investigate the effect of each material property change on the fatigue characteristics. The results are shown in Table 3. The variables m and C in the table are the fatigue ductility exponent and the fatigue ductility coefficient of Coffin-Manson's law, respectively, represented by the next equation:

$$\Delta\varepsilon_{in}^{m} \cdot N_f = C \,. \tag{9}$$

The creep properties have a larger influence on the fatigue characteristics compared with the yield stress, Young's modulus, and tangent modulus. The fatigue life changes are approximately 2.8 [%] at most with the yield stress, Young's modulus, and tangent modulus changes; however, the fatigue life changes 17 [%] with the creep exponent change.

Table 3. Material property changes and fatigue characteristics.

Material property changes (at 25°C)		Fatigue ductility exponent m	Fatigue ductility coefficient C	Thermal Fatigue life N_f [cycles]	Life deterioration [%]
Initial property		1.051	62.46	8788	0.00
Yield stress σ_y [MPa]	30→24	1.042	57.12	8702	0.99
Young's modulus E [GPa]	38.6→70	1.069	70.98	8545	2.77
Tangent modulus E_t [MPa]	700→350	1.052	62.96	8780	0.10
Creep exponent n	12.3→9.9	1.130	103.72	7326	16.64

5.3. The measurement of creep property changes by indentation tests
The actual creep property changes are measured in the solder bump of the BGA assembled board after the thermal cycle is loaded. Although the tensile test of the solder bump in the actual assembled board is impossible, Sargent et al. confirmed good correlation between the creep exponents and the time dependent results of the indentation test [11]. We also executed the indentation test to measure the creep property changes. Figure 5 shows the creep exponent measured by indentation tests at 25 [°C] and 75 [°C] after the thermal cycle stress with -40/80 [°C], -40/125 [°C], and -40/150 [°C] temperature ranges. The initial properties are the results of Sakai's measurement of bulk solder [8]. The properties after thermal cycle loading are calculated with the change ratio of our measurement results. The

approximated curves are also indicated in Figure 5 by considering the changes as exponential. The creep exponent degradations are observed and the changes are larger in the thermal cycle test with the higher maximum temperature.

5.4. The fatigue characteristics considering the measured creep property changes

The fatigue characteristic changes are calculated with the suggested simulation using the measured creep exponents. The creep exponents at the temperatures -35 [°C], 75 [°C], and 125 [°C] are assumed to change according to the thermal cycle with the same exponential ratio and creep exponents obtained at room temperature (25[°C]).

The fatigue characteristics considering the creep exponential changes are shown in Figure 6 with the fatigue characteristics calculated with constant material properties. The fatigue characteristics of the SnAgCu solder joint estimated in the suggested simulation show changes such as the fatigue ductility exponent from 1.05 to 1.19 and the fatigue ductility coefficient from 62.2 to 135.2. The values calculated by considering the creep property changes are closer to the reported fatigue ductility exponent and coefficient, respectively [12].

Figure 7 shows the crack behavior expected by the simulation and the thermal cycle test results at the conditions of temperature range -40/150 [°C], -40/125 [°C], and -40/80 [°C]. The simulation results with and without considering the creep property changes are both shown. The results of the condition
-40/150 [°C] are, of course, corresponding because the crack propagation rate is defined based on the experiment result. The simulation results by considering the creep property changes for the other two conditions show better expectations than those without considering the material property changes, although they are about 10% on the safer side.

6. Summary

The crack simulation to estimate the fatigue life of solder joints in electronic devices considering the strain distribution changes due to crack initiation and the material property changes due to material microstructural changes under thermal cycle stress is constructed. The suggested crack initiation and propagation simulation clarifies that the creep property influence on the fatigue life is larger than are the influences of the yield stress, Young's modulus, and tangent modulus. The actual creep property changes in the solder joints in the BGA assembled board are measured with indentation tests. The creep exponents decreased by the thermal cycle, and the degradation of the sample loaded with the

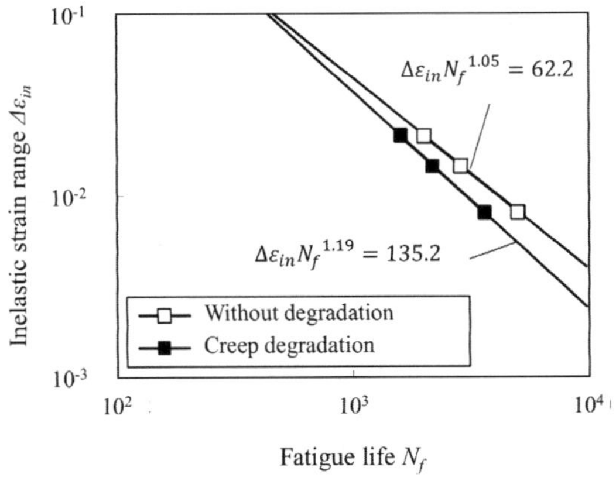

Figure 6. Fatigue characteristics under thermal cycle stress evaluated by thermal cycle stress model.

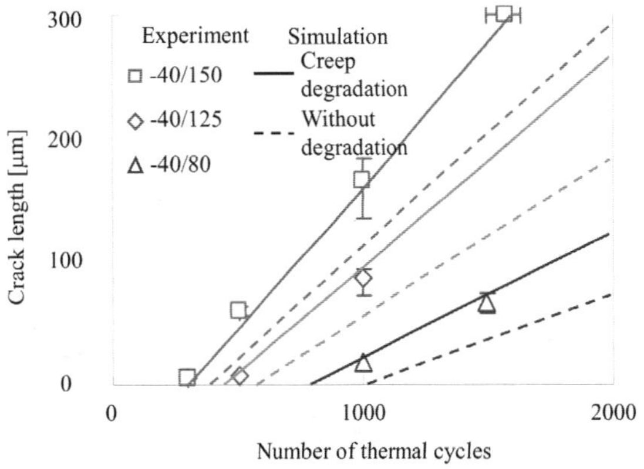

Figure 7. Simulated crack initiation and propagation with and without creep degradation compared with the experiment.

higher temperature thermal cycle is more significant. The fatigue characteristics are estimated with the measured creep property changes in the suggested simulation. The fatigue ductility exponent and the coefficient become closer to the reported values than the simulation result with constant material properties. The crack behaviors are simulated with the thermal cycle test of temperature ranges -40/150, -40/125 [°C], and -40/80 [°C]. The estimation results considering the creep property changes are better than those with a constant value.

From the results described above, the fatigue ductility exponent m and the fatigue ductility coefficient C are affected by the creep property changes under thermal cycle loading. The fatigue life simulation can be constructed without measuring the material property changes by estimating the changes from the stress-strain distribution in a thermal cycle analysis.

Acknowledgments
I would like to thank the Kobe Test Center of ESPEC Corp. and the Senju Metal Industry Co., Ltd. for the experiments and the micro-soldering education division of the Japan Welding Engineering Society for their financial support.

References
[1] Coffin L F and Wesley R P 1954*Trans. ASME* A Study of Effects of Cyclic Thermal Stresses on Ductile Metals **76** 931-950
[2] Manson S S 1966 *Thermal Stress and Low-Cycle Fatigue* McGraw-hill Book Company
[3] Kariya Y and Suga T 2007 *Fatigue Frac. Eng. Mater. Struct.* Low-cycle fatigue properties of eutectic solders at high temperatures **30** 413–419
[4] Terasaki K, Nagano K Miura, H Nakatsuka T 2001 *Mate* Stress Simulation Techniques to Fatigue Life Prediction in Electronic Devices **7** 441-446
[5] Matsushima M, Shishihara Y, Matsunami H, Fukumoto S and Fujimoto K 2010 *Proc. ICEP* The Material Deterioration and Microstructure Changes by the Thermal Load and the Effects on the Fatigue Life against Vibration Load 561-566
[6] Wiese S and Wolter K J 2004 *Microelectronics Reliability* Microstructure and creep behavior of eutectic SnAg and SnAgCu solders **44** 1923-1931
[7] Nishimura Y, Qiang Y and Maruoka T 2009 *Proc. EPTC* Reliability evaluation of fatigue life for solder joints in chip components considering shape dispersion **11** 838-845
[8] Sakai H, Nishimura K, Motegi M, Itoh N and Sakuyama S 2002 *Mate* The prediction of thermal fatigue life of lead-free solder joints **8** 443-448
[9] Ohata M and Toyoda M 2004 *Sci. and Technolo. of Advanced Materials* Damage Concept for Evaluating Ductile Cracking of Steel Structure Subjected to Large-Scale Cyclic Straining **5** 241-249
[10] Michiba K, Hiramatsu H, Hasegawa H, Matsumura H and Toyoda M 1996 *journal of the Japan Welding Society* Effect of Plastic Restraint on Strain at Crack Tip During Ductile Fracture Process : Study on Evaluation of Stable Crack Growth Resistance of Mis-matched Weld Joint **14** 414-421
[11] *International Organization for Standardization*, Road vehicles – Environmental conditions and testing for electrical and electronic equipment, ISO 16750-4
[12] Sargent P M and Ashby M F 1992 *Materials Sci. and Technolo.* Indentation creep **8** 594-601
[13] Nishimura K, Motegi M, Sakai H, Kawase K, Sakuyama S, Imamura K and Morimune K 2001 *Journal of Japan Institute of Electronics Packaging* Lead Free Solder BGA Fatigue Life Analysis **4** 416-420

ECO-MATES 2011 IOP Publishing
Journal of Physics: Conference Series **379** (2012) 012030 doi:10.1088/1742-6596/379/1/012030

Study on the low energy & pollution manufacturing of micro cutting tools by powder injection molding process

Young-Moo Heo[1,#], Gun-Hee Kim[1], Sung-Ho Chang[1] and Geun-An Lee[1]
[1]Molding & Forming Technology R&D Group, Korea Institute of Industrial Technology, 7-47 Songdo-Dong, Yeonsu-Gu, Incheon, 406-840, South Korea

E-mail: ymheo@kitech.re.kr

Abstract. In this paper, micro cutting tools were manufactured by the powder injection molding process. Most of cutting tools are manufactured by bulk-molding and grinding methods but, the fabrication of micro cutting tools is very difficult because of their minute flute shapes and cutting edges. Therefore, a powder injection molding (PIM) process was used to fabricate the green part of a micro cutting tool with zirconia mixer feedstock and the de-binding and sintering processes were performed. Besides, the grinding processes can be dropped by PIM. Finally, the micro cutting experiment using the newly manufactured tool by PIM was executed for verifying the utility of manufactured tool.

1. Introduction

As the trends of sub-miniature and slim products are recently accelerating in many industries, including machining, electric appliance, electronics, and bioengineering industries, interest in MEMS and high precision micro processing technologies is increasing. Especially in future-oriented industries such as information technology (IT), miniature/sub-miniature fuel cells, and bio-chips, micro-processing technology is emerging as an essential element, and many studies are being conducted on micro processes for the prototyping and mass production of products and parts that have sub-miniature functional micro patterns and shapes [1-4].

In the initial stage, micro processing technology was mainly the simple application of semiconductor manufacturing processes, but with the development of manufacturing technology for high precision processing equipment and tools, the applications of micro cutting technology in related industries are increasing. The quality of this cutting technology, which is based on the mechanical material removal mechanism, may be poorer than that of chemical or electrochemical processing methods. However, due to such advantages as accurate control of processing dimensions, 3-D micro shape processing, and diversity of processing materials, micro cutting is very useful for the manufacturing of micro injection and press molds and light metal parts that require mechanical hardness in MEMS.

In micro end milling, which is a representative micro cutting process, the micro endmill is the core element. The micro endmill determines the limits of the shape size that can be processed, and the geometry and properties of tools, such as smooth discharge of chips and wear resistance, have a direct effect on the shape processing quality. At present, most micro endmills are made of hard materials such as tungsten carbide and ceramics. Hard materials are appropriate for tools for their thermal characteristics and hardness, but as they are typically difficult-to-cut materials, high production costs are required to achieve various geometries of endmill, such as helix and relief angles. Furthermore, as micro endmills are manufactured through multiple processes, such as ion beam process, molding, turning and grinding, they cause such production process problems as increased energy of production process and the generation of by-products and dust from the turning and grinding processes.

On the other hand, manufacturing micro tools through powder injection molding (PIM) using metal and ceramic powders is advantageous for high quality mass production, and can reduce production energy consumption compared to existing tool manufacturing processes, and fundamentally solve the by-product problem by omitting the turning and grinding processes. Therefore, in this study, micro cutting tools that provide ease of manufacturing of tool blades and better structural hardness were designed and manufactured through the PIM process using zirconia ceramic materials with excellent wear resistance and hardness. Furthermore, the efficiency of the manufactured micro tool was verified through a micro cutting experiment with Cu as an electrode material.

2. Design of micro cutting tool and powder injection mold

2.1 Design of micro cutting tool

In this study, two types of micro cutting tools, flat-type and ball-nosed type, were designed by reviewing the existing micro tools and considering the characteristics of the micro cutting process. The flat-type tool was designed with a diameter of 500 μm, a blade length of 1 mm (double the tool diameter), and a tool relief angle of 5 degrees, to minimize cutting resistance at the bottom surface of the tool during processing. The ball-nosed type tool was designed with a radius of 250 μm and the same blade length as that of the flat type (double the tool diameter). Figure 1 shows the designed tool shapes.

To examine the micro cutting effectiveness of the designed tools, prototype tools were manufactured using hard materials before the PIM molds for the micro cutting tools were designed. The prototype tools with hard materials were designed with the same shape and geometry as the tools designed using turning and grinding processes (Figures 2 and 3). Tool geometry measurement and visual observation using a tool microscope found that the tool shape reproducibility was relatively good and the tool manufacturing error was 0.02 mm or less in diameter.

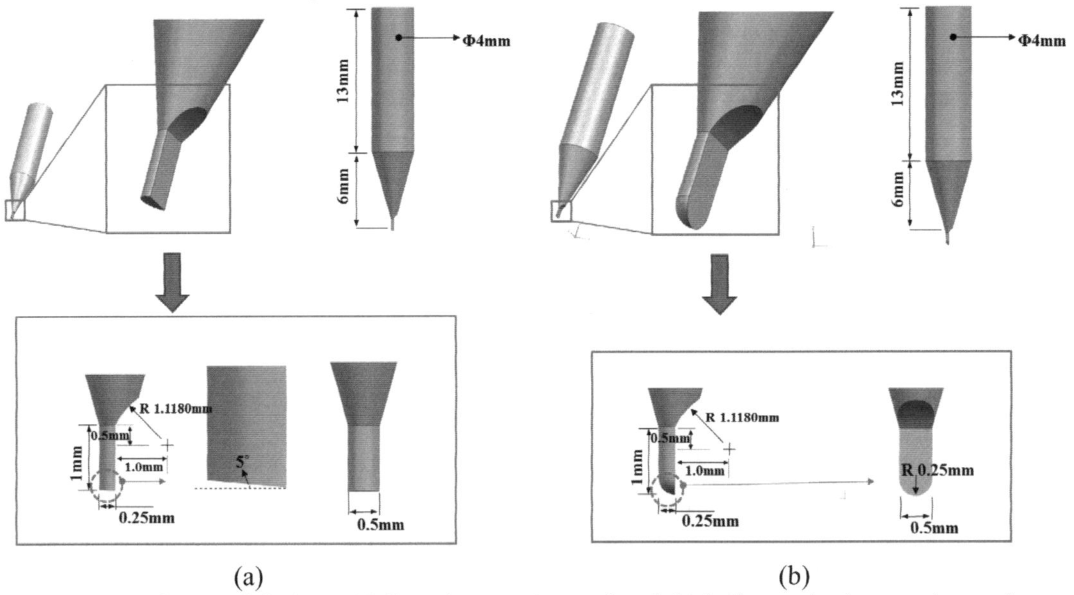

Figure 1. Micro tool designs. (a) flat micro cutting tool and (b) ball-nosed micro cutting tool.

Journal of Physics: Conference Series **379** (2012) 012030 doi:10.1088/1742-6596/379/1/012030

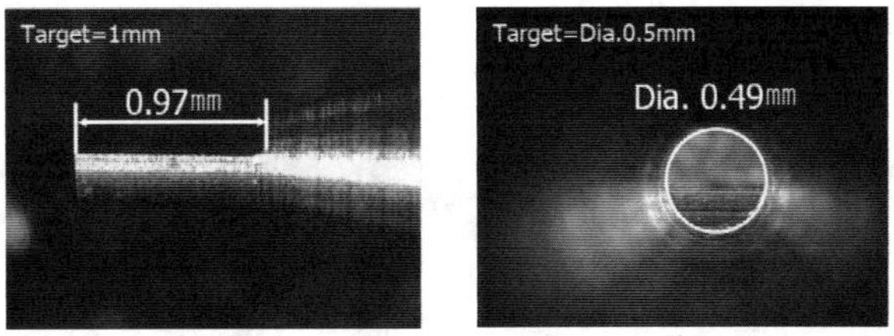

(a) (b)

Figure 2. Prototype flat micro cutting tool. (a) tool edge length and (b) tool diameter.

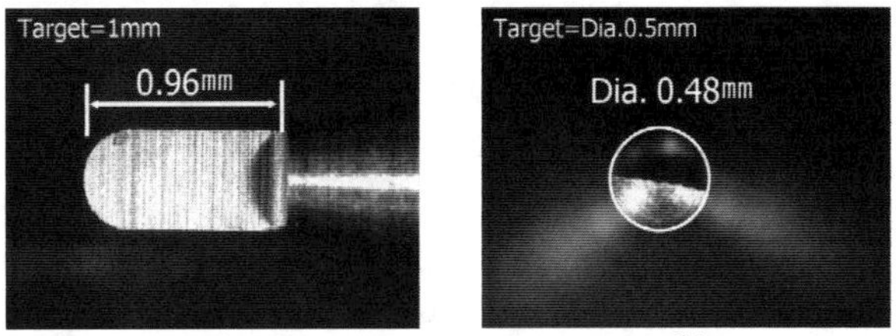

(a) (b)

Figure 3. Prototype ball-nosed micro cutting tool. (a) tool edge length and (b) tool diameter.

Table 1. Detailed conditions for micro cutting experiments.

Item	Conditions
Tool	1. Proto-type (flat) WC tool 2. Proto-type (Ball-nosed) WC tool 3. Dia. 500 ㎛ 2-flute endmill
Specimen	6:4 brass [Cu 60%, Zn 40%]
Cutting feature	Micro rob [width 200 ㎛]
Cutting velocity	30m/min [31,847 rpm]
Feed per tooth	0.94 ㎛/tooth [60mm/min]
Depth of cut	200 ㎛ [20 ㎛/step x 10steps]
Cutting environment	Wet cut with cutting lubrication oil

Table 1 shows the conditions for micro cutting experiment using the prototype tools. The cutting experiment was performed with commercial micro endmills under the same processing conditions to compare the processing results. As a result, it was found that the micro tool shapes designed in this study could perform the micro cutting process and provide relatively good micro-slot bottom surface quality. However, the processing quality of the ball-nosed type prototype tool was poorer than that of the flat-type tool, as it generated many cusps. Thus, it appears to be poorer than the general commercial hard endmills with respect to bur generation (Figure 4).

Figure 5 shows the width measurements of micro slots that were processed using the ball-nosed type prototype tool. The measurements revealed a shape processing error of up to 1.2 μm. The processing precision was relatively good, but as mentioned above, the bur generation quality was somewhat poor when the tools designed in this study are used. However, as micro cutting tools manufactured with ceramic materials through PIM have relatively good wear resistance, their utility as rough grinding tools is expect to be high.

(a) (b) (c)

Figure 4. Micro cutting features. (a) micro endmill (s-flutes), (b) prototype flat micro cutting tool, and (c) prototype ball-nosed micro cutting tool.

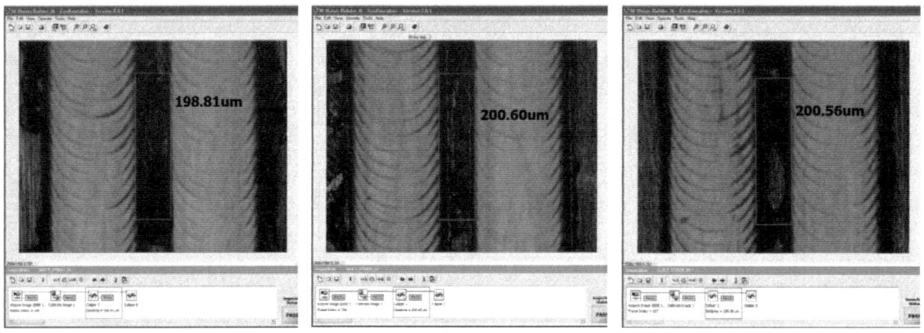

(a) (b) (c)

Figure 5. Measurements of micro slot widths. (a) entrance position, (b) center position, and (c) end position.

2.2 Design of powder injection mold

In this study, PIM molds were manufactured for molding the designed cutting tools. The general mold structures were designed to be similar to general plastic injection molds, and the flow system was designed in consideration of the PIM process characteristics (Figure 6). As PIM products contract through debinding and sintering after molding, the final product is smaller than the first molded (green) part. Thus, the molding shape size inside the core was increased by a factor of 1.3267 by considering the contraction characteristics of the debinding and sintering processes as well as the contraction characteristics of the feedstock materials used in the PIM process.

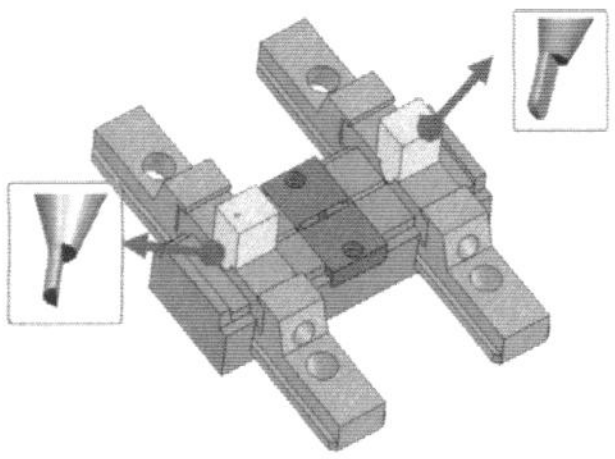

Figure 6. 3D design of powder injection mold for micro cutting tools.

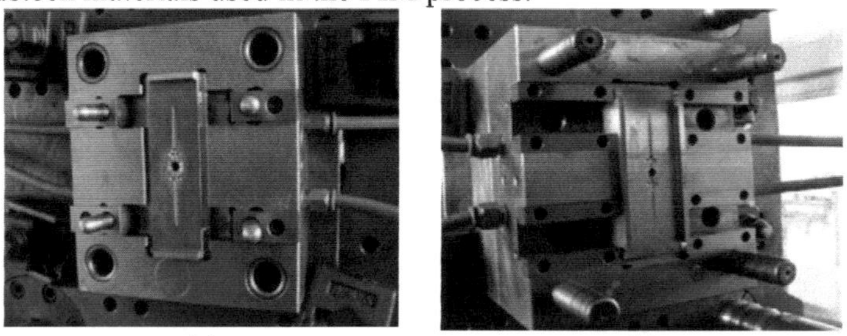

(a) (b)

Figure 7. Fabricated powder injection mold. (a) upper part and (b) bottom part.

3. Fabrication of micro cutting tools by PIM process

3.1 Powder injection molding

To manufacture the micro cutting tools using the designed molds, zirconia ceramic material was chosen as the feedstock. Table 2 shows the characteristics of the chosen material. This material was found to be appropriate for molding micro cutting tools as it had excellent thermal expansion , hardness, and wear resistance properties. Table 3 shows the molding conditions for manufacturing micro cutting tools and Figure 8 illustrates the shapes of the injection molded micro cutting tools.

Table 2. Material properties of zirconia ceramics.

Physical characteristics	
Crystallite size [nm]	26
Granule size [㎛]	60
Bulk density [g/cm^3]	1.2
Process characteristics	
Green density [g/cm^3]	2.50
Sintered density [g/cm^3]	5.50
Bending strength [MPa]	1900
Fracture toughness	5.0
Hardness [Hv10]	1400

Table 3. Powder injection molding conditions.

Item	Conditions
Injection temperature [℃]	170
Mold temperature [℃]	70
Injection time [sec]	0.92
Injection pressure [MPa]	37
Packing pressure [MPa]	37
Packing time [sec]	3
Cooling time [sec]	25

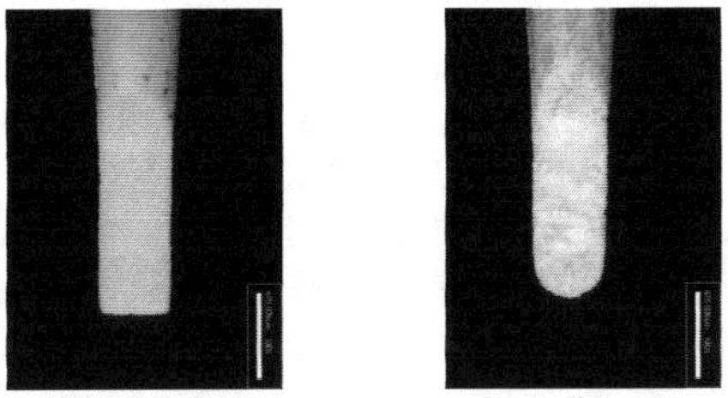

(a) (b)

Figure 8. Molded parts of micro cutting tools. (a) flat micro cutting tool and (b) ball-nosed micro cutting tool.

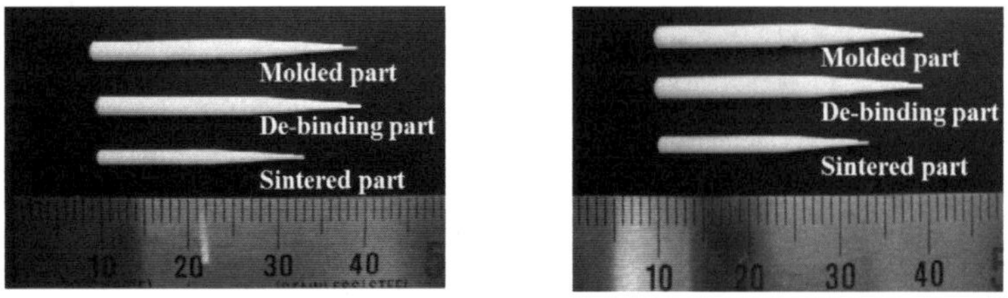

(a) (b)

Figure 9. Fabricated micro cutting tools. (a) flat micro cutting tool and (b) ball-nosed micro cutting tool.

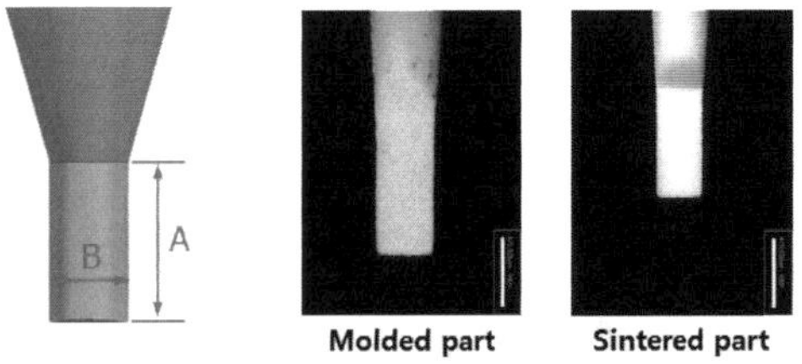

	A[mm]	B[mm]
Designed value	1.000	0.500
Molded part	1.351	0.609
Sintered part	1.037	0.480

(a) (b) (c)

Figure 10. Inspection of flat micro cutting tool. (a) inspection item, (b) photo of molded and sintered parts, and (c) datasheet of tool geometries.

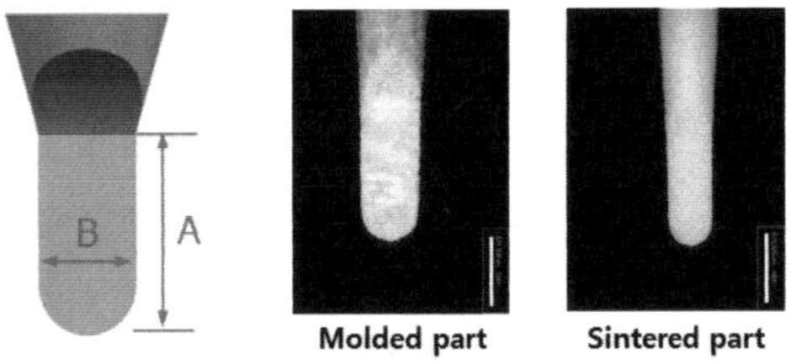

	A[mm]	B[mm]
Designed value	1.000	0.500
Molded part	1.370	0.613
Sintered part	1.055	0.485

(a) (b) (c)

Figure 11. Inspection of ball-nosed micro cutting tool. (a) inspection item, (b) photo of molded and sintered parts, and (c) datasheet of tool geometries.

3.2 Debinding and sintering process

The debinding process of the injection molded body (green part) in this study consisted of first catalysis debinding and second thermal debinding. The first catalysis debinding used n-hexane solvent at a temperature of 50°C and holding time of 12 hours. The second thermal debinding process consisted of raising the temperature to 250°C at 2°C per min, holding for 2 hours at 250°C, raising the temperature to 450°C at 1°C per min, holding for 2 hours at 450°C, raising the temperature to 700°C at 1°C per min, and holding for 1 hour at 700°C.

After debinding, a sintering process was performed to manufacture the final tool. For sintering, a sintering machine that could sinter at temperatures up to 1700°C was used. The sintering process consisted of raising the temperature to 1550°C at 100°C per hour, holding for 2 hours at 1550°C, and cooling to room temperature at 200°C per hour. Figure 9 shows the final micro cutting tools that were fabricated through sintering.

To test the dimensional precision of the tools manufactured through the injection molding-debinding-sintering process, measurements were taken using a tool microscope. As shown in the measurement results in Figure 10, the maximum shape error of the sintered part of the flat-type tool was 0.037 mm compared to the designed value, and the size of the sintered part was reduced by about

Journal of Physics: Conference Series **379** (2012) 012030 doi:10.1088/1742-6596/379/1/012030

22% compared to the molded part. Figure 11 shows the measurement results of the ball-nosed tool. The maximum shape error was 0.055 mm and as with the flat-type tool, the size of the sintered part was reduced by about 22% compared to the molded part.

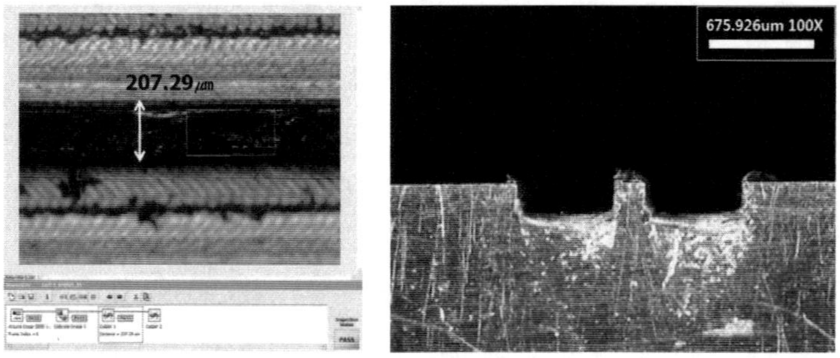

(a) (b)

Figure 12. Inspection of micro cutting feature fabricated by the flat micro cutting tool. (a) measurement of rib width and (b) sideview image of the micro feature.

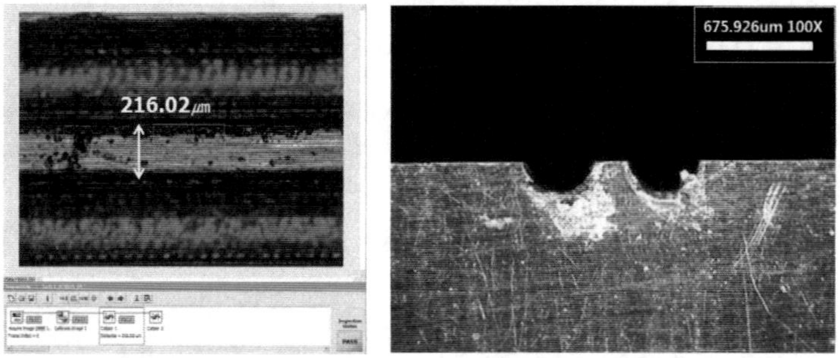

(a) (b)

Figure 13. Inspection of micro cutting feature fabricated by the ball-nosed micro cutting tool. (a) measurement of rib width and (b) sideview image of the micro feature.

4. Micro cutting experiments and results

To analyze the processing characteristics of ceramic micro cutting tools fabricated through PIM, the same experiment as the micro cutting experiment of the prototype tools was conducted. After the cutting experiment, the processed shape was examined and the processed micro rib width was measured. For tool wear analysis, the tool blade images were analyzed before and after processing.

Figure 12 shows the measurements of the micro rib shape that was processed with the flat-type micro tool. The reproducibility of the processed shape was good, but the maximum slot width processing error was 7.29 μm. The cause of this excessive shape processing error seems to be that, through the debinding and sintering after PIM, the tool was bent vertically, in the longitudinal direction of the tool, which worsened the runout error which occurs during tool setup. Figure 13 shows the measurements of the shape processed with the ball-nosed micro cutting tool. The shape processing quality is poorer and the processing error is greater than those of the flat type. The reason for this seems to be that the ball-nosed micro cutting tool has a different cutting load on the tool at different cutting depths, which causes irregular tool deformations during the process such as tool bending and runout.

Figures 14 and 15 show blade images before and after processing with the micro cutting tools that were fabricated through PIM. As shown in these figures, there was almost no tool wear. This excellent wear resistance was made possible by fabricating ceramic tools through PIM, and the wear resistance is better than that of hard materials. Micro cutting in general has the shortcoming of uneven processing quality for large or long area processes due to tool wear, but ceramic single-blade tools with excellent wear resistance will be more advantageous for rough grinding processes that have relatively large cutting volumes and loads. As a result, the cutting load is reduced in the post-processing, which reduces processing energy consumption and allows excellent processing quality. Above all, the reduction of the fabrication process and the omission of turning and grinding processes for micro cutting tools can reduce production energy consumption and establish a clean production process.

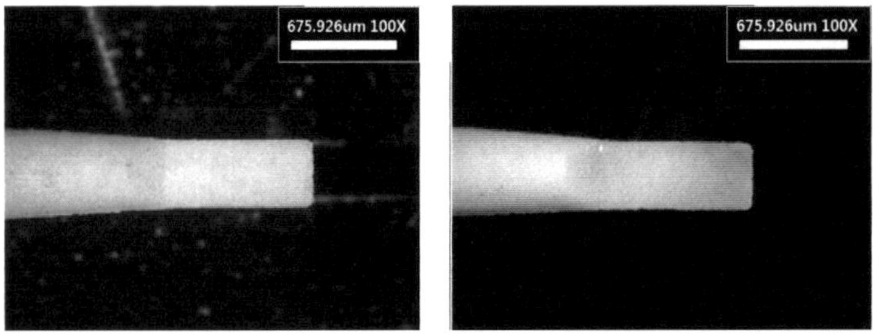

(a) (b)

Figure 14. Tool wear image comparison of the flat micro cutting tool. (a) new micro cutting tool and (b) used micro cutting tool.

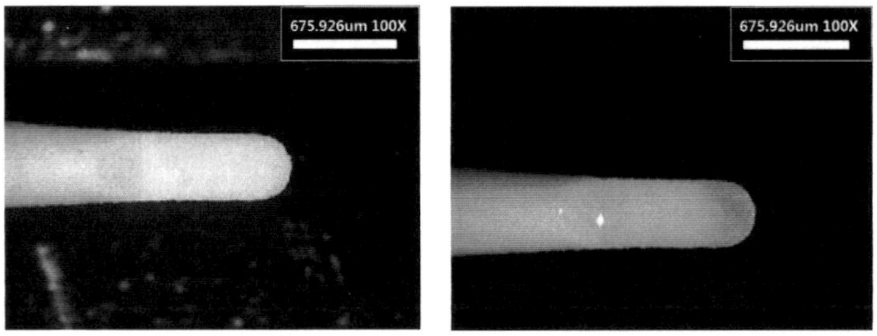

(a) (b)

Figure 15. Tool wear image comparison of the ball-nosed micro cutting tool. (a) new micro cutting tool and (b) used micro cutting tool.

5. Conclusions

In this study, micro cutting tools were fabricated through PIM with zirconia ceramic material with excellent wear resistance and hardness, and their efficiency were verified through cutting experiments. For this purpose, micro cutting tools were designed which are appropriate for the PIM process and offer relatively high hardness. Furthermore, the tool designs were verified through the manufacturing of prototype tools and cutting experiments. The PIM molds were designed and manufactured by considering the final tool designs, and the micro cutting tools with zirconia ceramic material were finally manufactured through the molding-debinding-sintering process. The micro cutting experiments with the fabricated tools revealed somewhat poor processing precision due to tool bending in the manufacturing process. However, as the tool has excellent wear resistance, it will be very useful as a rough grinding tool for large area micro cutting processes.

Micro cutting tools are generally manufactured through multiple production factories. This causes high production costs and energy consumption, as well as by-products due to turning and grinding. However, as the proposed micro cutting tool manufacturing process with PIM in this study can mass produce micro cutting tools only through molding, debinding, and sintering, not only the product cost and energy consumption can be reduced, but also the turning and grinding processes can be omitted.

Acknowledgments
This study has been financially supported by the Ministry of Knowledge Economy (MKE) as an 'Advanced Technology Center Project'.

References

[1] Asad A B M A, Masaki T, Rahman M, H S Lim and Y S Wong 2007 *J. Mat. Proc. Tech.* **192~193** 204
[2] J Chae, S S Park and Freiheit T 2006 *Int. J. Mach. T. Manuf.* **46** 313
[3] Ko S L, Y M Baron and Park J I 2007 *J. Mat. Proc. Tech.* **187~188** 19
[4] Aurich J C, Engmann J, Shueler G M and Harberland R 2009 *CIRP Ann.-Manuf.* **58** 311

Journal of Physics: Conference Series **379** (2012) 012031 doi:10.1088/1742-6596/379/1/012031

Characterization of inductively coupled RF plasmas for plasma-assisted mist CVD of ZnO films

Kosuke Takenaka, Yusuke Okumura and Yuichi Setsuhara

Joining and Welding Research Institute, Osaka University, 11-1 Mihogaoka, Ibaraki, Osaka 567-0047, Japan

E-mail: k_takenaka@jwri.osaka-u.ac.jp

Abstract. A plasma-assisted mist chemical vapor deposition (CVD) system with a high-density inductivity coupled RF plasma (ICP) source has been developed for low-temperature and high-rate deposition of zinc oxide films. In this paper, characterization of an ICP for plasma-assisted mist CVD is reported. It was found that the plasma density measured with a cylindrical Langmuir probe was as high as 1.2×10^{11} cm^{-3} and the electron temperature was 1.0 - 2.5 eV for an Ar plasma. The gas temperature in the ICP determined from molecular optical emission spectroscopy was 1450 K at an RF power of 1000 W for an Ar/air plasma. These results demonstrate that a plasma sustained by an inductively coupled RF plasma source has sufficient enthalpy for vaporization of mists and also has a high density for efficient dissociation of precursors and generation of oxygen radials during ZnO film deposition using plasma-assisted mist CVD.

1. Introduction

In recent years, zinc oxide (ZnO) has attracted considerable attention due to its diversity of applications. ZnO is a wide band semiconductor with a direct band gap of 3.37 eV at room temperature, which makes it a possible candidate for optical and electrical devices, such as ultraviolet emitters and detectors, transparent conductive oxides (TCOs), heterojunction transistors, thin film transistors and sensors [1-3]. For application to transparent conducting electrodes, the high electrical conductivity, high transparency to visible light, and material abundance make ZnO a very promising alternative to tin-doped indium oxide (ITO). Deposition of high-quality ZnO films for TCOs is attracting attention as one of the key technologies for fabrication of silicon thin-film solar cells to produce low-carbon energy. ZnO thin films have been deposited using various methods including dry processes such as magnetron sputtering [4-7], reactive evaporation [8], pulsed laser deposition (PLD) [9,10], metal organic chemical vapor deposition (MOCVD) [11,12], molecular beam epitaxy (MBE) [13,14], and wet processes such as spray pyrolysis [15-17], sol-gel [18] and mist CVD [19-21]. The dry deposition techniques are suitable for obtaining high-quality crystalline thin films; however, the deposition rate is low compared with that of wet processes. While wet deposition techniques contribute to system simplicity and high-rate deposition, the controllability of these processes is generally worse than that of dry processes. Furthermore, substrate heating is essential for pyrolysis in wet processes. Therefore it is difficult for conventional wet processes to deposit ZnO films on polymer materials since polymer materials are heat-sensitive.

In order to address these problems, we have developed a plasma-assisted mist CVD method that uses a high-density inductively-coupled RF plasma source, allowing low-temperature and high-rate

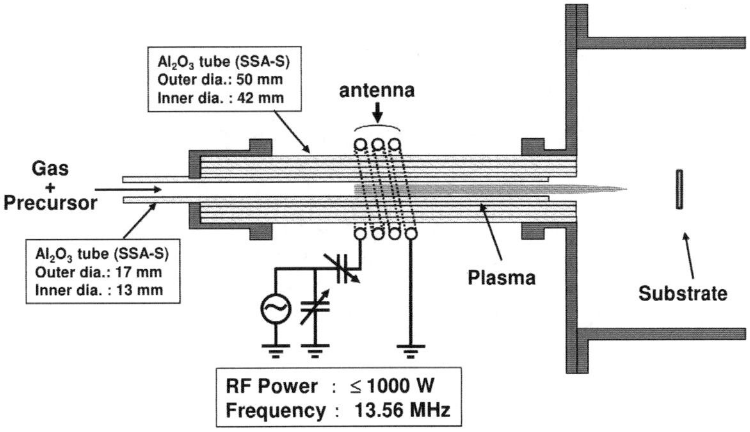

Figure 1. Schematic illustration of plasma-assisted mist CVD reactor.

deposition. Plasma-assisted mist CVD is a combination of mist CVD [19-21], which has the advantages of treating mist particles like gas, and plasma-assisted deposition. In this method, a water and/or alcohol solution of a metallic salt of zinc is ultrasonically atomized to form mist particles of the solution, and the particles are then transferred by a carrier gas onto the substrate, forming a ZnO film promoted by chemical reactions due to plasma irradiation. It is possible for this method to control both surface reactions and chemical reactions in the gas phase, allowing low-temperature and high-rate deposition of ZnO films.

We consider that in plasma-assisted mist CVD, the gas temperature together with the plasma density are among the key parameters determining the properties of the deposited films since they influence vaporization of the mists, gas-phase reactions of radicals and surface reactions of the precursors.

In this study, characterization of inductively coupled RF plasmas for plasma-assisted mist CVD was carried out in terms of plasma density and gas temperature.

2. Experimental

A schematic diagram of the experimental setup for the plasma-assisted mist CVD reactor is shown in Fig. 1. A water-cooled copper coil antenna (three turns) is looped around an alumina discharge tube with an outer diameter of 50 mm and an inner diameter of 42 mm. The antenna is coupled to a 1000 W RF power generator at 13.56 MHz via a matching network. Ar and Ar/air mixture gases are supplied at a flow rate of 2.5 slm and 2.5 slm/12.5 sccm, respectively, from upstream of a stainless tube with an outer diameter of 18 mm and an inner diameter of 15 mm. In the actual deposition process, a $Zn(CH_3COO)_2$ solution was used as the zinc oxide source. The $Zn(CH_3COO)_2$ solution was supplied by ejection of droplets of $Zn(CH_3COO)_2$ solution from a nozzle in the vacuum chamber. The total pressure was kept constant at 10-100 Pa. The Si and glass substrates were located in the downstream region at 300 mm from the copper coil antenna.

The plasma density and electron temperature were measured with a cylindrical Langmuir probe at a distance of 195 mm from the antenna. Optical emission spectroscopy (OES) analysis was carried out using a high-resolution fiber optic spectrometer (Ocean Optics HR-4000).

3. Theoretical background for spectral analysis

In this study, we analyzed the first and second positive systems of a nitrogen molecule to estimate the vibrational and rotational temperatures. Generally, the line intensity of the radiative transition of a molecule $(n', v', J') \rightarrow (n'', v'', J'')$, where n is the electronic state, v is the vibrational quantum number and J is the rotational quantum number, can be expressed as

$$I(n', v', J' \rightarrow n'', v'', J'') = hvA(n', v', J' \rightarrow n'', v'', J'')N_{n', v', J'}, \quad (1)$$

where $I(n', v', J' \rightarrow n'', v'', J'')$ is the radiant flux of the emission, h is Planck's constant and $N_{n', v', J'}$ is the number density of molecules in the upper states.

Concerning the transition probability $A(n', v', J' \rightarrow n'', v'', J'')$, we have [22]

$$A = \frac{64\pi^4 v^3}{3hc^3 g_{n'}} \frac{1}{2J'+1} \sum_{J'J''} \left|\bar{R}_{J'J''}\right|^2 q_{v'v''} S_{J'J''}), \tag{2}$$

where c is the velocity of light, $\sum_{J'J''} \left|\bar{R}_{J'J''}\right|^2$ is the transition moment, $g_{n'}$ is the statistical weight of the electronic state n', $q_{v'v''}$ is the Franck–Condon factor and $S_{J'J''}$ is the Hönl–London factor.

When we assume a Boltzmann distribution for the vibrational and rotational states, the number density of molecules in the upper states $N_{n';v';J'}$ is rewritten as

$$N_{n', v', J'} = N_{n'} \exp\left(-\frac{E_{vib}}{kT_{vib}}\right)(2J'+1)\exp\left(-\frac{E_{rot}}{kT_{rot}}\right), \tag{3}$$

where $N_{n'}$ is a constant independent of v and J, k is the Boltzmann constant, T_{vib} and T_{rot} are the vibrational and rotational temperatures, respectively, and E_{vib} and E_{rot} are the vibrational and rotational energy levels, respectively.

Considering the above equations, we can describe the unique rotation-vibration spectrum as a function of vibrational and rotational temperatures. Therefore, we can determine these temperatures by fitting the calculated spectrum to that measured experimentally.

In this study, we used the second positive system (2ndPS) bands for the estimation of the gas temperature. The 2ndPS appear frequently in ordinary discharges. This system corresponds to a transition between the electronic states $C^3\Pi_u$ and $B^3\Pi_g$.

To calculate the spectrum of the 2ndPS, we neglected the spin splitting; the line strengths for the P, Q, and R branches are combined. Therefore, we considered three rotational transitions and two vibrational transitions, namely, $(v', v'') = (0, 2)$ and $(1, 3)$.

The transition probability in the second positive system $A\frac{C_{v'J'}}{B_{v''J''}}$ is written as

$$A\frac{C_{v'J'}}{B_{v''J''}} = A\frac{C_{v'}}{B_{v''}}\frac{S_{J''}}{6(2J'+1)}, \tag{4}$$

where $S_{J''}$ for the P, Q, and R branches is given as [23]

$$P(J+1) = 6(J+1) - 10/(J+1), \tag{5}$$

$$Q(J) = 10/J + 10/(J+1), \tag{6}$$

and

$$R(J-1) = 6J - 10/J. \tag{7}$$

4. Results

To investigate the characteristics of the plasma source for plasma-assisted mist CVD of ZnO films, the plasma density and electron temperature for argon gas (Ar 100%) at a distance of 195 mm from the

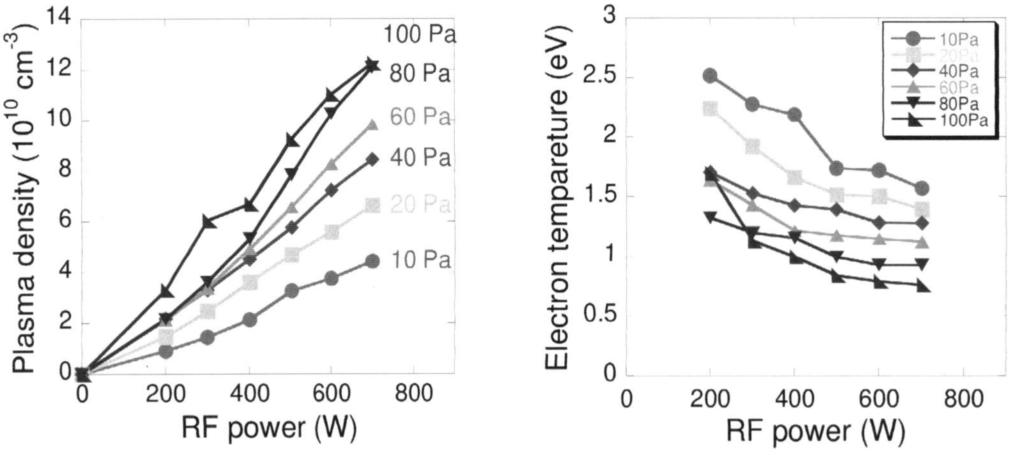

Figure 2. Plasma density and electron temperature.

antenna were measured. Figure 2 shows the plasma density and electron temperature of plasma sustained by an inductively coupled plasma source for plasma-assisted mist CVD. The plasma density and electron temperature became as high as 10^{11} cm^{-3} and 1.0 - 2.5 eV at a distance of 195 mm from the antenna. This result indicates that an inductively coupled plasma source can generate a high-density plasma for efficient deposition of ZnO films using plasma-assisted mist CVD.

To measure the gas temperature, we introduced an air-in-Ar plasma and performed optical emission spectroscopy measurements of the N_2 second positive system. Figure 3 shows a typical optical emission spectrum in the range of 300 - 900 nm for an Ar/air plasma sustained at an RF power of 1000 W. The optical emission spectrum contains electronic spectral lines of the first positive system ($B^3\Pi_g$ and $A^3\Sigma_g^+$) and the second positive system ($C^3\Pi_u$ and $B^3\Pi_g$). As shown in Fig. 4, the optical emission spectrum in the range of 700 - 900 nm contains atomic spectral lines from the oxygen radials that contribute to oxidation of ZnO films and removal of the organic impurities attributed to precursors. Figure 5 shows the spectrum of the N_2 second positive system for $v' - v'' = \Delta v = -2$. In this study, to determine the vibration and rotational temperature, we used the electronic spectrum of the second positive system ($C^3\Pi_u$ and $B^3\Pi_g$) in the range of 370 - 385 nm. Figure 6 shows comparisons between

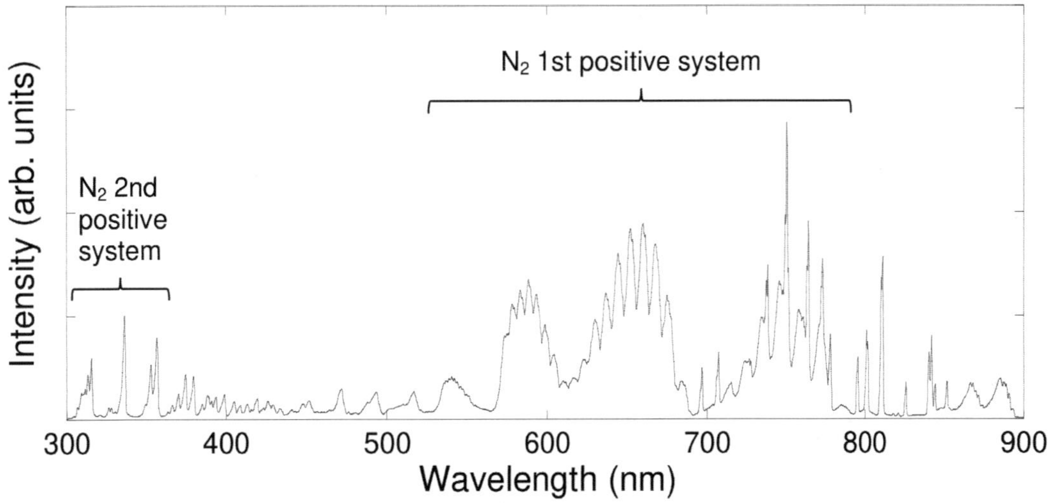

Figure 3. Typical optical emission spectrum of Ar/air plasma.

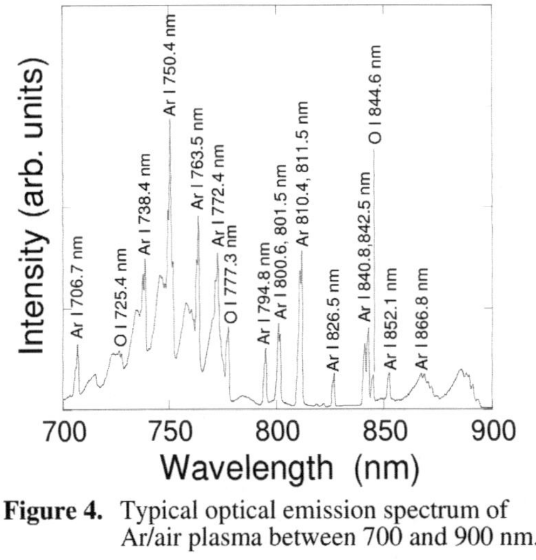

Figure 4. Typical optical emission spectrum of Ar/air plasma between 700 and 900 nm.

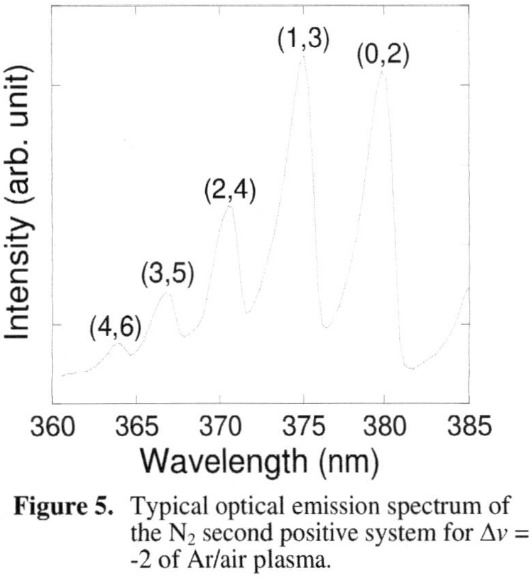

Figure 5. Typical optical emission spectrum of the N_2 second positive system for Δv = -2 of Ar/air plasma.

Journal of Physics: Conference Series **379** (2012) 012031 doi:10.1088/1742-6596/379/1/012031

theoretically calculated and experimentally measured spectra of the second positive system ($C^3\Pi_u$ and $B^3\Pi_g$) for $v = 0$ and 1. In this study, we measured the band spectra under steady state conditions at the center of the reactor, where rotational and translational relaxation may be accomplished [24]. Therefore, it is considered that the rotational temperature is almost equilibrated with the gas temperature [25]. Figure 7 shows the rotational and vibrational temperatures obtained from the second positive system. The rotational and vibrational temperature range from 1200 K to 1450 K and from 4600 K to 6900 K, respectively. The rotational temperature increases slightly with increasing RF power, while the vibrational temperature increases linearly up to 600 W and then remains almost constant. The rotational temperature is higher than the vibrational temperature. It is found that an increase in RF power generally causes an increase in rotational temperature. This is because the collision frequency between ions and neutral particles increases as the plasma density increases, and a larger amount of kinetic energy is transferred to neutral particles. Since the gas temperature is assumed to be close to the rotational temperature, the measured rotational temperatures are considered to be quite reasonable. Therefore, the gas temperature is found to be 1200 K to 1450 K. The results indicate that plasma sustained by an inductively coupled plasma source has sufficient enthalpy for

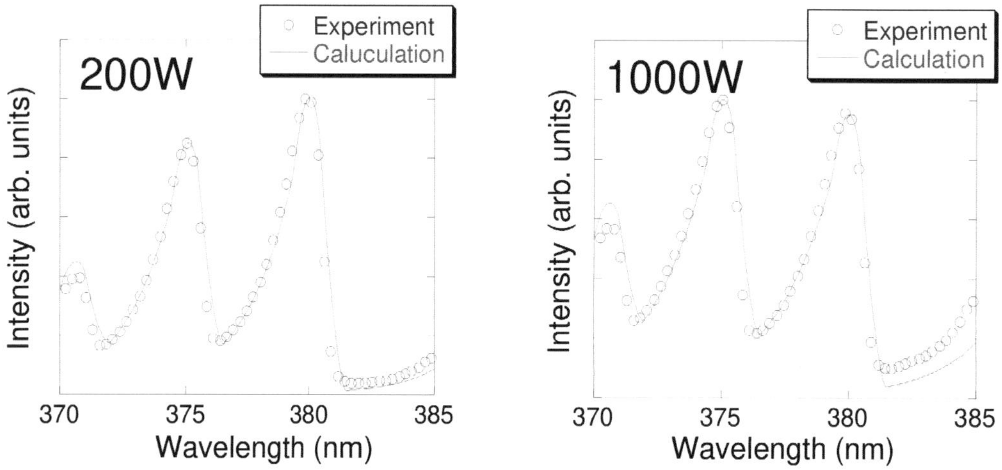

Figure 6. Examples of comparisons between spectra calculated theoretically and measured experimentally at RF powers of 200 W and 1000 W.

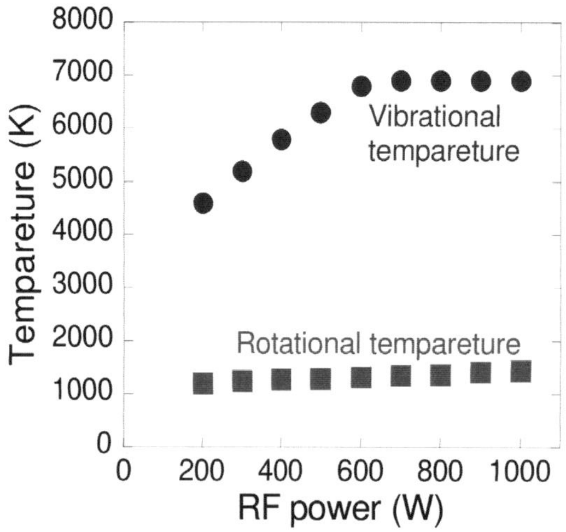

Figure 7. Rotational and vibrational temperatures measured experimentally as a function of RF power.

efficient vaporization of mists, including precursors, during plasma-assisted mist CVD.

5. Summary

Inductively coupled RF plasmas for plasma-assisted mist CVD have been characterized. It was found that the plasma density measured with a cylindrical Langmuir probe was as high as 1.2 x 10^{11} cm^{-3} and the electron temperature was 1.0 - 2.5 eV for Ar plasma. The gas temperature in the inductively coupled plasma determined from optical emission spectroscopy of N_2 molecular spectra was 1200 - 1450 K for an Ar/air plasma. These results indicate that a plasma sustained by an inductively coupled source has a high density for dissociation of precursors and generation of oxygen radials during ZnO film deposition, and has sufficient enthalpy for efficient vaporization of mists during plasma-assisted mist CVD.

References

[1] Mang A, Reimann K and Ruhenacke St 1995 *Sol. Stat. Commun.* **94** 251
[2] Tang Z K, Wong G K L, Yu P, Kawasaki M, Ohtomo A, Koinuma H and Segawa Y 1998 *Appl. Phys. Lett.* **72** 3270
[3] Pearton S J, Norton D P, Ip K, Heo Y W and Steiner T 2005 *Prog. Mater. Sci.* **50** 293
[4] Pei Z L, Sun C, Tan M H, Xiao J Q, Guan D H, Huang R F, and Wen L S 2001 *J. Appl. Phys.* **90** 3432
[5] Lu J G, Ye Z Z, Zeng Y J, Zhu L P, Wang L, Yuan J, Zhao B H and Liang Q L 2006 *J. Appl. Phys.* **100** 073714
[6] Kim K K, Niki S, Oh J Y, Song J O, Seong T Y, Park S J, Fujita S and Kim S W 2005 *J. Appl. Phys.* **97** 066103
[7] Minami T, Miyata T, Ohtani Y, and Mochizuki Y 2006 *Jpn. J. Appl. Phys.* **45** L409
[8] Ma J, Ji F, Zhang D H, Ma H L and Li S Y 1999 *Thin Solid Films* **357** 98
[9] Lu J G, Zhang Y Z, Ye Z Z, Zhu L P, Wang L, Zhao B H and Liang Q L 2006 *Appl. Phys. Lett.* **88** 222114
[10] Yang H, Li Y, Norton D P, Pearton S J, Jung S, Ren F and Boatner L A 2005 *Appl. Phys. Lett.* **86** 172103
[11] Moriyama T and Fujita S 2005 *Jpn. J. Appl. Phys.* **44** 7919
[12] Losurdo M, Giangregorio M M, Sacchetti A, Capezzuto P, Bruno G, Malandrino G and Fragala I L 2006 *J. Mater. Res.* **21** 1632
[13] Tang Z K, Wong G K L, Yu P, Kawasaki M, Ohtomo A, Koinuma H and Segawa Y 1998 *Appl. Phys. Lett.* **72** 3270
[14] Tanaka H, Fujita S and Fujita S 2005 *Appl. Phys. Lett.* **86** 192911
[15] Dutta A and Basu S 1993 *Mater. Chem. Phys.* **34** 41
[16] Paraguay D F, Estrada L W, Acosta N D R, Andrade E and Miki-Yoshida M 1999 *Thin Solid Films* **350** 192
[17] Pawar B N, Jadkar S R and Takwale M G 2005 *J. Phys. Chem. Solids* **66** 1779
[18] Jiménez-González A E, Urueta J A S and Suárez-Parra R 1998 *J. Crystal Growth* **192** 430
[19] Lu J G, Kawaharamura T, Nishinaka H, Kamada Y, Ohshima T and Fujita S 2007 *J. Crystal Growth* **299** 1
[20] Kamada Y, Kawaharamura T, Nishinaka H, and Fujita S 2006 *Jpn. J. Appl. Phys.* **45** L857
[21] Lu J G, Fujita S, Kawaharamura T, Nishinaka H, Kamada Y and Ohshima T 2006 *Appl. Phys. Lett.* **89** 262107
[22] Cacciatore M, Capitelli M and Gorse C 1982 *Chem. Phys.* **60** 141
[23] Phillips D M 1975 *J. Phys.* D **8** 508
[24] Naio K, Sakamoto T, Matsuura H and Akatsuka H 2005 *J. Adv. Oxid. Technol.* **8** 25
[25] Captitelli M, Ferreira C M, Gordiets B F and Osipov A I 2000 *Plasma Kinetics in Atmospheric Gases* (Berlin: Springer)

Temperature dependence for nitridation of aluminum films by sub-atmospheric pressure discharge

A Saikubo[1], **Y Ohtsuka**[1] **and Y Ueda**[1]
[1]Graduate School of Engineering, Osaka University,
2-1 Yamadaoka, Suita, Osaka 565-0871, Japan

E-mail: saikubo@st.eie.eng.osaka-u.ac.jp

Abstract. Characteristics of aluminum (Al) films exposed to helium/nitrogen (He/N_2) mixture plasma under sub-atmospheric pressure were investigated using X-ray photoelectron spectroscopy (XPS), powder X-ray diffraction (XRD), and scanning electron microscopy (SEM). After nitridation, the Al-N component became a dominant constituent element on the Al films through Al-N bonding due to exposure to He/N_2 mixture plasma. The content of the Al-N component with respect to temperature was almost constant at more than 0.6 up to 573 K. However, the Gaussian width of the Al-N component corresponding to disorder of aluminum nitride (AlN) crystal tended to become narrower with the elevation of the temperature. At a high temperature of 673 K, where the nitridation process may be easily promoted, an AlN 0002 reflection peak appeared in the XRD pattern, and crystals composed of collections of tiny white dots started to be observed in SEM images.

1. Introduction

Nondestructive inspection with an electromagnetic acoustic transducer (EMAT) employing permanent magnets is expected to detect cracks and wastage in pipes of a fast-breeder reactor under a moderate temperature of about 473 K [1]. Recently, in order to achieve more safety and reliability in the reactor, an ultrasonic inspection technique is strongly required under the high-temperature environment of 823 K during reactor operation.

A novel EMAT equipping the piled electronic circuits with a layer of insulation instead of permanent magnets has been proposed [2, 3]. However, the polyimide films widely used for insulating coatings are limited to a temperature under about 523 K.

Aluminum nitride (AlN) is well known to show excellent properties, such as high resistivity and high thermal conductivity [4, 5]. The properties of AlN films have been clarified using various techniques, such as magnetron sputtering, RF sputtering, and ECR resonance techniques at a low pressure condition below about 1 Pa [6-8]. However, experiments have been rarely performed for the aluminum nitridation process under high pressure, though the high deposition rate of silicon has been obtained using an atmospheric pressure plasma process technique [9]. If the mechanism of the aluminum nitridation process for the interaction between the high-density plasma and the surface is elucidated, it might be suitable for the coating of electronic circuits.

In this study, the basic growth process of the AlN film fabricated by exposing aluminum (Al) films to helium/nitrogen (He/N_2) mixture plasma under sub-atmospheric pressure was investigated from the viewpoint of the substrate temperature. Characteristics of these films were investigated using X-ray photoelectron spectroscopy (XPS), powder X-ray diffraction (XRD) and scanning electron microscopy (SEM).

2. Experimental

Al films used in the nitridation experiments were made in advance as follows. Al films were deposited onto a copper-tungsten substrate by an RF wave magnetron sputtering apparatus. In order to remove adsorption molecules such as water and oxygen molecules from the wall of the reaction chamber, the chamber was evacuated to a pressure of 1×10^{-4} Pa using turbo-molecular pumps. A disk of pure Al (99.99% in purity) 3 inch in diameter and 2 mm thick was set as a target material. Argon (Ar) gas as a source gas was accurately introduced into the chamber using a gas mass flowmeter. The deposition process of the Al film by Ar sputtering was performed under a pressure of 7×10^{-1} Pa at room temperature. During sputtering, the flow rate of the Ar gas was kept constant at 6 sccm. The prepared Al films were taken out from the magnetron sputtering apparatus at atmospheric pressure.

Next, nitridation of the deposited Al films was performed using an atmospheric pressure plasma apparatus with helium/nitrogen (He/N_2) mixture plasma under sub-atmospheric pressure, as shown in Figure 1. After fixing the Al films on the substrate holder, the reaction chamber was evacuated to a pressure of 1×10^{-4} Pa to remove oxygen influence. The Al film was heated by a resistive heater mounted inside the sample holder, which was kept within ±5 K of set values of 373 K, 473 K, 573 K and 673 K. Mixture gases of He and N_2 as source gases were accurately introduced into the chamber using a gas mass flowmeter. The Al film was exposed to He/N_2 mixture plasma at a pressure of 1×10^3 Pa, where the gap length between the RF electrode and the substrate holder was kept at several millimeters. During discharge, the flow rate of He and N_2 gases were kept at constant values of 50 sccm and 20 sccm, respectively.

The XPS measurement of the Al films exposed to He/N_2 mixture plasma was performed using an X-ray photoelectron spectrometer (Shimadzu Co. Ltd./Kratos Analytical Ltd.; AXIS-165). A magnesium target as an X-ray source was employed. An X-ray beam with an energy of 1253.6 eV obtained by an applied voltage of 12 kV and a current of 10 mA was irradiated on the sample. Photoelectrons emitted from the sample were detected by a hemispherical electron analyser. The pass energy employed was 40 eV. In this system, the total resolution has been estimated to be ~850 meV. The vacuum of the analysis chamber was kept below 1×10^{-7} Pa. Photoelectron spectra in an Al 2p region of the films were measured by detecting the photoelectrons emitted from the surface region of the samples.

The XRD measurement of the Al films exposed to He/N_2 mixture plasma was performed using a powder X-ray diffractometer (Rigaku Co. Ltd.; RINT-2000). A copper target as an X-ray source was employed. An X-ray beam with an energy of ~8 keV obtained by an applied voltage of 40 kV and a

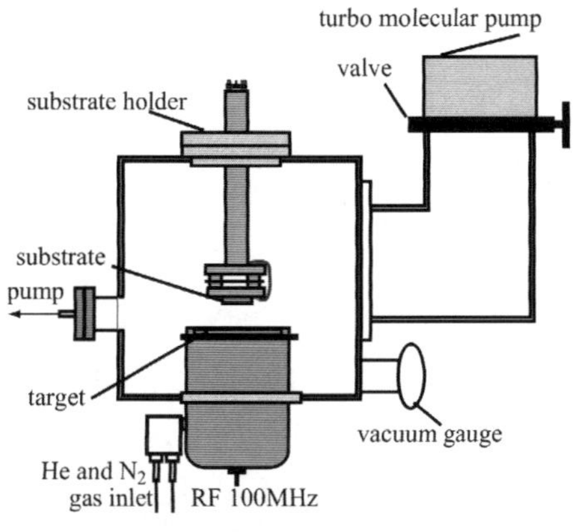

Figure 1. Schematic of atmospheric pressure plasma apparatus.

current of 30 mA was monochromatized by means of a graphite monochromator and was irradiated on the sample. Two Soller slits each were positioned in the incident X-ray beam in front of the divergent slit and in the diffracted X-ray beam in front of the graphite monochromator. The diffracted X-ray beam intensity was recorded using a NaI scintillation counter mounted on the 2θ arm. Diffraction patterns were obtained by scanning the 2θ in the angular range from 31° to 42°.

The surface morphology of the Al films exposed to He/N_2 mixture plasma was observed using SEM (Hitachi High-Technologies Co.; S-2150). An electron beam emitted thermally from an electron gun equipped with a tungsten filament was accelerated at an acceleration voltage of 20 kV in the section of an electrostatic lens. The accelerated electron beam was focused by condenser lenses to a spot of ~5 nm in diameter and irradiated on the sample. SEM images of the films were obtained at 3000 magnification through monitoring reflection electrons emitted by scanning the electron beam.

3. Results and discussion

Figure 2 shows a typical Al 2p core-level photoelectron spectrum for the Al film exposed to He/N_2 mixture plasma. In the Al 2p photoelectron spectrum, a broad peak containing several components was observed. The spectrum is divided into four peaks with consideration of chemical shift after Shirley background subtraction. A dominant peak observed at 73.4 eV is assignable to Al-N bonding [10-12].

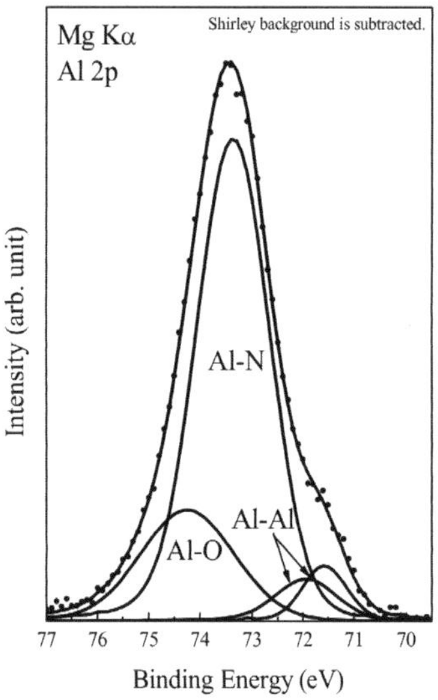

Figure 2. An Al 2p core-level photoelectron spectrum for an Al film exposed to He/N_2 mixture plasma.

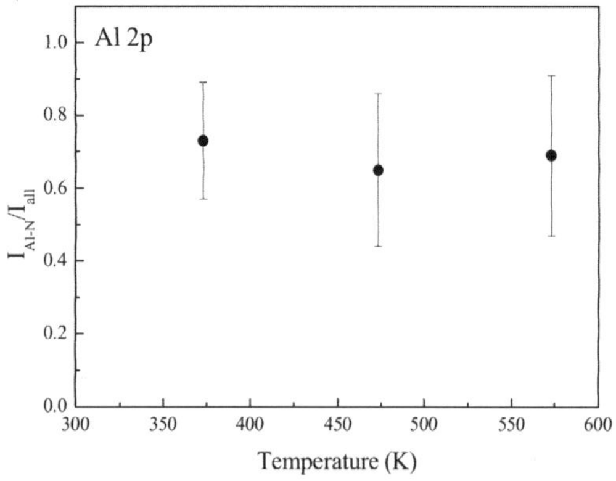

Figure 3. $I_{Al\text{-}N}/I_{all}$ ratio as a function of temperature.

Peaks obtained in a lower binding energy region of 71.6 eV and 71.9 eV are assignable to Al-Al bonding corresponding to metallic aluminum [13]. Another peak centered at a higher binding energy region of 74.2 eV is ascribable to Al-O bonding originating from native oxide [14, 15]. It is noted that a typical spectrum for native aluminum oxide/aluminum is only observed before nitridation (not shown).

The ratio of the area intensity for the Al-N component to that for all divided components (I_{Al-N}/I_{all}) is plotted as a function of the temperature in Figure 3. The I_{Al-N}/I_{all} ratio is almost constant at more than 0.6. However, at 673 K, the Al 2p spectrum was difficult to detect because of the surface charge-up effect.

On the other hand, the Gaussian width of the Al-N component derived from the Al 2p spectra is plotted as a function of the temperature in Figure 4. The Gaussian width tends to become gradually narrower with temperature, suggesting formation of crystalline AlN.

Figure 5 shows X-ray diffraction patterns for the Al films nitrided at temperatures from 373 K to 673 K. For the film fabricated at 673 K, an AlN 0002 reflection peak at 35.5° starts to appear, though its reflection peak width is broad.

Figure 6 shows the surface morphology of the as-deposited film and the films exposed to He/N_2 mixture plasma at 473 K and 673 K. Especially, Figure 6 (d) displays the magnification of the area indicated by the square in Figure 6 (c). On the as-deposited film, the collection of spherical particles, which would be the Al particles, was observed. No change of the surface morphology arose up to 473 K. However, the surface morphology was drastically varied at 673 K. Tiny white dots, the characteristic phenomena of charge-up of insulating materials, were observed. This phenomenon is in good agreement with results from the XPS measurement and XRD measurement.

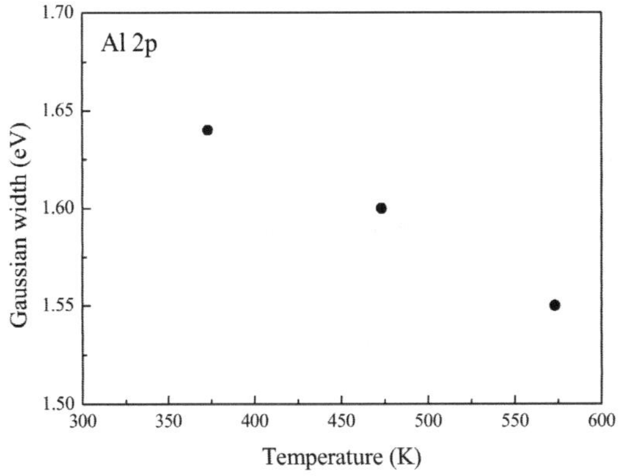

Figure 4. Gaussian width of an Al-N component in Al 2p spectra as a function of temperature.

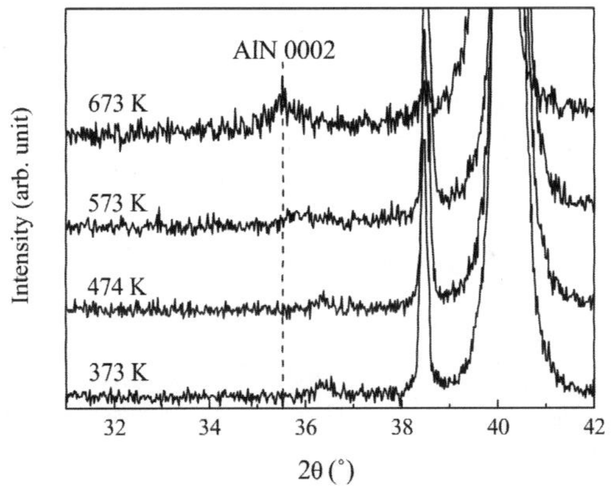

Figure 5. X-ray diffraction patterns for Al films nitrided at temperatures from 373 K to 673 K.

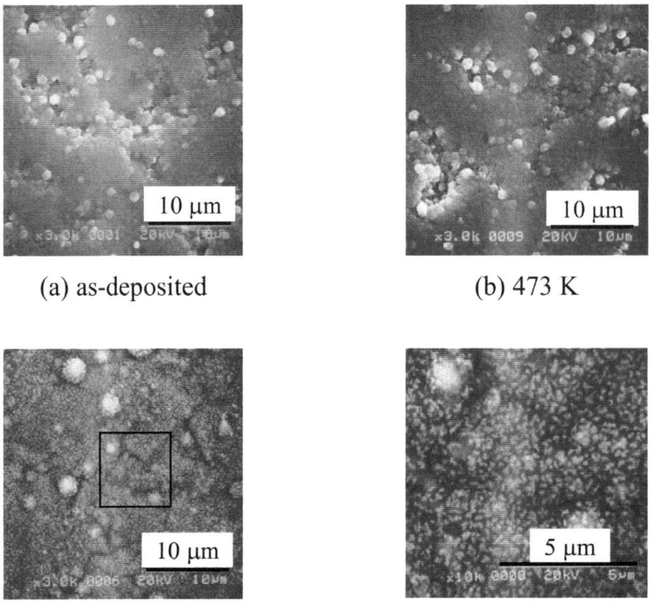

(a) as-deposited (b) 473 K

(c) 673 K (d) magnification

Figure 6. The surface morphology of Al films after nitridation by He/N_2 mixture plasma.

It is revealed that the Al film exposed to He/N_2 mixture plasma is dominantly composed of the Al-N component by progress of the nitridation. In spite of the fact that the I_{Al-N}/I_{all} ratio is more than 0.6 at temperatures from 373 K to 573 K, the surface charge-up phenomenon hardly occurred. However, at 673 K, a significant charge-up phenomenon is observed for XPS and SEM, as observed in insulating materials. It is considered that the disordered components decreased and the crystalline AlN increased on the film by the sufficient thermal energy of 673 K. Moreover, from the results of XRD patterns, it is indicated that the AlN crystal itself has a long-period structure with increasing temperature, though grain size of the AlN crystal is still small. Therefore, a temperature of more than 573 K would be required in order to form the AlN crystal with both a long-period structure and a large grain size.

4. Conclusions

The characteristics of Al films exposed to He/N_2 mixture plasma under sub-atmospheric pressure were investigated using XPS, XRD and SEM. The films exposed to He/N_2 mixture plasma are composed of Al-Al, Al-N and Al-O components. The I_{Al-N}/I_{all} ratio is almost constant up to 573 K. However, the Gaussian width of the Al-N component tends to become narrower with the elevation of the temperature. From the XRD patterns, the weak and broad AlN 0002 peak indicating growth of the AlN crystal is observed at 673 K. Moreover, on the Al film after nitridation at 673 K, a collection of tiny white dots appeared over the surface of the film. It is revealed that a temperature of more than 573 K would be required in order to form the AlN crystal with both a long-period structure and a large grain size.

Acknowledgments

The present study is the result of "Core R&D program for commercialization of the fast breeder reactor by utilizing Monju" entrusted to the University of Fukui by the Ministry of Education, Culture, Sports, Science and Technology of Japan (MEXT) and re-entrusted to Osaka University by the University of Fukui in FY2011.

References

[1] Xu Y, Tagawa A, Ueda M, Yamashita T, Ohtsuka Y, Osafune K and Nishikawa M 2005 W Arnold and S Hirsekorn eds. Kluwer Academic/Plenum Publishers Dordrecht & New York *Acoustical Imaging* **27** pp 175-184
[2] Ohtsuka Y, Yoshimura T and Ueda Y 2007 *Proc.2007 IEEE Ultrasonic Symp.* pp 1609-1612
[3] Ohtsuka Y 2011 *IEEJ Trans. Fundam. Mater.* **131** 900
[4] Talyansky V, Vispute R D, Ramesh R, Sharma R P, Venkatesan T, Li Y X, Salamanca-Riba L G, Wood M C, Lareau R T, Jones K A and Iliadis A A 1998 *Thin Solid Films* **323** 37
[5] Cai K F, McLachlan D S, Sauti G and Mueller E 2005 *Solid State Sci.* **7** 945
[6] Yin M L, Li M, Zou C W, Liu C S, Guo L P and Fu D J 2008 *J. Phys. D* **41** 085407
[7] Cheng H, Sun Y, Zhang J X, Zhang Y B, Yuan S and Hing P 2003 *J. Cryst. Growth* **254** 46
[8] Soh J-W, Jang S-S, Jeong I-S and Lee W-J 1996 *Thin Solid Films* **279** 17
[9] Kakiuchi H, Matsumoto M, Ebata Y, Ohmi H, Yasutake K, Yoshii K and Mori Y 2005 *J. Non-Cryst. Solids* **351** 741
[10] Taylor J A and Rabalais J W 1981 *J. Chem. Phys.* **75** 1735
[11] Manova D, Dimitrova V, Fukarek W and Karpuzov D 1998 *Surf. Coat. Technol.* **106** 205
[12] Rosenberger L, Baird R, McCullen E, Auner G and Shreve G 2008 *Surf. Interface Anal.* **40** 1254
[13] Sporken R, Thiry P A, Petit E, Pireaux J J, Caudano R, Ghijsen J, Johnson R L and Ley L 1987 *Phys. Rev. B* **35** 7927
[14] Laidani N, Vanzetti L, Anderle M, Basillais A, Boulmer-Leborgne C and Perriere J 1999 *Surf. Coat. Technol.* **122** 242
[15] Duez N, Mutel B, Dessaux O, Goudmand P and Grimblot J 2000 *Surf. Coat. Technol.* **125** 79

Laser processing for strengthening of the self-restoring metal-elastomer interface on a silicone sheet

Kiyokazu Yasuda
Dept. of Materials, Physics, & Energy Engineering, Nagoya University,
Furo-Cho, Chikusa, Nagoya, Aichi 464-8603, Japan

E-mail: yasuda@numse.nagoya-u.ac.jp

Abstract. A self-restoring microsystem is a unique concept which realizes the sensing functionality and robust interface which mechanically and electrically connects a deformable object such as a human body with printed electronic devices. For this purpose, the formation of conductive wiring on an elastomer substrate was attempted using the nickel ink printing process. Before the wiring process, surface patterning of a silicone sheet by a galvano-scanned infrared laser was conducted for the enhancement of interface adhesion of the metal deposit and polymer. Characterization of the fabricated pattern was conducted by optical microscopy. The novel method was successfully demonstrated as a fabrication of selective patterns of metal particles on self-restoring MEMS.

1. Introduction

With the progress in the field of nursing care as well as the advanced medical and engineering fields, the measurement of biological signals have been actively employed for the man-machine interface, brain-machine interface, and myoelectric hand [1-6]. Meanwhile, the developments of sensors and accurate measurements are needed without any restrictions. The sensors detect temperature, blood flow pulsating sound pressure, electromyography (EMG), heart telegraph signal, and brain wave which occur due to human activity. Since they are small and have weak signals with variations, the measurement requires high accuracy and sensitivity [7, 8]. As the technology requires such signal measurements to be achieved, it is important to consider the structure composed of new eco-friendly and bio-compatible materials and their interfaces for connecting electronic devices to a flexible human body electronically and mechanically [9,10].

In this paper, the fine structure of the device such as a thermistor is considered as a temperature sensor on a polymeric silicone elastomer material which is a promising bio-material interface. Such flexible polymer base devices are stretchable (typically several tens of percent deformation in length) and they can recover to their initial shape according to the human body motion because of their elastic and surface attaching nature, namely known as self-restoring characteristics. However, the circuit pattern printing on a hydrophobic surface such as silicone is one of the problems for creating stable patterns. Figure 1 shows the line patterns that became discontinued to isolated droplets after printing.

In order to solve such a problem, the effect of fine irregularities of the surface texturing treatment to the polymer substrate by a CO_2 laser was investigated. In addition, the influence of the contact angle of pure water as an indicator of the printed wiring conductor formed subsequently on the substrate was investigated. Finally, a trial to print a selection of the dipping in the solution after the pattern of dispersed nickel particles to the surface by the CO_2 laser was conducted.

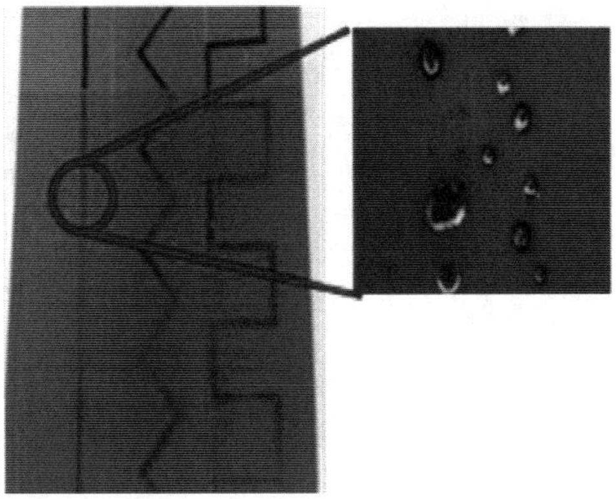

Figure 1. Hydrophobic property of a silicone sheet.

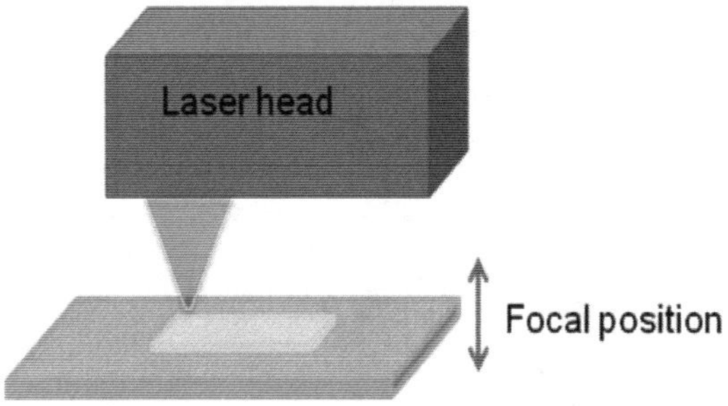

Figure 2. Surface modification of a silicone sheet by a scanning CO_2 laser.

2. Experimental

In the experiment, a silicone polymer (Nitto Denko, #66, thickness: 0.5 mm) was cut to a length of 25 mm as a sheet substrate. A micro-focused scanning CO_2 laser system (Keyence ML-9110, max. output power: 12 W, wavelength: 10.6 μm) was used for the surface pretreatment of the polymer sheet substrate as shown in figure 2.

As for the irradiation area (10 mm × 8 mm) by a CO_2 laser marker, 100 columns by 56 rows of micro lines (width: 150 μm) were drawn repeatedly. The irradiation time was 30 s. The laser output was varied from 120 mW to 720 mW for the surface treatment. In order to maintain the planarity, the substrate was fixed on a flat slide glass and a laser beam was focused on the polymer surface. The patterned surface of the substrate was observed by optical microscopy and FE-SEM. The contact angle of a distilled water droplet on the irradiated surface was measured to examine the wettability and consequently, the relevant index of adhesiveness of post-formed conductive pattern prints. The silicone sheet was used as the substrate material since it is considered to have the closest bio-compatibility to human activities among the rubber polymer materials. The physical properties of the silicone elastomer sheet are shown in Table 1. A solution of nickel particles (Mitsui Mining & Smelting Co., Ltd, average diameter: 15 nm) in deionized water was prepared as the printing material.

Table 1. Physical properties of a silicone sheet.

Tensile strength (MPa)	Elongation (%)	Adhesion force (N/19mm)
8.2	660	39
Resistivity (Ω cm)	**Contact angle (water) (°)**	
2.8×10^{15}	105	

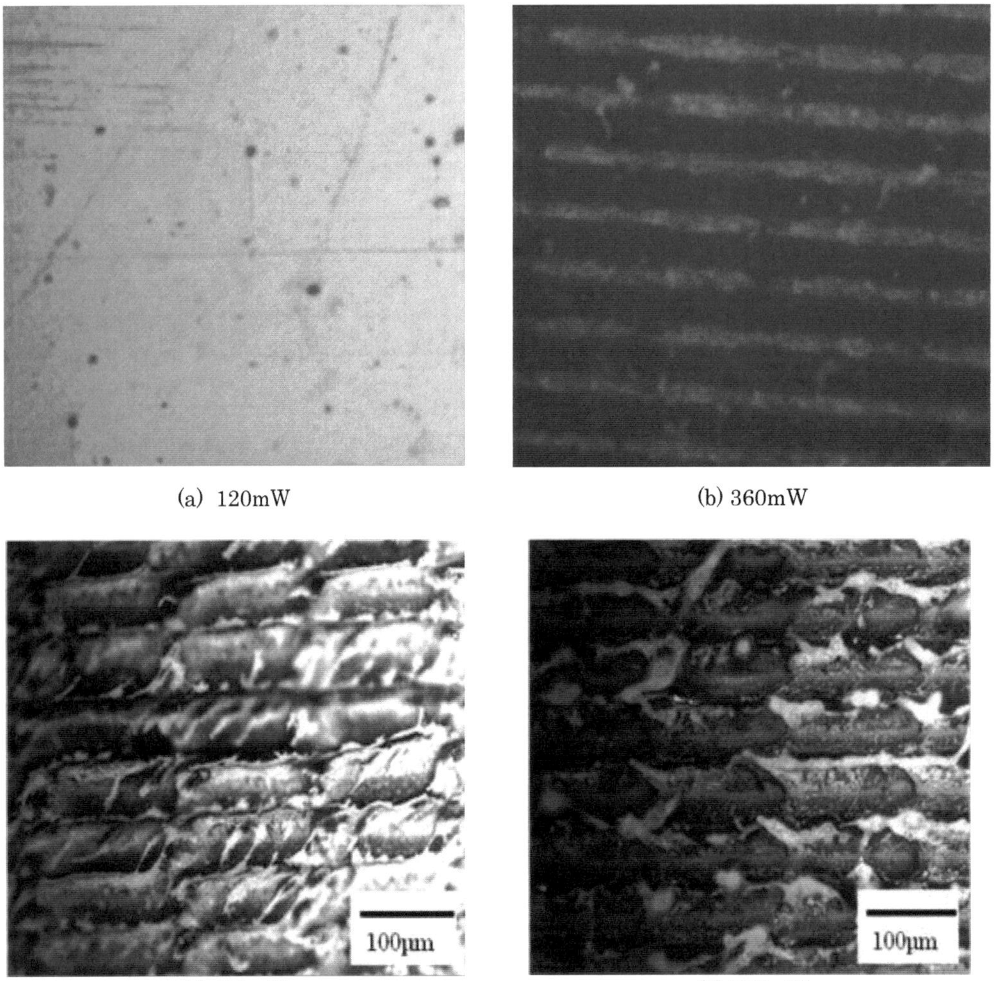

(a) 120mW (b) 360mW

(c) 600mW (d) 720mW

Figure 3. Optical image of the silicone surface after scanning CO_2 laser patterning.

3. Results and discussion

The typical surface texture after the laser irradiation of line patterns is shown in figure 3. The laser irradiation power was varied from 120 mW to 720 mW. The line pattern was drawn in the area of 10 mm x 8 mm. According to the optical microscopic observation, no apparent modification of the surface state was observed in the case of a low laser output power of below 240 mW. With the increase of the laser output above 360 mW, the area turned into a white mass on the surface of silicone. Fiber-like structures were also identified along with the formation of a fine white stripe texture line for the CO_2 laser irradiation area. In the case of over 600 mW, the debris (white area) was deposited along the irradiated area. As the irradiated power was increased, the silicone surface was significantly altered and exhibited fine structures. This fine structure was considered to be fabricated due to the thermal decomposition of the smooth surface of the silicone polymer to nanoparticles as debris. The periodical pattern of the surface was found to be deformable even after severe elongation of the sheet which might be suitable for a bio-interface since self-restoring property was required.

Journal of Physics: Conference Series **379** (2012) 012033 doi:10.1088/1742-6596/379/1/012033

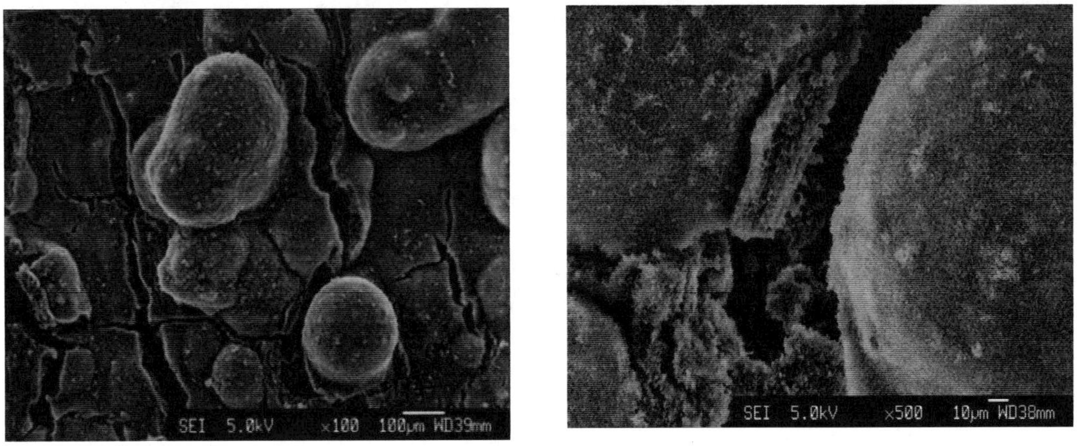

(a) (b)

Figure 4. SEM microphotograph of the silicone sheet surface after laser scanning.

Figure 4 is SEM microphotographs of the laser irradiated area. From this observation, it is clear that the irradiated area on the surface had a disordered microstructure with particulates and small grooves. The temperature increase by CO_2 laser irradiation was estimated over the melting point of silicone. This fact indicated that the origin of the particulates-like mounds was the melting mass of silicone. They seemed to evolve from the original sheet surface and adhere firmly. Since the temperature change was very abrupt, the volume shrinkage during cooling caused open grooves on the surface. Moreover, in the magnified photo of figure 4(b), a fibrous structure in the nanometer scale on the rough surface in the micrometer scale was observed. All those changes of the surface topography greatly cause the increase of the net area of the silicone surface after the laser irradiation. It can be emphasized that this hierarchical structure was easily produced by only the thermal laser treatment which enabled the formation of arbitrary patterns.

Figure 5 is the measurement result of the contact angle of deionized water droplets on the laser-irradiated region of the sheet material. There was no significant change in the surface wetting characteristics in the case of laser power less than 240 mW. The values of contact angle were almost the same level as the initial condition before the laser irradiation. The silicone rubber used in this experiment has a thermal stability up to approximately 523 K. From the fact that there was no change in both the morphology and wetting characteristics, it is clear that the surface retains its initial chemical composition as polymeric silicone. For the case of the medium laser power, especially at 360 mW, the variations in contact angle values were detected which infer the instability of the surface condition. As shown in figure 4, the local surface properties should be different between the white line area and non-irradiated area. The wettability measured by means of the contact angle of water droplets larger than these line pattern pitches reflects the instability of the measured contact angles.

The values of the contact angle in the cases of 480 mW and 600 mW were approximately 20°-30° larger than those of the original silicone rubber which had high water repellency (contact angle: 110°). This slight increase suggests that the periodic deep 3D structures of line patterns enhance water repellency. In contrast, greater wetting with small contact angles of 16°-40° was observed due to high hydrophilicity in the case of the high laser power condition of 720 mW. The steep change of wetting property is believed to result from structural changes in the surface texture by increasing laser power as well as the formation of a reaction substance having a hydrophilic group on the top surface by the thermal decomposition generated during the laser irradiation.

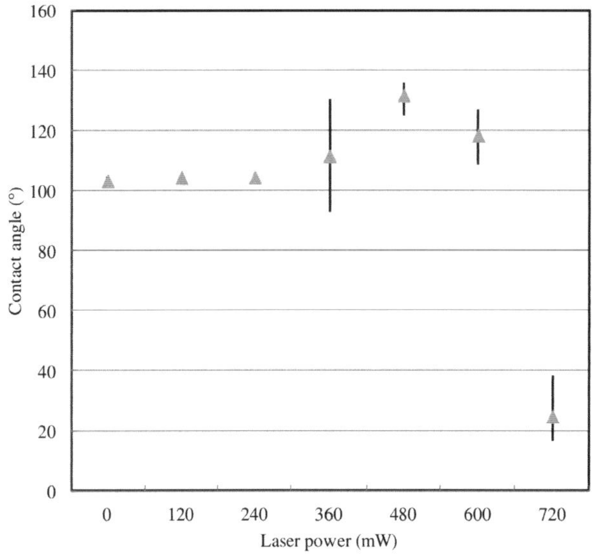

Figure 5. Laser power vs. contact angle of water.

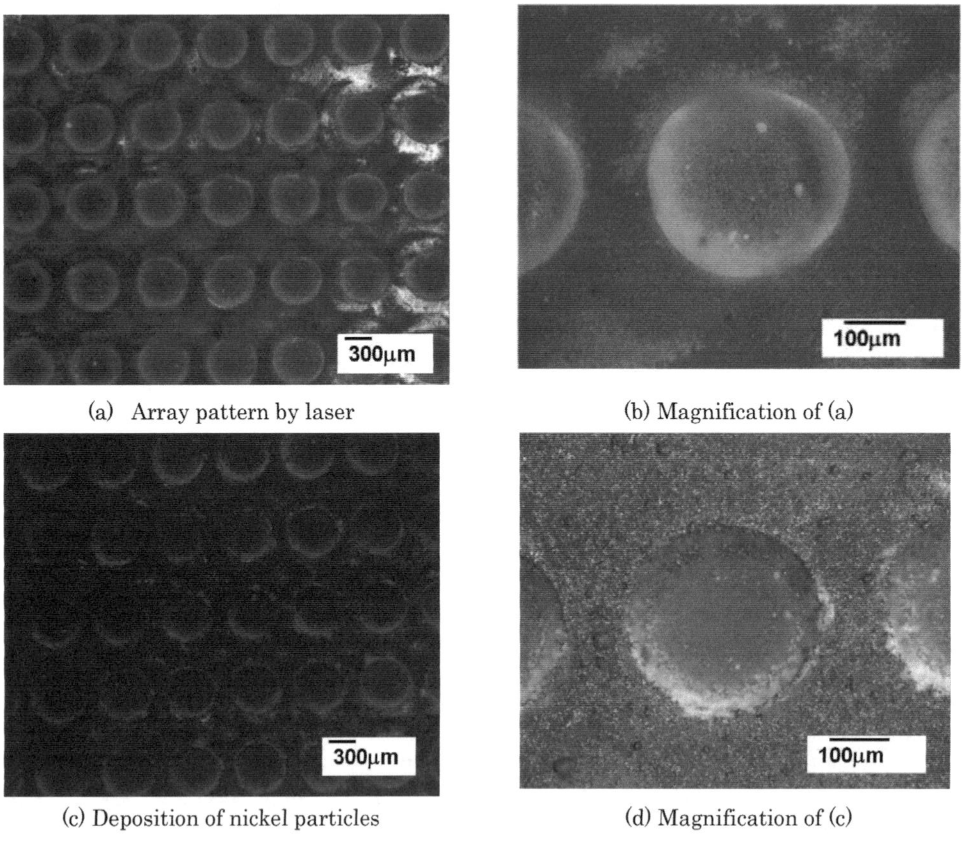

(a) Array pattern by laser (b) Magnification of (a)

(c) Deposition of nickel particles (d) Magnification of (c)

Figure 6. Nickel particles printing on the silicone polymer sheet after laser patterning.

After the modification process by CO_2 laser irradiation, the printing of nickel particles was conducted. Figure 6 showed nickel printing images observed under the optical microscope on the silicone polymer sheet. Before printing, the silicone polymer was irradiated by the CO_2 laser so that an array dot pattern (300 μm in diameter) formed. The laser treated silicone was immersed in the water solution of nickel particles. As shown in figures 6(c) and (d), the nickel particles were printed only on the surface area except on the array dot area where the laser was focused. The top surface of the laser irradiated dot area tightly attracts water molecules which might prevent nickel adhesion, while nickel particles reside on the other area, although water molecule can escape by a repulsive force due to surface high hydrophobic property. It was concluded that the selective printing was achieved by only the arbitrary template patterning by laser irradiation and post simple immersion in a liquid solution of metal particles.

4. Conclusions

By increasing the output of CO_2 laser irradiation on the surface of the silicone rubber, different texture structures on the surface were observed. In particular, very high hydrophilicity property appeared in the case of an output power of 720 mW. Using a scanning CO_2 laser process, the self-restoring silicone sheet exhibited hydrophobic ($\geq 100°$) property at the medium laser power condition, while showing a steep transition to exhibiting hydrophilic (16°-40°) property at the high laser power condition. The abrupt change of surface property from hydrophobic to hydrophilic coincides with the morphology change from the periodic regular structure to the disordered structure with the increase of the laser intensity. It is suggested that the flexible silicone polymer sheet is selectively printable by the simple immersion of the nickel particle solution using the prior scanning CO_2 laser irradiation. The proposed laser process is considered to be applicable to such self-restoring polymeric sheet devices where the interface between the metal and polymer should be strengthened.

References

[1] Birbaumer N 2006 *Psychophysiology* **43** 517
[2] Wolpaw J R 2002 *Clinical Neurophysiology* **113** 767
[3] Santhanam G, Ryu S I, Yu B M, Afshar A and Shenoy K V 2006 *Nature* **442** 195
[4] Ganguly K, Dimitrov D F, Wallis J D and Carmena J M 2011 *Nat. Neurosci.* **14** 662
[5] Chin T 2010 *Jpn. J. Rehabil. Med.* **47** 33
[6] Parker P, Englehart K and Hudgings B 2006 *J. Electromyography and Kinesiology* **16** 541
[7] Hattori T, Sato T, Minato K, Nakamura H and Yoshida M 2008 *Trans. Jpn. Soc. Med. Biol. Eng.* **46** 268
[8] Ito D, Ozeki M, Nakamura Y, Sakurazawa S, Toda M and Akita J 2008 *Human interface* **10** 107
[9] Sekido N, Yasuda K and Takai O 2011 *Proc. 17th Symp.Microjoing Assembly Technol. in Electronics* 257
[10] Yasuda K, Sekido N and Takai O 2011 *Proc.Int. Conf. Electronics Packaging (ICEP) 2011* 981

Greener manufacturing: Superplastic-like forming

Jun Liu[1], Ming-Jen Tan[1,*], Anders E W Jarfors[2], Samuel C V Lim[2], Kai-Soon Fong[2], Sylvie Castagne[1]
[1] School of Mechanical & Aerospace Engineering, Nanyang Technological University, Singapore 639798, Singapore
[2] Singapore Institute of Manufacturing Technology, Singapore 638075, Singapore

[*] E-mail: mmjtan@ntu.edu.sg

Abstract. Conventional superplastic forming (SPF) is normally conducted at a slow forming rate (10^{-3} s^{-1} or slower) and high temperatures (typically 773 K or higher for aluminum alloys), which is not attractive for wider applications. Recently, SPF has revived as an attractive process in the automotive and aerospace industries, especially for forming of excellent precision, large, and complex-shaped work pieces. A hybrid superplastic forming process has exploited the usage of superplastic forming for faster times, lower temperatures, and also for use by non-superplastic materials by designing a process that combines stamping and gas blow forming in one operation to establish an energy-saving technology. With recent developments in these areas, the light-weight, yet inexpensive aluminum alloys are promising candidates for green manufacturing. In comparison with the conventional SPF process, the forming time has been shortened (typically from 30 min to 8 min), and the forming temperature has been lowered down (from 773 K to 673 K). Furthermore, as a greener manufacturing process, it uses the non-superplastic grade materials, which is more compatible with existing manufacturing processes with less material preparation time and expense cost.

1. Introduction

Superplastic forming (SPF) is a high-temperature process by which sheet metals are deformed by gas pressure in a single-side die cavity to produce parts of complex shape by taking advantage of the superplastic flow behavior of the material. Conventional superplastic forming was conducted at slow forming rate (10^{-3} s^{-1} or slower) and high temperature (typically 773 K or higher for aluminum alloys), which have limited it for wider applications. SPF can also be combined with a joining operation involving diffusion bonding (SPF-DB) between several sheets of material [1]. New emerging superplastic developments also include superplastic tube forming [2]. No matter how the conventional SPF has been developed, it is only a perfect fit for the low production volumes. Recently, SPF has revived as an attractive process in the automotive and aerospace industries, especially for forming of excellent precision, large, and complex-shaped work pieces in one operation. With these developments, the light-weight, yet inexpensive aluminum alloys are promising candidates for green manufacturing.

In this paper, the superplastic-like forming process has been examined, expanding the usage of SPF for non-superplastic materials, lower temperatures, and faster times by designing the process that combines drawing and gas blow forming to establish an energy-saving technology.

2. Superplastic-like forming

Most of the prior SPF processes were conducted at a high temperature near the melting point, which also introduced many problems, such as reduced tool life, high heating demands etc. In addition, the mechanical properties of the product are poor due to higher amounts of cavitation introduced. Another big disadvantage is the slow forming rate. Cycle times vary from 30 min to 2 hr in the conventional SPF [3-6].

2.1. Materials usage

For industrial applications, the superplastic materials most commonly associated with conventional SPF are aluminum (5083, 7475) and titanium (Ti6Al4V) alloys with applications in the aerospace and automotive sectors. Other materials such as INCONEL 718 or magnesium alloys (AZ31B, ZK10) are also progressively entering the market. These kinds of materials largely complies with the SPF process requirements, such as fine grain size, high structure stability with respect to grain growth, and high purity of the structure. SPF5083 products have been implemented on various Boeing commercial airplanes [7]. Usually, the superplastic materials are thermal-mechanically prepared to meet the requirements for superplastic forming. However, for some alloys such as AA6xxx series [8], it is complicated and difficult to be processed, which also have not yet suited the commercial application for SPF. Therefore, the cost and preparation of raw materials is very high.

In order to reduce the material cost, the non-superplastic grade of AA5083 was employed in the hybrid SPF process. The lightweight and inexpensive alloy was only rolled and had a maximum elongation-to-failure of 93% [9]. Results indicate that it is possible to use the hybrid SPF process to form into 3-stepped die cavity at 673 K. Dry graphite was applied as lubricant, and evenly sprayed on both surfaces of the aluminum pieces. Compared to the superplastic materials, the non-superplastic alloys offer cost savings in the range of 20-50% [10]. As a greener manufacturing technology, the using of non-superplastic grade materials make it more compatible with existing forming process with less material preparation time and expense cost.

2.2. Process development

The technique studied in this paper is a combination of drawing (mechanical pre-forming) and superplastic forming (gas blow forming). This process, hereby named "superplastic-like forming", will overcome some of the problems that are encountered during the conventional superplastic forming. It will take advantages of the benefits of both drawing and gas blow forming. Initially, the sheet was mechanically drawn into the die cavity, and as a result some materials at the flange area flowed into the deformation area as shown in Figure 1. The two-stepped punch with length of 43 mm was used in the drawing phase. The punch dimension was designed in accordance with simulation and die cavity geometry. Experimental work has also shown that even a small amount of material can significantly improve the following forming performances not only in cycle times but also for forming temperature [11]. The pressure cycle was applied on the sample to keep the target forming strain rate at 0.002 s^{-1}. The whole gas blow forming process was conducted in only 8 minutes.

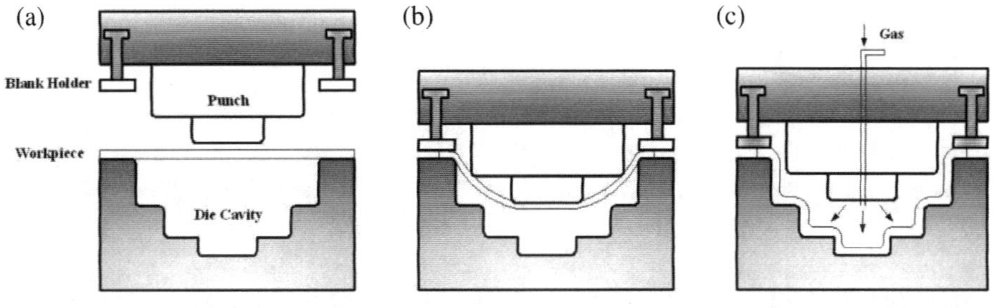

Figure 1. Forming sequence of the hybrid SPF of non-superplastic AA5083 sheet (a) clamping stage, (b) drawing and sealing stage, and (c) gas blow forming stage [11].

2.3. Forming temperature

Normally the temperature for SPF is in the range of 0.5 to 0.75 times the melting temperature (T_m). At these temperatures, some of the materials may encounter some problems such as oxidation at the grain boundary that reduces the ductility (e.g. for titanium alloys (Ti6Al4V)) or grain growth that causes unstable forming behavior (e.g. for magnesium alloys (AZ31B) or aluminum alloys (AA6xxx)).

In the hybrid forming of AA5083, the level of bulging into the 3-stepped die cavity after mechanical pre-forming was obtained for a series of experiments conducted within the temperature range from 300 to 500°C, as shown in Figure 2. Comparing these bulge heights, there is little variance between the temperatures of 673 and 773 K [11]. Therefore, it indicates the possibility of forming the sheet metal at 673 K. The changes of the microstructural features, such as less grain growth, less cavitation and dynamic recrystallization, can be obtained at the relatively lower temperatures. Furthermore, lower forming temperature will also inevitably lead to a lower carbon foot print by establishing an energy-saving technology.

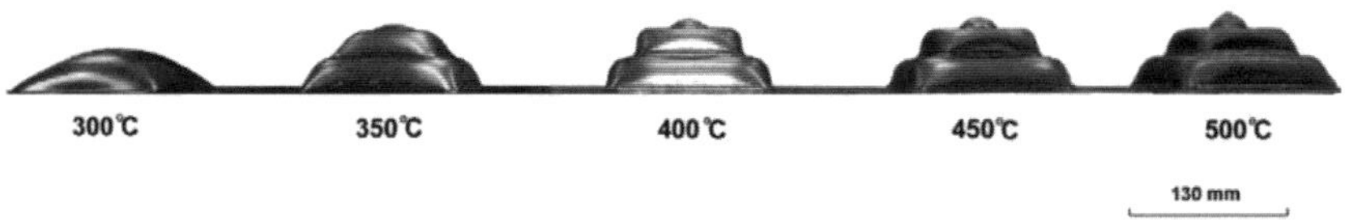

Figure 2. Bulge height profiles of formed parts at different forming temperatures[11].

2.4. Forming time

One of the reasons that hindered the conventional superplastic forming process not widely applied in industries is the low production rate. The typical forming time for superplastic process lies in between 30 to 120 min depending on the parts (the strain rate is in the range of 10^{-5} to 10^{-3} s^{-1}). This limitation implies that the technique is only suitable for low product volumes, e.g. for aerospace, prototype parts or niche cars. In the hybrid SPF process, the gas blow forming time has been shortened to 8 min. The pressure cycle obtained from prediction at a constant strain rate of 0.002 s^{-1} is plotted in Figure 3.

It is commercially viable to apply the SPF to high volume products such as automotive parts or aerospace components. Considered as essential in hybrid SPF, it utilizes the stretch-forming at a constant thickness, and also adopts advantages of superplastic forming of excellent precision and little springback. Therefore, the new design of hybrid process has revived as an attractive technology, but with time and energy savings.

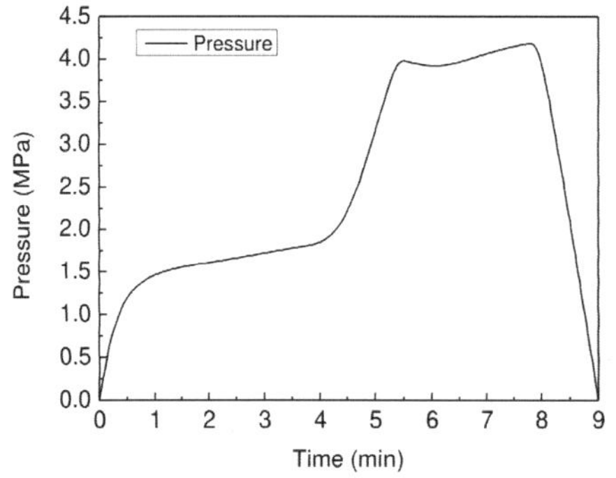

Figure 3. Pressure schedule predicted at a forming strain rate of 0.002 s^{-1}.

3. Finite element modeling (FEM)

The flow stress behavior of AA5083 during the hybrid superplastic forming has been represented by the power law given in ref. [12]. Figure 4 shows the thickness distributions at the end of forming phases. The simulation followed the experimental sequence and has been divided into two processing phases. The punch, which was modelled as a rigid tool, firstly moved down at a constant speed until it reached a total displacement of 43 mm. In this phase, the new punch has been changed into a two-stepped punch using the FEM design. For the second phase of gas blow forming, the punch was inactive and the workpiece was deformed due to the applied pressure only.

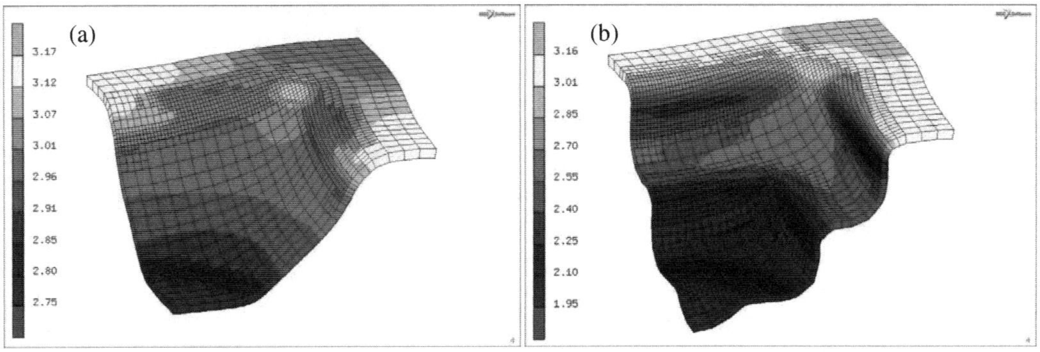

Figure 4. The predicted thickness contour of the workpiece in the phase of (a) mechanical pre-forming, and (b) gas blow forming.

The influences of process parameters, including of holder force, friction and punch geometry, have also been studied to facilitate process [12]. The conventional flow stress and the physically based material model related to the underlying behavior will be implemented in the simulation as well.

4. Summary

Non-superplastic grade of AA5083 has been used in the hybrid superplastic forming. As a greener manufacturing technology, it conducts the forming process at faster forming rate and lower temperature in one hybrid operation. It utilizes both the advantages of drawing and gas blow forming, and makes it more attractive for industry applications by forming the non-superplastic grade materials. The non-superplastic material is a promising candidate to be used in this green manufacturing with less cost, time, and energy consuming. Therefore, it is commercially viable to apply the hybrid SPF to high volume production, especially for the energy-saving technology to establish a more sustainable low carbon earth. Finite element modeling is a convenient method that can be used to facilitate the process design.

References

[1] Xun Y W, Tan M J 2000 *J. Mater. Process. Tech.* **99** 80-85
[2] Barnes A J 2007 *J. Mater. Eng. Perform.* **16** 440-454
[3] Kleiner M, Geiger M, Klaus A 2003 *CIRP Annals - Manuf. Tech.* **52** 521-542
[4] Osada K 1997 *J. Mater. Process. Tech.* **68** 241-245
[5] Ingelbrecht C D 1985 *J. Mater. Sci. Lett.* **4** 1021-1025
[6] Sorgente D, Scintilla L, Palumbo G, Tricarico L 2010 *Int. J. Mater. Form.* **3** 13-19
[7] Hefti L 2007 *J. Mater. Eng. Perform.* **16** 136-141
[8] Troeger L P, Starke E A 2000 *Mater. Sci. Eng., A* **277** 102-113
[9] Liu J, Tan M J, Jarfors A E W, Aue-u-lan Y, Castagne S 2010 *Mater. Design.* **31** S66-S70
[10] Friedman P A, Luckey S G, Copple W B, Allor R, Miller C E, Young C 2004 *J. Mater. Eng. Perform.* **13** 670-677
[11] Liu J, Tan M J, Aue-u-lan Y, Jarfors A E W, Fong K S, Castagne S 2011 *Int. J. Adv. Manuf. Tech.* **52** 123-129
[12] Liu J, Tan M J, Castagne S, Aue-u-lan Y, Fong K S, Jarfors A E W 2010 *Key Eng. Mater.* **447-448** 437-441

Laser micro-machinability of borosilicate glass surface-modified by electric field-assisted ion-exchange method

S Matsusaka, T Kobayakawa, H Hidai and N Morita
Department of Mechanical Engineering, Chiba University
1-33 Yayoi-cho, Inage-ku, Chiba 263-8522, Japan

E-mail: matusaka@faculty.chiba-u.jp

Abstract. In order to improve the laser micro-machinability of borosilicate glass, the glass surface was doped with metal (silver or copper) ions by an electric field-assisted ion-exchange method. Doped ions drifted and diffused into the glass substrate under a DC electric field. The concentration of metal ions within the doped area was approximately constant because the ion penetration was caused by substitution between dopant metal and inherent sodium ions. Nanosecond ultraviolet laser irradiation of metal-containing regions produced flat, smooth and defect-free holes. However, the shapes of holes were degraded when the processed hole bottoms reached ion penetration depths. A numerical analysis of ionic drift-diffusion behaviour in glass material under an electric field was also carried out. The calculated results for penetration depth and ionic flux showed good agreement with the measured values.

1. Introduction

It is well known that glass materials have excellent mechanical hardness, chemical stability and optical transparency. Micro-machining of glass materials, therefore, has become a key technology for various applications such as micro-machines, micro-opto-electro-mechanical systems (MOEMS) and analytical chemistry. However, mechanical or laser micro-machining of glass materials faces some problems with processing quality due to their brittleness and optical transparency. Therefore, chemical wet-etching techniques using hydrofluoric acid (HF) are frequently adopted in glass processing industries. As it is widely known, hydrofluoric acid is very harmful to human bodies and recycling of waste solutions is also difficult. For the reduction of the environmental burden, a novel processing method for glass materials is necessary.

The improvement of laser machinability is a promising approach for precise processing of glass materials. In fact, some research on the development of laser machinable glass has been carried out [1-5]. Typical laser machinable glass contains nanometer-sized metal clusters, such as silver or copper, in the glass matrix. Metal nano-clusters play two important roles during laser irradiation; they cause an increase in light absorption and mechanical strengthening of glass materials. This kind of laser machinable glass is usually prepared by an ion-exchange process in a molten salt bath that contains silver or copper ions. This 'wet' ion-exchange method is very simple and provides excellent laser micro-machinability to glass materials [1-4]. However, a long exchange time is required to obtain a large penetration depth of metal ions, because the process is driven by the substitution between alkali and metal ions at the glass surface. Also, penetrating metal ions are non-uniformly distributed in glass materials because the penetration is caused only by their ionic self-diffusion [4,6].

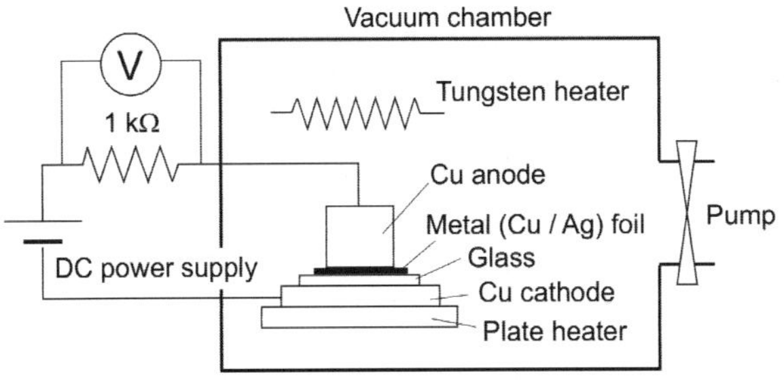

Figure 1. Experimental setup for electric field-assisted ion-exchange.

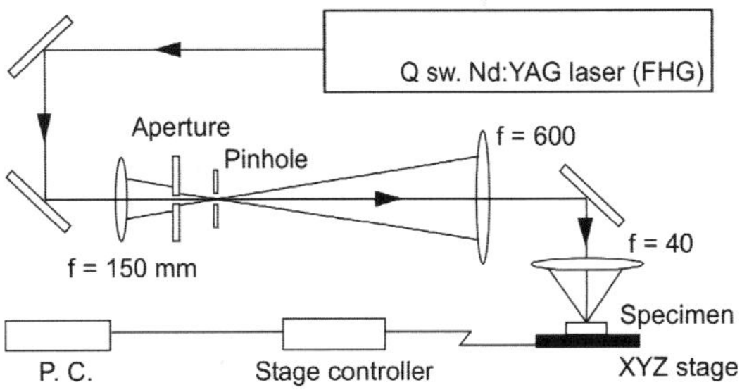

Figure 2. Laser micro-machining system.

In the present study, the authors propose a method for improving the laser micro-machinability of glass materials by electric field-assisted ion-exchange (FAIE). In the FAIE process, metal ions are supplied from a metal foil or film placed on the glass surface under an applied DC voltage. Because the process is electrically driven, metal penetration is more rapid and deep in comparison with the conventional wet ion-exchange process in a molten salt bath. In this report, the penetration behaviour of metal ions is experimentally and numerically examined. The laser micro-machinability of silver- and copper-doped borosilicate glasses prepared by FAIE is also investigated.

2. Experimental

Borosilicate glass slides (Schott, BOROFLOAT 33, 25 mm x 25 mm x 1.1 mm) and pure silver or copper foils (> 99.9 %, 20 mm x 20 mm x 10 μm) were used in the experiment. The glass slides and metal foils were rinsed in acetone, ethanol and distilled water before the experiment. A glass slide and metal foil were placed between two copper electrodes in a high vacuum chamber ($<10^{-3}$Pa). The experimental setup for ion-exchange is shown in Figure 1. After the evacuation and temperature elevation, a DC voltage was applied using the metal foil as the anode. The applied voltage, ion-exchange temperature and time were 100 V, 623 K and 3.6 ~ 21.6 ks, respectively. After the ion-exchange process, the glass slide was detached from the metal foil and rinsed in distilled water. The penetration depth and cross-sectional distributions of metal ions were measured by electron probe micro-analysis (EPMA: JEOL, JXA8900).

The laser micro-machining system consisted of a fourth harmonic generator of a Nd:YAG laser (Continuum, Surelite I-20, wavelength: 266 nm) and a PC-controlled translation stage (Suruga Seiki, KS201-20HD). In Figure 2, the laser micro-machining system is illustrated. The pulse duration and repetition rate of the laser beam were 6 ns and 20 Hz, respectively. All samples were irradiated on the front (ion-exchanged) surfaces by defocusing the laser beam to produce a spatially uniform (flat-top) energy distribution in atmosphere. The shapes and cross-sectional profiles of the processed holes were examined by scanning electron microscopy (SEM: JEOL, JSM-840A) and confocal laser scanning microscopy (Keyence, VK-8500), respectively.

3. Numerical analysis

We assume that only two mobile species A and B exist in the glass material, both of which are univalent cations. Symbols A and B denote metal (silver or copper) and sodium ions, respectively. When a DC electric field E is applied along the x-axis, the one-dimensional ionic drift-diffusion behaviour is expressed by the Nernst-Planck equation [6-8],

$$J_i = -D_i \frac{q}{kT} E C_i - D_i \frac{\partial C_i}{\partial x} \qquad (i = A, B), \tag{1}$$

where J is the ion flux, D is the self-diffusion coefficient, C is the ion concentration and q is the ionic charge. Because of local electrical neutrality, the total mobile cation concentration is assumed to be constant C_0, which is equal to the sodium ion concentration in un-doped glass, so that

Table 1. Material properties and calculation conditions.

Symbol / condition	Unit	Value
D_{Na}	m^2/s	5.0×10^{-15}
D_{Ag}	m^2/s	2.0×10^{-15}
D_{Cu}	m^2/s	1.0×10^{-16}
C_0	$1/m^3$	1.45×10^{27}
Temperature: T	K	623
Applied DC voltage: V_0	V	100
Boltzmann constant: k	J/K	1.381×10^{23}
Ionic charge: q	C	1.602×10^{-19}

$$C_A + C_B = C_0 . \tag{2}$$

Assuming that the potential drops at electrode-glass interfaces are negligible, the applied voltage V_0 between two electrodes is expressed by eq. (3),

$$\int_0^h E dx = V_0 , \tag{3}$$

where h is the thickness of the glass slide and V_0 is the applied DC voltage. Equations (1) and (2) were numerically solved under boundary condition (3) by the standard explicit finite-difference method. Material and physical properties used in the calculation are listed in Table 1. For simplicity, the concentration dependence of the self-diffusion coefficients was not taken into consideration in this calculation.

4. Results and discussion

4.1. Penetration behaviour of metal ions into glass substrates

Figure 3 shows typical silver and sodium ion distributions in the glass surface region obtained by EPMA. The field-assisted ion-exchange time of this sample was 21.6 ks. It is clearly indicated that the

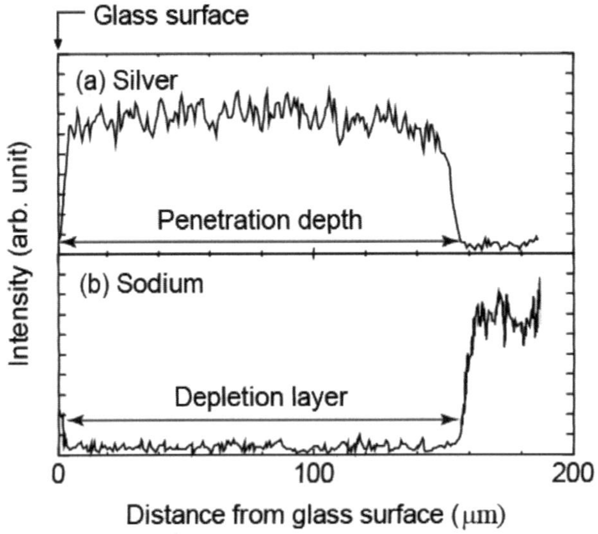

Figure 3. Cross-sectional ion distributions of (a) silver and (b) sodium in glass substrate measured by EPMA. Ion-exchange time was 21.6 ks.

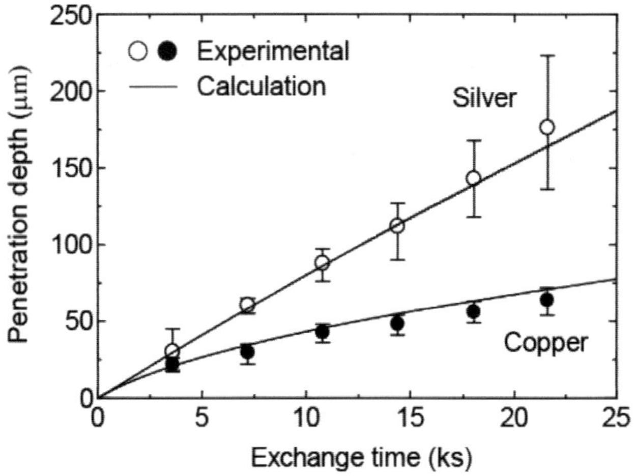

Figure 4. Change in ion penetration depths with exchange time.

silver concentration is approximately constant over the penetration area, and this is quite a contrast to that in silver-doped glass prepared by the wet ion-exchange method [4]. In the case of wet ion-exchanged glasses, the laser ablation rate (removed depth per laser shot) gradually decreased with distance from surface, because of the non-uniformities in ion concentration [4]. On the other hand, a constant ablation rate is expected for the ion-exchanged glasses produced by FAIE, because of their uniform ion distributions. Figure 3 also shows that a sodium depletion layer was formed at the glass surface, and its depth accurately accorded with the silver-doped depth. This result suggests that sodium ions in the glass substrate were fully substituted by penetrating silver ions. Although not shown here, similar results were obtained for copper-doped glasses.

The penetration depths of silver and copper ions for various exchange times obtained from EPMA profiles are indicated by white and black circles in Figure 4. Because the ionic mobility of silver in glass materials is much larger than that of copper [8], a difference in their penetration depths is seen in Figure 4. In a previous letter [5], we reported that the penetration depth of copper after 21.6 ks under similar experimental conditions was 27.2 μm, and this value is about half of the present value. The reason for the increase in penetration depth was the improvement of our experimental procedure, that is, surface cleaning and mirror polishing of two electrodes (anode and cathode) before every experiment. According to our previous result [5], the penetration depth of silver in wet ion-exchanged glass prepared in 1.0 % $AgNO_3$-$NaNO_3$ at 623 K after 21.6 ks was 8.8 μm, which was about one twentieth of that by FAIE after the same exchange time. This result suggests that the applied electric field is very effective for increasing the penetration depth, and therefore, for reducing the ion-exchange time.

In order to understand the penetration behaviour of silver and copper ions, the numerical analysis described in section 3 was carried out. The calculated penetration depths of silver and copper are shown by solid lines in Figure 4. As shown in Figure 4, the calculated results show good agreement with measured penetration depths. Especially, it should be noted that the geometrical difference between two penetration curves, i.e., linearly-increasing behaviour for silver and gradual saturation for copper, were accurately expressed by a simple simulation. Although the reason for the difference between the two penetration curves is now under investigation, we consider that the self-diffusion coefficient ratios $M_i = D_i/D_{Na}$ (i = Ag or Cu) have a large influence on the penetration behaviour. For validation of the calculation model, the numerically calculated and experimentally measured ionic fluxes after 21.6 ks were compared as listed in Table 2. The analytic and experimental ionic fluxes have comparable values within the range of experimental error, and therefore, the present calculation can reasonably describe the ion penetration behaviour.

Table 2. Comparison between numerically calculated and experimentally measured results for ionic flux after 21.6 ks of ion-exchange time.

Ionic flux [ions/m^{2}s]	Silver	Copper
Numerically calculated	1.00 x 10^{19}	2.95 x 10^{18}
Experimentally measured	1.09 x 10^{19}	2.04 x 10^{18}

4.2. Laser micro-machinability of metal-doped glasses

Figure 5 shows front views of laser ablated holes on silver- and copper-doped glasses, when the beam diameter and irradiated energy density were 300 μm and 10 J/cm^2, respectively. Cross-sectional profiles between the two white arrows are also shown in these micrographs. As shown in Figure 5, the bottoms of the ablated holes are remarkably flat and smooth before 680 irradiation pulses for silver and 200 pulses for copper-doped glasses. These flat bottom faces were assumed to be caused by the spatially uniform energy distribution of the laser beam. In Figure 5(a) to (c) and (d) to (f), the area of the flat bottom face decreased with increasing shot number. This is due to a decrease in the beam spot diameter with the progress of hole drilling, because the focal position of the laser beam was inside the glass substrate, i.e., the beam was underfocused. No large cracks or chips, which are often observed for laser irradiated un-exchanged glasses [5], were found in these holes. These results suggested that the input laser energy was effectively absorbed in the vicinity of the glass surface and rapid ablation of the surface layer occurred. On the other hand, the shapes of the 700 pulse hole for silver and 220 pulse hole for copper-doped glass were obviously different. The flat and smooth bottom faces were severely damaged and irregular-shaped structures appeared. As shown by the cross-sectional profiles in Figure 5, such deterioration of the bottom faces occurred at depths of 150 μm for silver and 60 μm for copper-doped glasses. These values approximately accorded with the penetration depths of silver and copper ions shown in Figure 4. Therefore, it was suggested that the defective structures at the hole bottoms were caused by the inherent brittleness of un-exchanged glass materials.

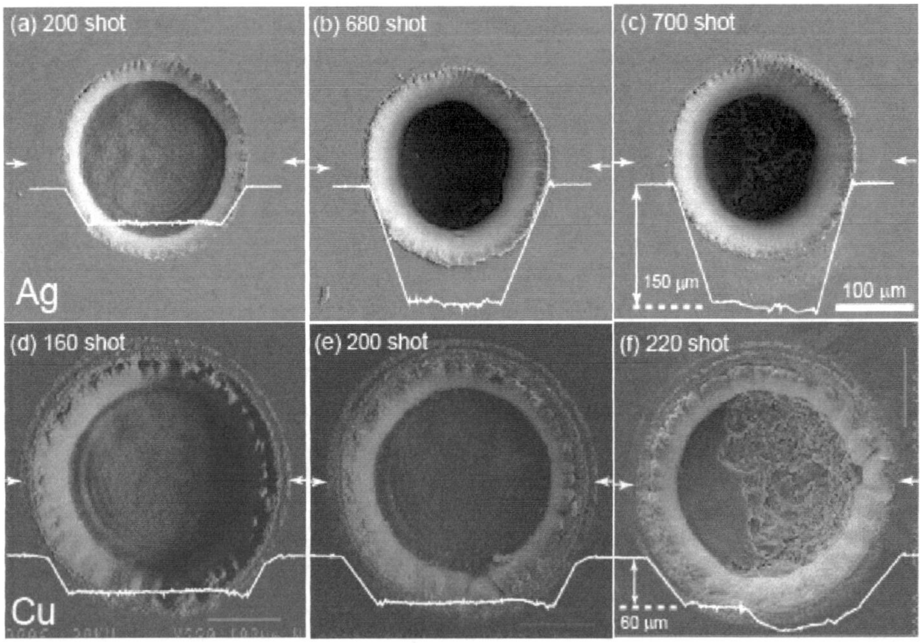

Figure 5. SEM micrographs of laser-ablated holes on metal-doped glasses. (a) 200, (b) 680 and (c) 700 irradiation shots on silver-doped glass. (d) 160, (e) 200 and (f) 220 shots on copper-doped glass. Irradiated energy density was 10 J/cm^2.

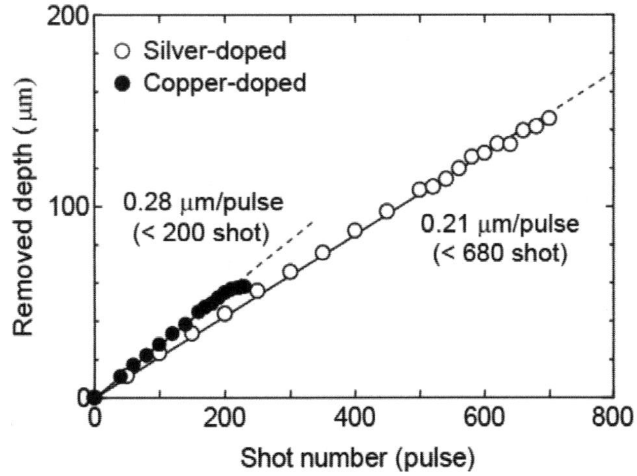

Figure 6. Change in removed depth with laser shot number for silver- and copper-doped glasses. Irradiated energy density was 10 J/cm^2.

Figure 6 shows the change in removed depth with laser shot number for silver and copper-doped glasses. As shown in Figure 6, the removed depths increased linearly with shot number, i.e., constant ablation rates (0.21 μm/pulse for silver and 0.28 μm/pulse for copper-doped glasses) were maintained over the ion penetrated area. We consider that these results are due to the constant ion concentrations using the FAIE process.

5. Conclusions

In the present study, a method of improving the laser micro-machinability of borosilicate glass by FAIE was proposed. UV nanosecond laser irradiation of silver or copper-doped glasses produced flat, smooth and defect-free holes. It was therefore found that metal doping of the glass surface was very effective for the improvement of laser micro-machinability. The concentrations of doped metal ions were approximately constant, and constant laser ablation rates were therefore achieved. However, the shapes of holes were degraded when the processed hole bottoms reached ion penetration depths. Therefore, a precise prediction method of ion penetration depth was strongly required for the high-accuracy fabrication of micro-optoelectronic devices. A simple numerical calculation using the finite-difference method was found to be sufficient to describe the ionic drift-diffusion behaviour in glass materials.

Acknowledgement

The authors gratefully acknowledge financial support by a Grant-in-Aid from the Ministry of Education, Culture, Sports, Science and Technology, Japan (No. 20760488).

References

[1] Tsunetomo K and Koyama T 1997 *Opt. Lett.* **22** 411
[2] Koyama T and Tsunetomo K 1997 *Jpn. J. Appl. Phys.* **36** L244
[3] Rosenbluh M, Antonov I, Ianets D, Kaganovskii Y and Lipovskii A. 2003 *Opt. Mater.* **24** 401
[4] Matsusaka S and Watanabe T 2008 *J. Mat. Proc. Tech.* **202** 514
[5] Matsusaka S and Watanabe T 2010 *Scripta Mater.* **62** 141
[6] Quaranta A and Gonella F 1995 *J. Non-Cryst. Solids* **192&193** 334
[7] Abou el leil M and Cooper A R 1979 *J. Am. Ceram. Soc.* **62** 390
[8] Oven R 2007 *J. Appl. Phys.* **101** 113113

Reactivity of CO_2 and H_2O on TiO_2 catalysts studied by gas phase FT-IR method and deactivation mechanism

Y Y Maruo, T Yamada and M Tsuda

NTT Energy and Environment Laboratories, 3-1 MorinosatoWakamiya, Atsugi, Kanagawa 243-0198, Japan

E-mail: maruo.yasuko@lab.ntt.co.jp

Abstract. The photoreduction of CO_2 to hydrocarbons is a sustainable energy technology that not only reduces emissions but also the need for alternative fuels. Wide band-gap TiO_2 catalysts are considered the most convenient candidates in terms of cost and stability. In this work, anatase TiO_2 powder was deposited on a silica filter paper, and photoreduction experiments were carried out in an FT-IR gas reactor using CO_2 gas with 50% relative humidity as the reactant under the irradiation of a high pressure Hg lamp. CH_4 and CO were detected as the CO_2 reduction products. However, there was a gradual decrease in the amounts of CH_4 and CO that were produced, seemingly because of the deactivation of the anatase TiO_2 surface. We analyzed the surface chemical states using XPS analysis to consider the cause of the TiO_2 deactivation, and found that the amounts of –OH group had increased on the TiO_2 surface after Hg lamp irradiation. Considering the analysis results, we assumed a surface deactivation mechanism.

1. Introduction

Increasing levels of CO_2 in the atmosphere constitutes the main cause of global warming. Recently, many efforts have been made to reduce CO_2 emissions through pre- or post- combustion CO_2 capture followed by compression and geological sequestration [1-7]. These processes are energy intensive and costly. Specifically, there are many uncertainties with regard to the long-term storage of CO_2 geological formations. An alternative and preferable approach is to recycle CO_2 as a fuel feedstock with energy input from cheap and abundant sources, for example solar energy. The photoreduction of CO_2 to hydrocarbons is a sustainable energy technology. This process utilizes ultraviolet (UV) and visible light as the excitation source, and the photo-excited electrons reduce CO_2 with H_2O on the catalyst surface and form energy-bearing products such as carbon monoxide (CO), methane (CH_4), methanol (CH_3OH), formaldehyde (HCHO) and formic acid (HCOOH) [8-14]. A variety of photocatalysts have been studied [15-18], and among them, wide band-gap titanium dioxide (TiO_2) catalysts are considered the most convenient candidates in terms of cost and stability [19-21]. TiO_2 forms three distinct polymorphs: rutile, anatase and brookite [22, 23]. Anatase was one of the most widely studied photocatalysts for environmental applications and there have been many papers dealing with the photoreduction of CO_2 on TiO_2 catalysts of anatase [10-14]. However, although it has many potential applications, the wider practical use of TiO_2 requires a photo-response extended to longer wavelengths and increased photo-efficiency achieved by retarding electron-hole recombination. In addition, focusing on a catalyst's photoreduction activity, it is reported that the amount of hydrocarbons produced gradually decreases, indicating TiO_2 deactivation [10, 24]. We obtained the same deactivation results in this work, and so we performed several kinds of bulk analyses (X-ray diffraction (XRD) and diffuse reflectance UV-vis spectra (UV-vis) of samples before and after Hg lamp exposure to consider the cause of the TiO_2 deactivation. However, we could not determine any difference between the two samples. Next, we performed a surface analysis of samples before and after Hg lamp exposure using X-ray photoelectron spectroscopy (XPS) to consider the cause of the TiO_2 deactivation, and obtained different chemical states on an anatase TiO_2 surface before and after Hg lamp exposure. We consider the cause of the TiO_2 deactivation using these results.

2. Experimental

2.1. Sample preparation

Anatase TiO_2 and 2-propanol were obtained from Kanto Chemical Co. Inc., and used without further purification. We used a silica filter paper (QR-100; Advantec Toyo) cut into 20×20 mm^2 squares as a substrate. We prepared a 10 ml 2-propanol suspension containing 0.1 g of anatase TiO_2. The silica filter papers were immersed in a glass petri dish filled with the suspension, and then 2-propanol was slowly evaporated in a dry nitrogen gas flow for five days. After that we obtained the samples of anatase TiO_2 deposited on the silica filter paper.

2.2. Sample characterization

Surface image of the samples were obtained using a scanning electron microscope (SEM) (SU1510; Hitachi) and a laser microscope (VK-8500; Keyence). XRD patterns of the powders were recorded with a MiniFlexII (Rigaku) instrument using Cu-Kα radiation. XPS analysis of the samples was performed with a PHI XPS5700 system using AlKα radiation (1486.6 eV). The photoelectron spectra were measured for C 1s, O 1s, and Ti 2p core levels. A diffuse reflectance UV-vis spectrometer (U-4100, Hitachi) was used to obtain the UV-vis spectra of the TiO_2 samples before and after Hg lamp irradiation.

2.3. Photocatalytic performance evaluation

We used a commercially available Fourier transform infrared spectroscopy (FT-IR) instrument (Nicolet iS10; Thermo Fisher Scientific) and designed an FT-IR reactor system to realize a photocatalytic reaction based on the FT-IR instrument. The system was not the flow system and was shown in Figure 1. The gas reactor was a glass cylindrical cell with two lengths of glass tube equipped with taps, and it was 32 mm in diameter and 100 mm long. We equipped both sides of the cell with BaF_2 windows, because BaF_2 has a high transmittance of near IR light up to about 1000 cm^{-1}. The light source for the photocatalytic reaction was a high pressure Hg lamp (UM-102, Ushio Inc.) and the output power was 100 W. First, we placed the TiO_2 deposited silica filter paper in the gas reactor. Next, compressed CO_2 (99.995%, Taiyonissann Inc.) controlled by a mass flow controller, was passed through a water bubbler to generate a mixture of CO_2 and water vapor (relative humidity: 50%). The CO_2 gas with 50% relative humidity was then introduced into the cylindrical gas reactor. We replaced the gas in the cylindrical gas reactor three times with generated CO_2 gas that has 50% relative humidity, and after that we closed the two taps of the cylindrical gas reactor. UV light was irradiated through the BaF_2 window using the high pressure Hg lamp. The light intensity of the lamp on the catalyst was 1.1 mW/cm^2 at 365 nm as measured by a spectroradiometer (USR-45; Ushio Inc.). Hourly FT-IR measurements were carried out, and the concentrations of the generated CO, CH_4, CH_3OH, HCHO, and HCOOH gases in the gas reactor were calculated from the absorption differences of 2165.5, 3017.5, 1032.5, 1105 and 2801 cm^{-1}, respectively. The CO and CH_4 calibration curves were obtained respectively using several kinds of diluted standard CO gas or CH_4 gas with compressed CO_2 gas. We obtained the CH_3OH, HCHO, and HCOOH calibration curves using several kinds of gases with specific concentrations. To obtain the gas with the intended concentrations, we injected the required amounts of each chemical reagent into 50 l Tedlar bags filled with CO_2.

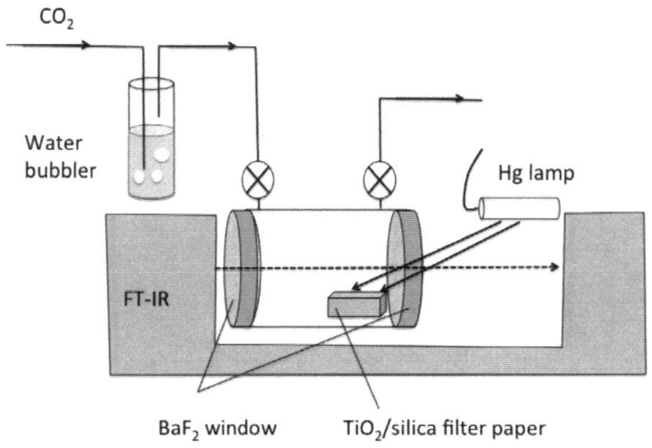

Figure 1. Experimental set-up.

3. Results and Discussion

3.1. Sample characteristics

The particle size of commercial anatase TiO_2 powder was checked using SEM images as shown in Figure 2. Anatase TiO_2 powder consisted of particles several ten of μm in diameter, which were aggregates of TiO_2 particles several hundred nm in diameter.

Figure 2. SEM images of commercial anatase TiO2; (a) ×350, (b) ×15.0K.

Laser microscope images of the silica filter paper and the TiO_2 deposited on it are shown in Figure 3. The TiO_2 powder was homogeneously deposited on the silica filter paper.

3.2. Photoreduction of CO_2

A background test was first performed on the gas reactor containing only the silica filter paper, and no CO_2 conversion products (for example CO or CH_4) were observed regardless of whether the light was on or off. We also carried out another background test on the gas reactor containing the TiO_2 sample deposited on the silica filter paper, and no CO_2 conversion products (for example CO or CH_4) were observed when the light was off. The conversion only occurred when TiO_2 was present together with Hg lamp irradiation. Figure 4 shows the product concentrations of CH_4 and CO in the gas reactor as a function of irradiation time for CO_2 photoreduction. The product concentrations of CH_3OH, HCHO and HCOOH were below the detection limit (1ppm) for the duration of the experiment. We carried out a second run after replacing the gas in the reactor with fresh CO_2 gas with 50% relative humidity. This process was repeated three times to ensure complete replacement. We found that the amounts of CH_4 and CO that were produced gradually decreased during the first run. In the second run, the product concentrations of both CH_4 and CO were small. CH_4 was identified as the major product using anatase TiO_2 catalyst. As shown in Figure 4 the measured concentration of CH_4 in the gas reactor increased with irradiation time and reached a peak concentration at around 120 min, and then gradually decreased. Similar results were reported by Y. Li et al. [24], namely that the yields of the CO_2 photoreduction products decreased once they had reached the maximum values. The reason for this is the deterioration of the photocatalytic activity of the anatase TiO_2 surface.

Figure 3. Laser microscope images; (a) silica filter paper, (b) TiO_2/Silica filter paper.

3.3. Characteristic of samples before and after UV irradiation

We measured the diffuse reflectance UV-vis spectra of the samples before and after Hg lamp irradiation to investigate the reason for TiO_2 deactivation. The reflectance was converted into the ratio (α) of absorption coefficient and scattering coefficient using the Kubelka-Munk equation, $\alpha=(1-R_\infty)^2/2R_\infty$, where R_∞ is the diffuse reflectance from a semi-infinite layer. The results are shown in Figure 5, and the band-gap energy of the samples before and after UV irradiation was estimated to be 3.31 ± 0.01 and 3.29 ± 0.01 eV, respectively. Considering the error range, two values are almost the same. An XRD measurement was carried out at the same time for the samples before and after Hg lamp irradiation. We used the same samples for both UV-vis and XRD measurements. We obtained the same diffraction pattern (anatase TiO_2 only) for both. On the basis of these results, we assumed that the cause of the deactivation would lie in the surface state, and not in the bulk state.

XPS is a valuable method for analyzing surface chemical states, therefore the XPS measurement was carried out for both non-irradiated and irradiated sample surfaces. The elemental concentrations calculated from the XPS intensities using sensitivity coefficients (Ti 2p:32.594, O 1s:13.099, and C 1s:5.214) (PHI Inc.) are summarized in Table 1. The calculated oxygen to titanium ratios were 2.24 and 2.28 for the non-irradiated and irradiated samples, respectively. The kinetic energy of a photoelectron from Ti 2p is about 1030 eV, and the mean free path of a 1030 eV photoelectron is about 1.4 nm [25]. If the surfaces of the two samples are terminated by oxygen atoms, the titanium atom is about 0.2 nm below the surface, because the bond distance of Ti—O is about 0.2 nm [26]. Considering the position of the Ti atom, the intensity of photoelectron emitted from Ti 2p levels is calculated 1.15 times, and based on this result, the calculated oxygen to titanium ratios are 1.94 and 1.98 for the non-irradiated and irradiated samples, respectively. This indicates that there are no stoichiometric changes on the surfaces of the two samples. In terms of carbon, the elemental concentration calculated from the XPS intensities increased by about 1.6 %. The origin of the carbon is the strong adsorption of the contaminant layer onto the surface [27], however the increased components would be related to the deactivation of anatase TiO_2.

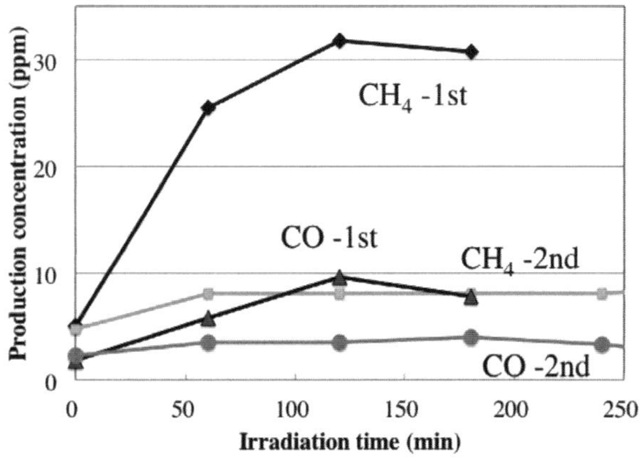

Figure 4. Time dependence of product concentration in gas reactor.

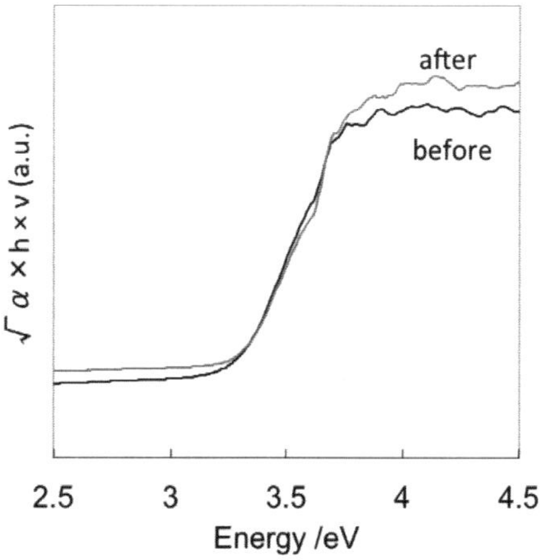

Figure 5. UV-vis spectra of samples before and after UV irradiation.

The XPS spectra for Ti 2p, O 1s and C 1s are shown in Figures 6, 7 and 8, respectively. The binding energy of the Ti 2p3/2 main peak was calibrated at 458.8 eV to correct the shifts caused by charging. No change was observed in the Ti 2p spectra after the Hg lamp irradiation except for the intensity change that was induced from the change in the amount of carbon on the TiO_2 surface. Shultz et al. reported that there are Ti^{3+} defects on a (110) rutile TiO_2 surface after UV irradiation, and this component has a lower binding energy of 1.7 eV than that of Ti^{4+} [28]. In our experiment, we could not detect a component with a lower binding energy of 1.7 eV, and observed a single peak of Ti^{4+} indicating that the chemical state of Ti had not changed after UV irradiation.

Table 1. Calculated atomic percentage using XPS intensities.

Sample	Atomic percentage (%)		
	Ti	O	C
Before irradiation	27.2	61.0	11.8
After irradiation	26.4	60.2	13.4

For the O 1s and C 1s spectra, the black line describes experimental data of the intensity of the signal in arbitrary units. The O 1s and C 1s spectra were resolved into individual peaks. The binding energies, FWHMs, counts and relative intensities of Ti 2p, O 1s and C 1s are summarized in Table 2. We set the counts of Ti 2p2/3 at 1.0 and calculated the relative intensities of the other components. The O 1s core level spectra had two peaks as shown in Figure 7. The peaks at approximately 530.0 and 531.3 eV are attributed to anhydrous oxide (Ti—O—Ti) and the hydroxyl group (—OH), respectively [27, 29, 30]. The relative intensity of anhydrous oxide (Ti—O—Ti) decreased by 0.02 and the relative intensity of the hydroxyl group (—OH) increased by 0.04 after Hg lamp irradiation. This indicates that the chemical state of anhydrous oxide (Ti—O—Ti) was converted into the chemical state of the hydroxyl group (—OH) by Hg lamp irradiation. Sasaki et al. and Yu et al. reported that UV irradiation in air increased the amount of the surface hydroxyl groups of TiO_2, and the source of the hydroxyl groups would be water in the air [31, 32]. Although our experiment was carried out in CO_2 gas with 50% relative humidity, the same formation of surface hydroxyl groups would occur between TiO_2 and water on the TiO_2 surface. The calculated percentage of oxygen in the hydroxyl group was 13.5% before irradiation and 16.2% after irradiation.

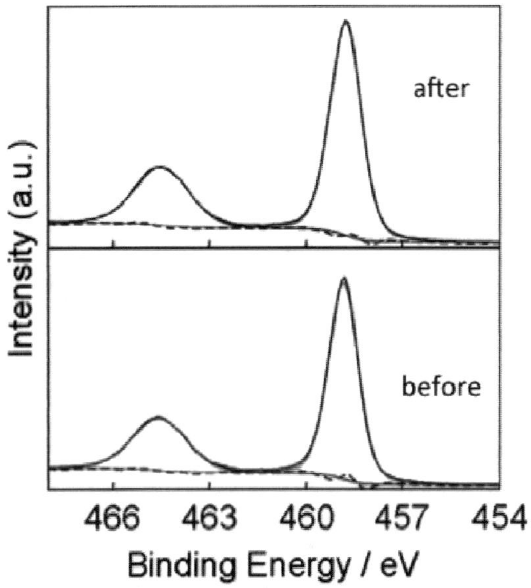

Figure 6. Change in Ti2p XPS spectra of TiO_2 before and after UV irradiation.

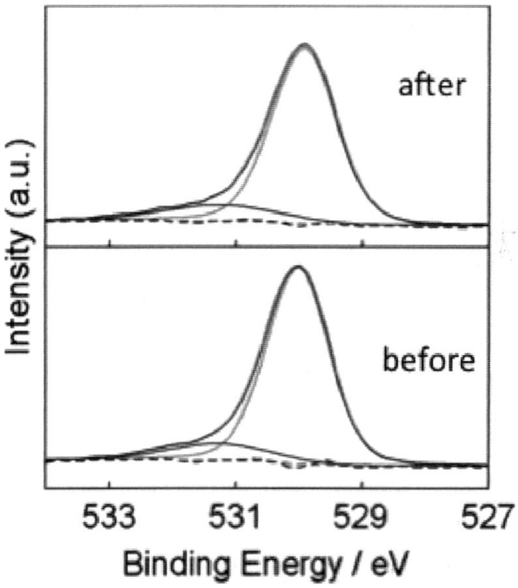

Figure 7. Change in O 1s XPS spectra of TiO_2 before and after UV irradiation.

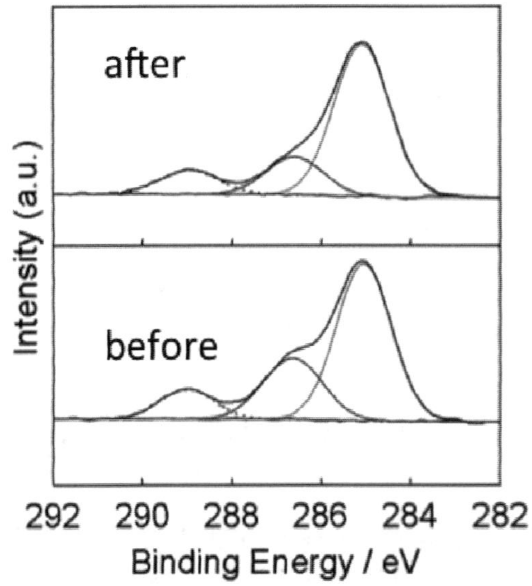

Figure 8. Change in C 1s XPS spectra of TiO_2 before and after UV irradiation.

Table 2. Binding energies, FWHMs, counts and relative intensities of Ti2p, O1s and C1s.

		Ti 2p		O 1s		C 1s		
Binding Energy (eV)		458.8	464.5	530.0	531.3	285.0	286.6	289.0
Before irradiation	FWHM(eV)	1.10	2.01	1.16	1.88	1.46	1.55	1.45
	Counts	31504	15238	38660	5228	2309	851	382
	Relative intensity	1.00	0.48	1.23	0.16	0.065	0.027	0.012
After irradiation	FWHM(eV)	1.15	1.98	1.21	2.16	1.48	1.49	1.55
	Counts	29659	14051	36012	5828	2498	652	410
	Relative intensity	1.00	0.47	1.21	0.20	0.084	0.021	0.014
Chemical species		Ti-O Ti2p 2/3	Ti-O Ti2p 1/2	Ti-O	Ti-OH	C	C-OH C-O-C	C=O

FWHM ; full width at half maximum.

The kinetic energy of a photoelectron from O 1s is about 955 eV, and the mean free path of a 955 eV photoelectron is about 1.3 nm [25]. If the sample surface is terminated by oxygen atoms, these atoms are present from the top of the surface to 0.2 nm below the surface, because the Ti—O bond distance is about 0.2 nm [26]. The photoelectron intensity from the O 1s is given as follows.

$$\boldsymbol{Io} = \int_0^{\infty} \boldsymbol{N0} \times \boldsymbol{e}^{-\frac{x}{1.3}} \boldsymbol{dx} \qquad \textbf{(1)}$$

In equation (1), N0 indicates the photoelectron intensity from the surface one layer. The photoelectron intensity of the irradiated TiO_2 was 41,840 (the sum of the O1s counts of 530.0 eV (36,012) and of 531.3 eV (5,828)), and the N0 was calculated as 32,190 by solving the eq. (1). Then the photoelectron intensity of O 1s from top of the surface to 0.2 nm below from the surface was calculated as 5,980, and the calculated intensity was almost the same of the photoelectron intensity of the hydroxyl group (—OH). For the non-irradiated TiO_2 surface, the photoelectron intensity was 43,890, and N0 was calculated as 33,760. Then the photoelectron intensity of O 1s from top of the surface to 0.2 nm below the surface was calculated as 6,280, and the experimentally obtained intensity of O 1s in the hydroxyl group (—OH) was about 80% of the calculated intensity. On the basis of these results, 80% of the oxygen in a non-irradiated TiO_2 surface would be in the hydroxyl state (–OH). The other 20% of the oxygen would be in an anhydrous oxide state (Ti—O—Ti), and this oxygen can be changed into the chemical state of the hydroxyl state (–OH) by Hg lamp irradiation, and then almost 100% of the irradiated TiO_2 surface would be terminated by a hydroxyl group (–OH).

The C1s core level spectra had three peaks as shown in Figure 8. The peaks at approximately 285.0, 286.6 and 289.0 eV are attributed to a hydrocarbon, an oxidant carbon such as alcohols and a carboxyl carbon, respectively [27, 29, 30]. The relative intensity of the oxidant carbon decreased by 0.006 and the relative intensity of the hydrocarbon increased by 0.019 after Hg lamp irradiation. Considering these results, we assumed that the carbon that existed near the oxygen in the (Ti—O—Ti) chemical state of the TiO_2 surface, changed into the carbon that existed near the hydrogen in the Ti—OH chemical state of the TiO_2 surface.

On the basis of these results, we proposed a model for the deactivation of the TiO_2 surface by Hg lamp irradiation as shown in Figure 9. 80% of the anatase TiO_2 surface is terminated by a hydroxyl group and this part would not work as a photocatalyst for the CO_2 photo-reduction reaction. The other 20 % of the oxygen in the anhydrous oxide state (Ti—O—Ti) can work as a photocatalyst for the CO_2 photo-reduction reaction. However, the oxygen in the (Ti—O—Ti) chemical state is not stable by the Hg lamp irradiation with H_2O, and the oxygen in the anhydrous oxide state (Ti—O—Ti) gradually changes into a hydroxyl group. After almost 100% oxygen of the TiO_2 surface had been terminated by a hydroxyl group, the TiO_2 changed into the deactivated state for the photo-reduction of CO_2.

Although more experiments are needed, the model shown in Figure 9 is reasonable considering the results we obtained in this experiment.

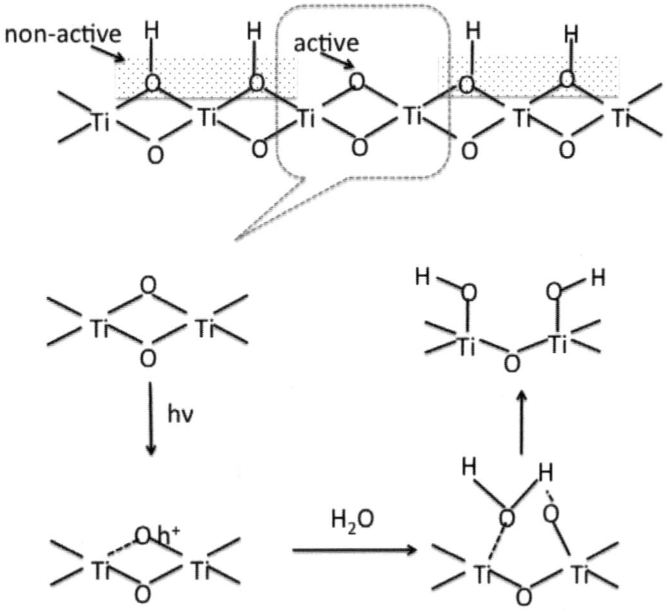

Figure 9. Surface deactivation models for the TiO_2 surface obtained by UV irradiation.

4. Conclusions

We found that anatase TiO_2 powder deposited on the silica filter paper was changed into a deactivated state by Hg lamp irradiation, and there was a gradual decrease in the amounts of CH_4 and CO that were produced by the CO_2 reduction. We analyzed the surface chemical states using XPS analysis to consider the cause of the TiO_2 deactivation, and found that the amounts of –OH group on the TiO_2 surface had increased after Hg lamp irradiation. On the basis of the analysis results, we assumed there to be a surface deactivation mechanism. However, when the (Ti—O—Ti) chemical state worked as a photocatalyst for the photo-reduction reaction of CO_2, the oxygen in the (Ti—O—Ti) chemical state was not stable as a result of the UV irradiation with H_2O, and the oxygen in the (Ti—O—Ti) chemical state gradually changed into a hydroxyl group that would not work as a photocatalyst for the photoreduction reaction of CO_2.

Acknowledgements

We thank Mr. Kunihiko Akaoka and Ms. Kaoru Yanase of NTT Energy and Environment Systems Laboratories for discussing about sample preparation methods.

References

[1] Balat M, Balat H and Oez C 2009 *Energ. Sources A Recovery Util Environ Eff* **31** 1473
[2] Romeo L M, Calvo E, Valero A and Devita A 2009 *Energy* **34** 1341
[3] Haszeldine R S 2009 *Science* **325** 1647

[4] Li H, Yan J and Yan J 2009 *Appl Energy* **86** 2760
[5] Tsuda T, Takeda M and Hosoya K 2010 *Corros Conf. Expo.* 2010 **3** 2339
[6] Florin N H, Blamey J and Fennell P S 2010 *Energy Fuels* **24** 4598
[7] Zeman F 2010 *J. Environ. Eng.* **136** 756
[8] Anpo M, Yamashita H, Ikeue K, Fujii Y, Ichihashi Y, Zhang S G, Park D R, Ehara S, Park S-E, Chang J-S and Yoo J W 1998 *Studies in Surface Science and Catalysis* **114** 177.
[9] Ikeue K, Nozaki S, Ogawa M and Anpo M 2002 *Catalysis Today* **74** 241
[10] Halmann M, Katzir V, Borgarelolo E and Kiwi J 1984 *Solar Energy Materials* **10** 85
[11] Tseng I-H, Chang W-C and Wu J C S 2002 *App. Catal. B* **37** 37
[12] Koci K, Mateju K, Obalova L, Krejcikova S, Lacny Z, Placha D, Capek L, Hospodkova A and Solcova O 2010 *Appl. Catal. B* **96**239
[13] Lo C-C, Hung C-H, Yuan C-S and Wu J-F 2007 *Solar Energy Materials & solar cells* **91** 1765
[14] Yahaya A H, Gondal M A and Hameed A 2004 *Chem. Phys. Lett.* **400** 206
[15] Bak T, Nowotny J, Rekas M and Sorrell C C 2002 *Int. J. Hydrogen Energy* **27** 991
[16] Guan G, Kida T, Harada T, Isayama M and Yoshida A 2003 *Appl. Catal. A* **249** 11
[17] Guan G, Kida T and Yoshida A 2003 *Appl. Catal. B* **41** 387
[18] Kydd R, Scott J, Teoh W Y, Chiang K and Amal R 2009 *Langmuir* **26** 2099
[19] Koci K, Obalova L and Lacny Z 2008 *Chemical Papers* **62** 1
[20] Kitano M, Matsuoka M, Ueshima M and Anpo M 2007 *Appl. Catal. A* **325** 1
[21] Usubharatana P, McMartin D, Veawab A and Tontiwachwuthikul P 2006 *Ind. Eng. Chem. Res.* **45** 2558
[22] Coronado D R, Gattoron G R, Pesqueira M E E, Cab C, Coss R and Oskam G 2008 *Nanotechnology* **19** 145605
[23] Mo S-D and Ching W Y 1995 *Phys. Rev. B* **51** 13023
[24] Li Y, Wang W-N, Zhan Z, Woo M-H, Wu C-Y and Biswas P 2010 *Appl. Cata. B* **100** 386
[25] Seah M P and Dench W A 1979 *Surf. Interface Anal.* **1** 2
[26] Diebold U 2003 *Surf. Science Reports* **48** 53
[27] Ohtsu N, Masahashi N, Mizukoshi Y and Wagatsuma K 2009 *Langmuir* **25** 11586
[28] Shultz A N, Jang W, Hetherington III W M, Baer D R, Wang L-Q and Engelhard M H 1995 *Surface Science* **339** 114
[29] Gopel W, Anderson J A, Frankel D, Jaehnig M, Phillips K, Schafer J A and Rocker G 1984 *Surface Science* **139** 333
[30] Huang W, Zuo Z, Han P, Li Z and Zhao T 2009 *J. Elect. Spectro. Relat. Phenom.* **173** 88
[31] Sasaki N, Fujishima A, Watanabe T and Hashimoto 2003 *J. Phys. Chem. B* **107** 1028
[32] Yu J C, Ho W, Yu J, Hark S K and Iu K 2003 *Langmuir* **19** 3889

Kinetic study on CO_2 photoreduction by Re complexes

Y Ono, M Tsuda, Y Y Maruo and J Nakamura
NTT Energy and Environment Systems Laboratories, 3-1 Morinosato Wakamiya, Atsugi, Kanagawa 243-0198, Japan

E-mail: ono.yoko@lab.ntt.co.jp

Abstract. The photoreduction of CO_2 using Re(btp)$(CO)_3$Cl (btp: bathophenanthroline) was investigated in a CO_2-saturated DMF-triethanolamine solution. CO formation was observed during irradiation with 365-nm light. Meanwhile, UV-vis spectral changes suggested that Re(btp)$(CO)_3$Cl was degraded and its amount decreased during irradiation. The degradation of Re(btp)$(CO)_3$Cl could cause CO formation, and the observed CO amount was the sum of CO produced by CO_2 reduction and Re(btp)$(CO)_3$Cl degradation. Thus, in the present paper, we discuss how the net amount of CO produced by CO_2 reduction could be determined via a kinetic study during UV irradiation which considers the CO production originating from the Re(btp)$(CO)_3$Cl degradation process.

1. Introduction

The use of CO_2 in chemical [1-4] and biological [5-7] CO_2-fixation reactions has been generating great interests because increasing atmospheric CO_2 concentration is considered to be responsible for global warming. Recently, much attention has been focused on photochemical CO_2 reduction because it can both reduce CO_2 and produce chemicals such as CO, CH_3OH and CH_4 as fuels [8]. Rhenium (Re) complexes are well known to have the potential to act as photocatalysts that photochemically reduce CO_2 to CO and/or COOH in the presence of electron donors such as triethanolamine (TEOA) [8-18]. Among the Re complexes, Re(bpy)$(CO)_3$Cl (bpy = 2,2′-bipyridine) has been reported to selectively reduce CO_2 to CO under UV irradiation. Many researchers have focused on increasing the reaction efficiency of CO_2 photoreduction and investigating its reaction mechanism by substituting Cl with a certain ligand [12,14,15]. We found that the amount of CO produced by CO_2 reduction greatly increased when the bpy moiety was substituted with bathophenanthroline (btp). However, since Re complexes originally have CO as a ligand, elimination of the CO ligand should be considered. Thus, an analytical method for determining the net amount of CO produced by CO_2 reduction, which takes Re complexes degradation into consideration, is required.

In this study, we investigate the degradation of Re(btp)$(CO)_3$Cl by undertaking a kinetic study on the change in UV-vis absorption of a Re(btp)$(CO)_3$Cl solution during UV irradiation, and establish an analytical method to determine the net amount of CO produced by CO_2 reduction during UV irradiation, using Re(btp)$(CO)_3$Cl as a CO_2-reducing photocatalyst.

2. Experimental

2.1. General procedures

UV-vis absorption spectra were measured using a Shimadzu UV-2200 spectrophotometer. IR spectra were recorded using a Perkin Elmer System 2000 FT-IR spectrometer. Elemental analysis was performed using a VARIO EL III analyzer. APCI mass spectra were measured using a Waters LCT Premier XE instrument in the negative-ion mode. The CO produced by CO_2 photoreduction was analyzed using a Shimadzu GCMS-QP2010 Plus instrument equipped with an Rt-Molsieve 5A column (L = 30 m, ϕ = 0.32 mm, d = 30 μm).

2.2. Photochemical CO_2 reduction

Photochemical CO_2 reduction was performed using a previously reported method [15]. Figure 1 shows the experimental set-up for photochemical CO_2 reduction. The reaction took place in a quartz cell (vol. = 5 mL) containing 2.0 mL of a $Re(btp)(CO)_3Cl$ DMF-triethanolamine (TEOA) (5:1, v/v) solution (conc. = 0.05 mM) which was stirred at room temperature. The solution was bubbled with CO_2 for 2 min at a flow rate of 200 mL/min, and sealed with a silicon-Teflon septum. A mercury lamp (λ = 365 nm, I = 614 μW/cm^2) was used as a light source for the photochemical reaction. The reaction products were analyzed by GC-MS. During irradiation, the amount of residual $Re(btp)(CO)_3Cl$ was determined by measuring the absorption maximum at 290 nm.

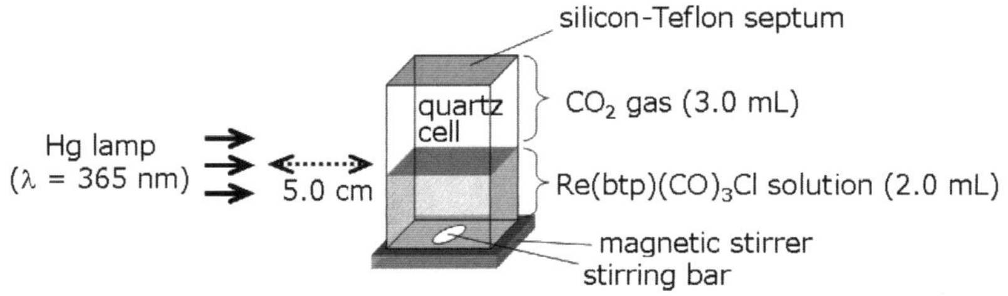

Figure 1. Experimental set-up for photochemical CO_2 reduction.

2.3. Materials

All solvents were of high purity and were used as received from Kanto Chemical Co. Rhenium pentacarbonyl chloride ($Re(CO)_5Cl$), 2,2′-bipyridine (bpy), and bathophenanthroline (btp) were purchased from Aldrich Chemical Co., Kanto Chemical Co., and Tokyo Chemical Industry Co., respectively. All reagents were used without further purification.

2.4. Synthesis

$Re(btp)(CO)_3Cl$ was synthesized as follows using a previously reported method [19] with some modifications. A solution of btp (249 mg, 0.75 mmol) in 100 mL of toluene was added to a solution of $Re(CO)_5Cl$ (271 mg, 0.75 mmol) in 150 mL of toluene at 110 °C. The mixture was kept at 110 °C and stirred vigorously for 4 h until the solution became a brown-orange color. The solvent was then evaporated in vacuum, and the brown-orange compound was isolated by column chromatography (H = 400 mm, ϕ = 25 mm) on silica using ethyl acetate as an eluent. After the solvent had been removed, the solid product was recrystallized from acetone-hexane. Yield: 65.0 mg (13.3%) Anal. Calcd for $C_{27}H_{16}ClN_2O_3Re$: C, 50.8; H, 2.5; N, 4.4; O, 7.5. Found: C, 51.1; H, 3.1; N, 3.9; O, 9.1. FT-IR (ν_{CO}/cm^{-1}): 2022, 1920, 1897 in CH_2Cl_2. UV-Vis (ε_{max}/nm (ε/M^{-1}cm^{-1}) : 284 (18880) in MeOH. APCI-MS (m/z): 638.0 (M^-) in MeOH.

3. Results and discussion

3.1. CO formation in the reaction cell

CO was formed when Re(btp)$(CO)_3$Cl in the CO_2-saturated DMF-TEOA solution was irradiated with 365-nm light, as shown in figure 2. CO formation increased linearly for about 60 min, and then the CO amount reached the saturation point at about 120 min. Since the observed CO might include CO from the degradation of Re(btp)$(CO)_3$Cl, the CO production will be discussed in detail in the following section by employing a kinetic study.

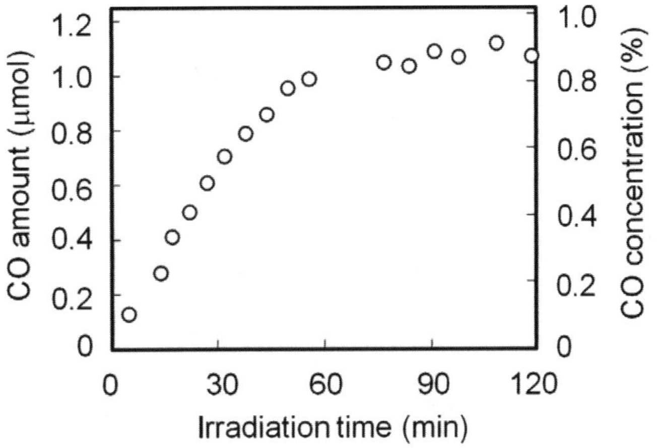

Figure 2. CO formation observed in a reaction cell during 365-nm light irradiation.

3.2. Determination of the degradation rate constant (k_d) of Re(btp)$(CO)_3$Cl

We measured the in-situ UV-vis absorption spectra of Re(btp)$(CO)_3$Cl in the CO_2-saturated DMF-TEOA solution during the photochemical reaction by irradiation with 365-nm-light. "o" in figure 3 show the absorbance at 290 nm, which was attributed to the π-π^* transition peak of the btp moiety of Re(btp)$(CO)_3$Cl. The absorbance intensity at 290 nm decreased during light irradiation, and further irradiation did not cause spectral changes. After removal of the light source, Re(btp)$(CO)_3$Cl retained its absorption intensity for more than 24 h. This results suggested that light irradiation induced Re(btp)$(CO)_3$Cl degradation. The amount of CO observed might include CO produced from Re(btp)$(CO)_3$Cl degradation. Therefore, we calculated the net amount of CO produced from CO_2 reduction by considering the CO production from Re(btp)$(CO)_3$Cl degradation using kinetic analysis. We assumed that this degradation involved CO-ligand elimination. We present the following reaction equation (1) for the degradation of Re(btp)$(CO)_3$Cl, with the assumption that the degradation proceeded by a first order reaction, and that the degradation of 1 mol of Re complexes produced 3 mol of CO, which was originally included as a ligand of Re(btp)$(CO)_3$Cl:

$$\mathrm{Re}(btp)(CO)_3Cl \xrightarrow{k_d} \mathrm{Re}'+3CO\,, \tag{1}$$

where k_d is the degradation rate constant, and Re′ represents a specific compound formed via the degradation. According to equation (1), the rate equation can be described by:

$$-\frac{d[\mathrm{Re}(btp)(CO)_3Cl]}{dt} = k_d[\mathrm{Re}(btp)(CO)_3Cl]. \tag{2}$$

Integration between the respective limits of {[Re(btp)(CO)$_3$Cl]$_{t=0}$, [Re(btp)(CO)$_3$Cl]$_{t=t}$}, and [0, t] gives:

$$[\mathrm{Re}(btp)(CO)_3Cl]_t = [\mathrm{Re}(btp)(CO)_3Cl]_{t=0} \times \exp(-k_d t), \tag{3}$$

where [Re(btp)(CO)$_3$Cl]$_t$ and [Re(btp)(CO)$_3$Cl]$_{t=0}$ represent the Re(btp)(CO)$_3$Cl concentrations at an arbitrary time and at t = 0, respectively. In the same way, the concentration of Re′ is represented by:

$$[\mathrm{Re'}]_t = [\mathrm{Re'}]_{t=\infty} \times \{1 - \exp(-k_d t)\}, \tag{4}$$

where [Re′]$_t$ and [Re′]$_{t=\infty}$ represent the Re′ concentrations at an arbitrary time and at $t = \infty$, respectively. Therefore, using the Lambert-Beer law to convert concentration to absorbance, the experimental value of the absorption intensity at 290 nm (Abs.$_t$) can be represented by the sum of the decrease in the Re(bpy)(CO)$_3$Cl absorbance and the increase in the Re′ absorbance by converting concentration to absorbance using the Lambert-Beer law as the following:

$$Abs._t = \varepsilon_1 \times [\mathrm{Re}(btp)(CO)_3Cl]_{t=0} \times \exp(-k_d t) + \varepsilon_2 \times [\mathrm{Re'}]_{t=\infty} \times \{1 - \exp(-k_d t)\}, \tag{5}$$

where t represents the light irradiation time (the time for attaching a light source is defined as t = 0), Abs.$_t$ is the observed absorbance at 290 nm at t, and ε_1 and ε_2 are the molar adsorption coefficients at 290 nm for Re(btp)(CO)$_3$Cl and Re′, respectively (ε_1 = 36000 M^{-1}cm^{-1} and ε_2 = 16400 M^{-1}cm^{-1}). In figure 3, the experimental values of the absorbance at 290 nm, represented by "o", could be well fitted to a bi-exponential function (equation (5)), as shown by the curve. The degradation rate constant (k_d) is obtained as:

$$k_d = 7.12 \times 10^{-4}\ \mathrm{s}^{-1}. \tag{6}$$

The theoretical absorbance values at 290 nm for Re(btp)(CO)$_3$Cl and Re′, which were calculated using equation (5), are also plotted in figure 3. It is clear that the amount of Re(btp)(CO)$_3$Cl decreased with increasing amount of Re′, followed by complete degradation of Re(btp)(CO)$_3$Cl.

The total observed CO formation (1.10 μmol) was sufficiently greater than the original amount of CO as ligands of Re(btp)(CO)$_3$Cl (0.30 μmol). Thus, it is considered that the CO formation originated from both the Re(btp)(CO)$_3$Cl degradation and CO_2 reduction. The inset of figure 4 summarizes the two different routes of CO formation. Under 365-nm light irradiation, Re(btp)(CO)$_3$Cl could act as a photocatalyst to reduce CO_2 to CO. On the other hand, light irradiation could also induce the degradation of Re(btp)(CO)$_3$Cl and adsorption of CO ligands.

Regarding CO produced by Re(btp)(CO)$_3$Cl degradation, the following equation could be derived:

$$[CO]_t = 3 \times [CO]_{t=\infty} \times \{1 - \exp(-k_d t)\}. \tag{7}$$

In equation (7), the value of 3 × [CO]$_{t=\infty}$ corresponds to the total concentration of CO produced by Re(btp)(CO)$_3$Cl degradation.

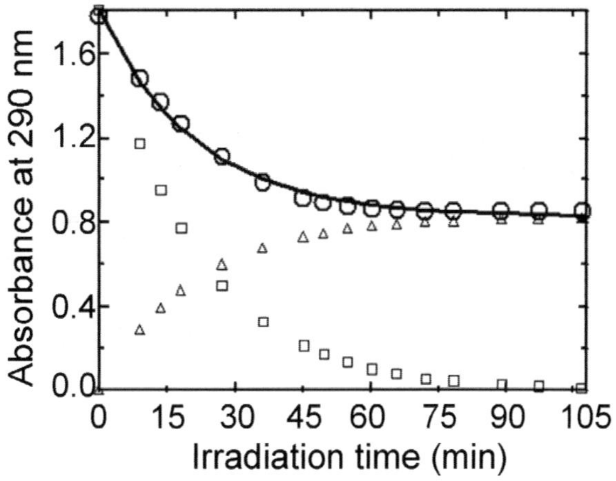

Figure 3. Absorbance at 290 nm for $Re(btp)(CO)_3Cl$ in the DMF-TEOA solution during 365-nm light irradiation. (o: experimental value, Δ :absorbance of Re′, □: absorbance of $Re(btp)(CO)_3Cl$, -: fitting function)

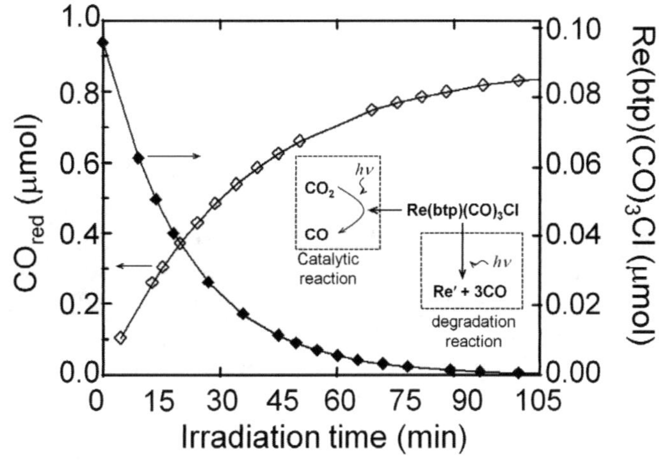

Figure 4. Amount of CO_{red} and $Re(btp)(CO)_3Cl$ vs. irradiation time plots (◊: amount of CO_{red}, ♦: amount of $Re(btp)(CO)_3Cl$, -: fitting function). The inset figure is a schematic illustration for CO formation through two different routes under 365-nm light irradiation.

3.3. Determination of the net amount of CO produced by CO_2 reduction

Taking the above discussion into consideration, if we assume that the CO_2 reduction equation is simply represented by equation (8), then the rate equation of CO formation can be described as equation (9):

$$\mathrm{Re}(btp)(CO)_3Cl + CO_2 \xrightarrow{k_{red}} CO_{red}\,, \tag{8}$$

$$\begin{aligned}\frac{d[CO_{red}]}{dt} &= k_{red}[CO_2][\mathrm{Re}(btp)(CO)_3Cl] \\ &= k_{red}[CO_2][\mathrm{Re}(btp)(CO)_3Cl]_{t=0} \times \exp(-k_d t) \\ &\approx a \times \exp(-k_{red} t),\end{aligned} \tag{9}$$

where k_{red} is the rate constant for CO_2-to-CO reduction, CO_{red} is CO produced by CO_2 reduction, and *a* is a constant when CO_2 concentration, $[CO_2]$, could be considered to be constant. Integration between the limits $\{[CO_{red}]_{t=0}, [CO_{red}]_t\}$ and $[0, t]$ gives the equation for the CO_{red} concentration, $[CO_{red}]$, at any time, *t*:

$$[CO_{red}]_t = [CO_{red}]_{t=\infty} \times \{1 - \exp(-k_{red} t)\}\,, \tag{10}$$

where $[CO_{red}]_{t=\infty}$ corresponds to the total amount of CO_{red} produced. We calculated the amount of CO produced by $\mathrm{Re(btp)(CO)_3Cl}$ degradation using equations (6) and (7), and determined the amount of CO_{red} by subtracting the amount of CO produced by $\mathrm{Re(btp)(CO)_3Cl}$ degradation from the experimental values for CO generation in the reaction cell at an arbitrary irradiation time. The results are plotted as a function of irradiation time in figure 4. These plots fit well to a mono-exponential function based on equation (10), and the following rate constant (k_{red}) and the total amount of CO_{red} produced ($CO_{red,t=\infty}$) were obtained as the fit parameters:

$$k_{red} = 4.20 \times 10^{-4}\ \mathrm{s}^{-1}, \tag{11}$$

$$CO_{red,t=\infty} = 8.73 \times 10^{-1} \pm 3.93 \times 10^{-6}\ \mu\mathrm{mol}. \tag{12}$$

The calculated amount of $\mathrm{Re(btp)(CO)_3Cl}$ using equation (3) is also plotted in figure 4, represented by "♦". It is clear that CO formation reaches the saturation point with decreasing amount of $\mathrm{Re(btp)(CO)_3Cl}$, depending on the irradiation time.

We determined the efficiency of the CO_{red} production on $\mathrm{Re(btp)(CO)_3Cl}$ to be 8.73 mol/cat-mol·2h, which was calculated by dividing $CO_{red,t=\infty}$ by the initial amount of $\mathrm{Re(btp)(CO)_3Cl}$ in the reaction solution. We took the degradation process of $\mathrm{Re(btp)(CO)_3Cl}$ into consideration, and improved the calculation accuracy of CO_{red}. If the degradation process could be suppressed such as by optimization of the light intensity and irradiation wavelength, $\mathrm{Re(btp)(CO)_3Cl}$ has the potential to be applied to possible CO_2 utilization systems.

4. Conclusions

CO_2 photoreduction was examined using $\mathrm{Re(btp)(CO)_3Cl}$ as a photocatalyst. The observed CO was found to be produced by both $\mathrm{Re(btp)(CO)_3Cl}$ degradation and CO_2 reduction. Assuming that the degradation involved CO-ligand elimination, we determined the net amount of CO produced by CO_2 reduction in a kinetic study by subtracting the CO produced by $\mathrm{Re(btp)(CO)_3Cl}$ degradation from the CO observed in the reaction cell at an arbitrary irradiation time. We improved the calculation accuracy of CO amount produced by CO_2 photoreduction. The CO production capacity resulting from CO_2 reduction using $\mathrm{Re(btp)(CO)_3Cl}$ was found to be 8.73 mol/cat-mol·2h.

References

[1] Aparicio P F, Baeza B B, Ruiz A G and Ramos I R 1998 *Surf. Sci. and Catal.* **114** 399
[2] Yogo K, Teng Y, Yashima T and Yamada K 2005 *Greenhouse Gas Control Technologies* **7** 2427

[3] Song C 2006 *Catal. Today* **115** 2
[4] Li J, Qi X, Wang L, He Y and Deng Y 2011 *Catal. Commun.* **12** 1224
[5] Michiki H 1995 *Energy Convers. Mgmt* **36** 701
[6] Usui N and Ikenouchi M 1997 *Energy Convers. Mgmt* **38** 487
[7] Otsuki T 2001 *Sci. Total Environ.* **277** 21
[8] Tanaka K 1995 *Adv. Inorg. Chem.* **43** 409
[9] Hawecker J, Lehn J M and Ziessel R 1983 *J. Chem. Soc., Chem. Commun.* 536
[10] Hawecker J, Lehn J M and Ziessel R 1986 *Helv. Chim. Acta* **69** 1990
[11] Ishitani O, George M W, Ibusuki T, Johnson F P A, Koike K, Nozaki K, Pac C, Turner J J and Westwell J R 1994 *Inorg. Chem.* **33** 4712
[12] Hori H, Johnson F P A, Koike K, Ishitani O and Ibusuki T 1996 *J.Photochem. Photobiol. A: Chem.* **96** 171
[13] Hori H, Ishitani O, Koike K, Kakeuchi K and Ibusuki T 1996 *Anal. Sci.* **12** 587
[14] Hori H, Johnson F P A, Koike K, Takeuchi K, Ibusuki T and Ishitani O 1997 *J. Chem. Soc. Dalton Trans.* 1019
[15] Hori H, Ishihara J, Koike K, Takeuchi K, Ibusuki T and Ishitani O 1999 *J.Photochem. Photobiol. A: Chem.* **120** 119
[16] Koike K, Okoshi N, Hori H, Takeuchi K, Ishitani O, Tsubaki H, Clark I P, George M W, Johnson F P A and Turner J J 2002 *J. Am. Chem. Sci.* **124** 11448
[17] Hori H, Takano Y, Koike K and Sasaki Y 2003 *Inorg. Chem. Commun.* **6** 300
[18] Takeda H and Ishitani O 2010 *Coord. Chem. Rev.* **254** 346
[19] Casper J V and Meyer T J 1983 *J. Phys. Chem.* **87** 952

CO_2 decomposition using electrochemical process in molten salts

Koya Otake[1], Hiroshi Kinoshita[2], Tatsuya Kikuchi[1] and Ryosuke O Suzuki[1,*]

[1] Division of Materials Science and Engineering, Faculty of Engineering, Hokkaido University, Kita 13, Nishi 8, Kita-ku, Sapporo, Hokkaido 060-8628, Japan
[2] Department of Mechanical Engineering, Fukushima National College of Technology, Iwaki, Fukushima 970-8034, Japan

E-mail: *rsuzuki@eng.hokudai.ac.jp

Abstract. The electrochemical decomposition of CO_2 gas to carbon and oxygen gas in LiCl-Li_2O and $CaCl_2$-CaO molten salts was studied. This process consists of electrochemical reduction of Li_2O and CaO, as well as the thermal reduction of CO_2 gas by the respective metallic Li and Ca. Two kinds of ZrO_2 solid electrolytes were tested as an oxygen ion conductor, and the electrolytes removed oxygen ions from the molten salts to the outside of the reactor. After electrolysis in both salts, the aggregations of nanometer-scale amorphous carbon and rod-like graphite crystals were observed by transmission electron microscopy. When 9.7 %CO_2-Ar mixed gas was blown into LiCl-Li_2O and $CaCl_2$-CaO molten salts, the current efficiency was evaluated to be 89.7 % and 78.5 %, respectively, by the exhaust gas analysis and the supplied charge. When a solid electrolyte with higher ionic conductivity was used, the current and carbon production became larger. It was found that the rate determining step is the diffusion of oxygen ions into the ZrO_2 solid electrolyte.

1. Introduction

Global environmental issues are currently regarded as one of the most urgent problems in the world, and the industrial sector accounts for a large fraction of CO_2 gas emissions [1]. The steel making industry emits a large amount of CO_2 gas, although it has been making great efforts to decrease the gas emission. Because the iron ore, Fe_2O_3, is reduced by coke, emission of a large amount of CO_2 gas from blast furnaces and converters is essential in the iron and steel making processes. The development of an innovative method is also required to decrease CO_2 gas emissions. However, there is only a few studies on the decomposition of CO_2 gas emitted from these furnaces. If all emitted CO_2 gas is efficiently decomposed, the amount of CO_2 gas emissions will decrease drastically.

As a new method to reduce metallic oxides such as titanium, vanadium, tantalum and nickel to their metallic state, a OS process consisting of electro-reduction of CaO and calciothermic reduction in a $CaCl_2$ bath has been studied [2-7]. In the present paper, a new method is proposed to decompose CO_2 gas to carbon and oxygen gas by applying the OS process to the reduction of gaseous oxides.

The reduction mechanism of the OS process for the case of titanium dioxide reduction was reported in previous papers [8-10]. At 1173 K, $CaCl_2$ can dissolve 20 mol% CaO and 2-4 mol% metallic Ca at equilibrium. CaO is ionized in the molten salt as presented in Equation (1). When an electrical voltage is applied between two electrodes, metallic Ca is produced near the cathode as shown in Equation (2).

The metallic Ca rapidly dissolves in $CaCl_2$ and the dissolved Ca then thermodynamically reacts with titanium dioxide near the cathode to produce metallic titanium and CaO as shown in Equation (3). Because CaO (by-product) is ionized again, the chemical reactions (Equations 1-3) are cycled in the furnace. Meanwhile, oxygen anions in the molten salt are removed as CO or CO_2 gas when a carbon anode is used as shown in Equation (4).

$$\underline{CaO}\ (l) = Ca^{2+} + O^{2-} \quad (1)$$

$$Ca^{2+} + 2e^{-} = \underline{Ca}\ (l) \quad (2)$$

$$2\underline{Ca}\ (l) + TiO_2\ (s) = Ti\ (s) + 2\underline{CaO}\ (l) \quad (3)$$

$$C\ (s) + nO^{2-} = CO_n\ (g) + 2ne^{-}\ (n = 1 \text{ or } 2) \quad (4)$$

When we apply this OS process to CO_2 decomposition, the solid TiO_2 can not be simply converted to CO_2 gas. The decomposition of CO_2 gas occurs at the surface of CO_2 gas bubbles, and it is important that the reductant Ca exists near the surface of the CO_2 gas bubbles. In the case of electrolysis, the reductant does not precipitate on the bubble, but on the cathodic surface. Therefore, it is essential that the reductant dissolves in the molten salt and reacts with the gas bubbles to decompose CO_2 gas. If this process can be realized, the exhaust CO_2 gas can be converted to useful substances. For example, carbon and oxygen gas produced from CO_2 gas can be used as raw materials in the iron and steel making industry.

In this study, two new proposals were introduced to the original OS process to decompose CO_2 gas to carbon and oxygen gas. The first proposal involves using LiCl-Li_2O molten salt as the solvent to prevent the occurrence of the Boudouard reaction, in which CO_2 gas preferentially reacts with carbon to form CO gas at high temperatures, as shown in Equation (5).

$$CO_2\ (g) + C\ (s) = 2CO\ (g) \quad (5)$$

$$(\Delta G^0_{923K} = 8.802\ kJmol^{-1}\ [11],\ \Delta G^0_{1173K} = -34.917\ kJmol^{-1}\ [11])$$

This reaction is undesirable since carbon, the desired product is consumed. When LiCl-Li_2O molten salt is used as the solvent, the operating electrolysis temperature can be decreased by about 250 K compared to the operating electrolysis temperature range 923-1173 K for $CaCl_2$-CaO molten salt. The latter salt has a wider solubility range of Ca and CaO, while LiCl can dissolve up to 4.2 mol% Li_2O at 923 K [12], and thus metallic Li has enough ability to decompose CO_2 gas to carbon and oxygen gas thermodynamically.

The second proposal is to use a ZrO_2 solid electrolyte. In the conventional OS process, carbon anode is used to remove oxygen ions from the molten salt to the outside of reactor. In this case, however, oxygen anions are removed as CO or CO_2 gas after the reaction with carbon. This is undesirable since we want the decomposition of CO_2 gas to carbon, not to CO and CO_2 gas. Therefore, the ZrO_2 solid electrolyte which is an oxygen ion conductor is applied as the anode. The oxygen anions in the molten salt diffuse through the ZrO_2 solid state electrolyte as shown in Equation (6) and the oxygen anions are removed as oxygen gas at the opposite side of the ZrO_2 anode as shown in Equation (7). When considering the diffusion rate of oxygen anions in a ZrO_2 solid electrolyte, a higher electrolysis temperature is desired. The decomposition mechanism of CO_2 gas is shown in Equations (6)–(10) when we apply the OS process in LiCl-Li_2O molten salt.

$$2O^{2-}(\text{in molten salt}) = 2O^{2-}\ (\text{in } ZrO_2 \text{ electrolyte}) \quad (6)$$

$$2O^{2-}\ (\text{in } ZrO_2 \text{ electrolyte}) = O_2\ (g) + 4e^{-} \quad (7)$$

$$\underline{Li_2O}\ (l) = 2Li^{+} + O^{2-} \quad (8)$$

$$Li^{+} + e^{-} = \underline{Li}\ (l) \quad (9)$$

$$4\underline{Li}\ (l) + CO_2\ (g) = C\ (s) + 2\underline{Li_2O}\ (l) \quad (10)$$

Another mechanism for carbon production in molten salt was reported [13-14], where the CO_2 gas dissolves as carbonate ions (Equation 11) and the electrochemical decomposition of carbonate ions occurs on the cathode as shown in Equation (12).

$$CO_2\ (g) + O^{2-} = CO_3^{2-} \quad (11)$$

$$CO_3^{2-} + 4e^{-} = C\ (s) + 3O^{2-} \quad (12)$$

In the present study, a large amount of oxygen ions exists in the molten salt due to the dissolution of Li_2O or CaO, and CO_2 gas is blown into the molten salt. Therefore, carbonate ions can be produced in the molten salt prior to electrolysis (Equation 11). The carbonate ions can receive electrons on the cathode, and carbon and oxygen ions are produced (Equation 12).

The purpose of this study is to decompose CO_2 gas to carbon and oxygen gas electrochemically in the molten salts using a ZrO_2 solid electrolyte. Another purpose is to investigate the rate determining step of this proposal in order to determine the key factor for an effective and faster CO_2 decomposition.

2. Experimental

$CaCl_2$ (>95 %, Wako Chemical Co.) was mixed with CaO (99.9 %, Kanto Chemical). The mixture was then calcined in air. The concentration of CaO was found to be 0.5 mol%. A magnesia crucible was filled with this mixture as the electrolyte (Figure 1). In a similar manner, LiCl (>99.0 %, Wako Chemical Co.) was mixed with 4.2 mol% Li_2O (>95 %, Wako Chemical Co.). A stainless steel pipe (6 mm in internal diameter) was used as the inlet of CO_2-Ar mixed gas and as the cathode. Two kinds of tubes made of the ZrO_2 solid electrolyte were used to remove oxygen ions in molten salts. These two tubes were produced by Nikkato Co. under the commercial name of ZR-8Y and YSZ-8. The electrical conductivity of YSZ-8 is 2.39 times larger than that of ZR-8Y at 923 K ($\sigma_{YSZ\text{-}8} = 0.79$ [$\Omega^{-1}m^{-1}$], $\sigma_{ZR\text{-}8Y} = 0.33$ [$\Omega^{-1}m^{-1}$]) as shown in Figure 2 [15]. These electrodes were attached to the cover of the stainless steel vessel using a silicone stopper. It was preferable that the inside of the ZrO_2 tube was applied with a platinum paste to provide electrical conductivity, but this was difficult to accomplish since the ZrO_2 tube was 500 mm long and its internal diameter was only 13 mm. As a result, carbon powder was used instead of a platinum paste. Since the outlet of the ZrO_2 tube was inside the reactor, oxygen gas which was produced at the internal surface of the ZrO_2 tube reacted with carbon produced in the molten salt, that is, the recombination of carbon with oxygen gas occurred. Therefore, it was important to make sure that the oxygen gas emitted did not make chemical contact with the molten salts.

After the salt was dehydrated, the $CaCl_2$-CaO and LiCl-Li_2O mixtures were slowly heated to 1173 K and 923 K, respectively, in an Ar atmosphere. LiCl-Li_2O molten salt has a higher vapor pressure in comparison with that of $CaCl_2$-CaO molten salt. Therefore, when LiCl-Li_2O molten salt was used, the electrolysis temperature was set at 923 K to prevent vaporization of the salt as much as possible. Next, 9.7 %CO_2-Ar mixed gas (CO_2 gas and Ar gas were blown at $5.0X10^{-8}$ m^3/s and $4.7X10^{-7}$ m^3/s, respectively) or 90 %CO_2-Ar mixed gas (CO_2 gas and Ar gas were blown at $6.0X10^{-8}$ m^3/s and $6.7\ X\ 10^{-8}$ m^3/s, respectively) was blown through the cathodic pipe into the molten salts. The mixed gas

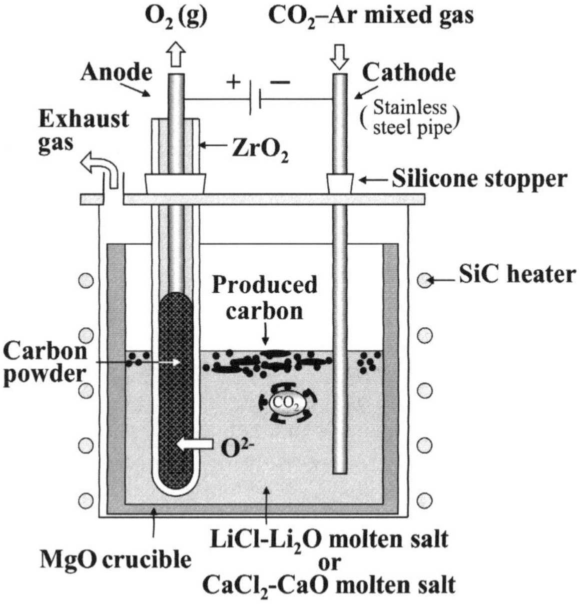

Figure 1. Experimental setup.

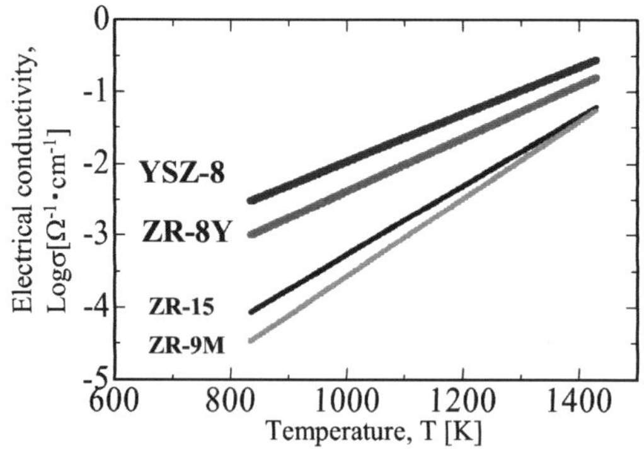

Figure 2. Temperature dependency of the electrical conductivity of each ZrO_2 electrolyte [15].

passed through the molten salts and the exhausted gas was introduced to the gas analyzers from the outlet. In the case when $CaCl_2$-CaO molten salt was used, electrolysis was initiated at 3.1 V after the salt was saturated with CO_2 gas, while in the case when LiCl-Li_2O molten salt was used, electrolysis was initiated at 3.2 V. The measured electrolysis voltage was larger than the theoretical decomposition voltage of the corresponding oxide, CaO or Li_2O, but smaller than that of the corresponding chloride, $CaCl_2$ or LiCl. During electrolysis of 9.7 %CO_2-Ar mixed gas, the concentration of CO_2 and CO gases in exhaust gas was measured at the outlet to investigate the decomposition behaviour of CO_2 gas using an infrared gas analyzer (YOKOGAWA Co.) which can measure gas concentrations up to 10 %. The gas chromatograph (GC-2014, SHIMADZU Co.) was used when 90 %CO_2-Ar mixed gas was blown. When the infrared gas analyzer was used, the concentrations of CO_2 and CO gases were able to be measured continuously, but the measurement by the gas chromatography was conducted about every $3.6X10^2$ s interval. After electrolysis with a constant cell voltage, the furnace was cooled down to room temperature in an Ar atmosphere after the cathodic pipe was raised to the upper part of the vessel. The black reduction products existed at the surface part of the solidified salt and around the cathode, and they were collected and washed with solvents in the following order, distilled water, acetic acid, distilled water, ethanol and acetone to completely remove the solidified salt residue from the reduction products. The products were observed using a transmission electron microscope (TEM, JEOL JEM-2000FX, acceleration voltage of 200 kV) and identified by X-ray diffraction (XRD, Philips Co.) using Cu-Kα radiation.

3. Results and discussion

3.1. Decomposition of 9.7 %CO_2-Ar mixed gas.

Table 1 lists the experimental conditions.

Figure 3(a) shows the time dependency of CO_2 and CO gas concentrations in the exhaust gas and the time dependency of current density during the electrolysis of 9.7 %CO_2-Ar mixed gas when LiCl-Li_2O molten salt and ZR-8Y were used, while Figure 3(b) shows those when $CaCl_2$-CaO molten salt and ZR-8Y were used. The current between the two electrodes were recorded every 10 s. The current density was determined by measuring the surface area of the electrode from the height of the solidified molten salt after cooling. The depth of the molten salt relates to the area where oxygen ions are transferred from the molten salt to the internal surface of the ZrO_2 tube. The effective surface area of the anode in the electrolysis of LiCl-Li_2O and $CaCl_2$-CaO molten salt was calculated as $2.45X10^{-3}$ m^2 and $1.63X10^{-3}$ m^2, respectively. The anodic current density was deduced from these surface areas.

To determine the optimal solvent, current density was measured when a constant voltage was applied to between the two electrodes. After the electrolysis was initiated, the concentration of CO_2 gas decreased, while the concentration of CO gas increased slightly, therefore, suggesting that the decomposition of CO_2 gas to carbon and oxygen gas occurred in the molten salts and the ZrO_2 solid electrolyte worked as an oxygen ionic conductor (as shown in Equations (6) and (7)).

$$4\underline{Li}\,(l) + CO_2\,(g) = C\,(s) + 2\underline{Li_2O}\,(l)\quad (\Delta G^0_{923K} = -555.760\ kJmol^{-1}\ [11]) \qquad (10)'$$

$$2\underline{Ca}\,(l) + CO_2\,(g) = C\,(s) + 2\underline{CaO}\,(l)\quad (\Delta G^0_{1173K} = -627.508\ kJmol^{-1}\ [11]) \qquad (13)$$

$$2\underline{Li}\,(l) + CO_2\,(g) = CO\,(g) + \underline{Li_2O}\,(l)\quad (\Delta G^0_{923K} = -273.479\ kJmol^{-1}\ [11]) \qquad (14)$$

$$\underline{Ca}\,(l) + CO_2\,(g) = CO\,(g) + \underline{CaO}\,(l)\quad (\Delta G^0_{1173K} = -331.212\ kJmol^{-1}\ [11]) \qquad (15)$$

Table 1. Experimental conditions, current efficiency, yield and average current during electrolysis.

Run	Molten salt	Temperature [K]	Gas concentration	ZrO_2 electrolyte	Current efficiency[%]	Yield[%]	Averaged current[mA]
1	LiCl-Li_2O	923	9.7%CO_2-Ar	ZR-8Y	89.7	29	75
2	$CaCl_2$-CaO	1173	9.7%CO_2-Ar	ZR-8Y	78.5	36	506
3	LiCl-Li_2O	923	90%CO_2-Ar	YSZ-8	26.7	24	178

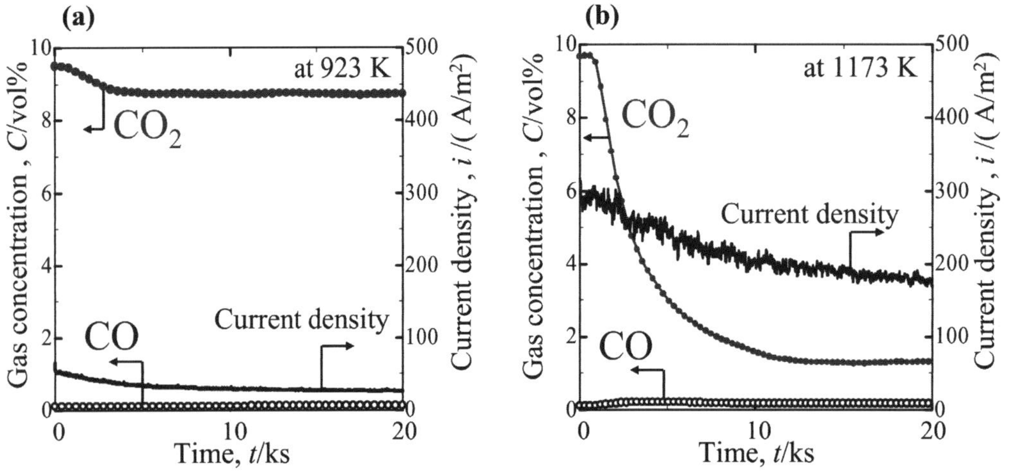

Figure 3. Changes in concentration of CO_2 and CO and current density (a) when LiCl-Li_2O and ZR-8Y were used and (b) when $CaCl_2$-CaO and ZR-8Y were used.

The decrease in CO_2 gas concentration for the case when $CaCl_2$-CaO molten salt was used is larger than that for the case when LiCl-Li_2O molten salt was used. This is because the reduction ability of metallic Ca is higher than that of metallic Li thermodynamically, although the difference in the free energy change is not significant as shown in Equations (10) and (13). Thus, the reason why the decrease of CO_2 gas concentration differs significantly would be due to the difference in electrolysis temperature, i.e., the decrease in CO_2 gas concentration depends on the diffusion rate of oxygen ions especially in the ZrO_2 solid electrolyte.

The Gibbs free energy change of the reaction between CO_2 gas and metallic Ca to produce carbon is lower than that of the reaction to produce CO gas. Similar relationship can be observed when metallic Li is used as a reductant as shown in Equations (10, 13-15). In particular, in the case when $CaCl_2$-CaO molten salt is used, CO gas tends to be produced by the Boudouard reaction as shown in Equation (5). However, even at a high electrolysis temperature, CO gas was not produced in both experiments (run 1 and 2) as shown in Figure 3. Therefore, in the case of $CaCl_2$-CaO molten salt, the fact that not only CO_2 gas, but also CO gas was decomposed can be explained by calciothermic reduction as shown in Equation (16).

$$\underline{Ca}\ (l) + CO\ (g) = C\ (s) + \underline{CaO}\ (l) \quad (\Delta G^0_{1173K} = -296.295\ kJmol^{-1}\ [11]) \qquad (16)$$

Current density gradually decreases as the electrolysis duration becomes longer. This behaviour is due to two reasons. The first reason is the decrease in the concentration of oxygen ions in the molten salts. When CO_2 gas as a source of oxygen ions is supplied to the molten salts, oxygen ions in the molten salts are continuously removed as oxygen gas from the internal surface of the ZrO_2 solid electrolyte by electrolysis. Therefore, the concentration of oxygen ions can be reduced to maintain a constant current during electrolysis. If the metallic oxide, Li_2O or CaO, is promptly supplied to the molten salt during electrolysis, a constant current would be maintained, i.e., the decrease in CO_2 gas concentration could be much more significant. Another reason is the decrease in effective surface area at the ZrO_2 solid electrolyte. Oxygen ions transfer from the molten salt to the ZrO_2 solid electrolyte, and then to the internal surface of the ZrO_2 tube, where oxygen gas is produced. However, as mentioned above, carbon powder was inserted inside the ZrO_2 tube instead of a platinum paste to provide good electrical conductivity due to experimental difficulties. However, the oxygen ions react with the carbon powder at the interface of the carbon and the internal surface of the ZrO_2 solid electrolyte, and produce CO and CO_2, which is similar to the anodic reaction in the original OS process as presented in Equation (4). Because the carbon is consumed, the effective surface area inside the ZrO_2 tube decreases during electrolysis. We plan to resolve these problems in the future and believe that it is possible to maintain a constant high current density.

In the present paper, two kinds of efficiencies are defined: current efficiency and yield. The former is the ratio of carbon production, which can be estimated from the decrease in CO_2 gas concentration and the increase in CO gas concentration, to carbon production, which can be calculated from the electric charge supplied during electrolysis. The yield is defined as the ratio of the amount of collected carbon from the furnace after electrolysis, to that of the amount of calculated carbon from the electric charge supplied during electrolysis. In these estimations, it was assumed that all the electric charge was consumed for the production of metallic Li or Ca, and that all the formed metallic Li or Ca were stoichiometrically used as a reductant of CO_2 gas.

When LiCl-Li_2O and $CaCl_2$-CaO molten salts were used with ZR-8Y electrolyte as the anode, the current efficiency were calculated to be 89.7 % and 78.5 %, respectively, as shown in Table 1.

In run 1, the decrease in CO_2 gas concentration is small as shown in Figure 3. However, the calculated value showed that the electric charge was used efficiently for the decomposition of CO_2 gas as listed in Table 1. Meanwhile, the yields remained at a low percentage of 29 % and 36 % in run 1 and 2, respectively. The yield of collected product could not be evaluated precisely because the particle size of produced carbon was very small, and not all the product could not be collected from the molten salt. A special technique to recover these fine particles is needed to apply this product as a reductant of iron ore in the iron and steel making processes, where the fine carbon can be effectively used as a pulverized coal in the blast furnace process.

3.2. Decomposition of 90 %CO_2-Ar mixed gas

The decomposition of 90 %CO_2-Ar mixed gas by electro-reduction at 923 K was performed at 3.2 V using LiCl-Li_2O molten salt and YSZ-8 (run 3). The electrical conductivity of YSZ-8 is 2.39 times larger than that of ZR-8Y at 923 K as shown in Figure 2 [15]. The concentrations of CO_2 and CO gases in the exhaust gas during electrolysis were determined from the gas chromatograph in Figure 4.

During the initial stage of the electrolysis, the concentration of CO_2 gas gradually decreased after application of the voltage between the two electrodes. In run 3, CO gas was produced even at the initial stage of the electrolysis and its concentration increased slightly during electrolysis.

The average current of run 3 is 2.37 times larger than that of run 1, although the other operating parameters were the same. In addition, the amount of collected carbon was 1.99 times larger and the concentration and flow rate of CO_2 gas increased from 9.7 % to 90 % and from $5.0X10^{-8}$ m^3/s to $6.0X 10^{-8}$ m^3/s, respectively. The increase in average current and amount of collected carbon does not depend on the variation in CO_2 gas concentration and flow rate, but on the difference in electrical conductivity of the ZrO_2 solid electrolyte used in this study. Therefore, it is considered that the rate determining step of this proposal is the oxygen ionic transfer in the solid electrolyte used as the anode. If the surface area of the ZrO_2 solid electrolyte becomes larger or the thickness of the ZrO_2 tube

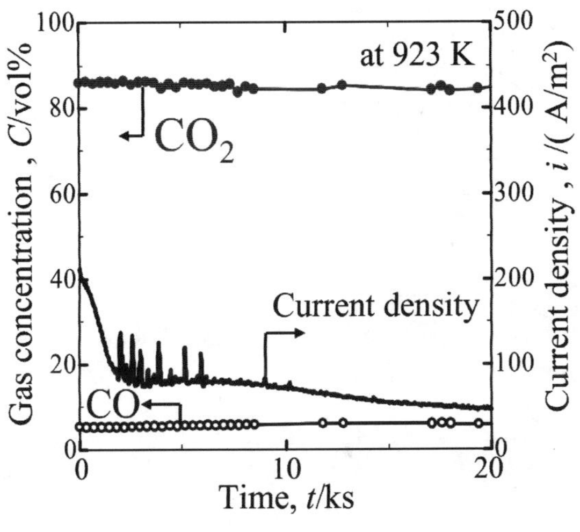

Figure 4. Changes in gas concentration and current density when LiCl-Li_2O and YSZ-8 was used and 90 %CO_2-Ar mixed gas was blown.

becomes thinner, the decomposition rate of CO_2 gas would increase and the carbon production rate would become much larger.

Although the average current increased when using YSZ-8 which has better electrical conductivity than ZR-8Y, the current efficiency decreased in run 3 compared with run 1 and 2. It is noted that the average current is proportional to the amount of metallic Li because it is produced by electrochemical reaction on the cathodic electrode (Equation (9)). It is desired that all the metallic Li reacts with CO_2 gas in the molten salt. When the flow rate of CO_2 gas exceeds the ability of metallic Li to decompose CO_2 gas, some CO_2 gas blown into the molten salt will be exhausted from the outlet without reacting. Therefore, it is reasonable that the current efficiency was lower in run 3 than those of run 1 and 2. In summary, it is important that CO_2 gas bubble remains in the molten salt for a longer time to react with the precipitated Li for obtaining higher current efficiency and larger carbon production.

3.3. The morphology of the produced carbon

After electrolysis, the reduction product adhered to the cathode, and some black powder was recovered from the solidified salt. They were washed to remove the solidified salt and then observed by TEM. The images of the carbon produced in the molten salts are shown in Figure 5.

Despite the types of salts used, two kinds of carbon materials were observed. One is the aggregations of fine particles of amorphous carbon (Figures 5 (a)-(c)). Their particle size was as large as 1 nm. The halo-pattern which is derived from the existence of an amorphous phase was observed as shown in Figure 5 (c). The d-spacing of these halo rings corresponded to the amorphous carbon.

Another is the rod-like graphite crystals (Figures 5 (d)-(f)). This material consists of aggregations of cubic-like graphite which are connected in a specific direction as shown in Figures 5 (d)-(f). The four diffraction rings derived from the diffractions of crystalline phases was also observed in Figure 5 (f).

Figures 6 (a) and (b) show the XRD pattern of the sample precipitated on the cathode. Most samples were recovered from the upper surface of the salt. XRD pattern of carbon was identified in both samples. In particular, the sample collected from the solidified salt contained graphite with a better crystalline symmetry. The diffraction peak at $2\theta = 26.6°$ can be assigned to the characteristic (002) diffraction line of graphitized carbon. The other peaks at about 42.5°, 44.5°, and 58.8° correspond to the (100), (101) and (004) reflections, respectively. Significant broadening of the diffraction peak at

Figure 5. TEM images of carbon obtained from the molten salts: (a) and (d) bright field, (b) and (e) dark field, and (c) and (f) electron diffraction.

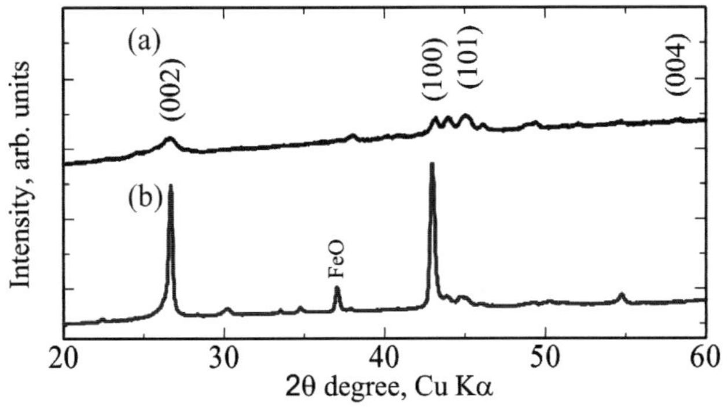

Figure 6. XRD pattern of products (a) precipitated on the cathode and (b) collected from the molten salts.

$2\theta = 26.6°$ was observed as shown in Figure 6 (a). This result shows that most part of the sample was amorphous. The XRD measurement results are in good agreement with the TEM observations.

4. Conclusions

CO_2 gas was decomposed electrochemically either in LiCl-Li_2O or $CaCl_2$-CaO molten salt. The OS process for the refractory oxide reduction was applied to CO_2 gas decomposition. The ZrO_2 solid electrolyte could remove oxygen ions from the molten salts when it was used as the anode. Gaseous oxides such as CO_2 and CO gases were electrochemically reduced in the molten salts. When $CaCl_2$-CaO molten salt was used at 1173 K, the decrease in CO_2 gas concentration was larger than that when LiCl-Li_2O molten salt was used at 923 K. The average current for the carbon production did not depend on either the concentration or the flow rate of CO_2 gas, but on the electrical conductivity of the ZrO_2 solid electrolyte. Therefore, we conclude that oxygen ionic transfer in the ZrO_2 solid electrolyte is the rate determining step. Despite the types of salts used, the aggregations of nanometre-scale amorphous carbon particles and rod-like graphite crystals were observed by TEM.

Acknowledgements

This work was financially supported by Grant-in-Aid for Scientific Research, #20360341, #21656193, and by JST-Advanced Low Carbon Technology Research and Development Program (ALCA).

References

[1] Ministry of the Environment in Japan 2010 *Japan's National Greenhouse Gas Emissions in Fiscal Year 2010 (Preliminary Figure)*
[2] Suzuki R O, Teranuma K and Ono K 2003 *Met. Mat. Trans* **34** 287
[3] Ono K and Suzuki R O 2002 *JOM Mem. J. Min. Met. Mater. Soc.* **54 (2)** 59
[4] Suzuki R O 2005 *J. Physics and Chemistry of Solids* **66** 461
[5] Suzuki R O, Tatemoto K and Kitagawa H 2004 *J. Alloys and Compounds* **385** 173
[6] Suzuki R O, Baba M, Ono Y and Yamamoto K 2005 *J. Alloys and Compounds* **389** 310
[7] Descallar-A R F, Kobayashi N, Kikuchi T and Suzuki R O 2011 *Electrochim. Acta* **56** 8422
[8] Hashizume J, Oka Y and Suzuki R O 2008 *Proc. of 2008 Joint Symposium on Molten Salts* 128
[9] Perry G and MacDonald L 1985 *J. Nucl. Mater.* **130** 234
[10] Zaitsev A and Mogutnov B 2001 *Metall. Mater. Trans.* **B32** 305
[11] Roine A 2007 *HSC Chemistry ver.6.12 Outotec Res. Oy, Pori, Finland*
[12] Usami T, Kurata M, Inoue T, Sims H, Beetham S and Jenkins J 2002 *J. Nucl. Mater.* **300** 15
[13] Van K, Groult H, Lantelme F, Dubois M, Avignant D, Tressaud A, Komaba S, Kumagai N and Sigrist S 2009 *Electrochim. Acta* **54** 4566
[14] Groult H, Kaplan B, Lantelme F, Komaba S, Kumagai N, Yashiro H, Nakajima T, Simon B, and Barhoun A 2006 *Solid State Ionics* **177** 869
[15] Nikkato Co. 1996 *Catalog No. 585 Zirconia solid electrolytes* 1

Development of a selective electroforming process for micro-sized probe tips used in LCD inspection machines

Jong Sun Kim[1], In Ki Min[2] and Jong Deok Kim[3]

[1] Incheon Technology Application Division, Korea Institute of Industrial Technology, 301-303, Buch TP, Ssangyong 3cha, Samjeong-Dong, Ojeong-Gu, Bucheon-Si, Kyunggi-Do, South Korea

[2] Department of Mechanical Engineering, Dankook University, 152, Jukjeon-Ro, Suji-Gu, Yongin-Si, Kyunggi-Do, South Korea

[3] Molds & Dies Technology R&D Group, Korea Institute of Industrial Technology, 7-47, Songdo-Dong, Yeongsoo-Gu, Incheon-Si, South Korea

E-mail: jdk@kitech.re.kr

Abstract. A probe station is an inspection apparatus that automatically transfers liquid crystal display (LCD) panels and separates defective LCD products using a micro-probe unit. However, as the number of pixels increases in LCD screens, the size of the inspecting probe unit has to decrease. There are many limits to the conventional methodology when it comes to fabricating micro-sized probes. In this study, a selective electroforming process was developed and a micro-probe tip was fabricated using this process. Selective electroforming is a high-precision fabricating process where a non-conductive barrier layer is deposited on a conductive wafer and the conductive parts are electroformed using only Ni. Finally, using the selective electroforming process, the micro-sized probe was fabricated with a shape error of less than 0.1%.

1. Introduction

A probe station is an apparatus that checks and separates defective samples of thin film transistor–liquid crystal display (TFT-LCD) panels at the inspection stage. Currently, the size of LCD panels is increasing, along with the resolution of the display; simultaneously, inspection stations are becoming bigger and more probe tips are needed. In addition, the probe tip requires a finer pitch and high dimensional accuracy to result in zero defects. In particular, the probe unit is a core part of the probe station, and consists of components such as a tip, alignment guide, printed circuit board (PCB), metal contactor, as shown in Figure 1 and Figure 2 [1].

LCD defect inspection is carried out by applying a current to the assembly of the LCD package of the color filter and the TFT glass, and checking defects by determining whether or not the TFT cell activates the light. The TFT electrodes and probe tips match one to one, so there needs to be the same number of probe tips as TFT electrodes. The probe unit requires conductance, position alignment, and tip strength for high accuracy and reliability. Conventional metal contactors in the micro-probe unit are manufactured using processes such as micro-machining, micro-electrical discharge machining (micro-EDM), laser micro-machining, chemical etching processes.[2] However, the pitch of the metal contactor patterns in the probe unit is becoming finer, so requiring greater dimensional accuracy to correspond to the high pixel density screens in LCDs; thus, conventional manufacturing processes for

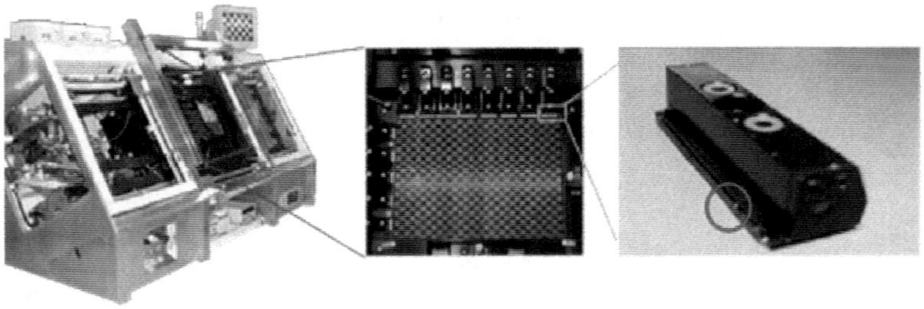

(a) Probe Station (b) Inspect Unit (c) Probe Unit

Figure 1. Composition of the probe station and probe tip.

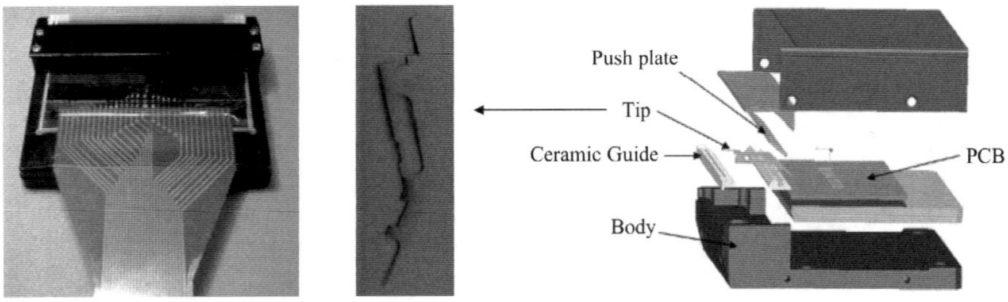

Figure 2. Blade-type probe unit and its construction.

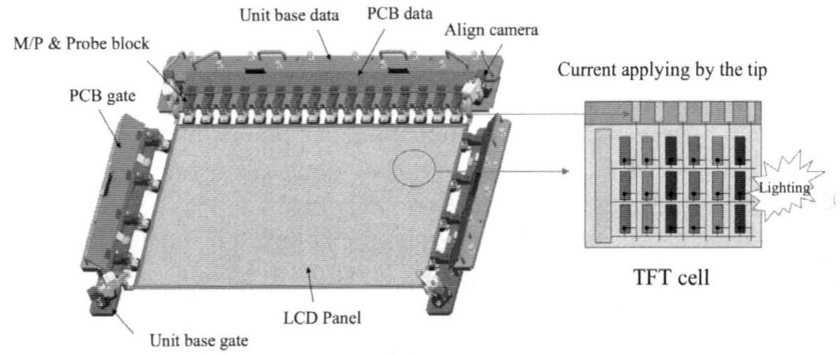

Figure 3. Principle of LCD inspection in the probe station.

probe tips have reached their limits. With conventional methods such as micro-EDM, laser machining is used to make the machining burrs and chemical etching is then used to produce an isotropic form, so the required precision cannot be achieved.

In the present study, a selective electroforming process was developed based on ultraviolet (UV) lithography, and the electroforming process was applied to the manufacture of probe tips. To draw a selective region, a negative photoresist was coated on an STS316 substrate and a metal contactor was fabricated through an exposure, development, and electroforming process [3, 4].

2. Experiment

2.1. Design and structural analysis

Depending on the shape of the probe, there are various types of tip, with the blade type being generally used. The tip in the probe unit has a two-dimensional geometry, and therefore has many merits with regard to processes such as machining and etching. The blade-type probe consists of a ceramic guide,

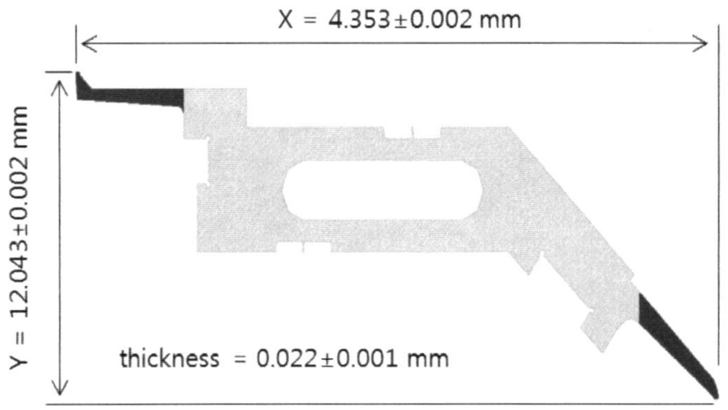

Figure 4. Tip design and target dimensions.

PCB, tip, body, and housing, as shown in Figure 2. The tip's role is to apply the current to the TFT cell in the probe station.

Figure 3 is a schematic representation of the principle of LCD inspection in the probe station.

Figure 4 shows the tip design. As a design guide, the tip should have high conductivity, strength, and reliability. In addition, X and Y deviations must be under 2 μm and the thickness deviation must be under 1 μm to avoid interference with the side walls.

2.2. Selective electroforming process and tip fabrication

Conventionally, the tip of the probe unit is fabricated by the normal chemical etching process used in the field of PCB making in order to save costs, and this process is based on metal substrates such as Cu, Be-Cu, and Be-Ni. However, there are limits to dimensional accuracy, such as shape errors due to undercutting and porous surfaces due to the isotropic etching characteristics and machining burrs in the system. In addition, the repeatability for re-fabrication is very low because it is difficult to maintain a constant concentration or mixing ratio of chemical components in every process. Thus, in this study, a selective electroforming process was investigated. Selective electroforming uses a photolithography process, with which the advanced polymer structure is formed. On a conductive substrate, the tips are fabricated by forming polymer structures using a photoresist and electroforming from the substrate bottom to the structure surface using nickel. This nickel substrate can be utilized over 5 times as a master mold. While building the metal structures on the substrate with nickel, the current density and processing time are the main factors in the electroforming process [1,2,5].

To form a micro-probe tip structure on the substrate (STS316L), spin coating was carried out uniformly on the cleaned substrate with the photoresist. For the photolithography, an SU8-2025 photoresist and PM-acetate developer from MicroChem Corporation were used. For the spin coating and baking, a HANA200D coater from Won Corporation and a hot plate were used, respectively. For exposure, a PC-500MFN UV lamp from Ushio Inc. was used, which discharges 365 nm I-line UV light every second.

Because the thickness of the tip has been decided, the speed of spin coating is a critical parameter in this step. After pretreatment by the Hexamethyldisilazane process (HMDS process) to improve the quality of the substrate surface, a polymer tip structure that was 25 mm in height was obtained at 1750 rpm (Table 1).

By exposing the photoresist-coated substrate to UV light through a probe tip patterned mask, the master structure was transferred and the substrate of the tip structure was formed after development. The photoresist was soft baked for 7 minutes at 95°C and illuminated with UV light for 12 seconds at a power of 200 mJ/cm^2. The post-exposure bake process needed to take place directly after exposure. Furthermore, the development process was carried out in PM-acetate at a temperature of 25°C for 5 minutes. All conditions for photolithography are listed in Table 2.

Table 1. Processing conditions for photoresist coating and pretreatment.

STEP	1	2	3	4	5
Chemical	HMDS	HMDS	SU8-2025	SU8-2025	SU8-2025
Process	Expansion	Main spin	Dispense	Expansion	Main spin
Spin speed (rpm)	500	2500	1	500	1750
Time (sec)	5	7	10	5	35

This study was conducted using an HG/1000 series electroforming machine (Digital Matrix Co., Ltd). This consists of a cleaning and silver spray booth, an electroforming booth, a passivation booth, and a current and temperature controller. Figure 5 shows the electroforming system used in this study. After pretreatment of the master surface with sodium dichromate ($Na_2Cr_2O_7$) and potassium dichromate ($K_2Cr_2O_7$) to ensure adhesion between the master and Ni layers, a selective electroforming process was performed on the master substrate whose fabrication was described above. In the electroforming process, the most important point is the height of the electroformed material; the height of the polymer structure (H) must be greater than the probe tip height h ($H > h$). Some preliminary trials were conducted and optimal electroforming conditions were determined. Figure 6 shows failed samples that were obtained owing to residual stress during demolding under unstable conditions. The electroforming was conducted over 30 minutes, which was divided into four intervals to prevent internal residual stress, as shown in Figure 7. Figure 8 presents the results under optimal electroforming conditions. The full process of selective electroforming is illustrated in Figure 9.

Figure 5. Construction of the electroforming machine.

Table 2. Processing conditions for photolithography.

No.	Process	Chemical	Time	Temp.	Comments
1	Photoresist coating	SU8-2025	-	-	-
2	Soft bake	-	7 min.	95℃	Hot plate
3	Exposure	-	12 sec.	-	200 mJ/cm^2
4	Post exposure bake	-	5 min.	95℃	Hot plate
4	Development	PM-acetate	7 min.	25℃	-
5	Hard bake	-	10 min.	150℃	Hot plate

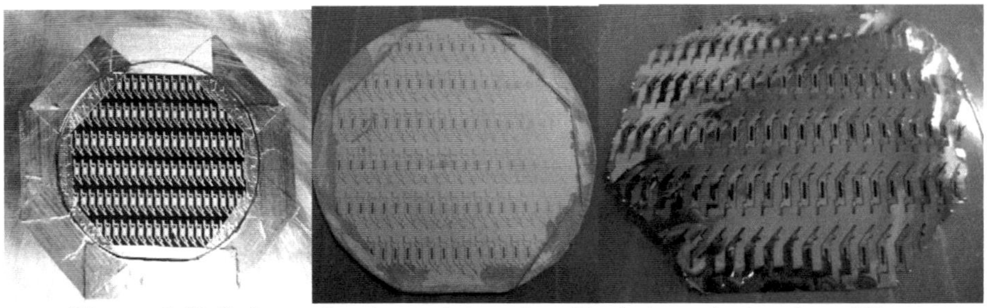

Figure 6. Failed samples due to internal residual stress during demolding.

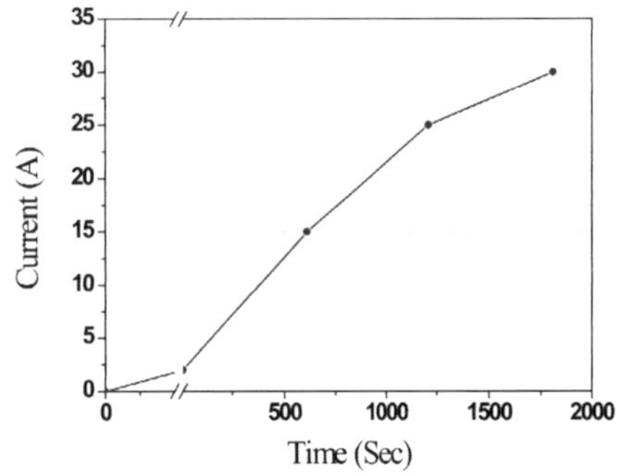

Figure 7. Current settings for electroforming.

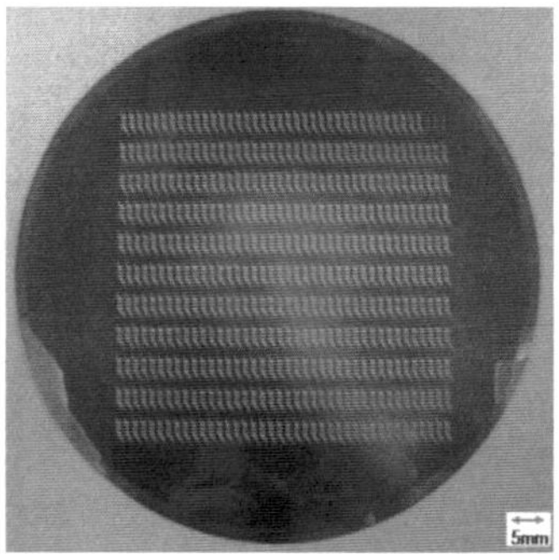

Figure 8. Electroformed probe tip under optimal conditions.

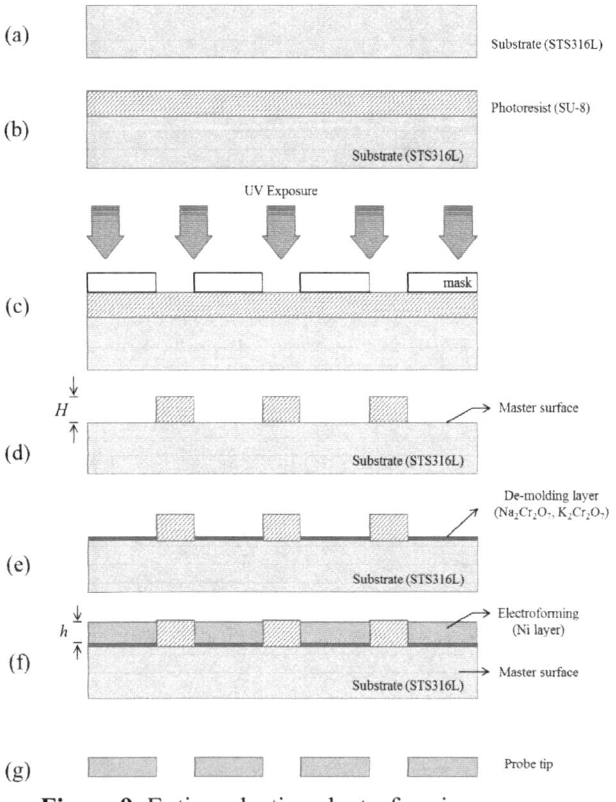

Figure 9. Entire selective electroforming process.

3. Results and discussion

In this study, the tip of a micro-probe unit was fabricated through a selective electroforming process based on UV lithography. The electroformed probe tip is shown Figure 10. In addition, Figure 11 shows the tip-assembled probe unit. For dimensional analysis, a three-dimensional measuring machine (VIEW-300) from VIEW Micro-Metrology was used. In Figure 12, the average dimensional accuracy and standard deviation were 4.353±0.001 mm in the X-direction and 12.043±0.0009 mm in the Y-direction. In addition, the thickness distribution was 0.022±0.0008 mm. In Figure 13, a red-green-blue (RGB) lighting test is shown on a TFT-LCD panel using the electroformed tips.

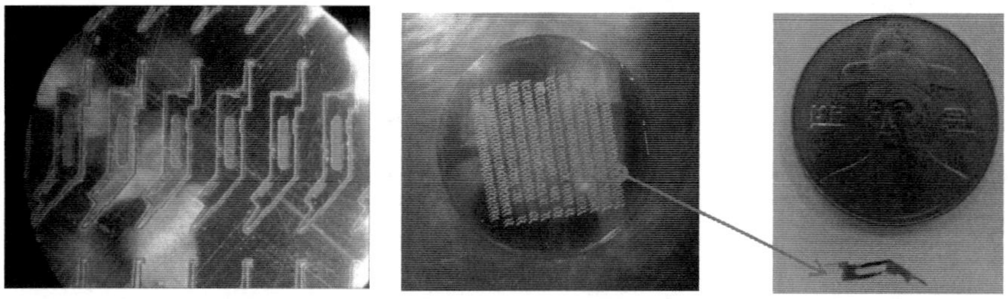

Figure 10. Tip of the micro-probe unit fabricated using the selective electroforming process.

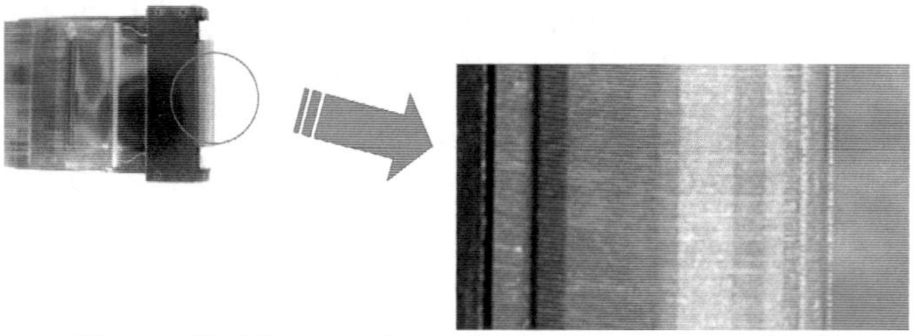

Figure 11. Micro-probe unit (the tip was assembled).

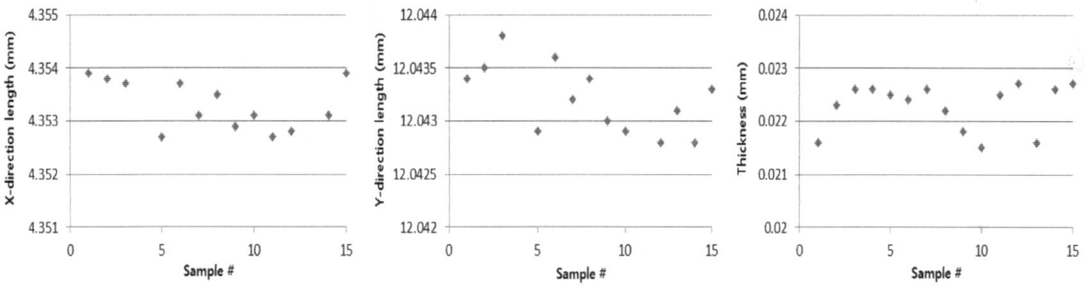

Figure 12. Dimensional accuracy deviation of micro-probe units.

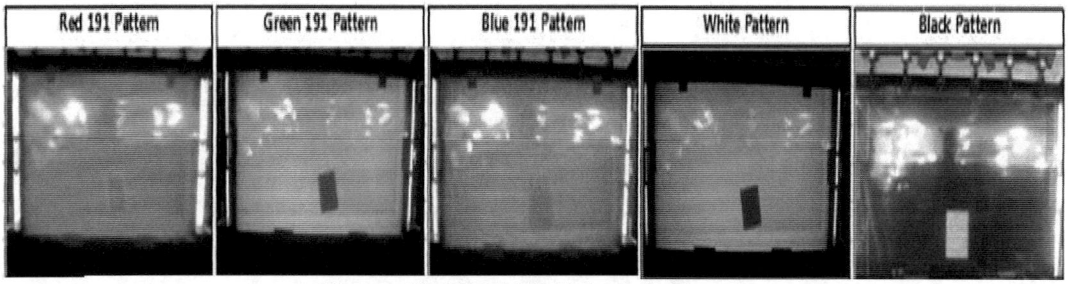

Figure 13. TFT-LCD light emitting test.

4. Conclusions

In this research, a micro-probe tip was fabricated through a selective electroforming process and high dimensional accuracy was obtained. The selective electroforming process could replace conventional machining techniques such as micro-machining, laser micro-machining, and chemical etching. It will also offer advantages in terms of reducing costs and manufacturing cycle time compared to classical machining processes.

Acknowledgments

This work was supported by the Korea Institute of Industrial Technology and the Ministry of Knowledge Economy Korea (manufacturing of precision electroforming tips using micro-molds, Project No. 10S-E2-0017).

References

[1] Bong S J and Chul L W 2008 *SICE Annual Conference* 2B04-2
[2] Menz W, Mogr J, and Paul O 2004 *Microsystem Technologies* (WILEY-VCH Publishing Company) chapter 3
[3] Saile V 2009 *LIGA and its Applications* (Wiley-VCH Publishing Company) chapter 2
[4] Hwang T Oh, J S Hong, J P Nam, G Y Bae, A H Son, S I Lee, G H Sung, H K Lee, and Y Nam 2012 *Carbon* **50** 612
[5] Prokop J Eberl, C Funk, M Prüfe, P Lorenz, J Piotter, V Welz, and M Ritzhaupt-Kleissl 2011 *Microsystem Technologies: Sensors, Actuators, Systems Integration* **17** 281
[6] Hong H S, Seo M H, Lee S, Hong S J, Suk H G, and Ahn J H 2011 *Current Applied Physics* **11** S289

Cluster model investigation of chemical bonding of elements in silica and water

Wataru Takahara

Division of Materials and Manufacturing Science, Graduate School of Engineering, Osaka University, 2-1 Yamada-oka, Suita, Osaka 565-0871, Japan

E-mail: takahara@mapse.eng.osaka-u.ac.jp

Abstract. Chemical bonding of different elements in the periodic table is investigated by the DV-Xα molecular orbital method for silicon oxide cluster and aqueous cluster models. The molecular orbital calculations reveal the distinctiveness of phosphorus-oxygen and calcium-oxygen bonds in the silica based cluster model. The former shows very strong covalency and the latter shows negative covalency. Because silica is ubiquitous in the Earth's crust, these characteristics seem to be important for the selection of elements in natural eco-systems.

1. Introduction

For development of an ecological system, it is important to understand the distribution of elements and the selection of elements in nature [1, 2]. There are great nanoscopic eco-machines [3] and nature provides clues for materials science and engineering. Earth's surface layer is divided into solid, liquid and gaseous parts, called the lithosphere, hydrosphere and atmosphere, respectively. The biosphere is present on the Earth's surface. The lithosphere consists mainly of soil and rock. The hydrosphere consists mainly of water in oceans, rivers and lakes. In the Earth's crust, the most abundant element is oxygen and the second most abundant element is silicon [4]. Silica and silicate minerals are dominant in rocks. Silica and silicates have building units of tetrahedral SiO_4 where four oxygen atoms surround a central silicon atom. Water is familiar but has complex properties. It is a solvent and forms complex solutions in oceans, rivers and lakes.

Natural eco-systems seem to be built around Si-O and H-O bond networks, because Si-O networks are dominant in the lithosphere and H-O networks are dominant in the hydrosphere. Thus, an investigation of the properties of elements in Si-O and H-O networks can form the basis for understanding eco-systems. In this paper, the chemical bonding of elements is investigated by the DV-Xα molecular orbital method in a silica based cluster model [5, 6] and an aqueous cluster model with water molecules as ligands.

2. Computational method

DV-Xα molecular calculations were carried out using the SCAT code developed by Adachi et al. [7]. Calculations were based on the Hartree-Fock-Slater approximation with an exchange correlation between electrons based on the Slater local exchange-correlation potential V_{xc} given by

$$V_{XC} = -3\alpha\left[\frac{3}{8\pi}\rho(\vec{r})\right]^{1/3}, \qquad (1)$$

where $\rho(\vec{r})$ is the local electron density at a position $\vec{r}$ [7-9]. In this calculation, the parameter α is fixed at a constant value of 0.7 and the self-consistent charge approximation is employed. The molecular orbital wave function ϕ is expressed by a linear combination of atomic orbitals (LCAO) as

$$\phi_n = \sum_i C_i^n \chi_i , \quad (2)$$

where χ_i is a basis function. Numerical atomic orbitals are employed as the basis functions in the DV-Xα method.

Figure 1 shows the model clusters analyzed in the present work, (a) $M(SiO_4)_4$ and (b) $M(H_2O)_6$. In both the $M(SiO_4)_4$ and $M(H_2O)_6$ clusters, the central site atom is M, as indicated by arrows in Figure 1. The $M(SiO_4)_4$ cluster is based on crystalline α-quartz with a lattice constant a=0.491 nm (M-O distance=0.161 nm) [10] and the $M(H_2O)_6$ cluster is based on an octahedral aqua complex (M-O distance=0.140 nm). DV-Xα molecular calculations are performed for these model clusters, where M = H – Sn across the periodic table. The atomic positions are fixed in the calculations for all the cluster models. The atomic basis functions used in the present calculations are 1s for H and He, 1s-2p for Li – Ne, 1s-3d for Na – Cl, 1s-3p for Ar, 1s-3p, 4s and 4p for K and Ca, 1s-4p for Sc – Kr, 1s-4p and 5s for Rb and Sr and 1s-5p for Y – Sn. No point groups are used to produce symmetry orbitals. Self-consistency within 0.001 electrons is obtained for the final orbital populations. A Mulliken population analysis [11] is employed for evaluation of the net charge and the bond overlap population. Non-spin-polarized cases are considered in the molecular orbital calculations.

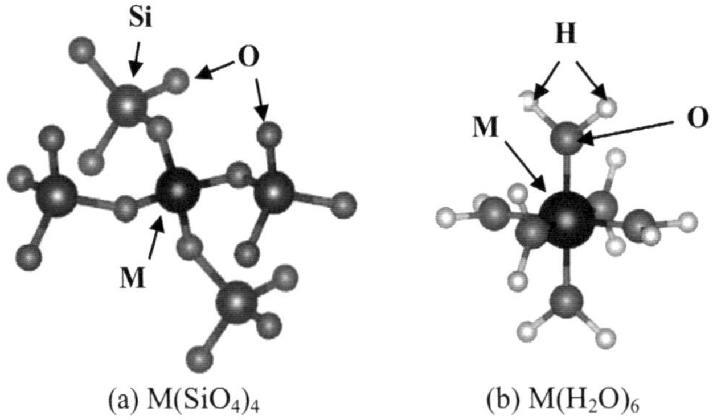

Figure 1. Cluster models of (a) $M(SiO_4)_4$ and (b) $M(H_2O)_6$, where M = H – Sn.

3. Results and discussion

Figure 2 shows the bond overlap population between the central atom M and neighboring O atoms in the cluster (a) $M(SiO_4)_4$ and (b) $M(H_2O)_6$, for M = H – Sn across the periodic table. In the $M(SiO_4)_4$ clusters, self-consistency within 0.001 electrons was not achieved in the calculations for M = V, Se and Ag. Self-consistency was achieved within 0.014 electrons for M = V, within 0.018 electrons for M = Se and within 0.037 electrons for M = Ag. In the $M(SiO_4)_4$ cluster models, the maximum value of the bond overlap population is observed for M = P, and minima are observed for M = Ca and Sr. An M atom is at the center of four SiO_4 tetrahedral units, and an M atom and four neighboring O atoms also form one MO_4 tetrahedral unit. In the present calculations, the M-O distances are set to be the same as for Si-O (0.161 nm). Of course, each stable structure for the M-O cluster is not necessarily tetrahedral, and both the coordination number with oxygen and the distance from oxygen are different from the case of Si-O for different elements in the periodic table. However, Figure 1 does show the individual characteristics of each element in Si-O tetrahedral networks.

The bond overlap population was derived from the DV-Xα orbitals in calculations considering only the Xα exchange correlation, without electron correlation. Although the finer details with respect to energy are rather sensitive to the way electron correlation is included, information concerning the wave functions is insensitive to electron correlation and such DV-Xα calculations can be considered reliable [7-9, 12, 13].

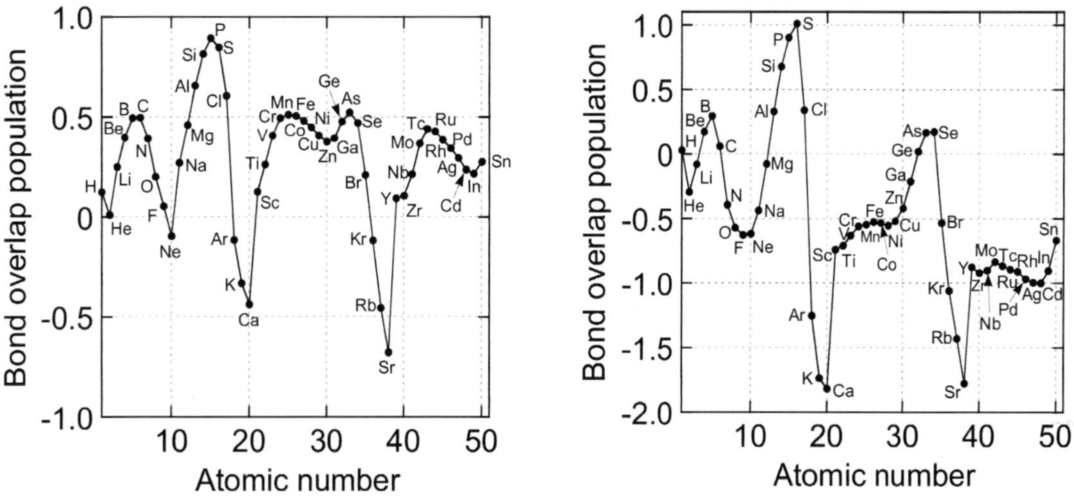

(a) $M(SiO_4)_4$ (b) $M(H_2O)_6$

Figure 2. Bond overlap population of M-O in (a) $M(SiO_4)_4$ and (b) $M(H_2O)_6$, where M = H – Sn. The M-O distances are 0.161 nm in $M(SiO_4)_4$ and 0.140 nm in $M(H_2O)_6$ respectively.

In the $M(H_2O)_6$ cluster model, the maximum value of the bond overlap population is observed for M = S, and minima are observed for M = Ca and Sr. The peak shapes are similar for the $M(SiO_4)_4$ and $M(H_2O)_6$ cluster models, but the maximum peak elements are different. The strength of the covalent bonding is estimated by the bond overlap population. The elements Si, P and S show strong covalency with O. On the other hand, the elements Ca and Sr show negative covalency. Especially in the case of $M(SiO_4)_4$ models, P and Ca seem to be distinctive elements.

It is interesting to note that Ca and P are also distinctive elements in biological systems. Ca acts as a messenger in cells and P is essential to all life, being a component of DNA and ATP [4]. In mammals, bone contain both Ca and P, often paired with each other. In the present calculations, although the bond overlap population curve for $M(H_2O)_6$ is similar to that for $M(SiO_4)_4$, the distinctiveness of Ca and P is clear for $M(SiO_4)_4$. In addition, a local minimum is observed for Zn in $M(SiO_4)_4$, and Zn is the second cell messenger [14]; however, this minimum is not observed in $M(H_2O)_6$.

Si-O networks are the dominant forms in the Earth and in the universe, the non-living world. In the Earth, they are found in the crust and mantle, as silica and/or silicate minerals. In the universe, they are found in cosmic dust, asteroids, terrestrial planets, and so on. Si-O networks are therefore ubiquitous. The essentiality of Si in life has been pointed out despite a failure to describe any mechanism of action [15, 16]. It seems that the characteristics of Ca-O and P-O bonds in Si-O networks are important for the selection of elements in eco-systems in nature.

4. Conclusions

The nature of chemical bonding with oxygen is investigated by the DV-Xα molecular orbital method for silica based cluster and water based cluster models. The effect of substitution at the central silicon site in α-quartz silica based tetrahedral cluster models is investigated for elements from hydrogen to tin in the periodic table. Central site substitution effects in clusters based on an octahedral aqua complex are also evaluated. The phosphorus-oxygen bonds show very strong covalency and calcium-oxygen bonds show negative covalency. This tendency is more clear in the silica based model than in the octahedral aqua complex model. The characteristics of phosphorus-oxygen and calcium-oxygen bonds in the silicon-oxygen networks seem to be important for element selection rules in natural eco-systems.

Acknowledgments

This work was supported in part by a Grant-in-Aid for the Global COE Program "Center of Excellence for Advanced Structural and Functional Materials Design" from the Ministry of Education, Culture, Sports, Science and Technology (MEXT) of Japan.

References

[1] Williams R J P and Frausto da Silva J J R 2000 *Coordin. Chem. Rev.* **200-202** 247
[2] Williams R J P 2006 *J. Inorg. Biochem.* **100** 1908
[3] Brunner E, Gröger C, Lutz K, Richthammer P, Spinde K and Sumper M 2009 *Appl. Microbiol. Biotechnol.* **84** 607
[4] Emsley J 2001 *Nature's Building Blocks* (Oxford: Oxford University Press)
[5] Takahara W and Minami F 2006 *Bull. Soc. DV-Xα* **19** 186
[6] Takahara W 2007 *Bull. Soc. DV-Xα* **20** 110 *(in Japanese)*
[7] Adachi H, Tsukada M and Satoko C 1978 *J. Phys. Soc. Jpn.* **45** 875
[8] Adachi H 1997 *Adv. Quant. Chem.* **29** 49
[9] Adachi H 1997 *Mater. Trans., JIM* **38** 485
[10] Villars P and Calvert L D 1991 *Pearson's Handbook of Crystallographic Data for Intermetallic Phases* Second Edition (Materials Park, OH: ASM International) p 4760
[11] Mulliken R S 1955 *J. Chem. Phys.* **23** 1833
[12] Morinaga M, Nasu S, Adachi H, Saito J and Yukawa N 1991 *J. Phys.: Condens. matter* **3** 6817
[13] Takahara W, Maeda H and Minami F 2003 *Adv. Quant. Chem.* **42** 275
[14] Yamasaki S, Sakata-Sogawa K, Hasegawa A, Suzuki T, Kabu K, Sato E, Kurosaki T, Yamashita S, Tokunaga M, Nishida K and Hirano T 2007 *J. Cell. Biol.* **177** 637
[15] Exley C 1998 *J. Inorg. Biochem.* **69** 139
[16] Exley C 2009 *Trends Biochem. Sci.* **34** 589

Separation mechanisms of reversible adhesive joining using an elastic body with variable elastic modulus

Y Sekiguchi[1], P Hemthavy[1], S Saito[2] and K Takahashi[1]

[1] Department of International Development Engineering, Tokyo Institute of Technology, I4-11, 2-12-1 O-okayama, Meguro-ku, Tokyo 152-8552, Japan

[2] Department of Mechanical and Aerospace Engineering, Tokyo Institute of Technology, I1-47, 2-12-1 O-okayama, Meguro-ku, Tokyo 152-8552, Japan

E-mail: sekiguchi.y.aa@m.titech.ac.jp

Abstract. Reversible adhesive joining that involves altering the elastic modulus of an elastic body is proposed by considering adhesion between a semi-infinite elastic body and a rigid body that has a slightly wavy surface with defects. Separation stress of adhered surfaces is theoretically expressed as a function of the elastic modulus. However, the expression of the separation stress varies depending on the condition of the maximum half contact width of the adhesion area and a parameter that depends on the elastic modulus, the work of adhesion, and the surface roughness. Therefore, separation mechanisms are categorized into six types by considering the maximum half contact width and the above parameter, and theoretically discussed. By utilizing an elastic body which has variable elastic modulus, the separation stress can be controlled and the elastic body can be repeatedly adhered to and separated from an object.

1. Introduction

Since adhesion is thermodynamically reversible, an elastic body can be repeatedly adhered to a rigid body. Adhesion of elastic bodies has the potential to be applied to thermal reversible joining in manufacturing and fabrication processes, for example. It can contribute to the simplification and reduction of industrial processes, making them more environmentally friendly. Also, the reuse of elastic bodies leads to a reduction in the amount of adhesive material used. So creation of a reversible adhesive system is important for environmental problems. However, for practical applications of adhesive joining, the control of the separation stress is necessary.

The separation stress of adhered surfaces depends on the elastic modulus [1], and the change in the separation stress has been experimentally measured for a shape-memory gel while varying the elastic modulus of the gel by heating and cooling it [2]. Adhesion can be utilized for realizing reversible joining. The adhesion of two elastic bodies with slightly wavy surfaces has been investigated theoretically [3] by extending the solution for a contact without adhesion [4]. The separation stress is expressed as a function of the elastic modulus [1].

It has been theoretically indicated that no separation occurs at the interface after two bodies make complete contact [1]. However, in practical situations, complete contact does not occur and adhered surfaces can be separated because of defects on the surface of the bodies, such as dents or cracks. It has been suggested that a discontinuous detachment transition occurs from complete contact to partial contact due to nucleation of a crack at the bottom of a valley [5].

Also, no adhesion area has been introduced as a defect at the bottom of a valley in an adhesion contact model [2]. Within this model, several separation mechanisms exist and the separation stress for each mechanism has a different expression. The condition of the separation mechanism is important for the control of the separation stress and it has been discussed theoretically by considering the relation between the external pressure and the contact width.

2. Adhesion model

2.1. Adhesion model with a sinusoidal surface roughness

Adhesion contact between a semi-infinite elastic body and a rigid body with a sinusoidal surface roughness is considered (Figure 1). The external pressure is defined as p, the half contact width as a, the amplitude of the surface roughness as h_0, and the wavelength of the roughness as λ, in which $h_0 \ll \lambda$. Young's moduli of the elastic body and the rigid body are given as E_1 and E_2 and Poisson's ratios of the bodies as ν_1 and ν_2, respectively. The elastic modulus E^* can be written as $1/E^* = (1-\nu_1^2)/E_1 + (1-\nu_2^2)/E_2$. Because $E_1 \ll E_2$, the elastic modulus can be given as $1/E^* = (1-\nu_1^2)/E_1$. The interface is assumed to be frictionless. The work of adhesion, $\Delta\gamma$, is the work required to separate a unit area of the adhered surfaces. Most real materials show inelastic behavior, such as viscoelasticity. So during a loading and unloading cycle, the effective work of adhesion decreases during loading and increases during unloading, which is known as adhesion hysteresis [6, 7]. But in this paper, a perfect elastic body is considered and adhesion hysteresis is not expected to occur. The relation between the external pressure and the contact width under equilibrium conditions has been obtained [1, 3] as

$$\frac{\tilde{p}}{\tilde{h}_0} = \sin^2(\pi a/\lambda) - \sqrt{\frac{2\Delta\tilde{\gamma}}{\left(\pi\tilde{h}_0\right)^2}\tan(\pi a/\lambda)} \tag{1}$$

$$= \sin^2(\pi\tilde{a}) - \sqrt{\Psi\tan(\pi\tilde{a})},$$

where the parameters are normalized as $\tilde{p} = p/\pi E^*$, $\tilde{h}_0 = h_0/\lambda$, $\Delta\tilde{\gamma} = \Delta\gamma/(\lambda E^*)$, and $\tilde{a} = a/\lambda$, and Ψ is given as

$$\Psi = \frac{2\Delta\tilde{\gamma}}{\left(\pi\tilde{h}_0\right)^2} = \frac{2\lambda\Delta\gamma}{\pi^2 h_0^2 E^*}. \tag{2}$$

This relationship is plotted in Figure 2(a) for several values of Ψ. For $3^{1.5}5^{2.5}/2^8 < \Psi$, there is no maximum or minimum point on the curve of Eq. (1). For $3^{1.5}/2^4 < \Psi < 3^{1.5}5^{2.5}/2^8$, $\tilde{p}/\tilde{h}_0$ at maximum and minimum points are both tensile stresses , i.e. negative pressures. For $0 < \Psi < 3^{1.5}/2^4$, $\tilde{p}/\tilde{h}_0$ at the minimum point is a tensile stress, whereas that at the maximum point is a pressure. The values of $\Delta\gamma$ for rubber and metals are about 0.01 and 1.0 J/m^2, respectively. The elastic modulus of the rubber is about 10^5 to 10^7 Pa. When considering representative wavelength and amplitude of surface roughness for λ as 10 mm and h_0 as 0.1 to 1 mm, the value of Ψ is obtained as $\Psi \approx 2 \times 10^{-6}$ to 2.

In Figure 2(b), the relation between the normalized external pressure and the normalized half contact width for $\Psi = 0.2$ is plotted. It has a minimum pressure (i.e. maximum tensile stress) at point A_0, a zero-crossing between the minimum and maximum pressures at A_1, and a maximum pressure at A_2. The pressure at A_3 is the same as that at A_0. The half contact widths at A_0, A_1, A_2, and A_3 are denoted by a_0, a_1, a_2, and a_3, respectively. When the elastic body starts contacting the rigid body at zero pressure, the contact width increases until equilibrium and it reaches A_1. When the elastic body partially contacts the rigid body (Figures 1(b) and (c)), the pressure changes following the equilibrium curve between A_0 and A_2.

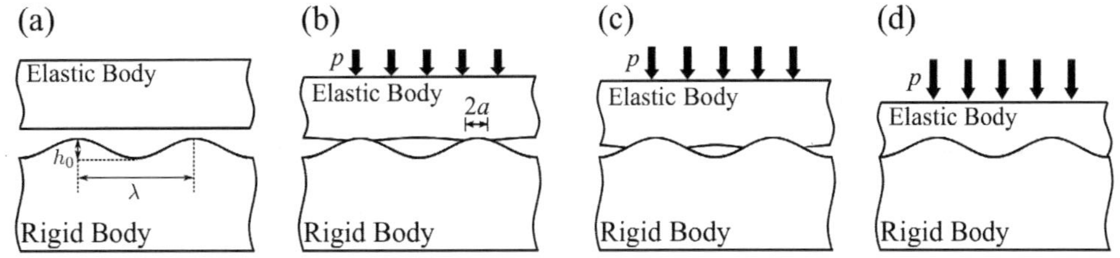

Figure 1. Adhesion contact model with a sinusoidal surface roughness on the surface of a rigid body: (a) an elastic body approaches the rigid body; (b), (c) partial contact at equilibrium with contact width $2a$; and (d) perfect contact.

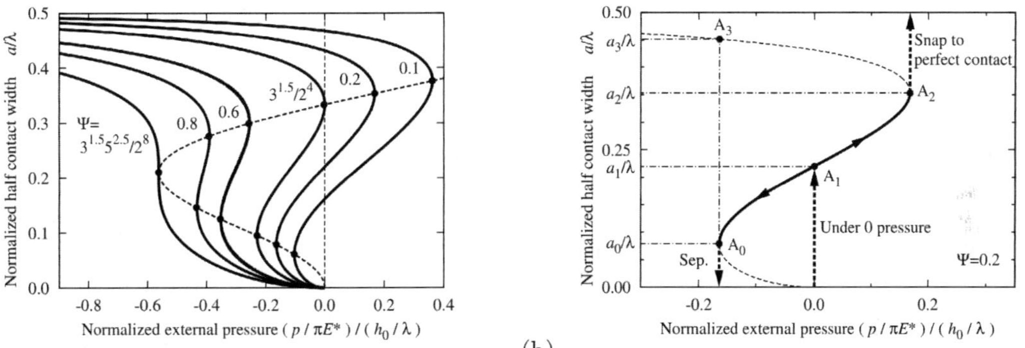

(a) (b)

Figure 2. Relation between the normalized half contact width and the normalized external pressure (a) for several Ψ and (b) when $\Psi = 0.2$.

A tensile stress is required to reduce the half contact width from a_1; it moves from A_1 to A_0 following the curve. The point A_0 is unstable. So when the half contact width reaches a_0, the elastic body separates from the rigid body. The tensile stress at A_0, i.e. the separation stress (denoted as $\sigma_{\text{Sep.}}$), can be well approximated [1] by

$$\breve{\sigma}_{\text{Sep.}} = \frac{3}{4}\left(\frac{\breve{E}^*}{\pi \tilde{h}_0}\right)^{1/3}, \tag{3}$$

where the separation stress and the elastic modulus are normalized as

$$\breve{\sigma}_{\text{Sep.}} = \sigma_{\text{Sep.}}/(\Delta\gamma/\lambda)$$

and

$$\breve{E}^* = E^*/(\Delta\gamma/\lambda),$$

respectively.

A pressure is required to increase the half contact width from a_1; it moves from A_1 to A_2 following the curve. The point A_2 is also unstable. When the half contact width reaches a_2, the elastic body completely contacts the rigid body; i.e., the normalized half contact width becomes $a/\lambda = 0.5$. After the surfaces make complete contact, the normalized half contact width remains $a/\lambda = 0.5$ even if the pressure is removed and the tensile stress is given.

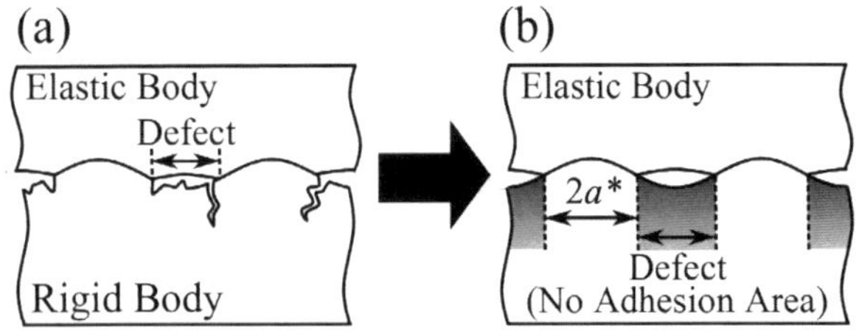

Figure 3. Schematic diagram of adhesion contact model with defects and a sinusoidal surface roughness on the surface of a rigid body: (a) cracks or dents exist on the surface and inhibit adhesion; (b) these defects are assumed to be areas where adhesion does not occur.

2.2. Adhesion model with defects and surface roughness

It has been theoretically predicted that no separation occurs at the interface when there are no defects on the surface and the surfaces are in complete contact [1]. However, body surfaces have defects, such as dents or cracks, in addition to surface roughness. These defects hinder bodies from completely adhering (Figure 3(a)) so that separation is present even for complete contact. Adhesion does not occur at defects. Therefore, no adhesion area forms on the surface of the rigid body, as shown in Figure 3(b). The maximum half contact width of the adhesion area is defined as a^*, as shown in Figure 3(b), i.e. the size of the defect is given as $\lambda - 2a^*$. The elastic body approaches the rigid body and makes contact. The half contact width then increases from a_1 to a^* during the loading process. After reaching a^*, a tensile stress is generated until separation occurs during unloading. As the tensile stress increases, the contact width decreases, following the curve given by Eq. (1) until separation occurs under certain conditions (Figures 4(a)-(d)). This is one of the separation mechanisms. However, for the other conditions, the contact width does not change until separation (Figures 4(e)–(h)). This is another separation mechanism.

The contact width is stable when total energy is a minimum. The total energy per unit depth (U_{total}) can be expressed as a sum of the elastic energy, the surface energy, and the mechanical potential energy, and it is normalized as $U_{\text{total}}/h_0^2E^*$ [2]. The relation between the normalized total energy and the normalized maximum half contact width for $\Psi = 0.2$ and $a^*/\lambda = 0.46$ is plotted in Figure 5(a). The unloading process starts from the point U_1 and it moves to the point U_2 with decreasing pressure. Between U_1 and U_2, the gradient of the energy curve is negative. Thus, the contact width does not decrease. However, the gradient at U_2 is zero. This point is unstable, and the contact width starts decreasing after U_2. Because the curve has no minimum on the left-hand side of U_2, separation occurs. Hence, the contact width is constant until separation occurs at U_2. This is referred to as separation with constant contact width. The separation stress at this condition is the tensile stress at U_2, and it can be obtained [2] as

$$\breve{\sigma}_{\text{Sep.}} = \sqrt{2\tan(\pi\tilde{a}^*)} \times \breve{E}^{*1/2} - \pi\tilde{h}_0 \sin^2(\pi\tilde{a}^*) \times \breve{E}^*, \tag{4}$$

where $\tilde{a}^* = a^*/\lambda$.

The relation between the normalized total energy and the normalized maximum half contact width for $\Psi = 0.2$ and $a^*/\lambda = 0.42$ is also plotted in Figure 5(b). a^*/λ in Figure 5(b) is smaller than that in Figure 5(a). Unloading starts from U_1 and the contact width is constant at $a = a^*$ until reaching U_2. The gradient of the curve at U_2 is zero. So U_2 is unstable and the contact width starts decreasing. However, the curve has a minimum point U_3 on the left-hand side of U_2. The contact width reaches equilibrium at the point U_3, but separation does not occur yet. The contact width subsequently decreases following the minimum points on the energy curve, i.e. the solid curve A_0U_3 in Figure 5(b), until separation occurs at A_0. This is referred to as separation with decreasing contact width. The separation stress for this condition is given by Eq. (3).

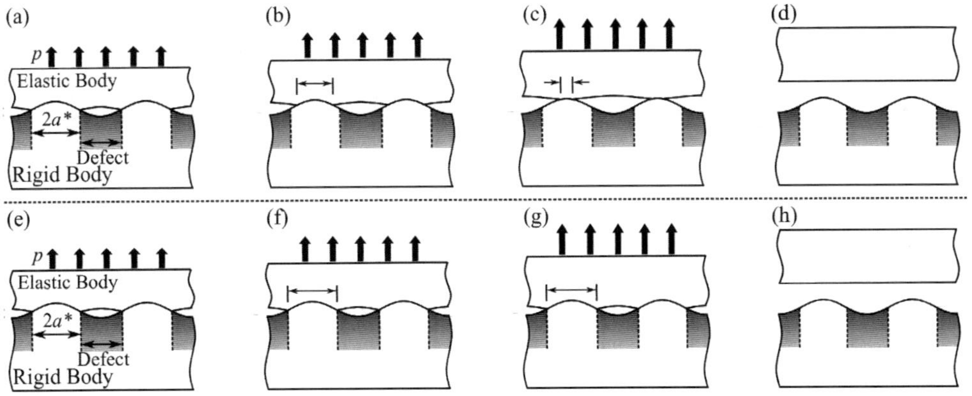

Figure 4. Adhesion contact model with a defect and a sinusoidal surface roughness on the surface of a rigid body: (a)–(d) separation with decreasing width and (e)–(h) separation with constant width.

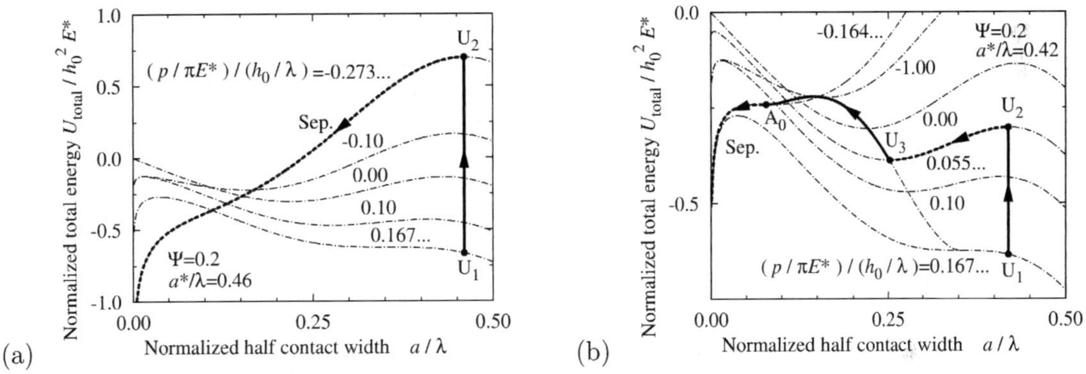

Figure 5. Unloading processes with the relation between the normalized total energy and the normalized half contact width when (a) $\Psi = 0.2, a^*/\lambda = 0.46$ and (b) $\Psi = 0.2, a^*/\lambda = 0.42$.

3. Conditions of separation mechanisms

3.1. Categorization of unloading process

The separation stress is expressed as a function of the elastic modulus by Eq. (3) or Eq. (4). Therefore, it is possible to control the separation stress by changing the elastic modulus by, for example, using a shape-memory gel with a Young's modulus in the range 2.2 $\times 10^5$ Pa (above 49°C) to 1.7 $\times 10^7$ Pa (30°C) [8]. However, two different expressions for the separation stress are obtained by considering the change in the contact width during unloading. It is thus necessary to clarify the conditions for each case.

The unloading process can be categorized into several types by considering Ψ and a^*/λ, as shown in Figure 6. First, the condition of Ψ is considered. Ψ is expressed as a function of E^* (Eq. (2)). Shape-memory gel has an elastic modulus that varies with temperature. By utilizing this kind of material, Ψ can be changed during processes. The stress curve changes with different Ψ, as shown in Figure 2(a). It can be divided into three ranges, $0 < \Psi < 3^{1.5}/2^4$, $3^{1.5}/2^4 < \Psi < 3^{1.5}5^{2.5}/2^8$, and $3^{1.5}5^{2.5}/2^8 < \Psi$, by considering the maximum and minimum points of Eq. (1).

Next, the condition of a^*/λ is considered for each range of Ψ. When $0 < \Psi < 3^{1.5}/2^4$, the maximum pressure, i.e. the pressure at A_2, is positive and the minimum pressure, i.e. the pressure at A_0, is negative. For this condition, the maximum half contact width can be divided into five range as $0 < a^*/\lambda < a_0/\lambda$, $a_0/\lambda < a^*/\lambda < a_1/\lambda$, $a_1/\lambda < a^*/\lambda < a_2/\lambda$, $a_2/\lambda < a^*/\lambda < a_3/\lambda$, and $a_3/\lambda < a^*/\lambda < 0.5$. These divisions are categorized as types A, B, C, D, and E, respectively, as shown in Figures 6(a)–(e). When $3^{1.5}/2^4 < \Psi < 3^{1.5}5^{2.5}/2^8$, the pressure at A_2 is negative and thus there is no zero-crossing, A_1. For this condition, type B has a maximum half contact width range of $a_0/\lambda < a^*/\lambda < a_2/\lambda$ and type C does not exist. Types A, D, and E are the same in the range $0 < \Psi < 3^{1.5}/2^4$. When $3^{1.5}5^{2.5}/2^8 < \Psi$, there is no maximum or minimum pressure in the curve given by Eq. (1). This condition is categorized as type F (Figure 6(f)). Thus, there are six conditions, as shown in Table 1.

3.2. Unloading process for each type

During loading, the contact width increases until it reaches $a/\lambda = a^*/\lambda$, i.e. the maximum half contact width, at the point U_1. This is the initial point of unloading. For types A, E, and F, the pressure varies horizontally from U_1 to U_2, as shown in Figures 6(a), (e), and (f). U_2 is the intersection between a horizontal line and the curve of Eq. (1). Between U_1 and U_2, the contact width remains constant. Separation occurs when the point U_2 is reached. These types are referred to as separation with constant contact width. The separation stress for types A, E, and F is expressed by Eq. (4). Point U_2 for type B and point U_1 for type C are on the equilibrium curve of A_0A_2, as shown in Figures 6(b) and (c). The pressure varies by following the curve after these points and separation occurs at point A_0. For type D, point U_2 is unstable and the contact width starts decreasing from U_2 until equilibrium and reaches point U_3, as shown in Figure 6(d). The pressure then follows the curve A_0A_2 until separation at A_0. When the pressure decreases by following the curve of Eq. (1), the contact width also decreases. Thus, types B, C, and D are classified as separation with decreasing contact width and the separation stress for these types is expressed by Eq. (3).

3.3. Condition of Ψ and a^*/λ for each type

The relation between Ψ and a^*/λ for the different types of separation is plotted in Figure 7. The upper side of the solid curve shows separation with constant contact width and the lower side shows separation with decreasing contact width. In addition, the dashed curves show the conditions for six types (A–F).

The solid curve between types A and B in Figure 7 shows the condition $a^*/\lambda = a_0/\lambda$ for $\Psi < 3^{1.5}5^{2.5}/2^8$. The dashed curves between B and D and between C and D show the condition $a^*/\lambda = a_2/\lambda$ for $\Psi < 3^{1.5}5^{2.5}/2^8$. These two conditions are given by

$$\Psi = \frac{2}{\breve{E}^*\left(\dfrac{\pi h_0}{\lambda}\right)^2} = 16\sin^3(\pi a^*/\lambda)\cos^5(\pi a^*/\lambda). \tag{5}$$

The dashed curve between B and C shows the condition $a^*/\lambda = a_1/\lambda$ for $\Psi < 3^{1.5}/2^4$. The solid curve between D and E shows the condition $a^*/\lambda = a_3/\lambda$ for $\Psi < 3^{1.5}5^{2.5}/2^8$. These two conditions are numerically obtained. The border line of F shows the condition $\Psi = 3^{1.5}5^{2.5}/2^8$.

4. Control of separation stress

4.1. Relation between separation stress and elastic modulus

To increase the elastic modulus, the separation stress is varied, as shown in Figure 8. The circles, ●, in Figure 8 represent the points where the expression for the separation stress changes. The

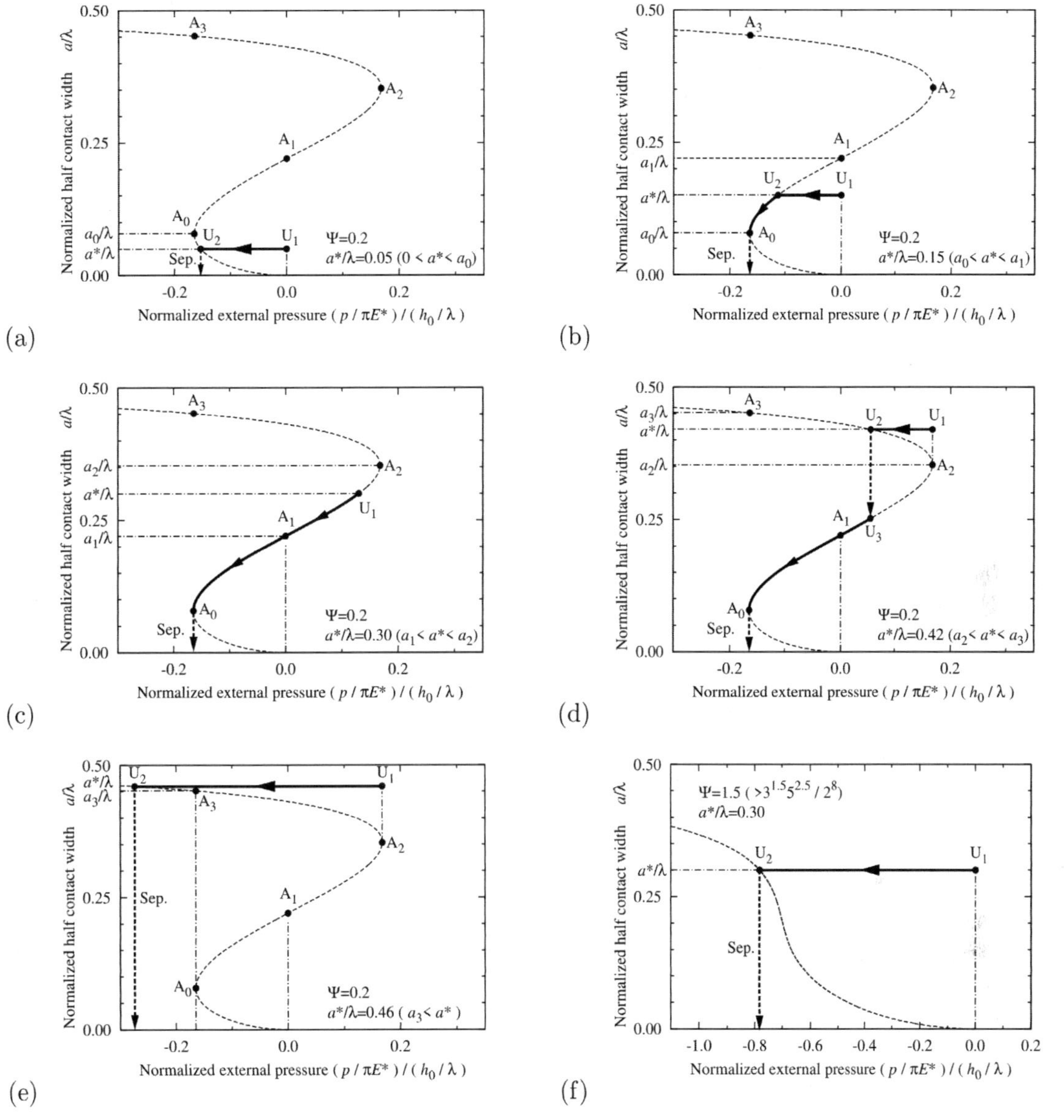

Figure 6. (a)–(f) Plots of the unloading processes of types A–F for the conditions (a) $\Psi = 0.2, a^*/\lambda = 0.05$, (b) $\Psi = 0.2, a^*/\lambda = 0.15$, (c) $\Psi = 0.2, a^*/\lambda = 0.30$, (d) $\Psi = 0.2, a^*/\lambda = 0.42$, (e) $\Psi = 0.2, a^*/\lambda = 0.46$, and (f) $\Psi = 0.1.5, a^*/\lambda = 0.30$ in the relation between the normalized half contact width and the normalized external pressure.

right- and left-hand sides of the circles show the separation stresses given by Eqs. (3). and (4), respectively.

The separation stress is in good agreement with the Griffith theory of brittle fracture when the amplitude and the elastic modulus are sufficiently small, where the separation stress of the Griffith theory can be expressed as

Table 1. Conditions of Ψ and a^*/λ for each type.

Type	Ψ	a^*/λ	Separation stress
A	$0 < \Psi < 3^{1.5}5^{2.5}/2^8$	$0 < a^*/\lambda < a_0/\lambda$	Eq. (4)
B	$0 < \Psi < 3^{1.5}/2^4$	$a_0/\lambda < a^*/\lambda < a_1/\lambda$	
	$3^{1.5}/2^4 < \Psi < 3^{1.5}5^{2.5}/2^8$	$a_0/\lambda < a^*/\lambda < a_2/\lambda$	Eq. (3)
C	$0 < \Psi < 3^{1.5}/2^4$	$a_1/\lambda < a^*/\lambda < a_2/\lambda$	
D	$0 < \Psi < 3^{1.5}5^{2.5}/2^8$	$a_2/\lambda < a^*/\lambda < a_3/\lambda$	
E	$0 < \Psi < 3^{1.5}5^{2.5}/2^8$	$a_3/\lambda < a^*/\lambda < 0.5$	Eq. (4)
F	$3^{1.5}5^{2.5}/2^8 < \Psi$	$0 < a^*/\lambda < 0.5$	

$$\sigma_{\text{Griffith}} = \sqrt{\frac{2E^*\Delta\gamma}{\pi(0.5 - \tilde{a}^*)\lambda}}, \tag{6}$$

and it is plotted as dashed-dotted line in Figure 8 for the case $a^*/\lambda = 0.46$.

a^*/λ is constant during unloading. Thus, when the elastic modulus changes during the processes, only Ψ changes and the condition moves vertically in Figure 7. The expression for the separation stress changes when the condition crosses a solid curve in Figure 7. Thus, the circles in Figure 8 are related to the solid curve in Figure 7.

4.2. Reversible adhesive joining process

A small elastic modulus is preferable for creating adhesion contact when the elastic body starts contacting the rigid body. Heating is required when using the shape-memory gel as the elastic body. After the bodies have adhered, the elastic modulus is increased to give strong adhesion, i.e. to increase separation stress. Thus, cooling is required for a shape-memory gel. However, when a^*/λ is close to 0.5 , for example when $a^*/\lambda = 0.42$ or 0.46 as plotted in Figure 8, the separation stress decreases just before the transition point between Eqs. (4) and (3) , i.e. the circles. Thus, the bodies can be separated by increasing the elastic modulus under these conditions,e.g. cooling the shape-memory gel. On the other hand, the separation stress does not decrease when a^*/λ is small, such as $a^*/\lambda = 0.15$. For this condition, it is necessary to reduce the elastic modulus for separation, e.g. heating the shape-memory gel. It is important to carefully discuss the relation between the separation stress, Ψ, and a^*/λ, especially near the transition points between Eqs. (4) and (3) using Figures 7 and 8, in terms of controlling the separation stress.

5. Conclusions

Adhesion contact between a semi-infinite elastic body and a rigid body is considered to determine the separation stress. The surface of the rigid body has a sinusoidal surface waviness and defects. By considering the total energy of the system and the relation between the external pressure and the contact width, the separation stress is obtained as a function of the elastic modulus. It is possible to separate adhered surfaces by changing the elastic modulus of the elastic body,

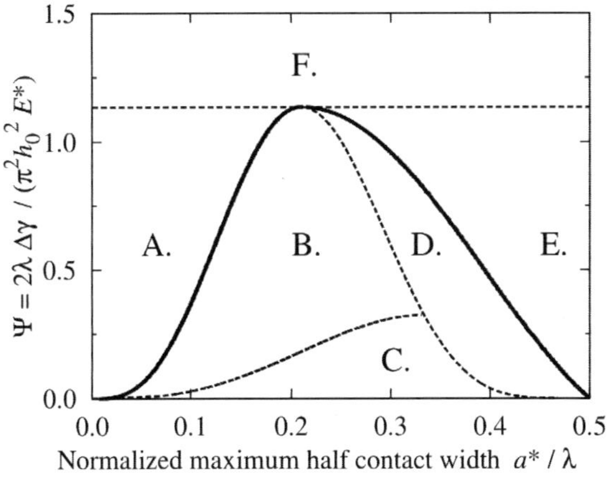

Figure 7. Condition of unloading processes in terms of the relation between Ψ and $a*/\lambda$. The upper and lower sides of the solid curve respectively represent separation with constant contact width and separation with decreasing contact width.

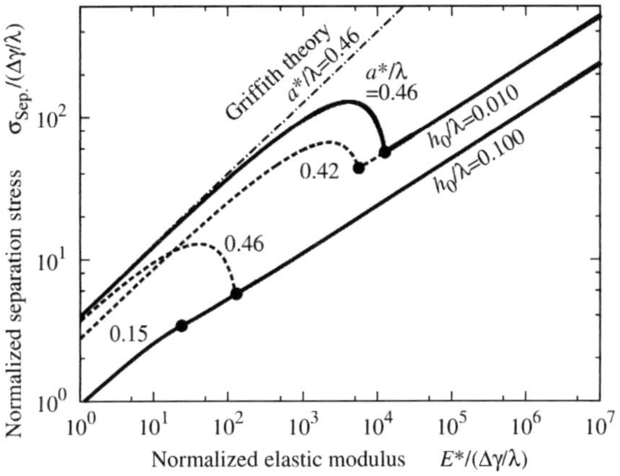

Figure 8. Normalized separation stress as a function of the normalized elastic modulus for $a^*/\lambda = 0.42$ and 0.46 with $h_0/\lambda = 0.010$ and $a^*/\lambda = 0.15$ and 0.46 with $h_0/\lambda = 0.100$. The left and right side of the circles, ●, represent separation with constant contact width and separation with decreasing contact width, respectively. The Griffith theory of brittle fracture for $a^*/\lambda = 0.46$ is also plotted.

e.g. shape-memory gel, during the processes. However, the expression for the separation stress changes with the maximum half contact width and a parameter that depends on the elastic modulus, the work of adhesion, and surface roughness parameters. The unloading process is categorized into six types by considering the change in the contact width during unloading. The conditions for each type are obtained by considering the relation between the maximum half contact width and the parameter. Reversible adhesive joining process is proposed and it is discussed by considering the curve of stress against elastic modulus.

References

[1] Takahashi K, Hemthavy P, Sriwijaya R R A, Lei L, Jatmiko K and Saito S 2008 *J. Japan Welding Soc.* **26** 292-297 in Japanese
[2] Sekiguchi Y, Lei L, Hemthavy P and Takahashi K 2010 *J. Adhesion Sci. Technol.* **24** 1819-1830
[3] Johnson K L 1995 *Int. J. Solids Structures* **32** 423-430
[4] Westergaard H M 1939 *J. Appl. Mech.* **6** A49-A53
[5] Persson B N and Tosatti E 2001 *J. Chem. Phys.* **115** 5597-5610
[6] Johnson K L 1998 *Tribology Int.* **31** 413-418
[7] Lei L, Takahashi K, Sriwijaya R R A and Hemthavy P 2009 *J. Japan Welding Soc.* **27** 196s-199s
[8] Kaneko K, Miyazaki T, Yamaoka K, Katayama Y, Matsuda A, Gong J P and Osada Y 1999 *Proc. SPIE* **3669** 199-208

Time dependence of adhesion contact between a rigid sphere and an elastic body

Pasomphone Hemthavy[1], Mohd Fahmi Bin Samir[1], Shigeki Saito[2] and Kunio Takahashi[1]

[1]Department of International Development Engineering, Tokyo Institute of Technology, I4-20, 2-12-1 O-okayama, Meguro-ku, Tokyo 152-8552, Japan

[2]Department of Mechanical and Aerospace Engineering, Tokyo Institute of Technology, I1-47, 2-12-1 O-okayama, Meguro-ku, Tokyo 152-8552, Japan

E-mail: pasomphone@ide.titech.ac.jp

Abstract. Adhesion contact between a rigid sphere and an elastic body with a flat surface is experimentally investigated. The time dependence of the contact radius on loading and unloading is observed. Based on the total energy of the JKR (Johnson, Kendall, and Roberts) contact, the hysteresis of the contact radius in loading-unloading trajectories is discussed. The correlation between the energy gradient and the velocity or rate of change in contact radius is introduced to discuss the time dependence of the adhesion qualitatively.

1. Introduction

Understanding thermodynamically reversible adhesion contact is one of the keys to saving the Earth's limited resources and to contributing to sustainable development. Some solutions to this issue can be found by carefully studying the adhesion phenomena between solids from the functions of living creatures. Recently, reversible adhesion using polymers is attracting researchers' interests, e.g. in the field of biomimicry technology, and is giving rise to some reports on mimicking gecko foot-hair structure for wall-climbing robots [1,2].

Adhesion contact behavior between elastic solids has been studied for several decades because of its significance in the areas of tribology, joining technology, and micro-manipulating technology [3]. One of the most popular adhesion tests involving the contact between a sphere and a flat surface is the theory of Johnson, Kendall, and Roberts (JKR theory) [4]. Based on the JKR theory, experimental studies on interfacial adhesion between polymers were carried out, which led to the observation of adhesion hysteresis [5,6]. Johnson later introduced the assumption of the energy release rate to characterize adhesion hysteresis [7]. Since the JKR theory deals with adhesion contact of elastic solids based on linear fracture mechanics, the time dependence of the adhesion contact is not taken into account.

The time dependence of adhesion contact is essential and practically important in order to utilize and control reversible adhesion. However, time dependent adhesion of elastic bodies has not been extensively studied.

In this study, we report an observation of JKR adhesion contact between a rigid sphere and an elastic body with a flat surface, and discuss the evolution of the contact radius during adhesion contact using the total energy of the JKR theory.

2. Adhesion contact model and theoretical background

The JKR contact between a rigid sphere and an elastic body with a flat surface used in this study is shown schematically in Figure 1. A displacement δ is externally given to press the rigid sphere of radius R against the surface of the elastic body, which gives rise to a circular contact area of radius a at the interface of the two bodies. $\delta = 0$ is defined at a virtual point contact, i.e. at the origin of the yr-coordinate system shown in Figure 1. For an appropriately small contact radius, i.e. the two bodies are assumed to contact under a light load, the spherical surface of the rigid body is approximated to be a paraboloid as

$$y = \frac{r^2}{2R} \,, \tag{1}$$

and is shown by a dashed line in Figure 1.

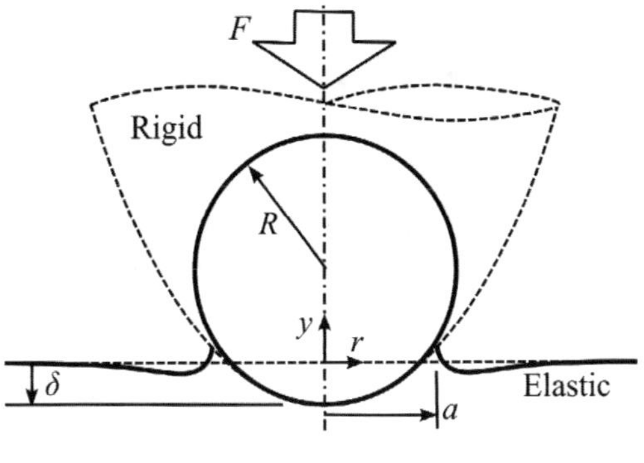

Figure 1. Schematic illustration of JKR contact between a rigid sphere and an elastic body with a flat surface. Displacement δ applied to the sphere by external load F gives rise to deformation of the elastic surface and contact radius a.

According to a previous work by Takahashi, one of the authors of this paper, the total energy E_{total} of the system is given as a function of the displacement and contact radius [1], expressed as

$$E_{\text{total}} = \frac{4E}{45(1-\nu^2)R^2} a^5 + \frac{Ea}{(1-\nu^2)} \left(\frac{a^2}{3R} - \delta \right)^2 - \pi a^2 \Delta\gamma, \tag{2}$$

where E and ν are the Young's modulus and Poisson's ratio of the elastic body, respectively, and $\Delta\gamma$ is the work of adhesion required to separate a unit area of adhered interface. Under constant displacement, the two bodies come into contact at equilibrium, from which the relation between the displacement and equilibrium contact radius can be obtained as

$$\delta = \frac{a^2}{R} - \sqrt{\frac{2\pi(1-\nu^2)\Delta\gamma\, a}{E}} \,. \tag{3}$$

Equation (3) can be used to determine the work of adhesion from measurements of the contact radius and applied displacement. However, in practical adhesion contacts, due to surface roughness, the viscoelastic behavior of elastomeric materials and so on, the work of adhesion appears different between loading and unloading. Following the assumption introduced by Johnson [5] that, during loading and unloading, an energy release rate G has to overcome the work of adhesion as well as some additional energy denoted by a fraction α of the elastic energy dissipation. Therefore, the work of adhesion $\Delta\gamma$ in Equation (3) should be replaced by

$$G = \frac{\Delta\gamma}{(1+\alpha)} \quad \text{on loading,} \tag{4}$$

$$G = \frac{\Delta\gamma}{(1-\alpha)} \quad \text{on unloading.} \tag{5}$$

3. Experimental

3.1. Preparation of samples

A transparent silicon rubber (Fusougomu) with dimensions of $35 \times 40 \times 10$ mm is used for the elastic body, of which the Young's modulus and Poisson's ratio are measured using a compact table-top universal tester (Shimadzu, EZ Test) to be $E = 45.5$ kPa, $\nu = 0.5$. For the rigid sphere, two steel balls (Sato Tekko, SUJ-2) of different radii, 12.5 mm and 15.08 mm, are used. Prior to each experiment, the samples are cleaned using an ultrasonic cleaner with ethyl alcohol followed by pure water.

3.2. Experimental setup and procedures

The experimental setup is shown schematically in Figure 2. The rigid sphere is fixed on an electronic balance (Sartorius, TE153S) to enable measurement of the applied load. The elastic body is attached to the bottom of a transparent acrylic beam connected directly to a motorized precision stage (Suruga Seiki, K701-20LMS). A crossed roller and goniometer stage (Suruga Seiki, B43-38N, B54-60UNR) are used to adjust the horizontal position and level of the elastic body. Loading and unloading are precisely controlled by a stepping motor on the motorized precision stage, which is manipulated by a series of digital pulses, with a resolution of $0.05\,\mu$m. An optical microscope (Keyence, VH-7000) is used to measure the radius of the contact area through the transparent elastic body.

To measure the work of adhesion, starting from initial displacement, loading/unloading is done by adding/reducing the displacement to the sphere step-by-step with a certain increment, up to a target displacement, in each of which the contact radius is measured as a function of a given displacement after it becomes stable or at equilibrium.

To measure the time dependence of the adhesion, after applying a target displacement with an average rate of loading/unloading of less than 0.5 mm/s, the evolution of the contact radius a is captured over time by a computer with a rate of 30 frames per second until the contact area becomes stable. The displacement, once given, is kept constant during the measurement. The contact radius is then measured manually, frame-by-frame, using a circle-fitting function of the optical microscope.

Note that all experiments are conducted in ambient conditions, and performed in an acrylic box on a vibration isolation table (Sigma Koki, TDI-129LA) to reduce the effects of dust, air currents, moisture and noise.

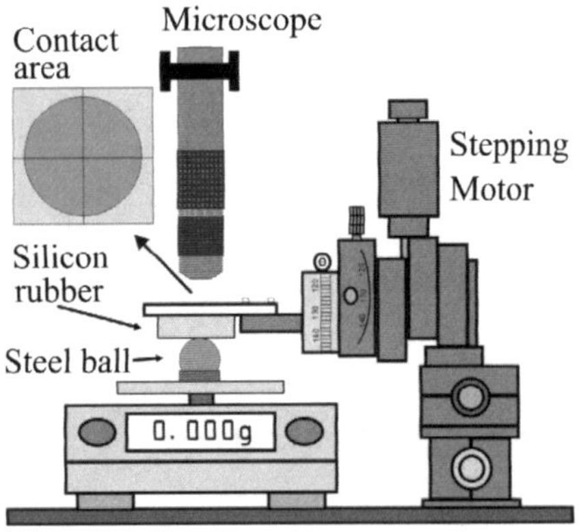

Figure 2. Schematic illustration of the experimental setup. The elastic body is brought into contact with the sphere by controlling the stepping motor. The contact radius is measured through the elastic body using an optical microscope.

4. Results and discussions

Experimental results of the loading/unloading cycles of the two spheres, R = 12.5 mm and 15.08 mm, on the elastic body are shown in Figures 3 and 4. Filled symbols (■) depict experimental data for the equilibrium contact radius a at an applied displacement δ on loading, whereas open symbols (○) depict those measured on unloading. As shown in the figures, adhesion hysteresis between loading and unloading can be observed in both experiments.

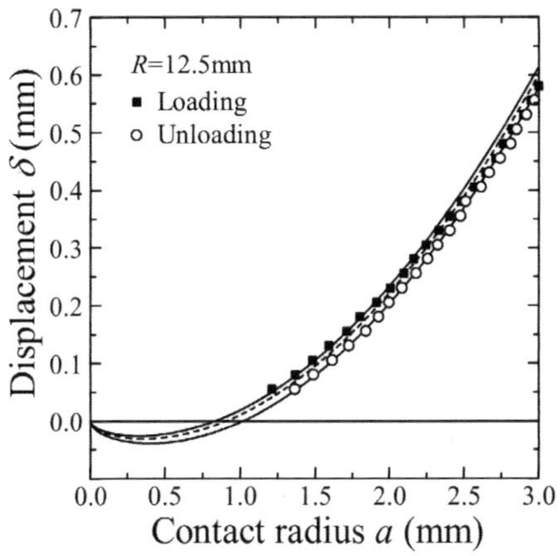

Figure 3. Experimental results of equilibrium contact radius as a function of displacement, used for estimating the work of adhesion $\Delta\gamma$ and energy release rate α, for R = 12.5 mm.

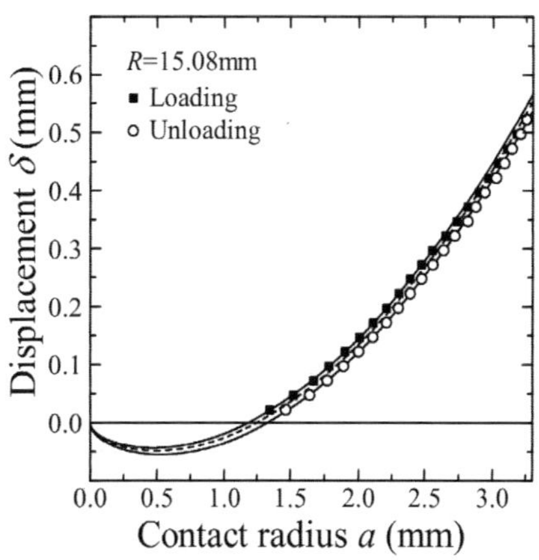

Figure 4. Experimental results of equilibrium contact radius as a function of displacement, used for estimating the work of adhesion $\Delta\gamma$ and energy release rate α, for R = 15.08 mm.

The work of adhesion $\Delta\gamma$ and the fraction α of the elastic energy dissipation can be determined by fitting the experimental data with Equations (3), (4) and (5). Taking into account the condition given in Equation (1), data pairs of δ and a that satisfy a/R < 0.18 are selected for fitting. Consequently, $\Delta\gamma$ and α are estimated as 0.0462 J/m^2 and 0.292 for R = 12.5 mm, and 0.0824 J/m^2 and 0.174 for R = 15.08 mm, respectively. To check the estimated results, theoretical curves are plotted as shown in Figures 3 and 4. The dashed curves do not take account of the energy dissipation, while the solid curves, on loading and unloading, fit the measurements very well for small contact radii that satisfy Equation (1). The hysteresis observed in this study is due to dry contact. However, hysteresis is also observed in lubricated contact as reported by Lei et al. (our research group) in a study on area contact using the same elastic material as that used in this study [8].

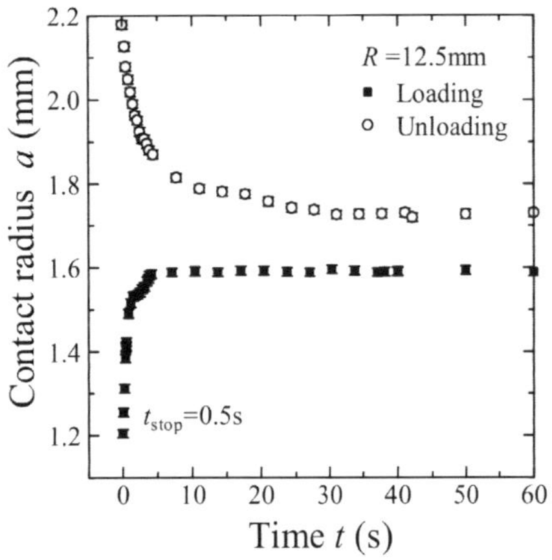

Figure 5. Measurements of the evolution of contact radius during adhesion. Filled symbols depict loading, and open symbols depict unloading. R = 12.5 mm.

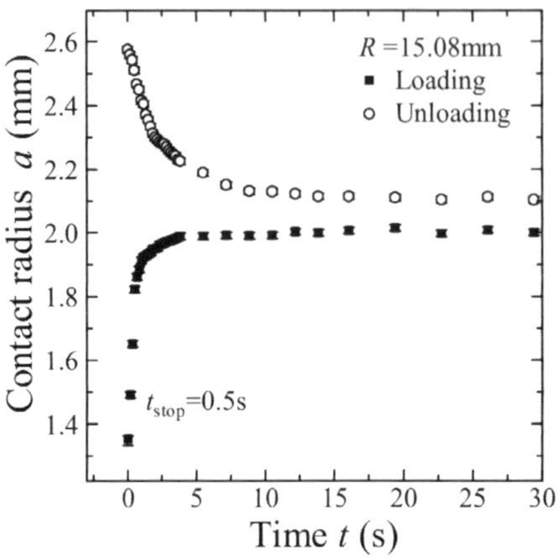

Figure 6. Measurements of the evolution of contact radius during adhesion. Filled symbols depict loading, and open symbols depict unloading. R = 15.08 mm.

Figures 5 and 6 show the change in contact radius against time, measured at δ = 0.130 mm for R = 12.5 mm, and at δ = 0.147 mm for R = 15.08 mm, respectively. Filled symbols depict loading, and open symbols depict unloading. Error bars in the measurements of contact radius are of the same order as the symbol size. The two different values of δ selected here are intended to obtain (almost) the same ratio a/R for each individual sphere. Although each target displacement is completely given by t_{stop} = 0.5 s from an initial contact, the contact radius, afterward, varies constantly up to 30 s to become stable. It is observed from both experiments that the velocity or rate of change in contact radius on loading is faster comparing to that on unloading.

Figure 7 shows the experimental results for the change in contact radius with time, measured at three different applied displacements, δ = 0.047 mm, 0.072 mm and 0.147 mm, for the sphere of R = 15.08 mm. Each target displacement is completely given by 0.5 s from initial contact. It is observed that the contact radius at each displacement changes in the same manner, but finally converges to a different equilibrium value.

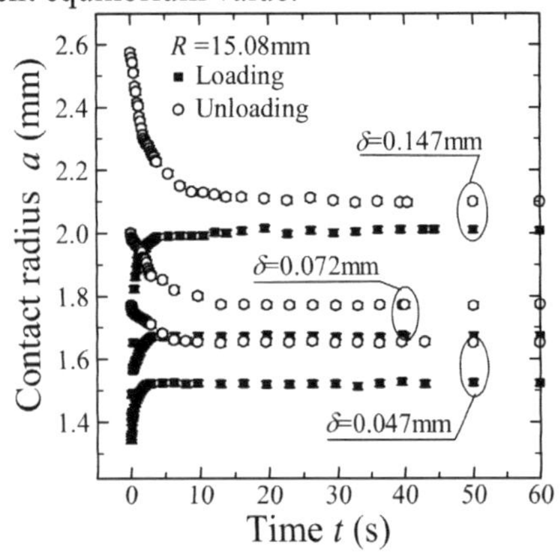

Figure 7. Measurements of the evolution of contact radius during adhesion, measured at different applied displacements, δ = 0.047, 0.072 and 0.147 mm, for R = 15.08 mm.

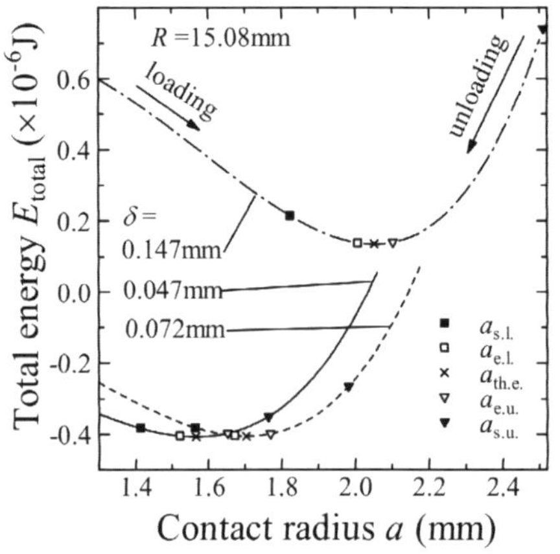

Figure 8. Relation between theoretical total energy and contact radius. Measured contact radii vary along each curve on loading and unloading; however, they cannot reach a stable equilibrium.

To give further consideration to these experimental results from the viewpoint of the total energy of the system, we plot the theoretical total energy of the system, Equation (2), for the three displacements. As shown in Figure 8, filled symbols depict starting or initial contact radius $a_{\text{s.l.}}$, $a_{\text{s.u.}}$, and open symbols depict equilibrium contact radius $a_{\text{e.l.}}$, $a_{\text{e.u.}}$. Rectangles represent loading, and triangles represent unloading. The cross depicts the thermodynamic equilibrium value of the contact radius $a_{\text{th.e.}}$. Precise values of the above-mentioned data are summarized in Table 1.

Table 1. Measured data for the contact radius at three different applied displacements, for the sphere of R = 15.08 mm. (Unit: mm)

R	δ	$a_{\text{s.l.}}$	$a_{\text{e.l.}}$	$a_{\text{th.e.}}$	$a_{\text{s.u.}}$	$a_{\text{e.u.}}$
	0.047	1.413	1.522	1.566	1.764	1.653
15.08	0.072	1.564	1.672	1.704	1.982	1.770
	0.147	1.823	2.007	2.052	2.510	2.103

It is observed from Figure 8 that the three theoretical energy curves are smooth and have a single thermodynamic equilibrium, i.e. stable equilibrium. Therefore, the contact areas at each applied displacement should attain the stable equilibrium. But in actuality, the contact radii from measurements on both loading and unloading cannot attain a stable equilibrium. There is one possibility that the existence of relatively small sub-roughness on the surface of the two bodies gives rise to the existence of an energy-gap on the interface that the contact area cannot overcome.

From the experimental data, we investigated the relation between the energy gradient $\partial E_{\text{total}} / \partial a$ and the velocity of change in contact radius $\partial a / \partial t$ for both spheres, as shown in Figures 9 and 10. It is observed that the energy gradient has a negative correlation with the velocity of change in contact radius, on both loading and unloading. As shown in Figure 10, the correlation is seen not to be influenced by the variation of applied displacement.

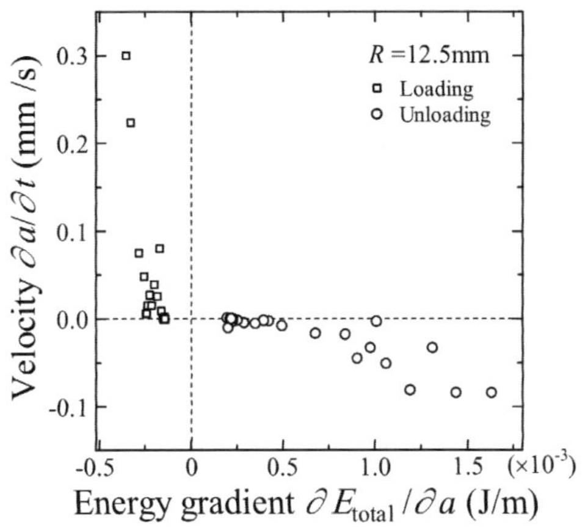

Figure 9. Negative correlation between energy gradient and velocity of change in contact radius can be observed.

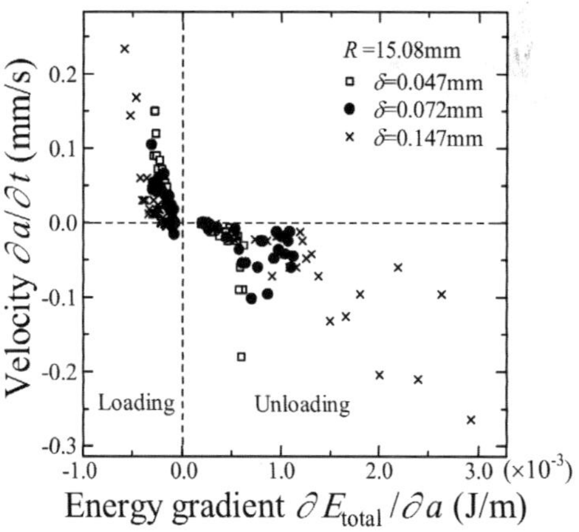

Figure 10. Negative correlation between energy gradient and velocity of change in contact radius can be observed. The correlation is seen to be independent of the applied displacement.

5. Conclusions

We experimentally investigated the adhesion contact between a rigid sphere and an elastic body with a flat surface. Hysteresis of the contact radius was observed between loading and unloading as a function of the applied displacement, from which the work of adhesion and the energy release rate could be estimated accurately. In the measurement of time dependence of the contact radius, it was observed that the contact radius on loading converges faster than that on unloading. Based on the total energy of the JKR theory, it was observed that the energy gradient has a negative correlation with the velocity of change in contact radius, on both loading and unloading.

References

[1] Geim A K, Dubonos S V, Grigorieva I V, Novoselov K S, Zhukov A A and Shapoval S Y 2003 *Nature mater.* **2** 461-63

[2] Kim S B, Spenko M, Trujillo S, Heyneman B, Santos D and Cutkosky M R 2008 *IEEE Trans. robotics* **24** 65-74

[3] Takahashi K, Mizuno R and Onzawa T 1995 *J. Adhesion Sci. Technol.* **9** 1451–64

[4] Johnson K L, Kendall K and Roberts A D 1971 *Proc. R. Soc. Lond. Ser.* A **324** 301-13

[5] Mangipudi V S, Huang E, Tirrell M and Pocius A V 1996 *Macromol. Symp.* **102** 131-43

[6] Ahn D and Shull K R 1996 *Macromolecules* **29** 4381-90

[7] Johnson K L 1998 *Tribol. Int.* **31** 413-18

[8] Lei L,Takahashi K, Rachmat A S R and Hemthavy P 2009 *Quarterly J. Jpn. Weld. Soc.* **27** 196-9

Influence of scale on hot rolling characteristics of Si-bearing steel sheets

K Hara[1], H Utsunomiya[1], R Matsumoto[1] and T Sakai[2]

[1] Division of Materials and Manufacturing Science, Graduate School of Engineering, Osaka University, 2-1 Yamada-oka, Suita, Osaka 565-0871, Japan
[2] Joining and Welding Research Institute, Osaka University, Ibaraki, Osaka, Japan.

E-mail: kenichiro.hara@mat.eng.osaka-u.ac.jp

Abstract. In hot rolling processes, surface oxide layer, scale, may influence friction and affect rolling characteristics. Si-bearing steel was rolled between 1273 K and 1473 K after oxidation for 0 s to 40 s at a reduction of 30%. In Si-bearing steel, Fe_2SiO_4 is formed in the scale and causes eutectic transformation between FeO at 1450 K. The rolling force is affected by the eutectic transformation of the scale. If the rolling temperature is just below the eutectic temperature, roll force scatters and curling of the sheet occurs. This is due to inhomogeneous formation of the liquid phase because of the transformation.

1. Introduction

In hot rolling processes, an oxide layer called 'scale' is formed on steel surface by oxidation at high temperature. The scale influences the friction and heat transfer between rolls and workpiece and affects rolling characteristics significantly.

Scale microstructure formed on steel surface is affected by alloying elements and heating conditions. Scale growth and microstructure is rather affected by added Si [1]. Si is widely used to harden steels or to decrease hysteresis loss in electrical steel. On mild steel, the major component of the scale is FeO (wüstite). On Si-bearing steel, an Fe_2SiO_4 (fayalite) layer is also formed in the scale. At 1450 K, fayalite causes eutectic transformation between wüstite [2]. Liquid phase is formed in the scale above 1450 K. In low carbon steel, scale acts as a lubricant [3]. However, the scale on Si-bearing steel sometimes increases friction and causes surface defects and curling during rolling. Influence of scale on hot rolling characteristics is supposed to be affected by the alloying elements and the scale structure.

In this study, Si-bearing steel sheets were rolled after oxidation around the eutectic temperatures. The scale on the as-hot rolled sheet was preserved by sprinkling glass powder and observed by a scanning electron microscope (SEM) and energy dispersive X-ray analysis (EDX). Roll force, surface temperature and internal shear deformation were measured.

2. Experimental

3 mm thick Si-bearing steel (1% Si) sheets were received. After homogenization at 1473 K for 1200 s, the scale was removed by machining the surface. The cold sheet was inserted into an electric tube furnace at hot rolling temperatures (1273 K, 1373 K and 1473 K) filled with argon gas. After 900 s, the atmosphere was changed from argon to air to allow oxide scale to grow for 0 s, 10 s and 40 s. After

the oxidation, the sheets were immediately rolled on a two-high laboratory rolling mill of which roll diameter was 100 mm. The thickness was reduced by 30% without lubrication. Sheet surface temperature just after the hot rolling was measured by an infrared radiation thermometer. It is assumed that the emissivity is equal to unity. Roll force during rolling was also measured by the load cell. After passing the roll bite, glass powder was sprinkled over the sheet to preserve the scale as hot rolled [4]. Sectioned longitudinal plane (TD plane) at the center of the length was observed with a SEM. In order to investigate the frictional effect, internal shear deformation caused by the friction from the rolls was also assessed by a pin inserted into a sheet before the hot rolling. For this purpose, before the hot rolling, a hole was drilled through the thickness at the center of the sheet. The hole was filled with a steel pin 2 mm in diameter. After the hot rolling, the sheet was sectioned precisely at the center of the width and the longitudinal section was polished to observe the deflection of the interface between the pin and the sheet.

3. Results

3.1. Scale thickness before hot rolling

The thickness of the scale formed on the steel during oxidization is measured on SEM micrographs of the specimens glass-coated without hot rolling. Thickness increases with oxidation time as shown in Figure 1. Scale grew even in the case without oxidation time (t=0 s) because the sheet was oxidized slightly when the sheet was put into and taken out from the furnace. At 1273 K, the scale thickness was less than 5 μm after 40 s oxidation. At 1373 K, the scale was brittle and detached from the steel substrate during cooling before SEM observation. The thickness was 13 μm after 10 s oxidation. Scale grew much faster when heated at 1473 K. The scale thickness was 21 μm at 10 s oxidation. In this case, the scale is adhered to matrix steel.

3.2. Scale morphology before hot rolling

Longitudinal sections of the sheets before the hot rolling (r=0%) after 10 s oxidation are shown in Figure 2(a). At 1273 K, the scale formed on steel surface is very thin. The thickness was about 3 μm. At higher temperature, the scale consists of two layers which show different contrasts. At 1373 K, just below the eutectic temperature between FeO and Fe_2SiO_4, the lower scale (subscale) which is just above the metal surface contains many voids and has porous structure. If the temperature is above the eutectic temperature at 1473 K, the lower scale supposed to be liquid because the interface between the scale and the steel is irregular before rolling.

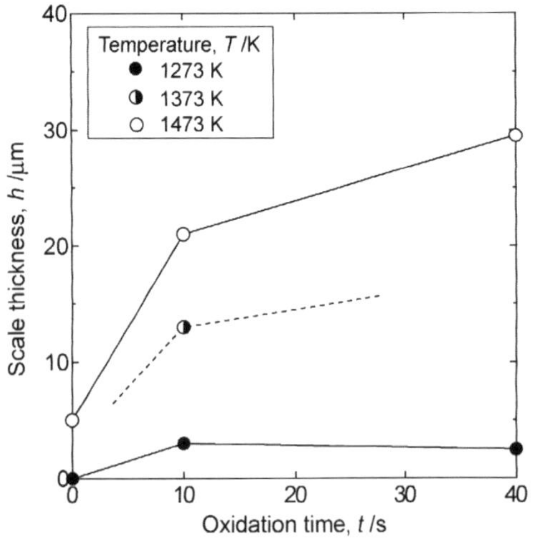

Figure 1. Scale thickness before rolling measured on SEM micrographs as a function of oxidation time and temperature.

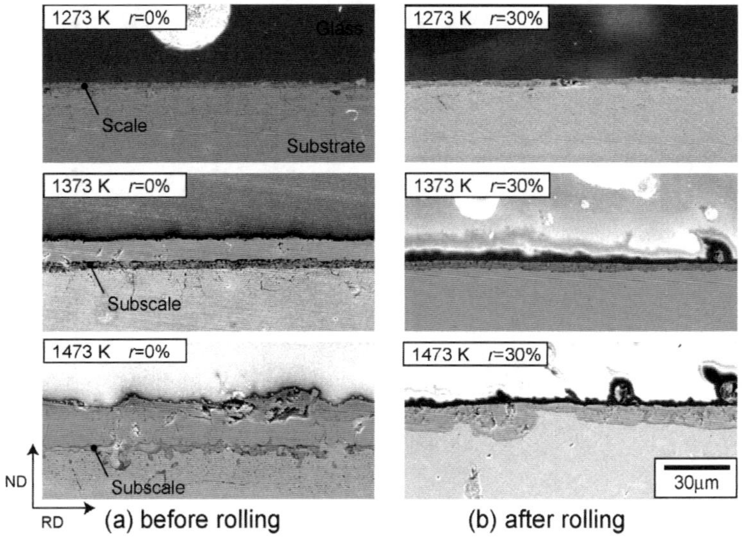

Figure 2. SEM micrograph on longitudinal section of the sheet oxidised for 10 s (a) before rolling and (b) after rolling at a reduction of 30% as a function of rolling temperature.

EDX analysis was performed to analyse the subscale. Figure 3 shows the subscale at the interface with the steel substrate and distribution of Si. This subscale is supposed to contain fayalite because Si is detected in both cases. Si is also detected near the surface of steel substrate. SiO_2 is supposed to be formed due to intergranular oxidation and internal oxidation [5]. At 1373 K, Si is detected only in the subscale and near the surface of steel substrate. At 1473 K, liquid subscale penetrates deeply into the upper scale.

3.3. *Scale observation after hot rolling*

Longitudinal sections of the sheets after 10 s oxidation rolled by 30% reduction in thickness are shown in Figure 2(b). Rolling temperature is defined as heating temperature in the furnace. At 1373 K, the scale thickness was relatively uniform over the surface. The sheet surface was covered with thin scale layer. The interface between the sale and the steel sheet was smooth. At 1373 K, just below the eutectic temperature between FeO and Fe_2SiO_4, the scale after hot rolling was very thin. When the sheet was reduced by 30% in thickness, curling of the sheet occurred and the scale was detached from the sheet surface and adhered to a roll surface. The scale left on the sheet surface was very thin and the interface between the scale and the steel is relatively smooth. At 1473 K, above the eutectic temperature, when the sheet was reduced by 30% in thickness, the sheet does not curl up or down and the scale deformed into non-uniformly. The subscale is distributed intermittently to the rolling direction.

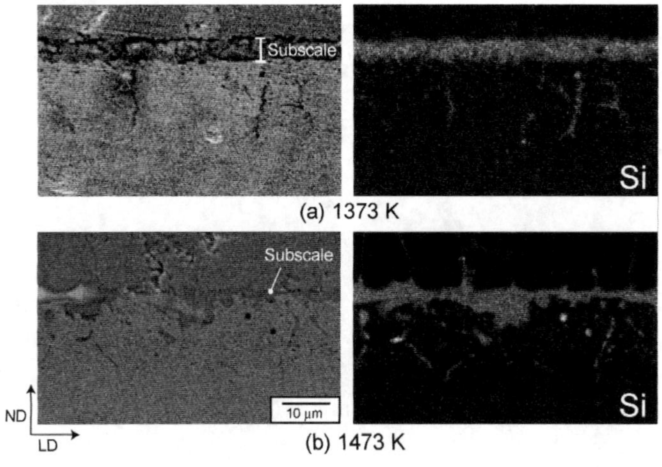

Figure 3. Longitudinal section of the sheet before rolling observed by SEM showing the morphology of the inner scale (subscale). Corresponding X-ray mappings show the distributions of Si.

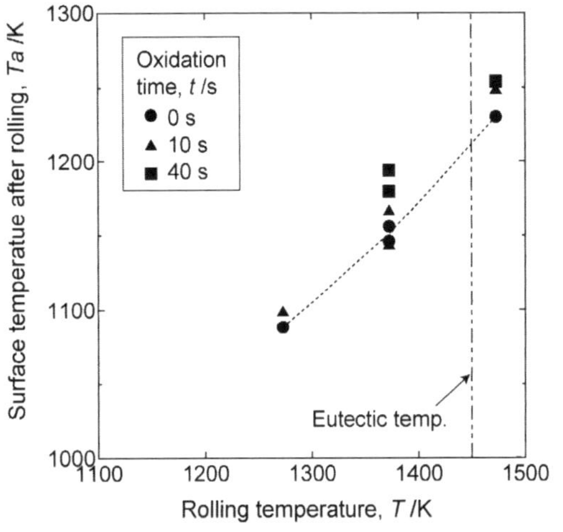

Figure 4. Surface temperature just after hot rolling measured by a radiation thermometer as a function of rolling temperature and oxidation time.

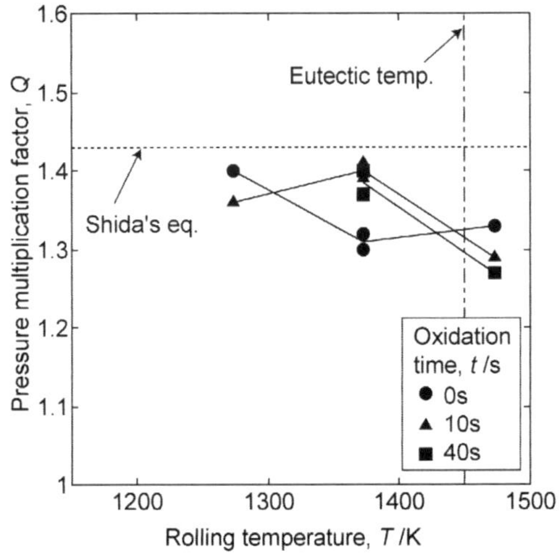

Figure 5. Pressure multiplication factor as a function of rolling temperature and oxidation time at a reduction of 30%.

3.4. Surface temperature

Surface temperature measured just after the hot rolling by a radiation thermometer is shown in Figure 4. The surface temperature decreases due to the heat transfer by touching the cold roll. For example, when a sheet was rolled at 1473 K after 0 s oxidation, the surface temperature decreased to 1230 K. Compared to the same reduction in thickness and rolling temperature, the surface temperature is higher when scale is thicker. The difference between 0 s and 40 s is about 30 K. The thermal conductivity of the scale, especially Fe_2SiO_4, is much lower than that of iron [6]. It is found that thick scale suppresses the heat transfer between roll and matrix steel.

3.5. Pressure multiplication factor

Pressure multiplication factor Q is rolling load normalized by the ideal rolling force. The factor is calculated by measured roll force using the following equation (shown in Figure 5).

$$P = Qk_{fm}\sqrt{R'\Delta h}w \quad (1)$$

where P: roll force, Q: pressure multiplication factor, k_{fm}: mean flow stress, R': flattened roll radius, Δh: thickness draft, and w: width.

Theoretical pressure multiplication factor by Shida's equation [7] assuming sticking friction is also shown. Below the eutectic temperature, the factor Q is higher when the oxidation time is longer. The Q value is close to the sticking condition. However, above the eutectic temperature, the factor Q is lower when the oxidation time is longer. It is supposed that above the eutectic temperature, friction supposed to be in the slipping condition.

3.6. Internal shear deformation

Figure 6 shows the longitudinal sections including pin/sheet interfaces in the 30% rolled sheets. The interface, which was vertical before rolling, deforms into a C-shape by the frictional shear stress acting on the surface. The slope of the pin/sheet interface decreases with increasing oxidation time when the rolling temperature is just below the eutectic temperature at 1373 K. The internal shear strain at the quarter plane in thickness increases from 0.60 to 0.76. If the sheet is rolled above the eutectic temperature at 1473 K, the slope increases with increasing oxidation time. The internal shear strain

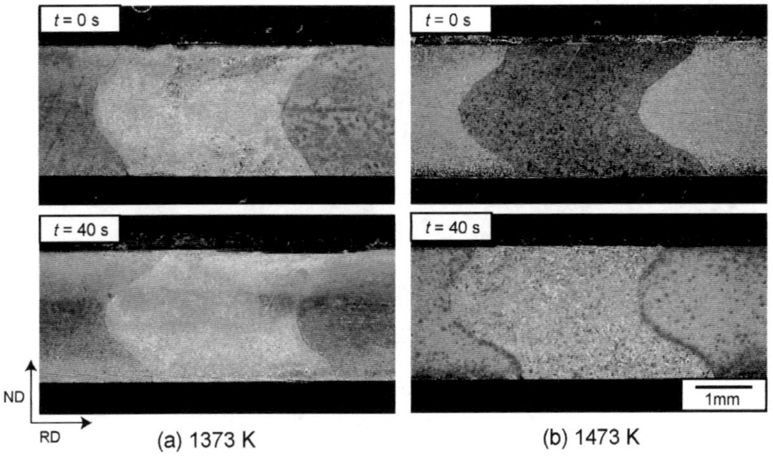

Figure 6. Deflection of embedded pin showing the shear deformation in the hot rolling. Longitudinal section after 30% rolling is shown as a function of rolling temperature and oxidation time.

decreases from 0.91 to 0.74. It is found that the internal shear deformation showing the frictional effect in hot rolling is much affected by the eutectic transformation of the scale.

4. Discussions

4.1. Scale morphology and adhesiveness

When the scale was formed just below the eutectic temperature at 1373 K, the scale was detached from the steel substrate during cooling before SEM observation. When the sheet was reduced by 30% in thickness, the scale was separated from the sheet surface and adhered to a roll surface. In this case, the subscale contains a lot of voids and has porous structure. It is supposed that the interface between the steel substrate and the scale is brittle just below the eutectic temperature due to the porous solid subscale. When the scale was formed above the eutectic temperature, the scale adheres tightly to the matrix steel before and after rolling. At this temperature, liquid phase is formed in subscale. This liquid subscale penetrates deeply into the upper scale. This subscale is supposed to increase the adhesion between the upper scale and the steel substrate to prevent the scale from detaching.

4.2. Effects on eutectic transformation of the scale to rolling characteristics

Figure 7 shows the supposed model of scale deformation during hot rolling of Si bearing steel sheet. In hot rolling of Si-bearing steel, rolling characteristics are sensitive to the eutectic transformation of the scale between FeO and Fe_2SiO_4.

If the rolling temperature is much lower than the eutectic temperature between FeO and Fe_2SiO_4, the amount of Fe_2SiO_4 is supposed to be small and have negligible effect. In this case, roll force decreases with scale thickness due to the relative sliding between the scale and the matrix steel as in the case of low carbon steel [3].

If the rolling temperature is just below the eutectic temperature, the scale was detached from the sheet surface and adhered to the roll surface when the sheet was reduced by 30% in thickness. Internal shear deformation is higher when the scale is thicker. The solid inner scale is supposed to increase the friction between the scale and the matrix steel. In this case, curling of the sheet occurs. This is due to the inhomogeneous formation of liquid scale phase because of the eutectic transformation. As shown in Figure 7(a), the subscale is solid below the eutectic temperature. However, when the sheet is hot rolled, the lower scale is supposed to be partially liquid due to the temperature rise related to friction or plastic work in rolling or non-uniform temperature distribution in heating. In this case, the sheet curls up or down because rolling is not stable due to non-uniform distributions of liquid scale.

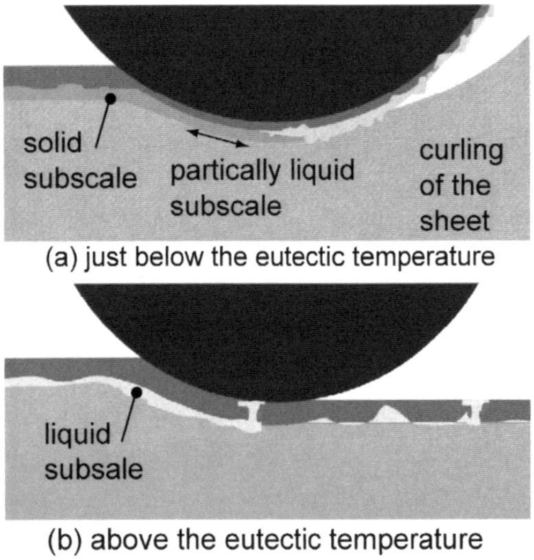

Figure 7. Proposed scale deformation model on hot rolling of Si-bearing steel sheet. (upper: just below the eutectic temperature, lower: above the eutectic temperature.)

If the temperature is above the eutectic temperature at 1473 K, the lower scale is distributed intermittently in the rolling direction after hot rolling. As shown in Figure 7(b), if rolling temperature is high and the scale has liquid phase during rolling, liquid lower scale flows freely and is squeezed out to the outermost surface. This liquid lower scale is supposed to reduce the friction between outer scale and matrix steel and decrease the internal shear deformation of the sheet.

5. Conclusions

In hot rolling of Si-bearing steel, rolling characteristics are sensitive to the eutectic transformation of the scale between FeO and Fe_2SiO_4. At 1373 K, just below the eutectic temperature, lower scale (subscale) which is just above the matrix steel shows porous structure. When the sheet is rolled at 30%, internal shear deformation increases with scale thickness. The sheet curls up or down due to non-uniform distributions of temperature and scale. When the rolling temperature is above the eutectic temperature at 1473 K, liquid phase is formed in the scale. This liquid lower scale flows freely and is squeezed out to the outermost surface during rolling. The internal shear deformation decreases with the scale thickness.

Acknowledgement

This work was partly supported by "Global COE program: Advanced Structural and Functional Materials Design" from MEXT of Japan.

References

[1] Fukumoto M, Hayashi S, Maeda S and Narita T 1999 *Tetsu to Hagané* **85** 878
[2] Verein Deutscher Eisenhüttenleute 1995 *Slag Atlas* (Düsseldorf :Verlag Stahleisen mbH) 79
[3] Hara K, Utsunomiya H, Sakai T and Yanagi S 2011 *Steel Research Int.* **82** 74
[4] Utsunomiya H, Doi S, Hara K, Sakai T and Yanagi S 2009 *CIRP Ann. - Manufact. Technol.* **58** 271
[5] Kusabiraki K, Watanabe R, Ikehata T, Takeda M, Onishi T and Guo X 2007 *ISIJ Int.* **47** 1329
[6] Takeda M, Onishi T, Nakakubo S and Fujimoto S 2009 *Materials transactions* **50** 2242
[7] Shida S 1966 *J. Jpn. Soc. Technol. Plast.* **7** 424

Improvement in rolling workability of Fe_3Al by high-speed rolling

A Muraoka[1], H Utsunomiya[1], R Matsumoto[1] and T Sakai[2]

[1] Division of Materials and Manufacturing Science, Graduate School of Engineering, Osaka University, 2-1 Yamadaoka, Suita, Osaka 565-0871, Japan
[2] Joining and Welding Research Institute, Osaka University, 11-1 Mihogaoka, Ibaraki, Osaka 567-0047, Japan

E-mail: akira.muraoka@mat.eng.osaka-u.ac.jp

Abstract. High-speed rolling is expected to improve the deformability of difficult-to-work alloys. In this work, high-speed rolling was applied to Fe_3Al, and the rolling workability was investigated. Fe-28%Al was prepared by plasma arc melting and multi-pass hot-rolled to a sheet 2 mm in thickness, followed by heat-treatment. Both the hot-rolling and heat-treatment were carried out at 1173 K. These specimens were rolled in a one-pass operation with a thickness reduction of 30 % at 1000 m/min in the temperature range from 573 to 973 K. Rolling at 5 m/min was also conducted for comparison. Sound sheets without defects or cracks were obtained even at 673 K in the high-speed rolling, while sound sheets were obtained only above 1173 K in the low-speed rolling. EBSD analysis showed that fine grains were formed either after the high-speed rolling at 873 K or after the low-speed rolling at 1073 K. Vickers hardness is not sensitive to the rolling speed below 773 K and decreased above 773 K and 1173 K in the high-speed rolling and low-speed rolling, respectively. The rolling workability of Fe_3Al was significantly improved by the high-speed rolling.

1. Introduction

Some intermetallic compounds have been expected for structural materials, *e.g.*, Ni_3Al, Ti_3Al, TiAl and Fe_3Al. In particular, Fe_3Al is an intermetallic compound which shows transformation from the $D0_3$ to B2 structures around 823 K. It has many advantages such as low cost, low density, high specific strength and good resistance to oxidation or sulfidation. Thus, it is an environmentally and economically friendly material. However, binary Fe_3Al shows very poor ductility as other intermetallics [1], which is a major barrier to its practical applications. So far, many works have been done in order to improve the ductility. The addition of Cr has been widely studied, since first proposed by McKamey *et al.* [1]. Other alloying elements are also effective for improvement in the ductility [2-4]. Nevertheless, the workability is not improved by such additional elements. For example, Fe-27.8%Al-5.0%Cr-0.1%Zr was rolled in a multi-pass operation with approximately 10 % reduction per pass at room temperature; edge cracking was noticed after about 40 % reduction [5]. In these cases, the rolling speed was no more than approximately 30 m/min. Because both alloying elements and rolling at low speeds are industrially inefficient, other efficient processing methods are required.

Recently, the authors [6, 7] reported that rolling faster than 200 m/min (3.33 m/s) enabled Mg alloys to roll in a one-pass operation with a heavy reduction of 50 %. This is because the time of contact between the sheet and the cold rolls is very short, which does not only prevent heat transfer to rolls during rolling, but also increase heat produced during rolling. Therefore, the temperature during rolling is higher than the initial rolling, and the grains remain fine due to the low heating temperature of the sheet before rolling and high Zener-Hollomon parameter deformation.

In this study, this high-speed rolling method is applied to Fe_3Al in expectation of improved rolling workability compared with conventional rolling with a low speed. Binary Fe-28%Al, where Al is richer than the stoichiometry, was used in order to evaluate only the influence of speed.

2. Experimental procedure

Fe-28%Al (at.%) ingot was prepared by plasma arc melting under argon atmosphere using pure iron and aluminum. The details can be found in [8]. The ingot was hot-rolled to a thickness of 2 mm in several pass operations at 1173 K, followed by homogenization at 1173 K for 1 h and furnace cooling. This hot-rolling process was carried out at 5 m/min with about 15 % reduction per pass. The sheet was cut into specimens with 25 mm length and 20 mm width. Two dummy steel sheets, whose lengths were 100 mm and widths and thicknesses were the same as the specimen, were welded to both the front and back of the specimen in the rolling direction by tungsten inert gas welding.

These sheets were rolled in a one-pass operation with a thickness reduction of about 30 % at 1000 m/min (16.7 m/s, high-speed) in the temperature range from 573 to 973 K at 100 K intervals. Hereafter, this preheating temperature is referred to as the 'rolling temperature'. Rolling at 5 m/min (8.33×10^{-2} m/s, low-speed) was also performed for comparison with that from 573 K to 1273 K. The rolls were coated with machine oil for lubrication, and the sheets were quenched into water immediately after the rolling. The high-speed rolling was operated on a rolling mill with rolls 530 mm in diameter which has water sprays after the rolls as shown in Figure 1, and the rolling mill with rolls 310 mm in diameter was used for the low-speed rolling.

The microstructure of the rolled sheets was examined by electron backscatter diffraction (EBSD) on the longitudinal section in the SEM. The specimens were mechanically polished and electropolished using a mixture of 7% H_2SO_4 and 93 % CH_3OH at 25 V and 273 K. The electron beam automatically scanned in 2 μm steps in the area of 2×1 mm^2. In the EBSD analysis, the crystal structure was assumed to be bcc. For the Vickers hardness test, ten points were measured around the mid plane of the rolled specimens using a load of 200 g at room temperature. The hardness of annealed specimens in the temperature range from 573 to 973 K was also measured.

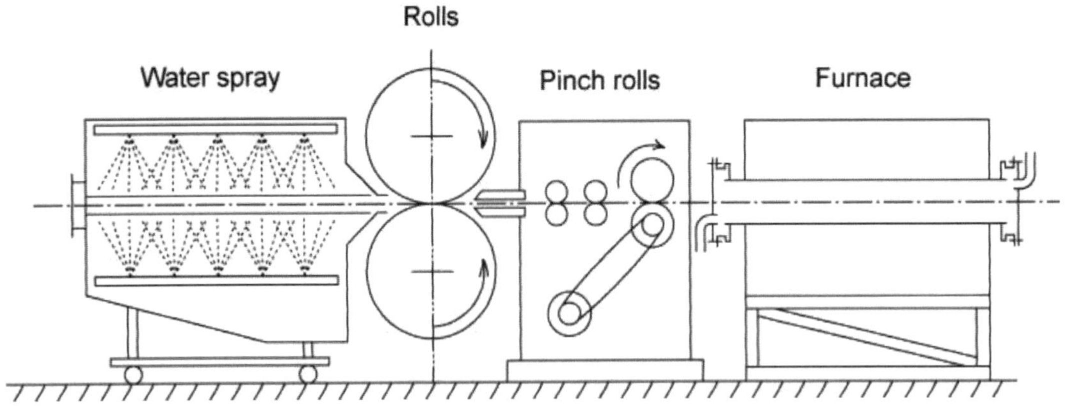

Figure 1. Schematic view of the high-speed rolling mill.

3. Results

3.1. Rolling workability

Figure 2 shows the appearance of rolled sheets, where r is the reduction in thickness of the specimen by rolling. In low-speed rolling below 1073 K, cracks and defects developed in the Fe_3Al specimen area of the sheets. Particularly, scissors cracks, where cracks initiate at edges and propagate in the rolling direction, developed at both 773 K and 873 K. Sound sheets without cracks and defects were obtained above 1173 K.

In contrast, sound sheets were obtained above 673 K in the high-speed rolling. Even at 573 K, the specimen had a minor crack on its surface, though the reduction in thickness of the high-speed rolled sheets is larger than that of the low-speed rolled ones. This indicates that high-speed rolling enables Fe_3Al to be rolled at a lower temperature, which was roughly 500 K lower.

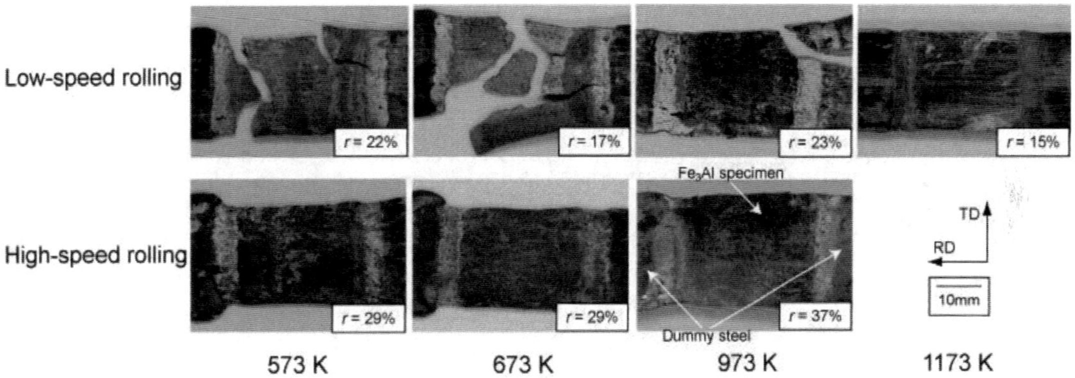

Figure 2. Appearance of low-speed/high-speed rolled sheets as a function of rolling temperature.

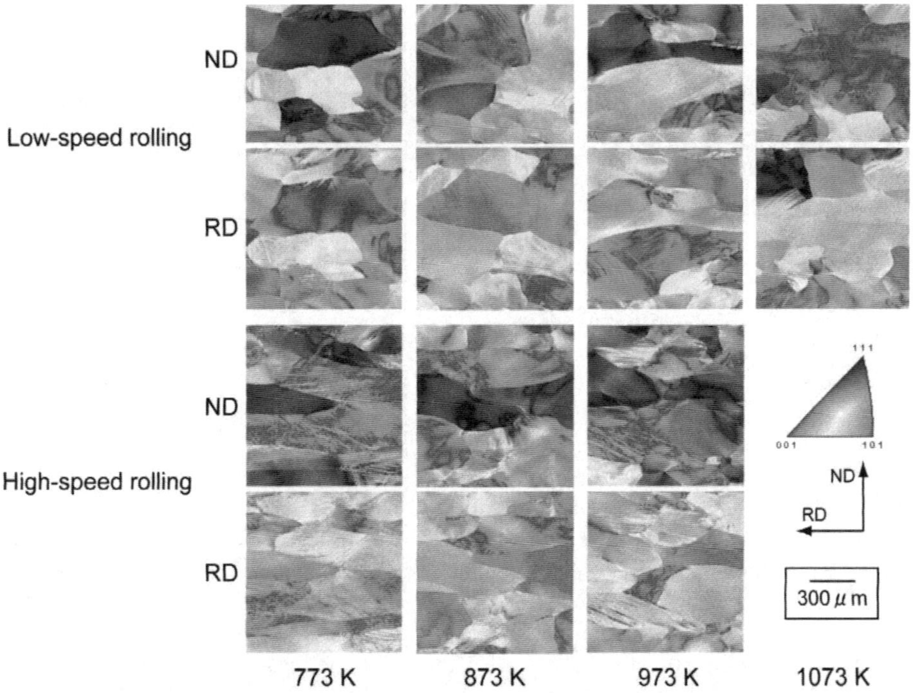

Figure 3. Inverse pole figure maps of low-speed/high-speed rolled sheets as a function of rolling temperature.

3.2. Microstructure and texture

Inverse pole figure maps of ND (top) and RD (bottom) orientations measured by EBSD analysis are shown in Figure 3. After the heat-treatment at 1173 K, equiaxed grains which have little preferred orientation were observed. In either the low-speed rolled sheets below 973 K or the high-speed rolled sheets below 773 K, grains elongated parallel to the rolling direction. In either the low-speed rolling at 1073 K or the high-speed rolling at 873 K, some fine grains believed to be recrystallized grains were formed. Grain size increased with rolling temperature due to grain growth during preheating in the furnace. The texture components of the rolled specimens were similar to those of typical bcc rolling textures, *i.e.*, α-fiber, <110>//RD and γ-fiber, <111>//ND.

3.3. Micro-hardness

Figure 4 shows Vickers hardness of the specimens before and after rolling. The hardness of the rolled specimens was not sensitive to the rolling speed below 773 K, and decreased above 773 K in the high-speed rolling and above 1173 K in the low-speed rolling. It means that high-speed rolling causes the softening temperature to decrease by 400 K. In the figure, hardness of the sheets which were heated in the same furnace, but not rolled is also shown. The hardness is much lower than that after rolling. It is also lower than that of the initial sheet shown by a dashed line. It means that work hardening in the hot rolling is significant. The difference between the hardness before and after the rolling decreases with heating temperature, particularly above 773 K in the high-speed rolling. In other words, dynamic restoration is apparent above 773 K.

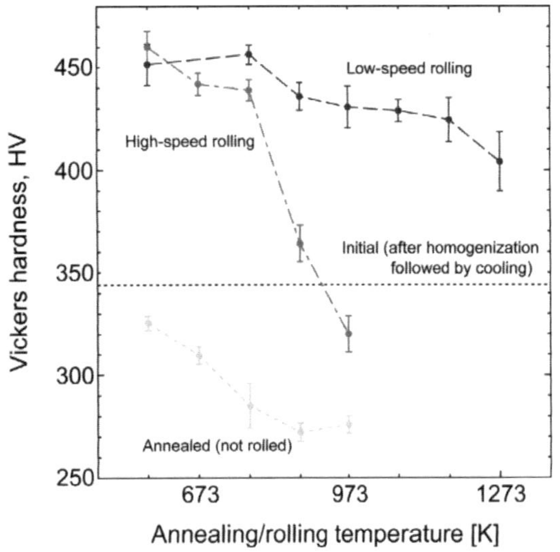

Figure 4. Vickers hardness before and after rolling as a function of rolling temperature.

4. Discussion

Fe-28%Al sheets were rolled in a one-pass operation with a thickness reduction of 30 % at 1000 m/min in the temperature range from 573 to 973 K. In the high-speed rolling, sound sheets were obtained even at a temperature 500 K lower than the lowest rollable temperature in the low-speed rolling. The hardness decreased above 1173 K in the low-speed rolling and 773 K in the high-speed rolling. The temperature of hardness drop was 400 K lower. Some fine grains were observed at 1073 K in the low-speed rolling and 873 K in the high-speed rolling.

These results can be explained consistently if occurrence of dynamic recrystallization in hot rolling is assumed. If dynamic recrystallization occurred at lower temperatures in high-speed rolling, it is a possible cause for the improvement in rolling workability. In the low-speed rolling, a gradual decrease occurred below 1173 K, resulted in a slight drop of the hardness. On the other hand, in the high-speed rolling, the hardness shows an abrupt drop above 773 K. Thus, the rolling workability of Fe_3Al may be improved by high strain rate deformation and dynamic recrystallization associated with it. When the sheet is in contact with the rolls, cracks may develop in low-speed rolling due to the temperature drop. It is supposed that the temperature during the high-speed rolling was higher than that during the low-speed rolling.

The dynamic recrystallization at lower temperatures in high-speed rolling is mostly attributed to higher temperatures during rolling. The temperature change in hot rolling is due to (i) heat generated by plastic work (ii) and friction, as well as (iii) heat loss to the rolls. The plastic work is mostly converted to heat. Flow stress increases with strain rate, so that (i) heat by plastic work increases with speed. A higher relative velocity between the rolls and the specimen produces higher frictional heat. Meanwhile, the high-speed rotating rolls prevent heat transfer to the rolls from sheets due to the short contact time.

If we assume that the thickness temperature distribution is constant, the temperature change of the specimen during rolling can be estimated using equation (1),

$$\Delta T = \Delta T_P + \Delta T_f - \Delta T_R \quad (1)$$

where ΔT is the total change in temperature during rolling, ΔT_P is the temperature rise due to plastic work, ΔT_f is the temperature rise due to friction and ΔT_R is the temperature drop due to contact with cold rolls. ΔT_P is calculated using the values of the processing heat, which is estimated by the mean flow stress (shown in Table 1), the equivalent strain, the density of the sheet (7057 kg/m^3), and the specific heat of the sheet (shown in Table 1). Similarly, ΔT_f is calculated using the value of the frictional heat, which is estimated by the assumed friction coefficient of 0.2, the flow stress, the speed of the rolls, the relative velocity and contact time between the rolls and sheet, as well as the thickness reduction, the density and the specific heat of the sheet. ΔT_R is calculated using the value of the conductive heat from the material to the rolls, which is estimated by temperatures of the sheet and the rolls, the contact time of the sheet and the rolls, the thermal conductivity of the sheet (shown in Table 1) and the rolls of 73.3 J/(msK), the density of the sheet and the rolls of 7870 kg/m^3, and the specific heat of the sheet and the rolls of 444 J/(kgK), and the thickness reduction. The details of this method can be found in [9, 10]. The temperature difference during the high-speed rolling and the low-speed rolling is estimated to be 80-250 K as shown in Table 1. Here, the flow stress is substituted by the yield stress of Fe-25%Al, and the values are extrapolated from [11]. The values of the specific heat and the thermal conductivity are extrapolated from [12, 13].

Table 1. Temperature difference during the high-speed rolling and the low-speed rolling.

Rolling temperature [K]	Mean flow stress [MPa]	Specific heat of the sheet [J/(kgK)]	Thermal conductivity of the sheet [J/(msK)]	Temperature during low-speed rolling [K]	Temperature during high-speed rolling [K]	Temperature difference [K]
573	110	807	16.8	493	578	85.3
673	170	814	16.9	561	682	121
773	340	820	17.0	633	814	181
873	270	827	17.0	694	893	199
973	90	833	17.1	730	968	248

Chen *et al.* [14] tested the mechanical properties of Fe-28%Al under a strain rate ranging from 10^{-4} to 1300 s^{-1} at room temperature, and concluded that the yield stress is sensitive to the strain rate. In this work, the values of strain rate were about 5.9 x 10^2 s^{-1} in the high-speed rolling and 2.6 s^{-1} in the low-speed rolling. According to Chen's data, the flow stress in high-speed rolling could be about 1.3 times as high as that in low-speed rolling. If we suppose that the work during rolling was completely converted to heat, the processing heat is at least 1.3 times higher in the high-speed rolling. Given that the rolling was carried out at an elevated temperature while Chen's work was done at room temperature, the influence may be more significant; however, it may be less than the experimental value of 500 K, which is the difference of the temperatures where the hardness decreased.

As shown in Table 1, the temperature during the low-speed rolling is 80-250 K lower than the rolling temperature, while the temperature during the high-speed rolling is as high as the rolling temperature. However, the sheet was successfully rolled at 673 K in the high-speed rolling, and the temperature during rolling is estimated to 682 K. If the factor which influences the workability was only temperature, the sheet could be successfully rolled in the low-speed rolling at 873 K, because the temperature estimated is 694 K.

Therefore, it is impossible to explain the improvement in rolling workability (< 500 K) by only the influence of temperature (< 250 K). Other factors, such as the grain size of the specimen, could affect rolling workability. In the case rollability is improved, the grain size is kept finer because the specimen can be held at a lower temperature before rolling. Materials having a finer grain microstructure are generally easier to deform. Thus, the rolling workability of Fe_3Al is further improved by high-speed rolling. Furthermore, the change in the deformation mode such as slip systems may occur. In order to clarify the improvement in workability, more studies are required.

5. Conclusions

Fe-28%Al sheets were rolled in a one-pass operation with a thickness reduction of 30 % at 1000 m/min in the temperature range from 573 to 973 K. Sound sheets without defects or cracks were obtained even at 673 K in the high-speed rolling, while sound sheets were obtained only above 1173 K with rolling at 5 m/min.

Grains of rolled sheets were elongated parallel to the rolling direction. After either the low-speed rolling above 1073 K or the high-speed rolling above 873 K, equiaxed recrystallization grains were observed.

The hardness is not sensitive to the rolling speed below 773 K, and decreased above 873 K and 1173 K in the high-speed rolling and low-speed rolling, respectively.

These results indicate that the rolling workability of Fe_3Al is significantly improved by high-speed rolling. The improvement in the workability of Fe_3Al by high-speed rolling is mainly due to a smaller temperature drop in hot rolling by a short contact time with rolls. The temperature difference between the high-speed rolling and the low-speed rolling is estimated to be 80-250 K.

Acknowledgement

This work was supported by the "Global COE Program: Center of Excellence for Advanced Structural and Functional Materials Design" from MEXT.

References

[1] McKamey C G, Horton J A and Liu C T 1988 *Scr. Metall.* **22** 1679
[2] Falat L, Schneider A, Sauthoff G and Frommeyer G 2005 *Intermetallics* **13** 1256
[3] Schneider A, Falat L, Sauthoff G and Frommeyer G 2005 *Intermetallics* **13** 1322
[4] Matsumoto N, Kaneno Y and Takasugi T 2007 *Tetsu-to-Hagané* **93** 400
[5] Morris D G, Muñoz-Morris M A and Gutierrez-Urrutia I 2010 *Mater. Sci.Eng. A* **528** 143
[6] Utsunomiya H, Sakai T, Minamiguchi S and Koh H 2006 *Magnesium Technology 2006* eds A A Luo, N R Neelameggham and R S Beals (San Antonio: TMS) pp 201-204
[7] Hamada G, Sakai T and Utsunomiya H 2010 *Adv. Mater. Res* **89-91** 227
[8] Yasuda H Y, Kouzai K, Kawamura Y and Umakoshi Y 2010 *ISIJ Int.* **50** 147
[9] Kokado J 1970 *Sosei-to-Kakou* **11** 816
[10] Ataka M 1972 *Seisankenkyu* **24** 225
[11] Nishio Y and Tanahashi T 2004 *Mater. Sci. Eng. A* **387-389** 973
[12] Rudajevová A and Šíma V 1997 *Mater. Res. Bull.* **32** 441
[13] Rudajevová A and Buriánek J 2001 *J. Phase Equilibria* **22** 560
[14] Chen M, Lin D, Xia Y and Liu C T 1997 *Mater. Sci. Eng. A* **239-240** 317

Evaluation of *r*-value of steels using Vickers hardness test

Kazutoshi Ohashi[1]**, Hiroshi Utsunomiya**[1] **and Ryo Matsumoto**[1]

[1] Division of Materials and Manufacturing Science, Graduate School of Engineering, Osaka University, 2-1 Yamadaoka, Suita, Osaka 565-0871, Japan

E-mail: kazutoshi.ohashi@mat.eng.osaka-u.ac.jp

Abstract. The plastic strain ratio (Lankford value, *r*-value) is an index of mechanical anisotropy of sheet metals. The Vickers hardness is measured to determine the *r*-value. Tensile and indentation tests of three kinds of commercial steel sheets are conducted in directions 0°, 45° and 90° to the rolling direction. A relationship is found between the *r*-value and impression aspect ratio by a hardness tester. The aspect ratio is defined as the ratio of the diagonal length in the longitudinal direction to that in the thickness direction.

1. Introduction

Many parts for automobiles and home electric appliances are made of steel sheets. In sheet metal forming processes, the *r*-value is an index of mechanical anisotropy [1] that shows good correlation with deep drawability [2,3]. Accurate measurement of *r*-values enables more precise control of processes and results in many benefits such as an improvement in yield efficiency in machining processes and a decrease in loss of material. The *r*-value is defined as the strain ratio (width strain/thickness strain) in tensile tests. The *r*-value is unity for isotropic materials because strains in both directions are equal. The *r*-value is higher than unity for anisotropic materials, which deform more in the width direction than in the thickness direction. It is known that the *r*-value is governed mainly by crystallographic texture. The *r*-value increases with <111>//ND texture components and decreases with <100>//ND texture components [1,4,5]. Therefore, sheet metals with high *r*-values for deep drawing are often cold-rolled under well-lubricated conditions to develop γ-fibers (<111>//ND).

The *r*-value, generally measured by tensile tests, reflects only the average anisotropy through the thickness. However, the *r*-value near the surface seems to be lower than that around the center owing to shear texture formed in the rolling process. For tensile tests, the materials tested should be a particular shape of flat sheet. It is important to evaluate the distribution of *r*-values in a tailored sheet or a sheet with a graded microstructure. Thus, establishment of other methods to measure the *r*-value is strongly desired.

A Vickers hardness test gives local mechanical information because the impression after the test has a micrometer size. The shape of the impression is normally independent of applied load. The test provides a hardness value that corresponds to the yield stress or the strength. Nakamura and Tozawa proposed that anisotropic information on materials can be obtained from the shape of the impression made by the indenter, mainly for silicon steels [6]. However, this could also be applied to other steels.

The two diagonal lengths of the impression after indentation are equal for isotropic materials, but not for anisotropic ones. The shape of the impression should be affected by anisotropy in flow stress. The shape of a micrometer-sized impression is influenced by the texture. In this study, three kinds of commercial steel sheets for deep drawing were tested to provide an estimation of the *r*-value of steel sheet from the shape of the impression formed by a Vickers hardness test.

2. Experimental procedure

2.1. Observation of microstructure
Three kinds of commercial cold-rolled steel sheets were used, (a) extra-low carbon steel, (b) ultra-low carbon steel, and (c) low carbon steel. The chemical compositions are shown in Table 1. Electron Back Scattering Diffraction analysis was conducted in a 200 μm x 200 μm area on the TD plane, by a field emission scanning electron microscope (FE-SEM; Japan Electro Optical Laboratory, JSM-6500F) with a slow-scan CCD camera. Specimens were sampled in 10 mm lengths. In the SEM observations, the accelerating voltage was 15 kV, the exposure current was 16 nA and the scanning pitch was 0.5 μm. An EDAX system (OIM Data Collection ver.5.2 and Analysis ver.5.2) was used.

Table 1. Chemical compositions of the steel sheets used (mass%).

Sheet	JIS	C	Si	Mn	P	S	Fe
A	SPC270C	0.019	0.01	0.14	0.012	0.0068	Bal.
B	SPC270D	0.0018	0.02	0.17	0.014	0.005	Bal.
C	SPC440	0.172	0.01	0.76	0.014	0.003	Bal.

2.2. Measurement of r-value by tensile test
Specimens for tensile tests were prepared so that the tensile axis was parallel to the rolling, transverse and diagonal directions of the three steel sheets. The gauge length and width were 20 mm and 10 mm, respectively. Uniaxial tensile tests were conducted with a cross-head speed of 1 mm/min. After uniform straining of 15% in tension, the specimen was unloaded and the *r*-value was calculated. The *r*-value is defined as the strain ratio (width strain/thickness strain)

$$r = \frac{\ln(w_1 / w_0)}{\ln(t_1 / t_0)}, \tag{2}$$

where w_0 and t_0 are the initial width and thickness of the gauge, respectively. w_1 and t_1 are the width and thickness of the gauge after deformation, respectively. As it is difficult to measure the strain in the thickness direction of a thin sheet accurately, it was calculated from the strains in the longitudinal and width directions using the volume constancy [7].

$$l_0 w_0 t_0 - l_1 w_1 t_1 = 0 \tag{3}$$

l_0 and l_1 are the initial gauge length in the tensile direction and that after deformation, respectively. Hence, the *r*-value was calculated by

$$r = \frac{\ln(w_1 / w_0)}{\ln(w_1 \cdot l_1 / w_0 \cdot l_0)}. \tag{4}$$

The planar average *r*-value is defined by r_0, r_{90} and r_{45} for a tensile axis along the rolling, transverse and diagonal directions, respectively, of the sheets.

$$\bar{r} = \frac{r_0 + r_{90} + 2r_{45}}{4} \tag{5}$$

The specimens were reloaded to measure the tensile strength and elongation. Mechanical properties of the steel sheets are shown in Table 2.

2.3. Vickers hardness test
Three faces of the initial sheets shown in Figure 1 were sectioned and polished to a mirror finish. After polishing, a Vickers hardness test was conducted with an indentation load of 100 gf and a duration of 15 s (number of tests: 20). The diagonal length of the impression in the thickness direction $d_1(d_{ND})$ and

that in the longitudinal (rolling / transverse / diagonal) direction $d_2(d_{RD}, d_{TD}, d_{DD})$ were measured. The Vickers hardness (HV) is defined as

$$H = 1.854 \frac{F}{d^2} \quad . \tag{6}$$

F is the indentation load (kgf) and d is the diagonal length (mm). d is generally an average of d_1 and d_2. To estimate the anisotropy, the aspect ratio α is defined as the ratio of the two diagonal lengths.

$$\alpha = \frac{d_2}{d_1} \tag{7}$$

The aspect ratio α is unity in isotropic materials and higher than unity in anisotropic materials that deform more in the longitudinal direction than in the thickness direction. To estimate the planar anisotropy of the aspect ratios, α_0, α_{90} and α_{45} were defined as aspect ratios in the rolling plane, transverse plane and diagonal plane, respectively. The planar average of the aspect ratio is defined as

$$\bar{\alpha} = \frac{\alpha_0 + \alpha_{90} + 2\alpha_{45}}{4} \quad . \tag{8}$$

Table 2. Mechanical properties of the steel sheets used.

Sheet	Angle to RD (deg.)	0.2% proof stress /MPa	UTS /MPa	Elongation (%)
A	0	246.5	336.4	49.2
	45	260.3	352.7	42.3
	90	258.5	342.0	52.0
B	0	163.7	290.6	62.3
	45	166.1	298.6	60.3
	90	163.8	291.9	62.7
C	0	319.2	462.5	44.9
	45	335.5	473.9	43.8
	90	317.8	461.6	44.6

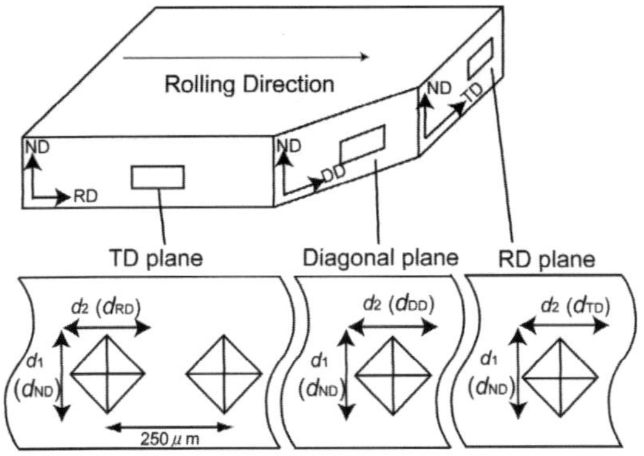

Figure 1. Indentation directions of Vickers hardness tests.

3. Results

3.1. Crystallographic texture

Inverse poll figure maps on TD sections of initial sheets are shown in Figure 2. Sheets A and B have equiaxed microstructure with a mean grain size of 15.6 μm and 13.8 μm, respectively. The grain size of sheet C was 4.8 μm with a random preferred orientation.

It is possible to determine texture quantitatively in Bunge ODF sections for $\phi_2 = 45$(deg.), as shown in Figure 3. The major texture components in sheet A are {111}<211> and the sub-components are {112}<110>. The major texture components in sheet B are {111}<110>. The major texture components in sheet C are {112}<110>, although the intensity is low. <111>//ND (= γ-fiber) components, which increase the deep drawability, are strongly developed in sheets B and A, showing strong intensity. The proportion of <110>//RD (= α-fiber), which increases the planar anisotropy, is relatively low in sheet B.

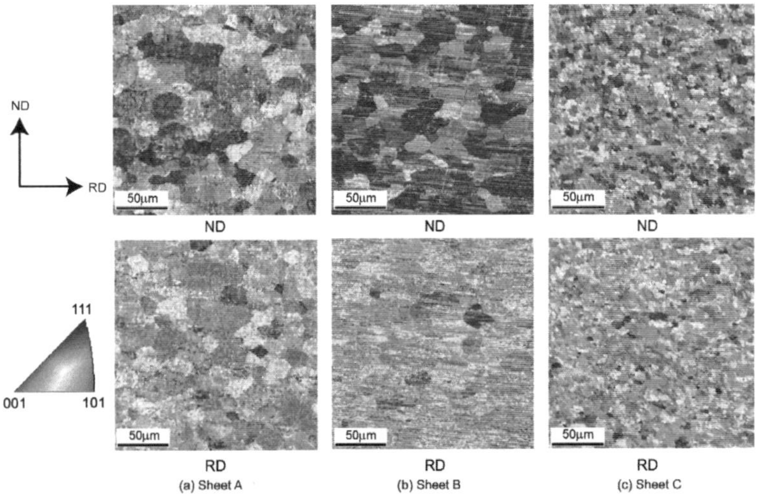

Figure 2. Inverse pole figure maps on TD sections of initial sheets.

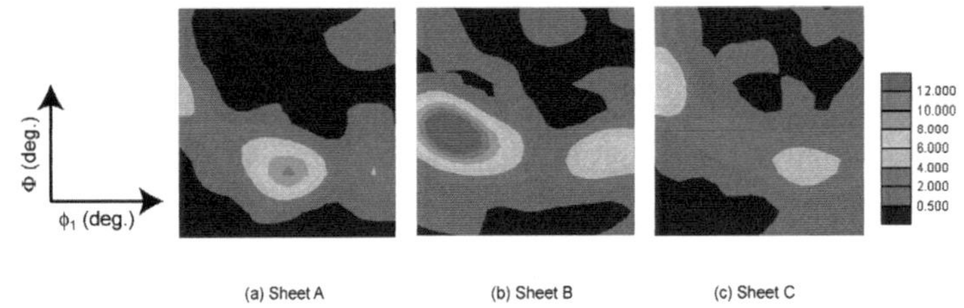

Figure 3. Bunge ODF sections for $\phi_2 = 45$(deg.).

3.2. *r*-value obtained by tensile test

Sheet B has the highest *r*-value, and sheet C has the lowest among the three sheets in Figure 4. This is reasonable if the intensities of γ-fibers in the ODF are considered. The *r*-value shows high planar anisotropy as in Figure 4. The planar anisotropy of the *r*-value in sheets A and C is higher than that in sheet B. This is in good agreement with the intensities of α-fibers from the ODF results.

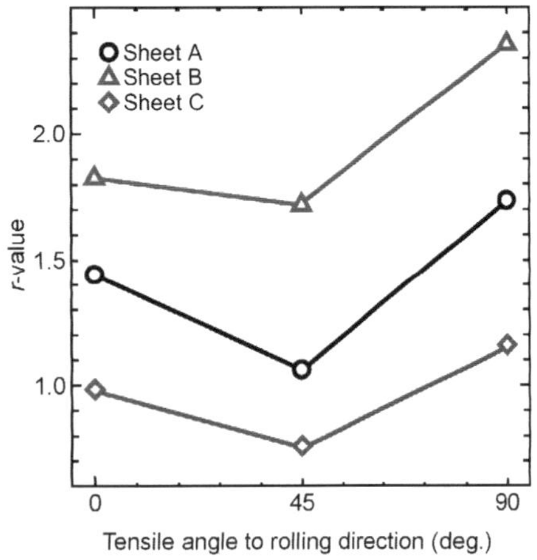

Figure 4. Relationship between *r*-value of carbon steels and tensile angle to rolling direction.

3.3. Vickers hardness

In Figure 5, the Vickers hardness shows lower planar anisotropy. The Vickers hardnesses of sheets A, B and C were 116HV, 100HV and 151HV, respectively. The average diagonal length calculated by Eq. (6) with these hardness values were 40.0 μm, 43.1 μm and 35.0 μm, respectively. The diagonal lengths are larger than the mean grain sizes of the sheets in Figure 2. The hardness reflects averaged mechanical properties of grains under the indenter.

The aspect ratio α shows high planar anisotropy in Figure 6. In sheets A and C, α_0 and α_{90} are higher than α_{45}. In sheet B, the aspect ratios are higher than unity in all directions. This result indicates that sheet B deformed more in the longitudinal direction.

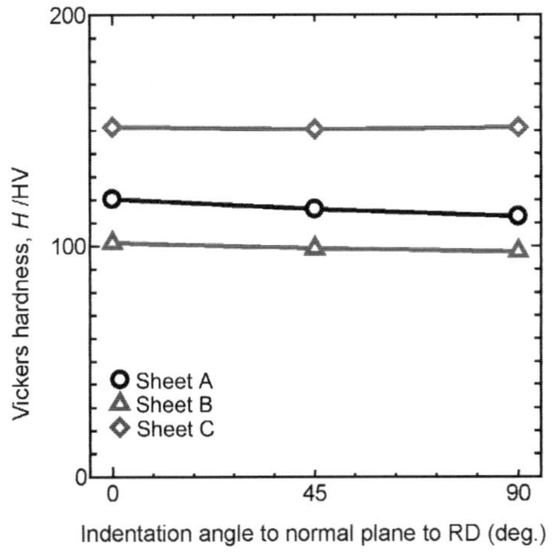

Figure 5. Relationship between Vickers hardness of carbon steels and indentation angle to the rolling direction.

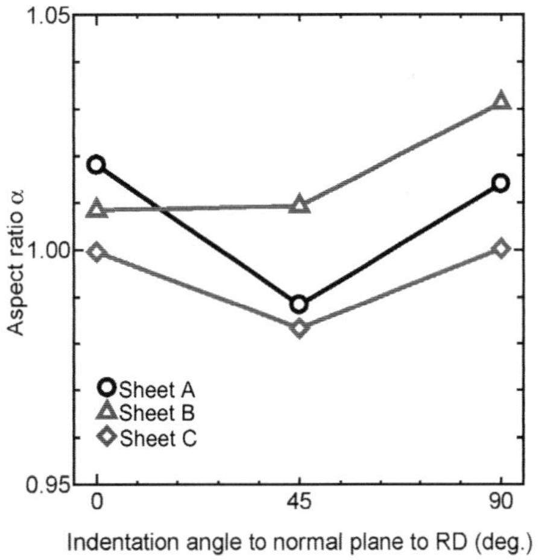

Figure 6. Relationship between aspect ratio of carbon steels by Vickers hardness test and indentation angle to the rolling direction.

4. Discussion

There is high planar anisotropy in the r-value and aspect ratio, but not in the Vickers hardness. The Vickers hardness defined by Eq. (6) is calculated from the average of d_1 and d_2. The diagonal lengths d_1 and d_2 may have low planar anisotropy because it reflects a property either in the thickness direction or another direction. It is supposed that flow stress in the thickness direction is much higher than in either direction on the rolling plane. To estimate the r-value, a parameter having a high dependence on anisotropy is required, such as the aspect ratio α defined by Eq. (7).

In this study, α^2 is used as the parameter that corresponds to the ratio of two hardness values calculated using only the diagonal length in the thickness direction and only in the longitudinal direction. Figure 7 shows the relationship between $\bar{r}$, the planar average r-value measured by the tensile test and $\bar{\alpha}^2$, the square of the planar average of the aspect ratio obtained by the Vickers hardness test. The coefficient of correlation between $\bar{r}$ and $\bar{\alpha}^2$ is 1.00. There is a good relationship between $\bar{r}$ and the impression geometry.

The r-value measured from the tensile test and the aspect ratio α obtained from the Vickers hardness test show high planar anisotropy, and as has been mentioned, seem to have some relationship. To discuss this relationship in more detail, the r-values, r_0, r_{45} and r_{90} are compared with the aspect ratios, α_0, α_{45} and α_{90}, respectively, under the condition that the indentation plane is normal to the tensile axis. From Figure 8, a linear relationship between the r-value determined from the tensile test and the aspect ratio α obtained from the Vickers hardness test is found as

$$r = 15.1\alpha^2 - 13.8. \quad (9)$$

The deformation in the thickness direction decreases with increasing r-value and with higher α as the shape of the impression becomes more rhombic, that is, the impression is longer in the longitudinal direction than in the thickness direction. Eq. (9) gives a better correlation than Nakamura's nonlinear relationship plotted as a curved line in Figures 7 and 8. Nakamura and Tozawa [6] proposed a nonlinear relationship between the r-value and the ratio of two Vickers hardness values calculated using the diagonal lengths in the thickness and width directions. They mainly used silicon steels with a relatively low r-value, so the equation deviates at high r-values.

In Figure 8, Eq. (9) does not pass through the point where both the r-value and aspect ratio are unity. This result is unexpected considering that both the r-value and aspect ratio should be unity for isotropic materials. This may be because the r-value may have a distribution through the thickness of sheet. In this study, the r-value was determined by the tensile test, while the Vickers hardness test was conducted only around the midplane. Additionally, this might be attributed to experimental errors caused by loose guiding in the rolling operation, sectioning of the specimen or tilting of the specimen in hardness test.

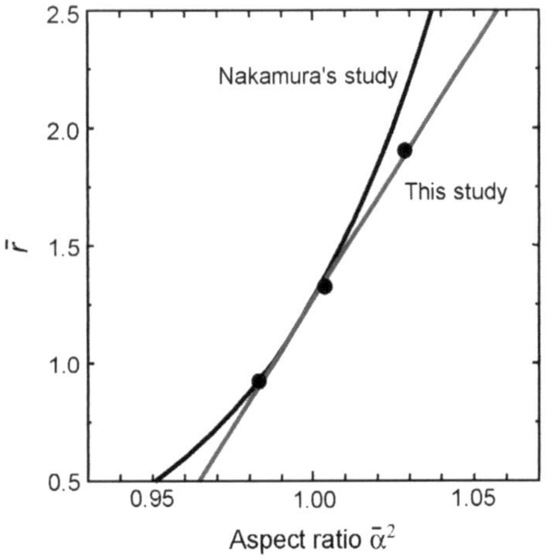

Figure 7. Relationship between planar average r-value obtained by tensile test and planar average aspect ratio obtained by Vickers hardness test.

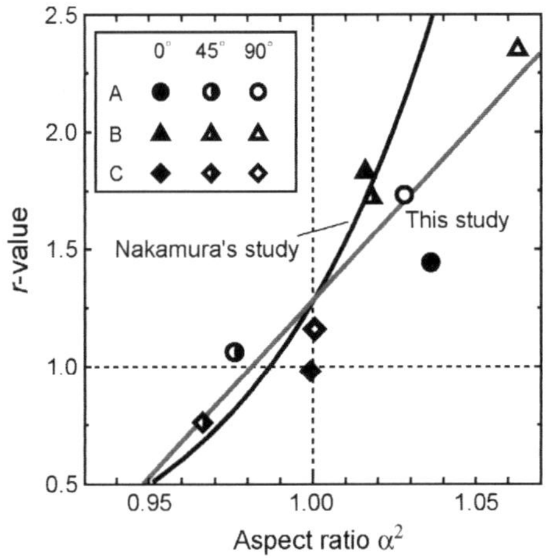

Figure 8. Relationship between r-value obtained by tensile test and aspect ratio obtained by Vickers hardness test.

5. Conclusions

Tensile and indentation tests of three kinds of commercial steel sheets were conducted in directions 0°, 45° and 90° to the rolling direction. A good linear relationship between the r-values determined by the tensile test and the aspect ratio α of the Vickers hardness test is found as $r = 15.1\alpha^2 - 13.8$. The aspect ratio α is defined as the ratio of the two diagonal lengths of the diamond-shaped impression.

By taking advantage of this equation, the r-value of steel can be locally evaluated.

Acknowledgements

This work was supported by "Global COE program: Center of Excellence for Advanced Structural and Functional Materials Design" from MEXT.

References

[1] Lankford W T *et al.* 1950 *Trans. A.S.M.* **42** 1197
[2] Whiteley R L, Wised D E and Bliekwede D J 1961 *Sheet Metal Ind.* **38** 349
[3] Ito K 1984 *Tetsu-to-Hagane* **70** 1881
[4] Lequeu Ph, Jonas J J 1988 *Metall. Trans. A* **19A** 105
[5] Inoue H, Inakazu N 1994 *J. Jpn. Inst. Light Metal* **44** 97
[6] Nakamura M and Tozawa Y 1977 *J. Jpn. Soc. Technol. Plast.* **18** 164
[7] Determination of plastic strain ratio ISO 10113 2006

Effect of columnar grain shape modeling on the microscopic stress distribution at the clad weld metal surface of Ni base alloy

Takahiro Sakai[1], **Yoshiki Mikami**[1] **and Masahito Mochizuki**[1]
[1] Division of Materials Manufacturing Science, Graduate School of Engineering, Osaka University, 2-1 Yamadaoka, Suita, Osaka 565-0871, Japan

E-mail: sakai@mapse.eng.osaka-u.ac.jp

Abstract. Recently, crystal plasticity simulations have been widely used to evaluate microscopic stress. However, the appropriate simulation model has not yet been determined. In this study, the microscopic stresses in the clad weld metal of Ni base alloy were simulated. A series of simulations using columnar grain shape models with different thicknesses indicated that a sufficient thickness is required for the evaluation of microscopic stress at the surface and that the adequate model thickness could be determined by the proposed approach. Additionally, a series of simulations using models with different column axes indicated that the column axis has little effect on the microscopic stresses at the surface of the metal.

1. Introduction

The increase in atmospheric carbon dioxide is a global problem. A large part of human carbon dioxide emissions comes from fossil-fuelled thermal power plants. Nuclear power generation can replace thermal power generation; however, the risk associated with nuclear power plants is a major concern. Recently, intergranular stress corrosion cracking (IGSCC) has been observed in Ni base alloy weld metal in nuclear power plants. The microscopic stress at the surface is an important factor. The stress concentrations near grain boundaries and triple junctions are caused by the different crystal orientations of neighboring grains. Therefore, a numerical simulation that considers the crystal orientations is needed to calculate the microscopic stress. The crystal plasticity finite element method (CP-FEM) is a numerical method that considers crystal orientation and has recently been widely used.

Although it is known that the simulation results are affected by the geometrical model used, the appropriate model for CP-FEM simulations has not yet been determined. To calculate the microscopic stress distribution at the surface of Ni base alloy weld metal, the appropriate model needs to be established. In particular, the grain cross sections and the crystal orientation distribution of a section can easily be measured using the electron backscatter diffraction (EBSD) technique. However, it is difficult to obtain accurate three-dimensional grain shapes. So, the effect of grain shape modeling on the simulation result should be investigated.

Generally, the crystalline structure of weld metal consists of columnar grains, which are slender in shape. In particular, columnar grains constituting clad weld metal are aligned as shown in Figure 1. So, the crystalline structure of clad weld metal can be characterized by the column axis, which is the longitudinal direction of the columnar grain. The column axis corresponds to the grain growth direction, and the grain growth direction depends on the welding conditions and the welding direction.

The purpose of this study is to investigate the effects of the model thickness and the column axis on the microscopic stress distribution at the surface of the clad weld metal of Ni base alloy. In this study, columnar grain shape models with different thicknesses and different column axes were generated based on the crystal orientation distribution measured using the EBSD technique. Then, the microscopic stresses at the surface of the clad weld metal of Ni base alloy were calculated using CP-FEM.

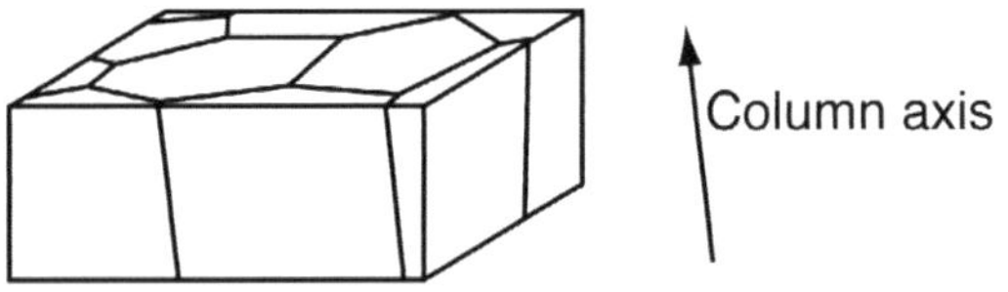

Figure 1. Illustration of the crystalline structure of clad weld metal.

2. Crystal orientation distribution at the surface of the clad weld metal of Ni base alloy

First, a clad weld specimen was fabricated. Two layers of Ni base alloy were deposited on low alloy steel by submerged arc welding using a band electrode. The welding conditions are shown in Table 1. The low alloy steel (SQV2A) was 226 mm in width, 410 mm in length, and 167 mm in thickness. The Alloy 82 electrode was 75 mm in width and 0.4 mm in thickness. The chemical composition of the low alloy steel is shown in Table 2, and the chemical composition of Alloy 82 is shown in Table 3. An overview of the clad weld joint is shown in Figure 2. The X direction was the welding direction, the Y direction was perpendicular to the welding direction, and the $Z = 0$ plane was the surface of the weld metal. A specimen for measuring the crystal orientations was cut from the center of the last pass as shown in Figure 2. Then, the crystal orientation distribution at the surface of the weld metal was measured using the EBSD technique. The inverse pole figure (IPF) map is shown in Figure 3. As seen in the figure, the <100> direction is mainly observed along the direction perpendicular to the surface of the weld metal. Additionally, the average grain diameter (equivalent to the average width of the columnar grains on the surface) was 2.5×10^2 μm.

Table 1. Welding conditions.

Power source	Welding current (A)	Arc voltage (V)	Welding speed (cm/min)	Wire extension (mm)	Preheating and interpass temperature (°C)
DC	1100–1300	26–28	18	40	100–150

Table 2. Chemical composition of the low alloy steel (mass%).

C	Si	Mn	Ni	Fe	Mo	P	S
0.2	0.27	1.47	0.66	Bal.	0.55	0.01	0.008

Table 3. Chemical compositions of Alloy 82 (mass%).

C	Si	Mn	Ni	Fe	Cr	Ti	Cb+Ta
0.012	0.02	3.45	72.56	0.83	19.83	0.61	2.61

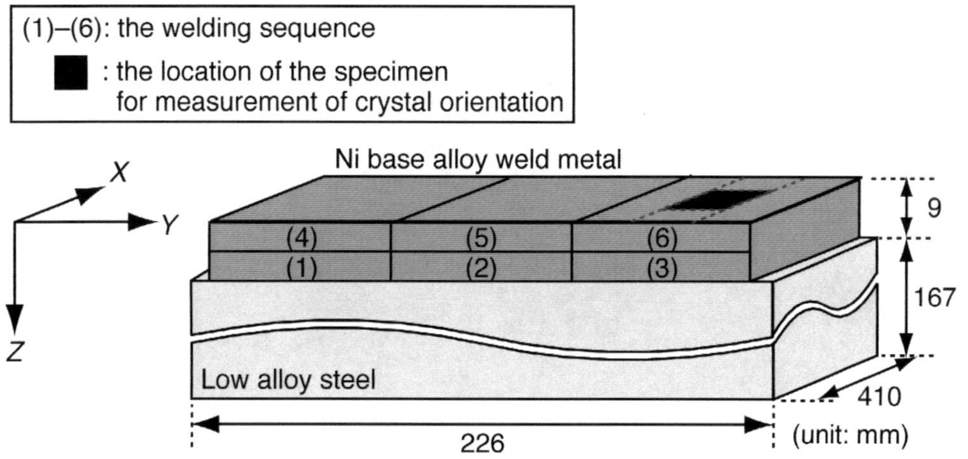

Figure 2. Geometry of the clad weld joint and location of the specimen for measuring the crystal orientation.

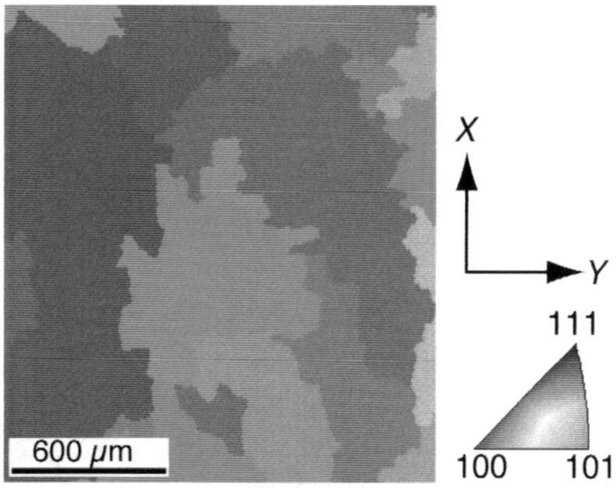

Figure 3. IPF map of the surface of the weld metal.

3. Numerical simulation model

Columnar grain shape models were generated with the same grain cross section on the surface and with different grain shapes in the *Z* direction. The grain cross section on the surface and the crystal orientation of each grain were based on the EBSD results.

In order to investigate the effect of the model thickness on the microscopic stress distribution at the surface of the metal, simulation models with different thicknesses were generated. The column axis was set parallel to the *Z* direction. The thicknesses of the simulation models were varied according to the following seven cases: 15, 75, 150, 450, 750, 1050, and 1500 μm. The simulation models with the different thickness are shown in Figure 4.

In order to investigate the effect of the column axis on the microscopic stress distribution at the surface of the metal, simulation models with different column axes were generated. The thickness of the simulation models was 750 μm. The column axis was defined by two angles: α and β. The angle between the column axis and the *X* direction was defined as α and the angle between the column axis and the *Y* direction was defined as β. These angles were varied according to the following five cases: (α, β) = (90°, 90°), (80°, 90°), (−80°, 90°), (90°, 80°), and (90°, −80°). The simulation models with the different column axes are shown in Figure 5.

A displacement corresponding to 5% macroscopic tensile strain was applied to each model in the *Y* direction. At this strain level, stress in the *Y* direction becomes comparable to the residual stress occurring in clad weld metal. The numerical simulation was performed using the commercial finite element software Abaqus and incorporated a UMAT subroutine to consider the crystal plasticity theory [1, 2]. In each simulation model, we confirmed that the characteristics of the microscopic stress distribution at the surface, for example, the location where stress concentration occurs, do not change qualitatively when the applied macroscopic tensile strain changes.

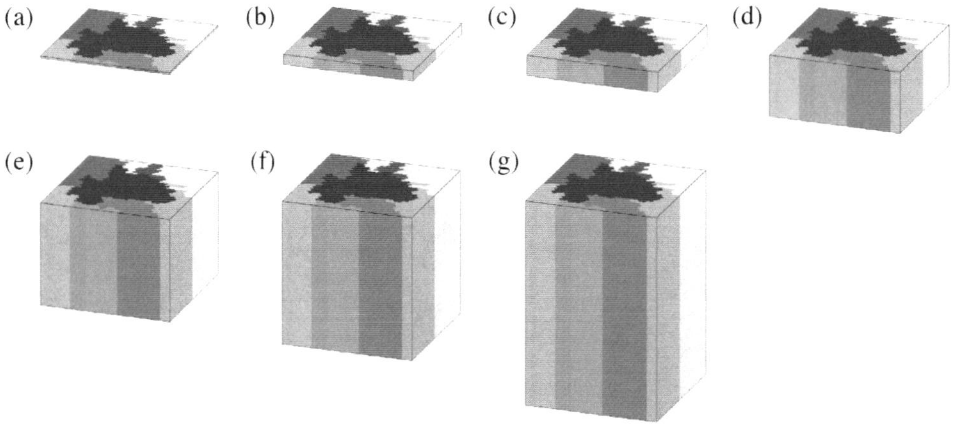

Figure 4. Columnar grain shape models with different thicknesses: (a) 15 μm, (b) 75 μm, (c) 150 μm, (d) 450 μm, (e) 750 μm, (f) 1050 μm, and (g) 1500 μm.

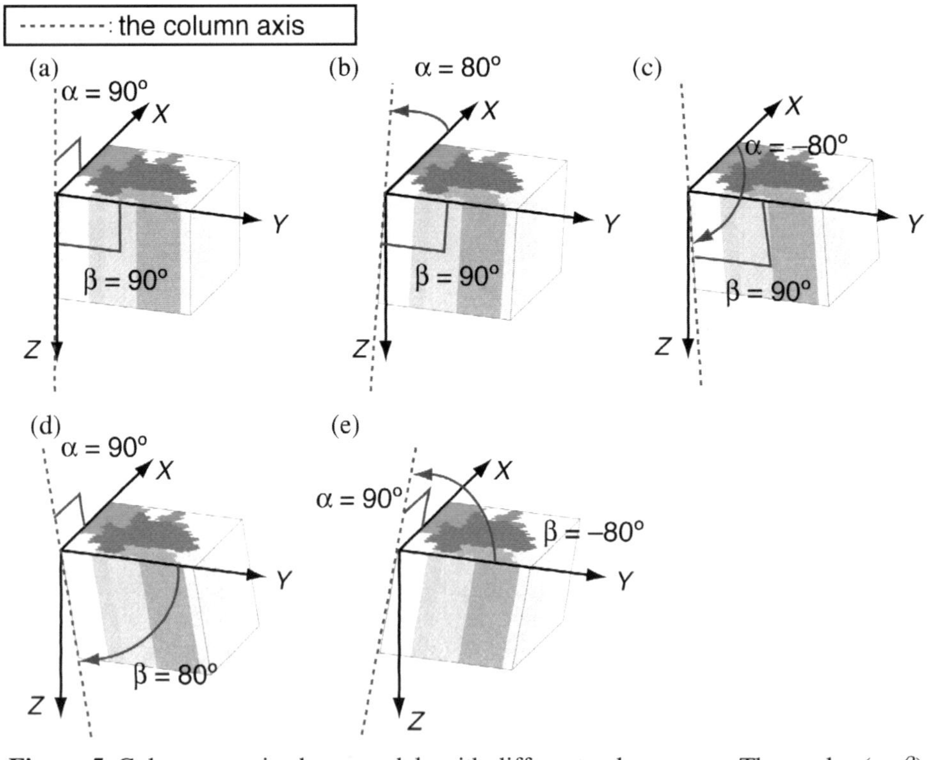

Figure 5. Columnar grain shape models with different column axes. The angles (α, β) are: (a) (90°, 90°), (b) (80°, 90°), (c) (–80°, 90°), (d) (90°, 80°), and (e) (90°, –80°).

4. Effect of grain shape modeling on the microscopic stress distribution at the surface

4.1. Effect of the model thickness on the microscopic stress distribution at the surface
The results of the series of simulations using models with different thicknesses are shown in Figure 6. When the thickness is less than 150 μm, the microscopic stresses near grain boundaries are greatly affected by the thickness of the simulation model. It is known that the surface of a thin plate is under a plane stress state. The more the thickness increases, the more the stress state at the surface deviates from a plane stress state. For a model thickness of less than 150 μm, it is considered that the model thickness greatly affects the microscopic stresses near grain boundaries at the surface due to the deviation from a plane stress state. This indicates that a sufficient thickness is required for the evaluation of IGSCC at the surface. The microscopic stress distribution becomes more similar as the model thickness increases. The threshold is approximately 750 μm, corresponding to three times the average width of the columnar grains.

4.2. Effect of the column axis on the microscopic stress distribution at the surface
The results of the series of simulations using the models with different column axes are shown in Figure 7. Although the column axes are different, the results are similar to each other. This indicates that the column axis has little effect on the microscopic stresses at the surface of the weld metal.

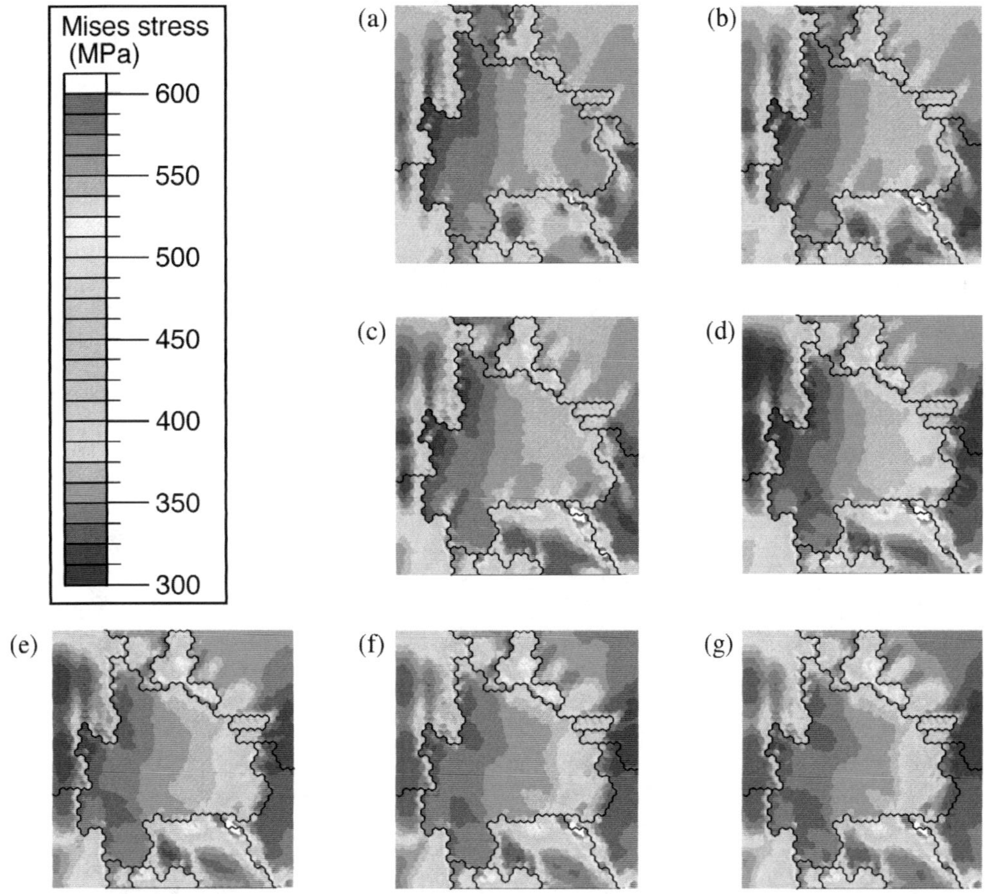

Figure 6. Microscopic stress distributions at the surface of the metal calculated using simulation models with different thicknesses: (a) 15 μm, (b) 75 μm, (c) 150 μm, (d) 450 μm, (e) 750 μm, (f) 1050 μm, and (g) 1500 μm.

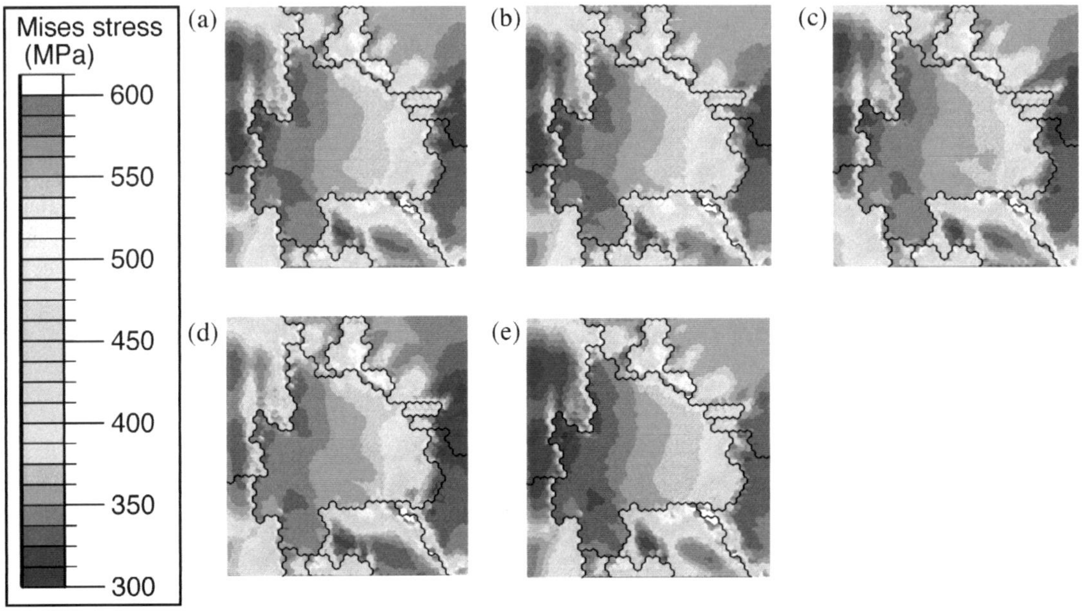

Figure 7. Microscopic stress distributions at the surface of the metal calculated using simulation models with different axes. The angles (α, β) are: (a) (90°, 90°), (b) (80°, 90°), (c) (−80°, 90°), (d) (90°, 80°), and (e) (90°, −80°).

5. Conclusions

In this paper, the microscopic stress distributions were calculated using CP-FEM with columnar grain shape models with different thicknesses and different column axes. Then, the effects of model thickness and column axis on the microscopic stress distribution at the surface of the clad weld metal of Ni base alloy were discussed. The conclusions of this paper are summarized as follows.

(1) When the column axis is perpendicular to the surface of the weld metal, a sufficient thickness is required for the evaluation of microscopic stress at the surface. Furthermore, the adequate model thickness for the simulation of microscopic stresses at the surface of the weld metal could be determined. This suggests that the adequate thicknesses of other grain models can also be determined by the proposed approach.

(2) The column axis had little effect on the microscopic stress distribution at the surface of the weld metal.

Acknowledgements

This study was partly supported by Priority Assistance for the Formation of Worldwide Renowned Centers of Research – The Global COE Program (Project: Center of Excellence for Advanced Structural and Functional Materials Design) from the Ministry of Education, Culture, Sports, Science and Technology (MEXT), Japan.

References

[1] Huang Y 1991 A user-material subroutine incorporating single crystal plasticity in the abaqus finite element program *Mech. Report* 178 (Cambridge, Massachusetts: Division of Applied Sciences, Harvard University)

[2] 2009 *Abaqus Analysis User's Manual Version 6.9* (Providence, RI, USA: Simulia)

Dissimilar joint characteristics of SiC and WC-Co alloy by laser brazing

K Nagatsuka[1], **Y Sechi**[2] **and K Nakata**[3]

[1] Graduate School of Engineering, Osaka University, Joining and Welding Research Institute, 11-1 Mihogaoka, Ibaraki, Osaka 567-0047, Japan
[2] Kagoshima Prefectural Institute of Industrial Technology, 1445-1 Oda, Hayato-cho, Kirishima, Kagoshima 899-5105, Japan
[3] Joining and Welding Research Institute, Osaka University, 11-1 Mihogaoka, Ibaraki, Osaka 567-0047, Japan

E-mail: nagatuka@jwri.osaka-u.ac.jp

Abstract. SiC and WC-Co alloys were joined by laser brazing with an active braze metal. The braze metal based on eutectic Ag-Cu alloy with additional Ti as an active element ranging from 0 to 2.8 mass% was sandwiched by the SiC block and WC-Co alloy plate. The brazing was carried out by selective laser beam irradiation on the WC-Co alloy plate. The content of Ti in the braze metal was required to exceed 0.6 mass% in order to form a brazed joint with a measurable shear strength. The shear strength increased with increasing Ti content up to 2.3 mass%Ti and decreased with a higher content.

1. Introduction

Brazing is suitable for joining dissimilar materials, particularly those that cannot be easily joined by fusion welding, such as ceramics and metals [1-12]. Regarding ceramic materials, silicon carbide (SiC) has several functional characteristics, such as a low thermal expansion coefficient, high thermal conductivity, good high-temperature mechanical properties, superior wear, corrosion, heat, and oxidation resistance, and semiconducting properties [10-15]. However, it is difficult to braze SiC to other materials without the use of an active braze metal or some kind of surface treatment because the wettability of the SiC and the molten braze metal plays a crucial role in the brazing of these materials [7-11, 16, 17]. The technique for the dissimilar brazing of SiC using an active braze metal is applicable to construction materials, cutting tools, ceramic heat exchangers, optical devices, and power devices [10, 11, 14, 18]. Much research on the dissimilar brazing of SiC to other materials has been carried out in recent studies, such as the brazing of SiC to Ni alloy [12, 19, 20], Nb, W [15], TiAl [21], Fe alloy [20], and graphite [2] and the brazing of carbon fiber reinforced SiC to Ti alloy [3, 22]. The behavior of an active element and a molten braze metal in the conventional furnace heating process is well known [13, 16-18, 20, 23-26].

The active element (e.g., Ti, Zr, or Cr) in the braze metal improves the wettability of the SiC and molten braze metal and reacts with the Si and C of the SiC to form stable compounds [13, 16, 18, 20]. The wettability improves with increasing active element content in the braze metal, and the increased wettability in turn increases the joint strength [5, 8, 18, 24]. However, several problems have remained in conventional brazing, such as material deterioration and thermal stress and strain in the joint [2-11, 15, 18, 22-24, 27]. Especially, depending on the brazing temperature and time, an increase in the active element content can cause a decrease in the joint strength regardless of the improvement in the wettability. This decrease in joint strength occurs because of the high thermal expansion coefficient mismatch between the SiC and the thick reaction layer [4-6, 10, 17, 18, 23]. These problems arise because conventional furnace brazing requires a long treatment time in order to ensure heating and cooling of the whole component.

Such problems can potentially be avoided in laser brazing, which is a novel brazing technique, as the heating and cooling times in this technique are short, and only the selected part of the component needs to be heated [7, 8, 11]. In addition, it contributes to a decrease in the energy consumption of the brazing process. The behaviors of the active element and molten braze metal during laser brazing differ from those during conventional brazing because of the short processing time and smaller selected heating area [7, 8]. In particular, the short processing time affects the diffusion and interfacial reaction processes at the interface of the SiC and the braze metal [3, 10]. However, little information has been reported regarding the effect of active element contents in the dissimilar laser brazing of ceramics and metals [7, 8]. Based on this background, the objective of this study was to determine the effect of Ti as an active element in the braze metal of a dissimilar laser-brazed joint of SiC to WC-Co alloy on the joint strength and the interface structure of the joint; WC-Co alloy was selected as a counter material owing to its low thermal coefficient and relatively high heat resistance.

2. Material and methods

Experiments were carried out using SiC blocks, WC-Co alloy plates, and nine types of Ag-Cu-Ti braze metal sheets with different Ti contents. The SiC block was classified as recrystallized silicon carbide (>99% SiC, porosity: 17%, NGK INSULATORS, LTD.) and had dimensions of 5 × 5 × 3.5 mm. The WC-Co alloy plate (ISO K10 grade, Mitsubishi Materials Corporation) was used as the substrate plate and had dimensions of 10 × 10 × 2 mm. Table 1 lists the chemical compositions of the braze metals. The braze metals were eutectic-type Ag-Cu alloy containing additional Ti as an active element, with the Ti content ranging from 0 to 2.8 mass%. The thickness of the braze metal was 0.1 mm, and its dimensions were fixed at 3 × 3 mm to ensure that it does not flow out of the joint interface. Before brazing, all materials were degreased by ultrasonic agitation in acetone for 10 min and dried in air.

Table 1. Chemical composition of the braze metals (mass%).

No.	Ag	Cu	Ti
1	72.0	28.0	0
2	71.5	28.2	0.3
3	71.5	28.1	0.4
4	71.3	28.1	0.6
5	71.2	27.9	0.9
6	70.9	27.8	1.3
7	70.2	28.1	1.7
8	70.2	27.5	2.3
9	69.6	27.6	2.8

Figure 1 shows a schematic illustration of the laser brazing apparatus. The specimen was shaped like a "top hat." A braze metal sheet was sandwiched between the SiC block (above it) and the WC-Co alloy plate (below it) and placed in a vacuum chamber. The top of the specimen was covered with a transparent quartz glass plate, which not only acted as a window of the vacuum chamber for laser beam irradiation but also fixed the specimen in place. The vacuum chamber was evacuated to less than 10^{-1} Pa, and then, Ar gas of 99.999% purity was pumped in to atmospheric pressure. This evacuation and substitution cycle was performed five times prior to brazing. During brazing, Ar gas continued to flow at a rate of 5 L/min. The YAG and LD lasers were coaxially transferred via optical fibers to the laser head unit and irradiated through the transparent quartz glass plate to the top side of the WC-Co alloy plate at an irradiation angle of 85°. Table 2 lists the laser brazing conditions, which were set according to JIS Z3261 BA-8 and previous studies [7, 8]. The WC-Co alloy plate around the SiC was subjected to laser irradiation for 36 s. During laser heating, the temperature of the WC-Co alloy plate was monitored by an R-type thermocouple inserted into a hole beneath the joint on the lower surface of the plate.

After brazing, a shear strength test of the joint was carried out using a precision universal tester operated at a cross-head speed of 0.5 mm/min. Five joints were tested for each of the nine types of braze metals shown in Table 1. Figure 2 shows a schematic illustration of the shear strength test. The WC-Co alloy plate was clamped, and the SiC was loaded by the steel jig. The shear strength was calculated as the maximum load divided by the joining area, which was estimated from the fractured surface. The brazing joint specimens were cross-sectioned and mounted in epoxy resin before being ground with #220 SiC paper and polished with diamond paste. Observations of the macrostructure and microstructure of the brazed interface and its elemental analysis were then performed using a scanning electron microscope (SEM), a transmission electron microscope (TEM), and an energy dispersive X-ray analyzer.

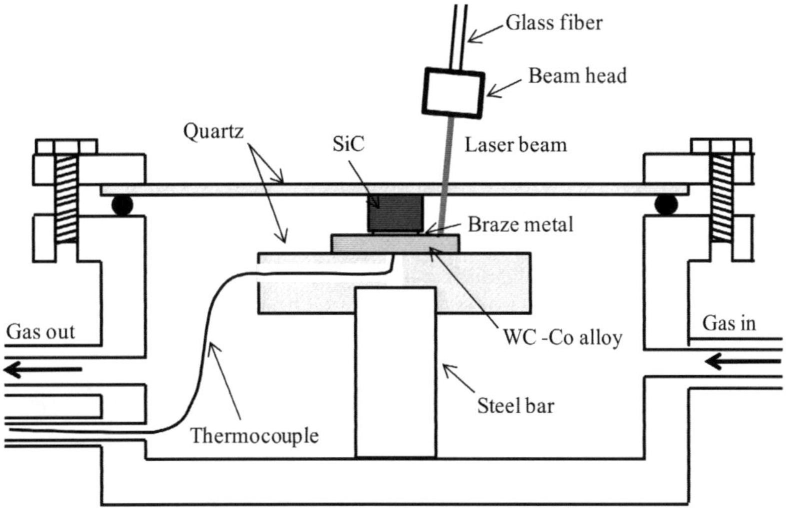

Figure 1. Schematic illustration of the laser brazing apparatus.

Table 2. Laser brazing conditions.

Pulsed YAG average output (W)	Pulsed YAG wave length (nm)	CW LD output (W)	CW LD wave length (nm)	Pulse frequency (Hz)	Scanning time (s)	Atmosphere
134	1064	20	808	100	36	Ar flow (5L/min)

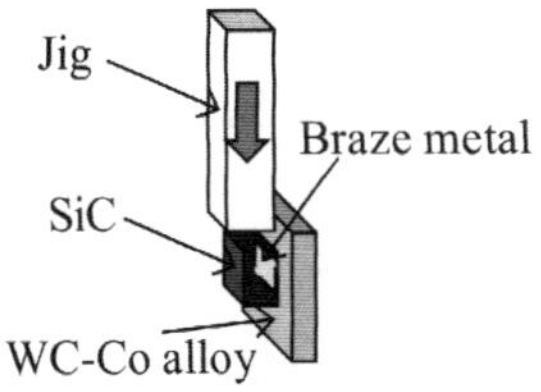

Figure 2. Schematic illustration of shear strength test.

3. Results and discussion

3.1. Strength of the brazed joint

Figure 3 shows a typical temperature profile of the WC-Co plate during laser brazing. The highest brazing temperature was 1102 K, which is above the melting temperature of the braze metal (approximately 1073 K). Macro cracking could not be observed in any of the brazed joints.

Figure 4 shows the effect of the Ti content on the shear strength of the brazed joint. The content of Ti in the braze metal was required to exceed 0.6 mass% in order to form a brazed joint with a measurable shear strength. Using the braze metal containing 0.4 mass%Ti, the SiC block and WC-Co alloy plate were joined; however, the shear strength was too low to measure. The shear strength increased as the Ti content increased up to 2.3 mass%. Figures 5 (a) and (b) show the fractured surface of the joint on the WC-Co alloy plate after the shear strength test using the braze metals containing 0.6 and 2.3 mass%Ti, respectively. The fracture occurred at the SiC block near the interface of the SiC block and the braze metal with 0.6 mass%Ti and almost at the SiC block with 2.3 mass%Ti. Fracture between the WC-Co alloy plate and the braze metal was not observed in any case. Figure 6 shows the effect of the Ti content on the area fraction of the fractured surface of the SiC block on the braze metal. The area fraction increased with increasing Ti content in the braze metal in a similar tendency as observed in Figure 4 for the range from 0.6 to 2.3 mass%Ti. Figure 7 shows the relationship between the area fraction of the fractured surface of the SiC block on the braze metal and the shear strength. This result indicates that the shear strength increases with increasing the area fraction. These mean that Ti as an active element increased the joint strength of the interface between the SiC block and the braze metal. The increase in shear strength with increasing the Ti content in the braze metal is possibly a result of the improvement in the wettability of the SiC and molten braze metal and the formation of stable intermetallic compounds when the Ti reacts with the Si and C in the SiC [13, 16, 18, 20]. The shear strength decreased from 2.3 to 2.8 mass%Ti, the reason for which is discussed later in Section 3.2: *Microstructure of the brazed joint*. The wide distribution of the shear strength of the joints can be explained by the wide distribution of the strength of the SiC block itself. The fracture strength of the SiC block was determined by the dimension of the pores and cracks in the sintered compact because SiC is a porous ceramic.

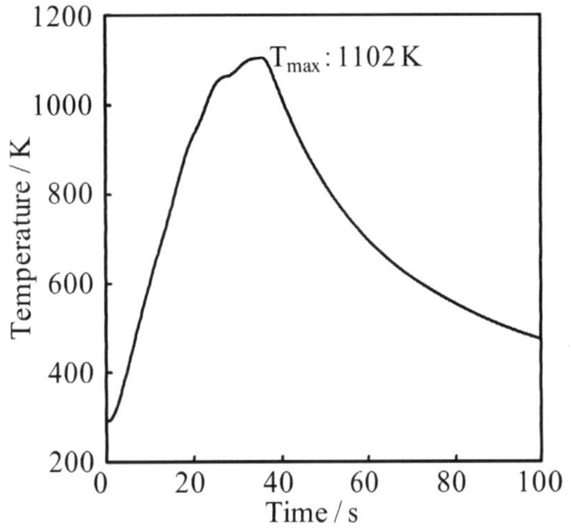

Figure 3. Typical temperature profile of the WC-Co plate during laser brazing.

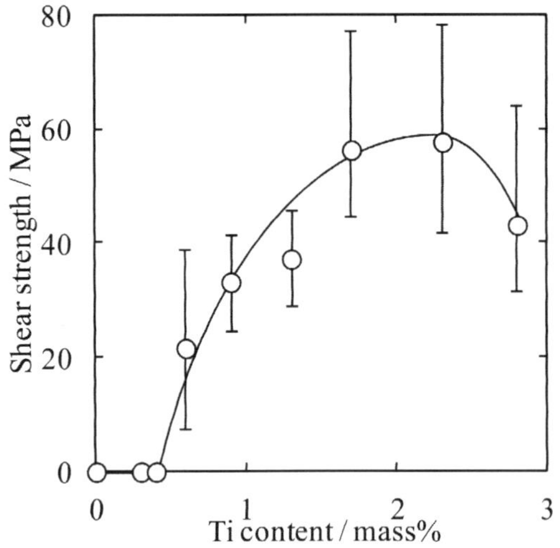

Figure 4. Effect of the Ti content on the shear strength of the brazed joint.

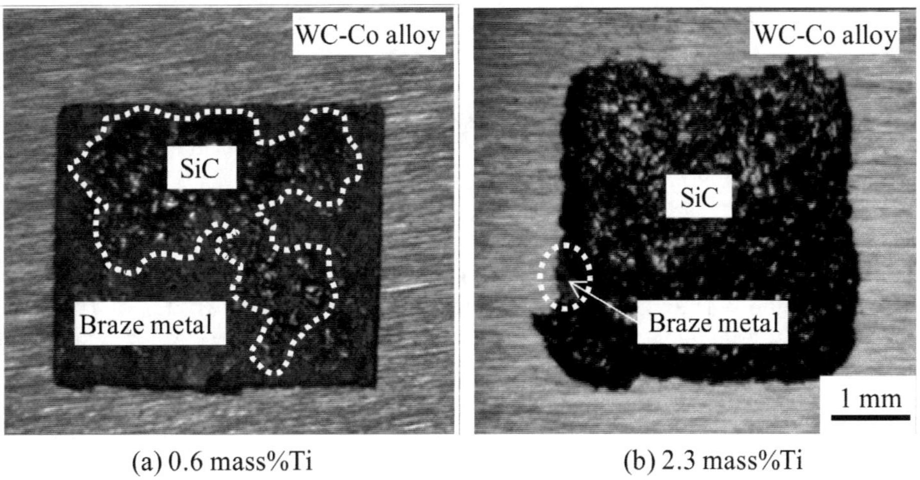

Figure 5. Fractured surface of the joint on the WC-Co alloy plate after shear strength test using the braze metals containing 0.6 and 2.3 mass%Ti.

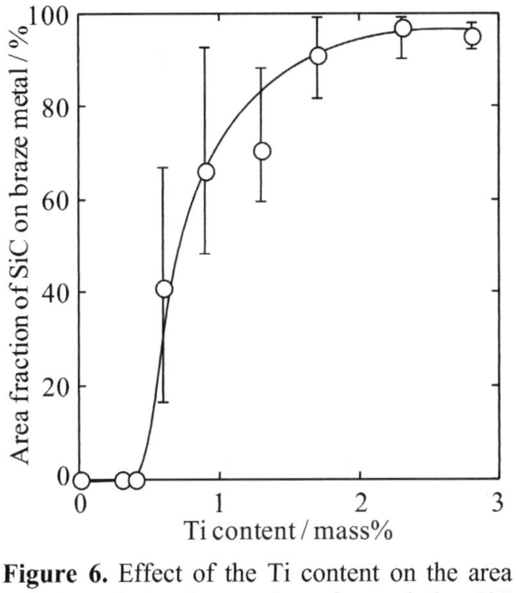

Figure 6. Effect of the Ti content on the area fraction of the fractured surface of the SiC block on the braze metal.

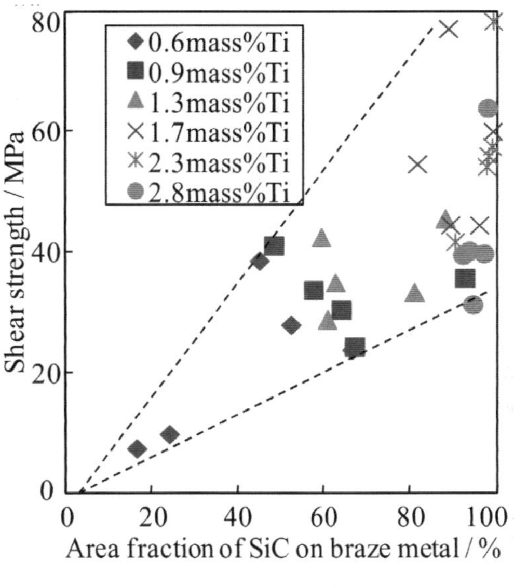

Figure 7. Relationship between the area fraction of the fractured surface of the SiC block on the braze metal and shear strength.

3.2. Microstructure of the brazed joint

Figures 8 (a) and (b) show the cross-sectional macrostructure of the joints using the braze metals containing 0.4 and 2.3 mass%Ti, respectively. Using the 2.3 mass%Ti braze metal, it was observed that the braze metal infiltrated into the void of the SiC blocks; however, this was not the case for the 0.4 mass%Ti. This showed that a sufficient amount of Ti as an active element in the braze metal improved the wettability of the braze metal. Figure 9 shows the cross-sectional microstructure and the element distributions of Si, Ti, Cu, and Ag at the brazed joints of SiC and the braze metal containing 0.4, 2.3, and 2.8 mass%Ti. Carbon could not be identified using the energy dispersive X-ray analyzer. An obvious concentration of Ti was observed at the interface between the SiC and each braze metal. This suggests that the Ti as an active element diffused from the braze metal to the interface so as to decrease the interfacial energy and reacted with the SiC during laser heating. The formation of Ti concentrated layers at the interface became more obvious with increasing Ti content in the braze metal, which resulted in an increase in the thickness and area of the Ti concentrated layer.

Figure 10 shows the TEM bright field image and element distributions of Si, Ti, Cu, and Ag at the interface between the SiC block and the braze metal containing 2.3 mass%Ti. The microstructure at the interface was divided into five areas depending on the dominant constitutional elements at each position, which are marked in the bright field image as follows: Si at position 1 (in SiC), Ti at position 2, Ti and Si at position 3, Ti and Cu at position 4, and Ag at position 5 in the braze metal. Figures 11 (a)-(d) show the selected area diffraction patterns of the areas corresponding to positions 2, 3, 4, and 5, respectively. These results reveal that the precipitated intermetallic compounds at the interface between the SiC and the braze metal were identified as TiC at position 2, Ti_5Si_3 at position 3, and Cu_4Ti at position 4. Ag was found to be the braze metal at position 5. These results indicated that the Ti concentrated layer formed at the interface consisted of multiple thin layers of TiC, Ti_5Si_3, and Cu_4Ti. The Ti in the braze metal has a high affinity for C and Si in SiC, and it would thus diffuse to the interface and form thin intermetallic compound layers. In addition, Cu as the major alloying element of the braze metal also has a high affinity for Ti, and thus, a Cu_4Ti compound was formed following the formation of TiC and Ti_5Si_3.

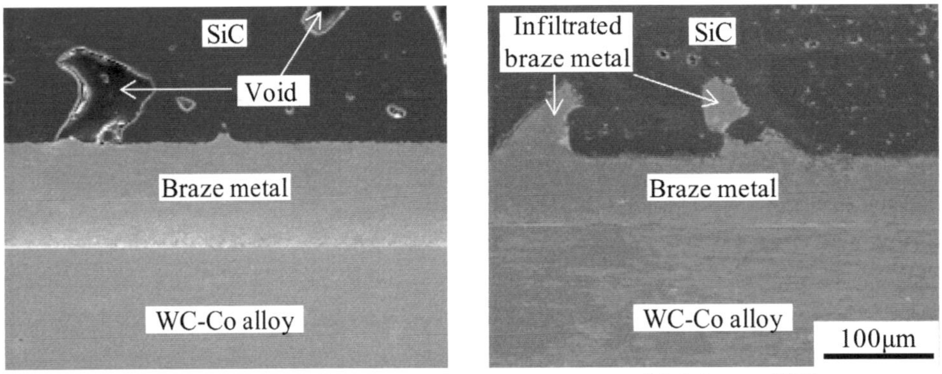

(a) 0.4 mass%Ti (b) 2.3 mass%Ti

Figure 8. Cross-sectional macrostructure of joints using the braze metals containing 0.4 and 2.3 mass%Ti.

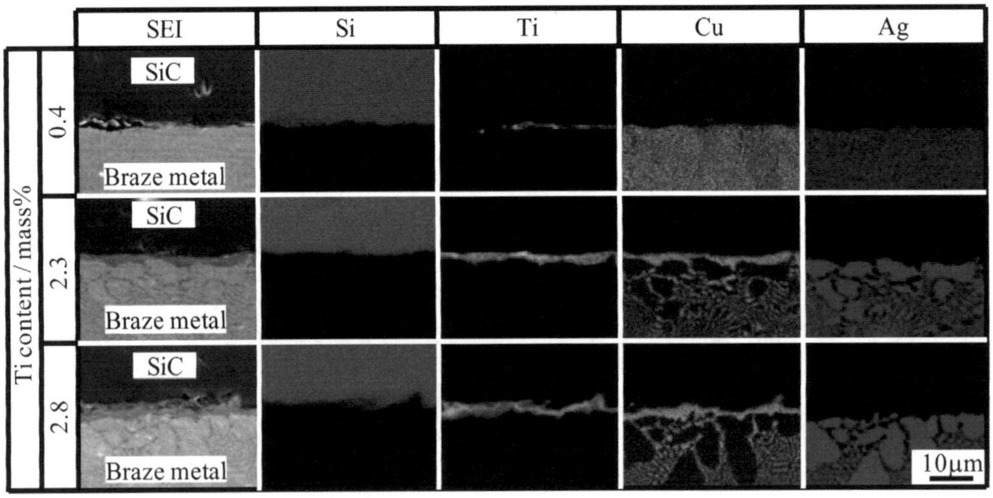

Figure 9. Cross-sectional microstructure and the element distributions of Si, Ti, Cu, and Ag at the brazed joints of SiC and the braze metals containing 0.4, 2.3, and 2.8 mass%Ti.

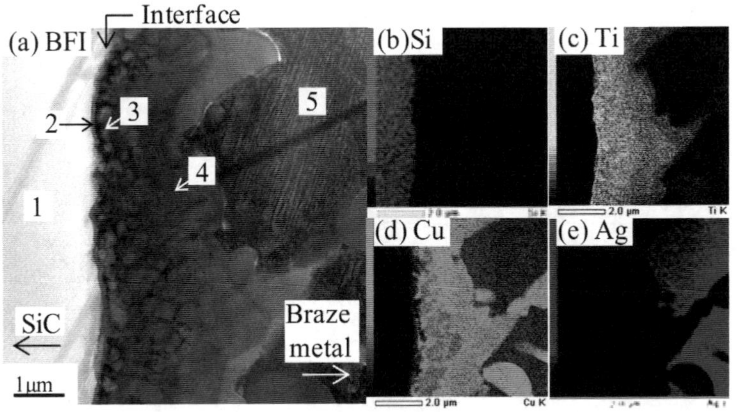

Figure 10. TEM bright field image and element distribution at the joint interface between the SiC block and the braze metal containing 2.3 mass%Ti.

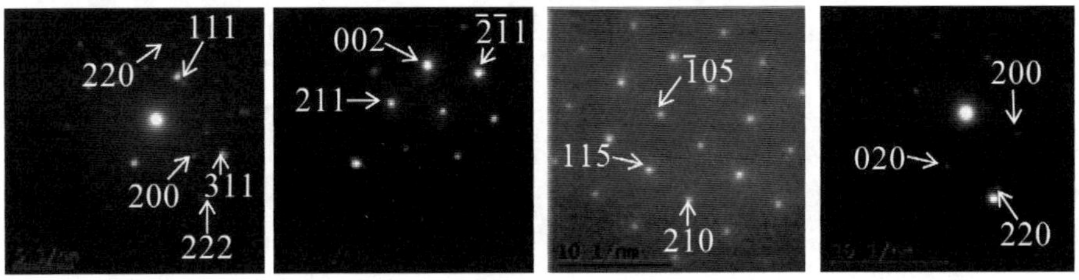

(a) TiC (b) Ti_5Si_3 (c) Cu_4Ti (d) Ag

Figure 11. Selected area diffraction patterns at (a) position 2, (b) position 3, (c) position 4, and (d) position 5 in the TEM bright field image in Figure 10.

Figures 12 (a)-(c) show the TEM bright field image at the joint interface between the SiC block and the braze metals containing 0.4, 2.3, and 2.8 mass%Ti, respectively. TiC, Ti_5Si_3, and Cu_4Ti were observed using the braze metals containing 0.4, 2.3, and 2.8 mass%Ti. The thickness of the compound layers increased with increasing Ti content in the braze metal. Table 3 lists the coefficients of thermal expansion of SiC, WC-Co alloy, and the reaction compounds [27, 28]. The coefficient of thermal expansion of SiC is similar to that of WC-Co alloy and different from that of the reaction compounds (TiC and Ti_5Si_3). Residual stresses in the SiC and reaction layers would occur in all of the joints [4-6, 10, 17, 18, 23]. Thus, it is considered that the decrease in the shear strength of the joint using the braze metal containing 2.8 mass%Ti would be caused by the high residual stress because of the formation of the thick reaction layers.

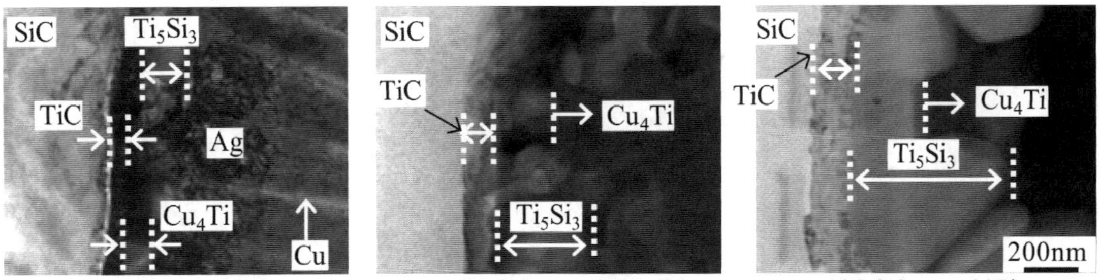

(a) 0.4 mass%Ti (b) 2.3 mass%Ti (c) 2.8 mass%Ti

Figure 12. TEM bright field image at the joint interface between the SiC block and the braze metals containing 0.4, 2.3, and 2.8 mass%Ti.

Table 3. Coefficient of thermal expansion (CTE) of SiC, WC-Co alloy, and reaction compounds.

	Coefficient of thermal expansion ($\times 10^{-6}$ K^{-1})
SiC	4.7
WC-Co alloy	5.4
TiC	7.4-8.6
Ti_5Si_3	11.0

4. Conclusions

The effects of Ti as an active element in the Ag-Cu-Ti braze metal of a dissimilar laser-brazed joint of SiC to WC-Co alloy on the joint strength and the interface structure of the joint were investigated. The following conclusions were drawn about the shear strength and the interface structure of the joint:

(1) Ti serving as an active element in the Ag-Cu braze metal enabled the joining of SiC to WC-Co alloy by heating for a short time using laser brazing. The content of Ti in the braze metal was required to exceed 0.6 mass% in order to form a brazed joint with a measurable shear strength.

(2) The shear strength of the laser-brazed joint increased with increasing Ti content in the range of 0.4 to 2.3 mass% in proportion to the area fraction of the fractured surface in the SiC, and a higher Ti content decreased the shear strength of the joint.

(3) A Ti concentrated layer was formed at the interface between the SiC and the braze metal using laser brazing. This layer consisted of TiC, Ti_5Si_3, and Cu_4Ti compound layers, and the thickness of these compound layers increased with increasing Ti content in the braze metal.

References

[1] Li J, Liu L, Wu Y, Zhang W and Hu W 2008 *Mater. Lett.* **62** 3135-8
[2] Mao Y, Li S and Yan L 2008 *Mater. Sci. Eng. A* **491** 304-8
[3] Xiong J H, Huang J H,Zhang H and Zhao X K 2010 *Mater. Sci. Eng. A* **527** 1096-101
[4] Zhang J, He Y M, Sun Y and Liu C F 2010 *Ceram. Int.* **36** 1397-404
[5] Wang Y, Qiu X M, Sun D Q and Yin S Q 2011 *Int. J. Refract. Metal. Hard Mater.* **29** 293-7
[6] Liu C F, Zhang J, Zhou Y, Meng Q C and Naka M 2008 *Mater. Sci. Eng. A* **491** 483-7
[7] Sechi Y, Tsumura T and Nakata K 2010 *Mater. Des.* **31** 2071-7
[8] Sechi Y, Takezaki A, Matsumoto T, Tsumura T and Nakata K 2009 *Mater. Trans.* **50** 1294-9
[9] Shibayanagi T 2010 *J. Jpn. Weld. Soc.* **79** 653-9
[10] Liu Y, Huang Z R and Liu X J 2009 *Ceram. Int.* **35** 3479-84
[11] Sudmeyer I, Hettesheimer T and Rohde M 2010 *Ceram. Int.* **36** 1083-90
[12] Xiong H, Mao W, Xie Y, Guo W, Li X and Cheng Y 2007 *Mater. Lett.* **61** 4662-5
[13] Lee H and Lee J 1992 *J. Mater. Sci. Lett.* **11** 550-3
[14] Chin H, Cheong K and Ismail A 2010 *Met. Mat. Trans. B* **41B** 824-31
[15] Naka M, Taniguchi H and Okamoto I 1990 *Trans. JWRI* **19** 189-94
[16] Nomura M, Ichimori T, Iwamoto C and Tanaka S 2000 *J. Mater. Sci.* **35** 3953-8
[17] Akselsen O M 1992 *J. Mater. Sci.* **27** 1989-2000
[18] Okamura H, Sakamoto M and Shida T 1990 *Quarterly J. Jpn. Weld. Soc.* **8** 272-9
[19] Hattali M, Valette S, Stremsdoerfer G, Treheux D, Ropital F and Mesrati N 2009 *J. Eur. Ceram. Soc.* **29** 813-9
[20] Prakash P, Mohandas T and Raju P D 2005 *Scr. Mater.* **52** 1169-73
[21] Liu H, Feng J and Qian Y 2000 *J. Mater. Sci. Lett.* **19** 1241-2
[22] Xiong J H, Huang J H, Wang Z P, Ban Y H, Zhang H and Zhao X K 2010 *Mater. Sci. Technol.* **26** 356-60
[23] Liu G, Balenza F, Muolo M, Passerone A and Liu G 2010 *J. Mater. Sci.* **45** 4299-307
[24] Liu G M, Zou G S, Wu A P and Zhang D K 2006 *Mater. Sci. Eng. A* **415** 213-8
[25] Lee H and Lee J A 1996 *J. Mater. Sci.* **31** 4133-40
[26] Peteves S D, Ceccone G, Paulasto M, Stamos V and Yvon P 1996 *JOM* **48-1** 48-52

[27] He T, Zhang J and Sun Y 2010 *Mater. Sci. Eng. A* **527** 2819-25
[28] Costa A K and Camargo Jr S S 2003 *Mat. Res.* **6** 39-42

Dissimilar laser brazing of h-BN and WC-Co alloy in Ar atmosphere without evacuation process

Y Sechi[1] , K Nagatsuka[2] and K Nakata[3]
[1] Kagoshima Prefectural Institute of Industrial Technology, 1445-1 Oda, Hayato-cho, Kirishima, Kagoshima 899-5105, Japan
[2] Graduate School of Engineering, Osaka University, 2-1 Yamadaoka, Suita, Osaka 565-0871, Japan
[3] Joining and Welding Research Institute, Osaka University, 11-1 Mihogaoka, Ibaraki, Osaka 567-0067, Japan

E-mail: sechi@kagoshima-it.go.jp

Abstract. Laser brazing with Ti as an active element in Ag-Cu alloy braze metal has been successfully applied to dissimilar joining of h-BN and WC-Co alloy in Ar (99.999% purity) gas flow atmosphere without any evacuation process. Good wettability of the braze metal with h-BN and WC-Co alloy were confirmed by the observation and structural analysis of the interface by electron probe micro-analysis and scanning acoustic microscopy. The oxidation of titanium was not observed and this showed that the laser brazing with titanium as an active element in braze metal could be performed even in an Ar gas flow atmosphere without an evacuation process using a high-vacuum furnace.

1. Introduction

The brazing process is used in many industrial fields for developing engineering structures and electronic devices [1-13]. This process has several advantages making it suitable for use in joining dissimilar materials and materials which are difficult to connect precisely; further, this process can be efficiently applied to the mass production of structures and devices. For the application of ceramic materials, it is frequently required that a ceramic material be bonded to metals. Moreover, the development of high-functionality products in recent years has led to a demand for a new joining process that can join different materials. However, certain problems are encountered while meeting the above requirements, such as the formation of joint defects due to thermal stress in the joint field and material deterioration due to heating in the brazing process.

Boron nitride has various functional characteristics. Especially, h-BN has a good thermal resistibility and solid-lubrication [14]. Meanwhile, it is difficult to braze h-BN to another material because of its low wettability. As a result, it is difficult to braze hexagonal boron nitride ceramics to other materials. Because wettability of a material affects joining characteristics, it is speculated that brazing of hexagonal boron nitride ceramics may be suitable to demonstrate laser brazing of dissimilar materials such as ceramics and a metal.

In general, the thermal expansion coefficient of a metal is higher than that of ceramics; this may result in the generation of a large thermal stress at the brazing joint, which leads to the formation of defects at the interface [15]. Tungsten carbide / cobalt alloy made by powder metallurgy has a low thermal expansion coefficient and high rigidity, which is a good complement as a structural material to ceramics. However, only a few studies have focused on the joining of boron nitride and tungsten carbide alloy or other metals [5, 8, 10] by brazing in a furnace that needs a long heating time.

Among brazing processes [16], laser brazing [10, 11, 13, 17-30] has good characteristics for a dissimilar joining process because of the possibility of short heating time and small heating area, and suppression of damage to the base materials without a furnace, in comparison with conventional furnace brazing.

Recently, laser brazing with the addition of Ti as an active element in Ag-Cu alloy braze metal has been successfully applied to dissimilar joining of h-BN and WC-Co alloy after the evacuation on the order of 10^{-1} Pa and Ar (99.999% purity) substitution to avoid oxidation of the active element of Ti [12], but a more simplified method without evacuation is needed for the practical use of dissimilar laser brazing. So this study describes the dissimilar laser brazing of h-BN and WC-Co alloy without an evacuation process and the characteristics of a dissimilar joint, and in order to investigate the characteristics of the dissimilar joint, cross-sectional observation, elemental and structural analyses of the joint interface, adhesion evaluation, and shear strength measurement were performed.

2. Experimental procedures

Commercially available tungsten carbide equivalent material classified with ISO K10 grade and high-purity h-BN made by the hot pressing method without using sintering additives were used in this work. Silver-copper-titanium alloy braze metal sheet included Ti as a major active ingredient for direct ceramic brazing with a thickness of 0.1 mm. Nominal compositions and properties are summarized in Tables 1 and 2. The size of the braze metal sheet was determined to cover 80% of the joint area of h-BN so that the melted braze metal would not flow out of the joint. Before brazing, the h-BN block, braze metal and tungsten carbide plate were degreased by ultrasonic agitation for 10 min in acetone and dried in air. Sample configuration was a top hat shape. A braze metal sheet was sandwiched with an h-BN block from the top and a tungsten carbide plate from the bottom in a vacuum chamber. A low pressure of 1.2 MPa was applied to prevent the workpiece from moving when the braze metal sheet was melted, and no gap adjustments for the joint were performed. A chamber with a 100 mm diameter was used in the brazing experiments, which had inner volume 145 ml. To avoid the oxidation of Ti, 99.999% purity Ar was flowed during laser brazing at a rate of 10 L/min without evacuation of the chamber.

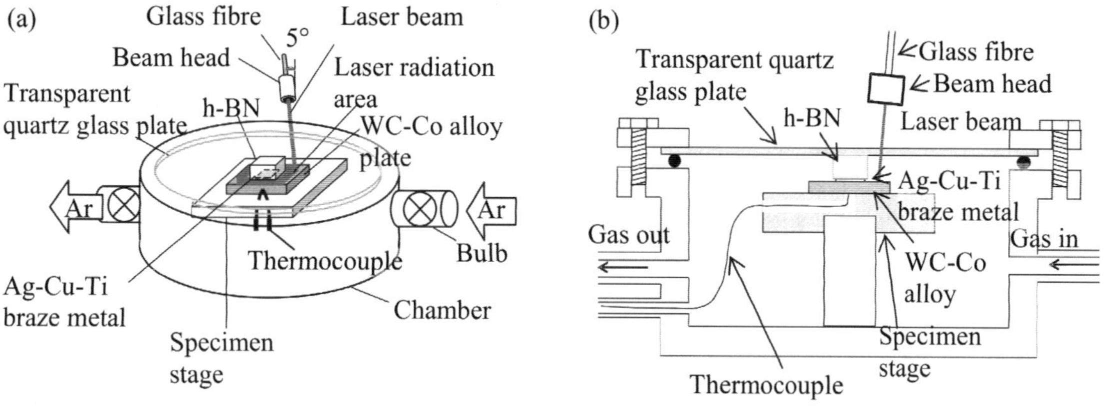

(a) perspective illustration (b) cross-sectional illustration

Figure 1. Schematic diagram of laser brazing.

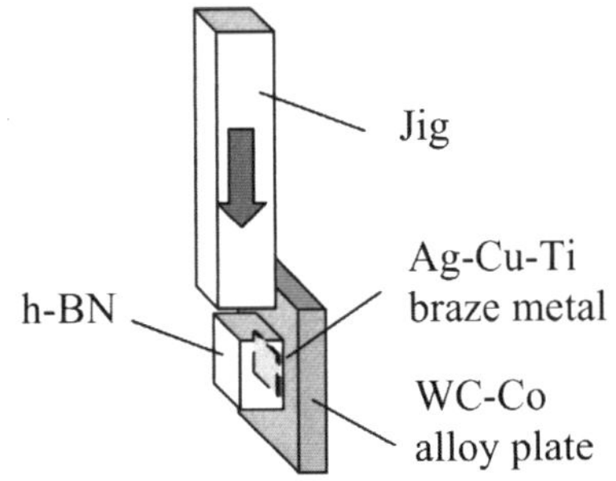

Figure 2. Schematic diagram of shear strength test.

Table 1. Materials used in this work.

Material	Nominal composition (mass %)	Density ($\times 10^3$ kg / m^3)	Relative Density (%)	Size (mm)
Tungsten Carbide	WC: 94, Co: 6	14.9	-	10*10*2
h-BN	h-BN > 99.993	1.93	82.5	5*5*3.5

Table 2. Chemical composition of Ag-Cu-Ti alloy braze metal.

Elements (mass %)		
Ag	Cu	Ti
70.2	bal.	1.7

Table 3. Laser heating condition.

Pulsed YAG Average Output (kW)		0.134
Pulsed YAG wave length (nm)		1064
CW LD Output (kW)		0.02
CW LD wave length (nm)		808
Pulse frequency (Hz)		100
Scanning speed (mm/s)	(1st run)	0.6
	(2nd run)	1.0
	(3rd run)	1.0
	(4th run)	1.0
Laser beam diameter (mm)		0.5

The setup of laser brazing was as shown in Figure 1. In (a), a perspective illustration is shown, and in (b), a cross-sectional illustration is shown to describe the schematic diagram of laser brazing. The sample was located in a small vacuum chamber. The topside of the specimen was covered with a transparent quartz glass plate, which also acted to fix the specimen. The generated YAG and LD lasers were transferred with an optical fibre to the laser head unit and radiated through the transparent quartz glass plate to the topside of the tungsten carbide plate, with a radiation angle of 85°. Laser scanning of the tungsten carbide substrate was performed around the h-BN block. The laser brazing condition is summarized in Table 3. The scanning speed of laser was 0.6 mm/s on the first run and 1.0 mm/s on the second run to the fourth run in order to heat the specimen efficiently in a circular manner. In this study, the laser scanning was performed on the tungsten carbide substrate around the h-BN block. Temperature during laser brazing was monitored at the bottom of the tungsten carbide plate using a thermocouple. Some of the samples were subsequently cross sectioned by a low-speed diamond saw with water cooling and mounted by epoxy resin, curing at room temperature for about 8-10 h, ground using SiC paper #120-#1200, and polished by 3 to 1 μm polycrystalline diamond to provide microstructural information. Cross-sectional observation and elemental analysis of the interface were performed using an electron probe micro-analyzer (JEOL Co. Ltd. JXA-8230) and an X-ray diffractometer (Bruker AXS Co. Ltd. D8: Co Kα). The diameter of the collimator was 50 μm, to determine crystallographic phase of the micro-region. Interfacial observation and estimation of the interface area were performed using a scanning acoustic microscope (Hitachi Kenki FineTech Co., Ltd. HSAM220), which can observe interfacial structure using ultrasonic echo radiated from the probe in a water tank. Some of the samples were placed in a shearing jig as shown in Figure 2 and stressed to destruction in a precision universal tester (Shimadzu Cooperation Autograph AGS-5kNB) operating at a crosshead speed of 0.5 mm/min. In order to compensate the effect of interfacial area on the shear strength test, shear strength was calculated from the maximum load divided by the interface area estimated from scanning acoustic microscopy, and its average shear strength was calculated using the Weibull distribution function [31, 32]:

$$\ln \ln(1-F)^{-1} = m \ln\sigma - m \ln \sigma_0, \quad (1)$$

where F is the cumulative failure probability, m is the Weibull modulus, and σ_0 is the characteristic strength. In this paper, the median rank method was used for computing cumulative failure probability F in Eq. (1) because of its good reliability despite the small number of samples:

$$F = (i - 0.3) / (n + 0.4), \quad (2)$$

where i is the rank of the observation and n is the number of samples. Average shear strength is as follows:

$$\mu = \sigma_0 \Gamma(1+1 / m), \quad (3)$$

where the Γ function is expressed as follows.

$$\Gamma(x) = \int_0^\infty e^{-t} t^{x-1} dt \quad (x > 0). \quad (4)$$

3. Results

3.1. Heating profile and appearance of a specimen

Figure 3 shows the typical profile of the bottom temperature of the WC-Co plate during laser brazing joining. The increase of temperature continued in approximate proportion to the gradient until irradiation of the laser on the fourth side. The maximum bottom temperature was around 900 K. After the heating, the bottom temperature decreased quickly to 620 K within 25 sec from the end of the fourth run of heating.

The maximum bottom temperature of the WC-Co alloy in this case was about 70 K lower than that of the case at the Ar gas flow rate of 5 L/min with an evacuation process [12], but both temperature profiles showed a similar trend.

Figure 4 (a) and (b) shows the appearance of the specimen and the scanning acoustic microscopy image at the interface of h-BN / Ag-Cu-Ti braze metal / WC-Co, respectively, after laser brazing. In (a), the white part of the square in the centre is an h-BN block, and the peripheral metallic lustre part is WC-Co. At the right side in (a), a Ag-Cu-Ti braze metal lump, which occasionally flowed out from the interface between h-BN and WC-Co, as indicated by a white arrow can be seen, and its colour kept a metallic lustre, which showed that a large amount of titanium oxide did not exist at the surface of the braze metal. The black area in the centre of (b) is Ag-Cu-Ti braze metal which was melted at the joint interface and no void was observed in the joint interface.

Figure 5 shows a typical cross-sectional SEM image at the interface of h-BN / Ag-Cu-Ti braze metal / WC-Co after laser brazing. The upper right side is h-BN and lower area is the WC-Co alloy plate. The gray layer between h-BN and WC-Co is Ag-Cu-Ti braze metal. As an active element in Ag-Cu alloy braze metal, Ti reacted with N in h-BN, and a continuous TiN layer was formed at the interface, which improved the wettability between ceramics and braze metal. Also, Ti reacted with C in WC-Co, and a continuous reaction layer was formed at the interface between braze metal and WC-Co. These results reveal a similar tendency to the former study with evacuation process [12]. Therefore, these observations are addressed in the former study [12].

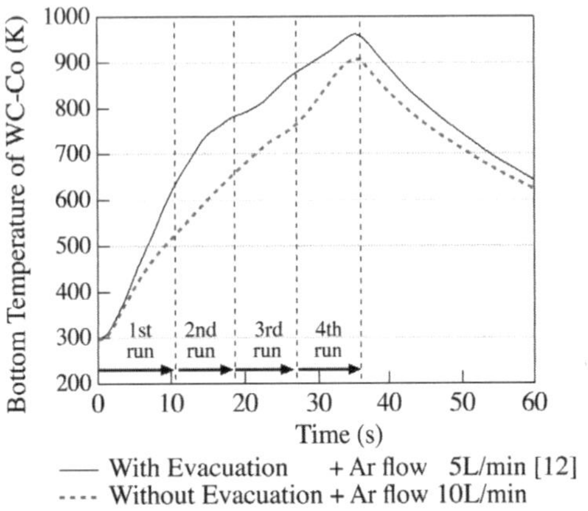

Figure 3. Heating profile of the bottom temperature profile of the WC–Co plate during laser brazing.

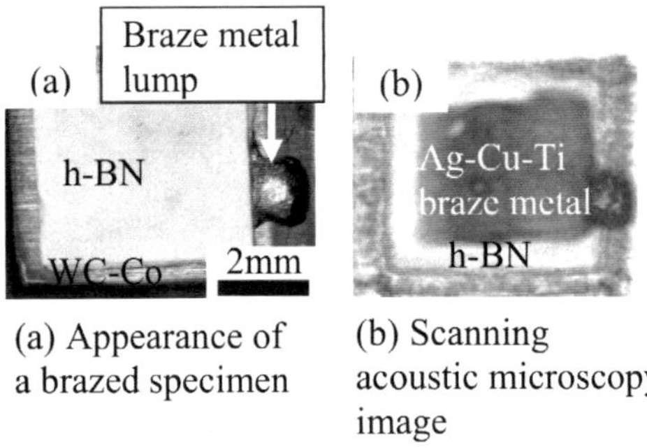

Figure 4. Appearance of a laser brazed specimen and interface observation of a h-BN / Ag-Cu-Ti braze metal / WC-Co interface using scanning acoustic microscopy.

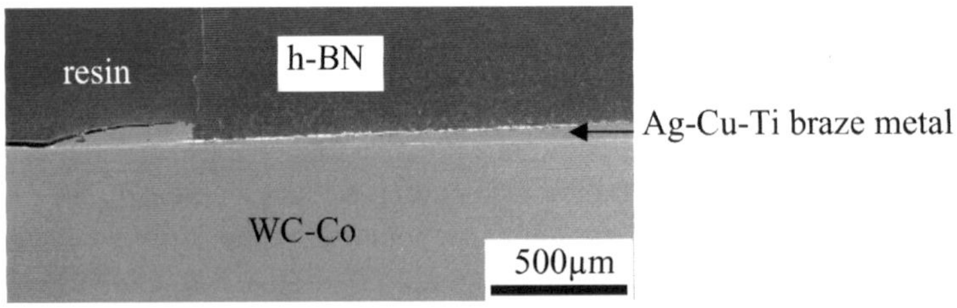

Figure 5. Cross-sectional SEM image of the specimen.

3.2. The surface condition of Ag-Cu-Ti braze metal

Figure 6 shows cross-sectional SEM observation of an edge of the h-BN / Ag-Cu-Ti braze metal interface and its map analysis using electron probe micro-analysis. In (a), the upper left area is h-BN as indicated, the lower whitish area shows the Ag-Cu-Ti braze metal, the upper right area is epoxy resin which is used to mount the specimen and there is a space between the Ag-Cu-Ti braze metal and epoxy resin because of the shrinkage of the curing epoxy resin. The resin side of the Ag-Cu-Ti braze metal corresponds to the surface of the melted braze metal, which overflowed from the h-BN/WC-Co interface and was exposed to the Ar atmosphere. In (b), the distribution of oxygen was limited to the epoxy resin area and was not observed in the Ag-Cu-Ti braze metal. In (c) and (d), the distributions of silver and copper were the same as that of the braze metal area, as shown in the SEM image. The distribution of titanium is shown in (e). In (f) and (g), the distributions of boron and nitrogen were the same as that of the h-BN area. In (e), most of the titanium existed at the interface between the h-BN and Ag-Cu-Ti braze metal, and overlap of the distribution of nitrogen and titanium was observed, as shown in (g), which suggested the formation of a reacted layer of Ti-N intermetallic compound [5]; and the rest of the titanium was in the bulk of the braze metal, which overlapped the concentration of copper shown in (d). In the high-concentration regions of titanium and copper, the concentration of silver was low in contrast with titanium and copper. These results showed a similar tendency to that of the former study with evacuation [12]. From (b) and (e), the distributions of oxygen and titanium were not overlapped, which was apparently distinguished due to the existence of the space between the Ag-Cu-Ti braze metal and resin. These facts show that the oxidation layer was not observed at the surface of the melted braze metal.

In (h), the concentration of tungsten was low, and its detection was at background level. In (i), the distribution of carbon was limited to the epoxy resin area. In (j), the concentration of cobalt was low, and its detection was at background level in the same way as tungsten.

Figure 7 shows surface SEM observation of Ag-Cu-Ti braze metal lump and its map analysis using electron probe micro-analysis. In (a), small white spots are considered as contamination on the surface of Ag-Cu-Ti braze metal from their forms. In (b), the concentration of oxygen was low and its detection was at background level, and a significant concentration of oxygen was not observed. In (c), (d) and (e), the distribution of silver, copper and titanium, which were elements of braze metal, were the same as that of the braze metal area as shown in the SEM image of (a). In the high-concentration regions of titanium and copper, the concentration of silver was low in contrast with titanium and copper. In (f) and (g), the concentrations of boron and nitrogen were low and their distributions were restricted to small white spots which were considered as contamination on the surface. In (h), (i) and (j), the concentrations of tungsten, carbon and cobalt, which were elements of WC-Co, were low and their detection was at background level.

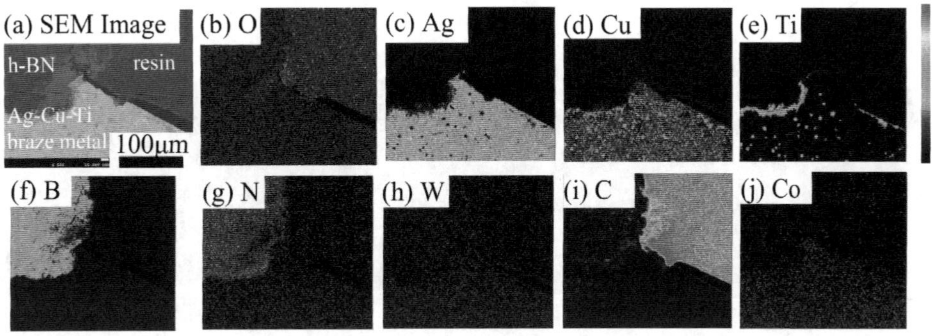

Figure 6. Map analysis of h-BN/Ag-Cu-Ti braze metal interface (at the edge).

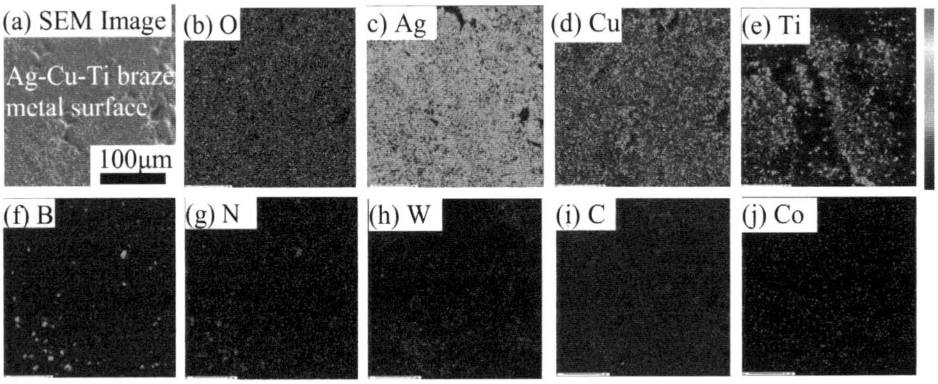

Figure 7. Map analysis of Ag-Cu-Ti braze metal surface (at the braze metal lump).

3.3. The interface condition between h-BN and Ag-Cu-Ti braze metal

Figure 8 shows cross-sectional SEM observation and its map analysis of the central area of an h-BN / Ag-Cu-Ti braze metal interface. In Figure 8 (a), the upper area is h-BN as indicated, and the lower gray area shows Ag-Cu-Ti braze metal. In (b), oxygen was not observed at the interface. Silver near the interface in the same area of (a) is shown in (c) and also copper is shown in (d). In (e), the distribution of titanium was concentrated at the interface. In (f) and (g), the distributions of boron and nitrogen were the same as that of h-BN area, as shown in the SEM image of (a). From these element distributions, titanium concentration near the interface was apparently observed, of which the thickness was about 2-5 μm. The interface between h-BN and Ag-Cu-Ti was very complicated. It is presumed that the molten Ag-Cu-Ti braze metal infiltrated into open pores of the h-BN surface. These results showed a similar tendency to that in the former study with evacuation [12]. In (h), (i) and (j), the distributions of tungsten, carbon and cobalt, which are elements of WC-Co, are shown. Concentrations of tungsten and cobalt were low, at background level. In (i), the concentration of carbon was observed in the ragged area of h-BN as shown in the SEM image of (a), which was considered to be contamination remaining from the polishing material. In (j), the concentration of cobalt was low, at background level.

Figure 9 shows XRD profiles of the samples at the interface of the h-BN / Ag-Cu-Ti braze metal without evacuation and with evacuation [12] in the central area of the interface. Among the strong peaks of h-BN, WC-Co and silver, which resulted from bulk ceramics, WC-Co and braze metal elements, respectively, some peaks from titanium nitride were found to exist at the interface. These weak peaks were indexed as TiN [2, 6, 10] in both cases. Oxide peaks such as TiO_2 were not observed even in the case without evacuation.

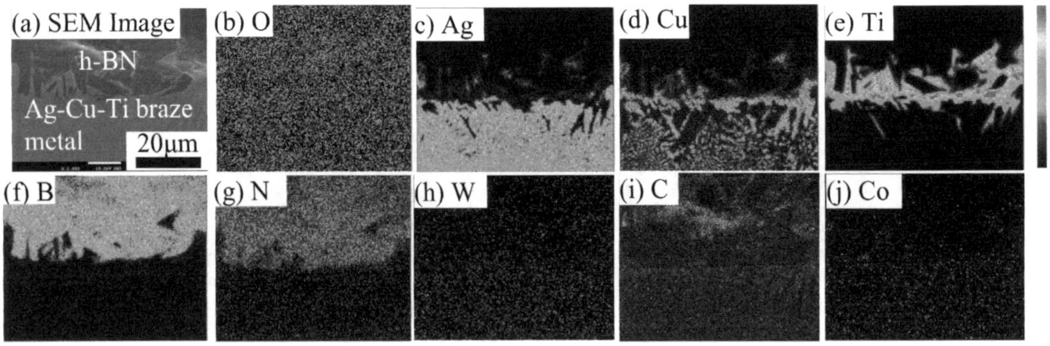

Figure 8. Map analysis of an h-BN/Ag-Cu-Ti braze metal interface (at the central area).

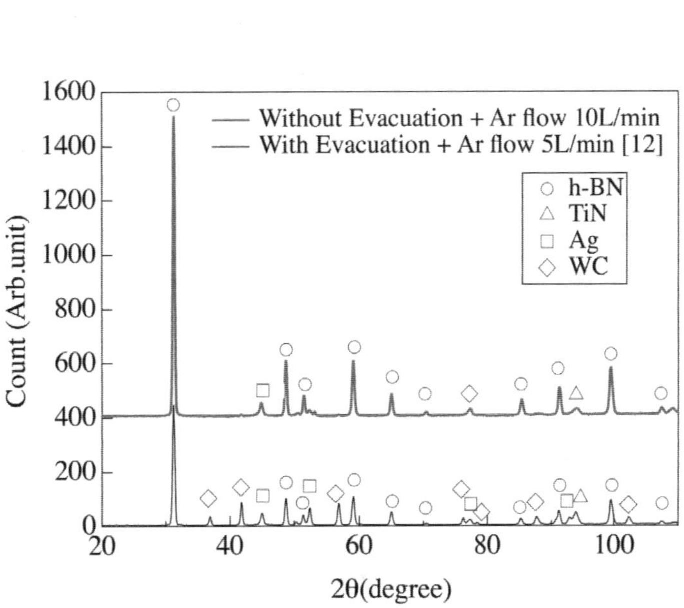

Figure 9. XRD profile of the h-BN/Ag-Cu-Ti braze metal interface.

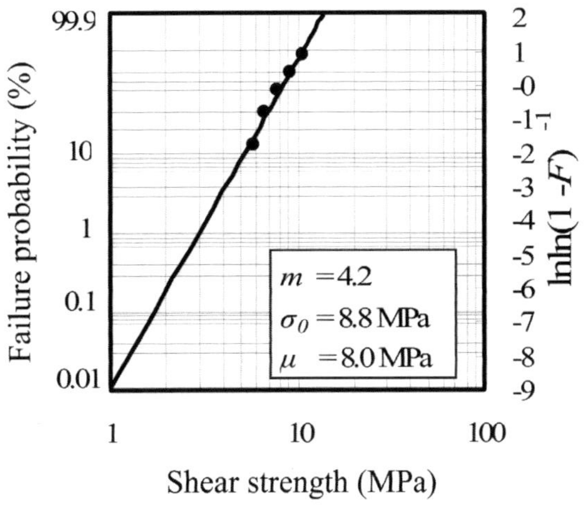

Figure 10. The Weibull distribution of shear strength test.

3.4. Shear strength test

After cross-sectional and non-destructive examinations, a shear strength test of the dissimilar joint was performed. Figure 10 shows the Weibull distribution of shear strength tests of h-BN / Ag-Cu-Ti braze metal and h-BN, which was plotted using the median rank method. In all tests, fracture occurred at h-BN side of the specimen near the interface. The difference between the maximum and minimum shear strengths was about 4-5 MPa. The distributions of shear strength were spread widely and could be approximated by the line formula as shown. The average shear strength μ was 8.0 MPa, which showed similar strength to that obtained by supplying Ar gas after evacuating the chamber before brazing [12].

4. Discussion

As shown in Figure 3, the maximum bottom temperature of WC-Co alloy in this case was about 70 K lower than for the case with an evacuation process of the chamber before brazing [12]. It is assumed that this decrease in temperature results from the difference in cooling effect with the flow rate of Ar. However, Ag-Cu-Ti braze metal was melted, and a reaction layer at the interface was formed as shown in Figures 4, 6 and 8. These facts indicate that the brazing condition in this study is sufficient to melt the Ag-Cu-Ti braze metal and make the joint.

Figure 4 shows that the Ag-Cu-Ti braze metal, which flowed out from the interface between h-BN and WC-Co, had a metallic lustre. This fact shows that a large amount of titanium oxide did not exist at the surface of the Ag-Cu-Ti braze metal, even when only Ar gas was supplied at a flow rate of 10 L/min without evacuating the chamber before brazing.

From Figure 6 (a), the upper right area of Ag-Cu-Ti braze metal was exposed to Ar gas. If the remaining amount of oxygen in Ar gas flow atmosphere was sufficient to react with titanium in the braze metal, titanium oxide would be formed, but the distribution of oxygen in (b) showed that the oxidation layer was not observed at the surface of the braze metal. In addition, from Figure 7 (b), the concentration of oxygen was low and its detection was at background level, and a significant concentration of oxygen was not observed in the surface elemental analysis. Thus, it is considered that the oxidation layer at the surface of the braze metal over sub-micrometre order, which is the minimum limit of detection of EPMA, was not observed. In the conventional dissimilar brazing process, a vacuum furnace is commonly used to avoid oxidation of Ag-Cu-Ti braze metal during a long brazing period. In contrast, the laser brazing process enabled the realization of a good braze metal joint without oxidation by supplying only Ar gas at a flow rate of 10 L/min, even without evacuating the chamber before brazing.

In Figures 6 and 8, the concentrations of tungsten, carbon and cobalt were low and it is considered that they have little or no effect on the interfacial reaction compared to the former study [12]. In Figure 7, the concentrations of tungsten, carbon and cobalt were low and their detection was at background level, and these concentrations are considered that they have little or no effect on the surface reaction of Ag-Cu-Ti braze metal.

Figures 8 to 10 indicate that the brazing behaviour [2, 34, 35] and the average shear strength were similar to those in the former study [12] even when supplying only Ar gas at a flow rate of 10 L/min without any evacuation process of the chamber before brazing.

5. Conclusions

(1) The laser brazing of h-BN and WC-Co alloy using Ag-Cu-Ti braze metal as an active element of titanium could be achieved only in a high-purity Ar gas flow atmosphere in a chamber without any evacuation process.

(2) Oxidation of titanium was not observed, and titanium existed as Cu-Ti composite in Ag-Cu-Ti braze metal near the edge of the braze metal.

(3) Average shear strength calculated from the interface area was 8.0 MPa, which showed similar strength to that obtained by using an evacuation process.

Acknowledgements

This work was supported by the Joint Research System of the Joining and Welding Research Institute, Osaka University.

References

[1] Nakao Y, Nishimoto K and Saida K 1989 *Trans of J.W.S.* **20** 66-76
[2] Nicholas M G, Mortimer D A, Jones L M and Crispin R M 1990 *J. Mat. Sci.* **25** 2679-89
[3] Nakao Y, Nishimoto K and Saida K 1990 *ISIJ Int.* **30** 1142-50
[4] Murakawa M and Takeuchi S 1991 *Mater. Sci. Eng.* A **140** 759-63
[5] Chattopadhyay A K and Hintermann H E 1993 *J. Mat. Sci.* **28** 5887-93
[6] Peteves S D 1996 *Ceram. Int.* **22** 527-33
[7] Fernandes A J S, Fonseca M J, Costa F M, Silva R F and Nazare M H 1999 *Dia. Rel. Mat.* **8** 855-8
[8] Felba J, Friedel KP, Krull P, Pobol I L and Wohlfahrt H 2001 *Vac.* **62** 171-80
[9] Huang S F, Tsai H L and Lin S T 2004 *Mater. Chem. Phys.* **84** 251-8
[10] Sechi Y, Takezaki A, Tsumura T and Nakata K 2008 *Smart Process Tech.* **2** 27-30
[11] Rohde M, Suedmeyer I, Urbanek A and Torge M 2009 *Ceram. Int.* **35** 333-7
[12] Sechi Y, Tsumura T and Nakata K 2010 *Materials and Design* **31** 2071-7
[13] Suedmeyer I, Hettesheimer T and Rohde M 2010 *Ceram. Int.* **36** 1083-90
[14] Biddulph R H 1985 *Proceedings of the 1st Europian Symposium on Engineering Ceramics* p.45-61
[15] Wlosinski W 1985 *Fueg. Keram. Gla. Met.* p.22-36
[16] Jenney C L, O'Brien 2001 *A Welding Handbook* Vol. I. (American Welding Society)
[17] Witherell C E and Ramos T J 1980 *Weld J.* **59** 267S-277S
[18] Hanebuth H, Hoffmann P and Geiger M 1994 *Proc. SPIE Int. Soc. Opt. Eng.* 2207: 146-153
[19] Whitehead D G and Foster R J 1995 *Assem. Auto.* **15** 17-19
[20] Dave V R, Carpenter R W, Milewski J O, and Christensen D T 2001 *Weld. J.* **80** 142.S-147.S
[21] Markovits T, Takacs J, Lovas A and Belt J 2003 *J. Mater. Process. Technol.* **143/144** 651-5
[22] Vollertsen F and Grupp M 2005 *Steel Res. Int.* **76** 240-4
[23] Saida K, Song W and Nishimoto K. 2005 *Sci. Tech. Weld. Join.* **10** 227-35
[24] Mathieu A, Cicala E, Matteie S, Gewvey D, Pontevicci S and Viala J C 2006 *Mater. Sci. Eng.* A **435/436** 19-28
[25] Li M G, Sun D Q, Qiu X M, Sun D X and Yin S Q 2006 *Mater. Sci. Eng.* A **424** 17-22
[26] Mathieu A, Matteie S, Grevey D, Deschamps A and Martin B 2006 *NDT E Int.* **39** 272-6
[27] Saida K, Song W and Nishimoto K 2006 *Sci. Tech. Weld. Join.* **11** 694-700
[28] Saida K, Song W and Nishimoto K 2007 *Design Int. Str. Adv. Mat. Join. Sol. Sta. Phen.* (Part B) **127** 301-6
[29] Takemoto T, Kawahito Y, Nishikawa H, Katayama S and Kimura S 2008 *J. Lig. Met. Weld. Cons.* **46** 300-8
[30] Saida K, Song W and Nishimoto K 2008 *Mat. Sci. For.* **580/582** 271-4
[31] Weibull W 1951 *J. Appl. Mech.* **18** 293-7
[32] Stanley P, Fessler H and Sivill A D 1973 *Proc. Brit. Ceram. soc.* **22** 453-87
[33] Chattopadhyay A K and Hintermann H E 1993 *J. Mat. Sci.* **28** 5887-93
[34] Benko E 1995 *Ceram. Int.* **21** 303-7
[35] Benko E, Bielanska E, Perevertelio V M and Loginova O B 1997 *Dia. Rel. Mat.* **6** 931-34

A study for high accuracy measurement of residual stress by deep hole drilling technique

Houichi Kitano, Shigetaka Okano and Masahito Mochizuki

Graduate School of Engineering, Osaka University, 2-1 Yamada-oka, Suita, Osaka 565-0871, Japan

E-mail: h-kitano@mapse.eng.osaka-u.ac.jp

Abstract. The deep hole drilling technique (DHD) received much attention in recent years as a method for measuring through-thickness residual stresses. However, some accuracy problems occur when residual stress evaluation is performed by the DHD technique. One of the reasons is that the traditional DHD evaluation formula applies to the plane stress condition. The second is that the effects of the plastic deformation produced in the drilling process and the deformation produced in the trepanning process are ignored. In this study, a modified evaluation formula, which is applied to the plane strain condition, is proposed. In addition, a new procedure is proposed which can consider the effects of the deformation produced in the DHD process by investigating the effects in detail by finite element (FE) analysis. Then, the evaluation results obtained by the new procedure are compared with that obtained by traditional DHD procedure by FE analysis. As a result, the new procedure evaluates the residual stress fields better than the traditional DHD procedure when the measuring object is thick enough that the stress condition can be assumed as the plane strain condition as in the model used in this study.

1. Introduction

Residual stresses are found in weld joints because of the localized heat input, the difference between the expansion coefficients of joint members, and the restraints around welded zones. In particular, high residual tensile stresses occur near welded zones. These stresses can affect the fracture and fatigue behaviours in welded structures[1]. Therefore, residual stresses in welded structures should be assessed in detail for the appropriate design, production, and maintenance. These are necessary for environmental consciousness of welded structures. For this purpose, the inner residual stress fields are as important as surface residual stress fields. The neutron diffraction technique is a well-known evaluation method of inner residual stress fields[2]. However, it is not always possible to conduct the evaluations with this technique easily because it requires huge experimental facilities and a stress-free sample. In addition, this technique cannot be applied to the residual stress evaluation of plates more than several tens of millimetres in thickness. Therefore, a new technique for evaluating inner residual stress fields is needed when the neutron diffraction technique cannot be applied.

Consequently, the deep hole drilling (DHD) technique has received much attention in recent years as a method for measuring through-thickness residual stresses[3-5]. The DHD technique measures the diametric change of a reference hole drilled thorough the component between before and after the residual stress release. The residual stress release is performed by trepanning a column of material containing the reference hole. The DHD technique has some advantages over the neutron diffraction technique, such as simpler test devices and procedures, applicability to thick plates, and in-field testing. Therefore, the DHD technique has merits that differ from those of the neutron diffraction technique.

However, some accuracy problems relevant to the appropriate design, production, and maintenance of structures occur when residual stress evaluation is performed by the DHD technique. The reason is the traditional DHD evaluation formula applies to the plane stress condition. In addition, the effects of plastic deformation produced in the DHD process are ignored. These can affect the accuracy of the evaluation results obtained by the DHD technique. Therefore, the traditional DHD evaluation procedure does not perform actual precise evaluation of the residual stress fields.

In this study, a new evaluation procedure is discussed for the high accuracy evaluation of residual stresses by the DHD technique.

2. Traditional DHD technique

Figure 1 provides a schematic illustration of the measurement method, which consists of following four experimental steps:

Step 1 A reference hole is drilled through the specimen with a gun drill.
Step 2 The initial diameters d_0 of this hole are measured accurately at several depths and angles using an air probe.
Step 3 A column of material containing the reference hole as its axis is coaxially freed using electro-discharge machining (EDM).
Step 4 The final diameters d of the reference hole are re-measured at the same depths and angular positions as those in step 2.

Let the direction of drilling the reference hole is denoted by z-axis, the depth from the top of the reference hole is denoted z, the angle from the x-axis is denoted by θ, (See Figure 1(e)) the differences between d_0 and d at depth z and angle θ is denoted by Δd and the normalized diametric distortion $\Delta d/d_0$ is denoted by $u_r(z, \theta)$. The relationship between $u_r(z, \theta)$ and remote stress fields σ_x, σ_y, σ_{xy} is given by the elastic solution for a hole in an infinite plate, as follows[6]:

$$u_r(z,\theta) = \frac{d(z,\theta) - d_0(z,\theta)}{d_0(z,\theta)} = \frac{\Delta d(z,\theta)}{d_0(z,\theta)} = -\frac{1}{E}[(1 + 2\cos 2\theta)\sigma_{\mathrm{x}} + (1 - 2\cos 2\theta)\sigma_{\mathrm{y}} + 4\sin 2\theta\, \sigma_{\mathrm{xy}}], \quad (1)$$

where E is Young's modulus. Therefore, three $u_r(z, \theta)$-θ relationships at each depth give one evaluation result of σ_x, σ_y, σ_{xy}.

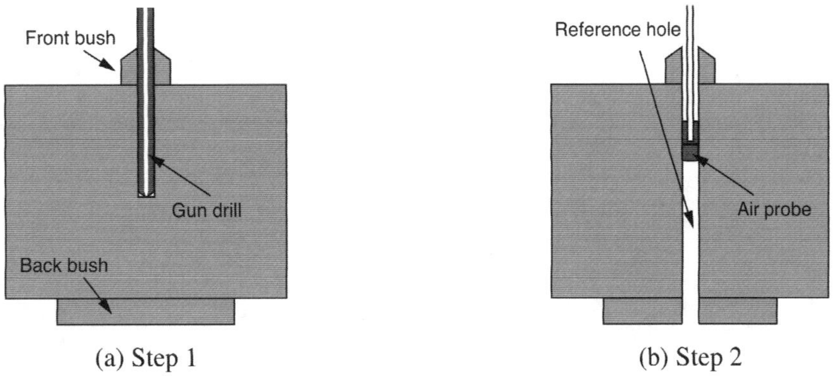

Figure 1. Steps in the DHD technique.

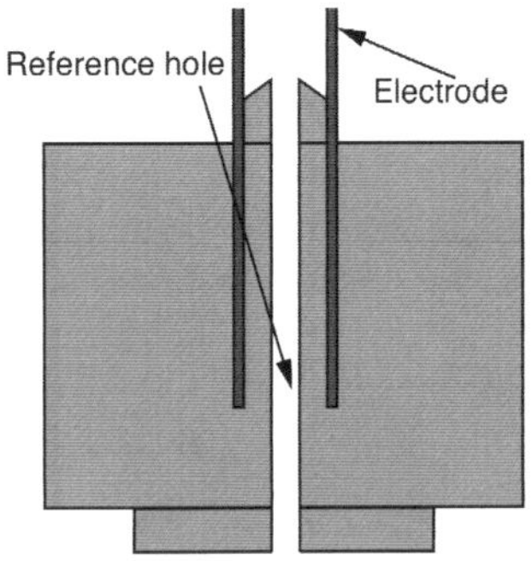

(c) Step 3

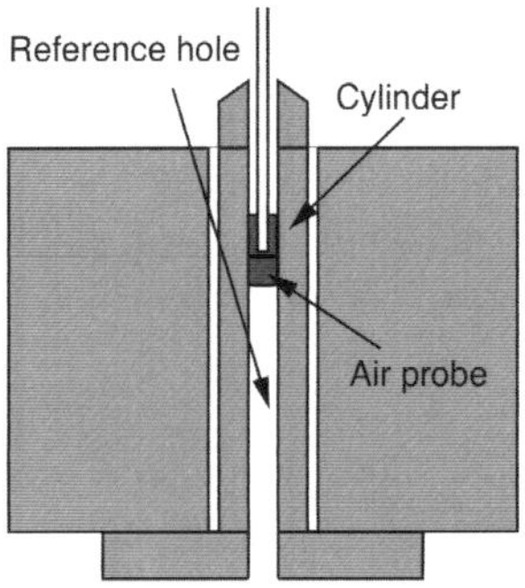

(d) Step 4

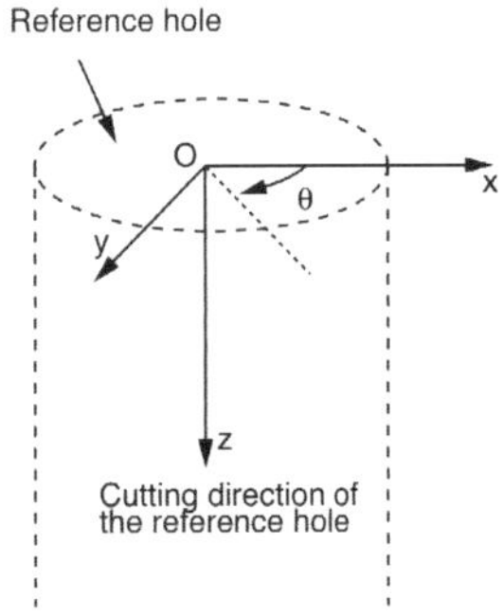

(e) Difinition of the each systems

Figure 1. Continued.

3. Modification of the evaluation formula

The traditional evaluation formula applied to the plane stress condition. However, when inner residual stress fields are evaluated, the stress condition around the reference hole is close to the plane strain condition. Therefore, the diametric change of the reference hole before and after the residual stress release is produced by the release of the in-plane residual stress field and the restraint of deformation in the axial direction. When the residual stress release under the plane strain condition is performed, the expansion of the cylinder is caused by the release from the stress, which is denoted by σ_z':

$$\sigma_z' = -\upsilon(\sigma_x + \sigma_y), \tag{2}$$

where υ is Poisson's ratio. Thus, the diametric change caused by σ_z' is calculated as follows:

$$\frac{d - d_0}{d_0} = \frac{\upsilon\sigma_z'}{E} = -\frac{\upsilon^2}{E}(\sigma_x + \sigma_y). \tag{3}$$

The relationship between $u_r(z, \theta)$ and the residual stress field is given by the elastic solution in the following equation:

$$u_r(z,\theta) = -\frac{1}{E'}[(1 + 2\cos 2\theta)\sigma_x + (1 - 2\cos 2\theta)\sigma_y + 4\sin 2\theta\, \sigma_{xy}] - \frac{\upsilon^2}{E}(\sigma_x + \sigma_y), \tag{4}$$

where $E'=E/(1-\upsilon^2)$.

4. FE analysis condition

Numerical analysis is used for evaluating the effects of plastic deformation produced in the DHD processes. Figure 2 shows the object used for finite element analysis. This is the 1/4 model and

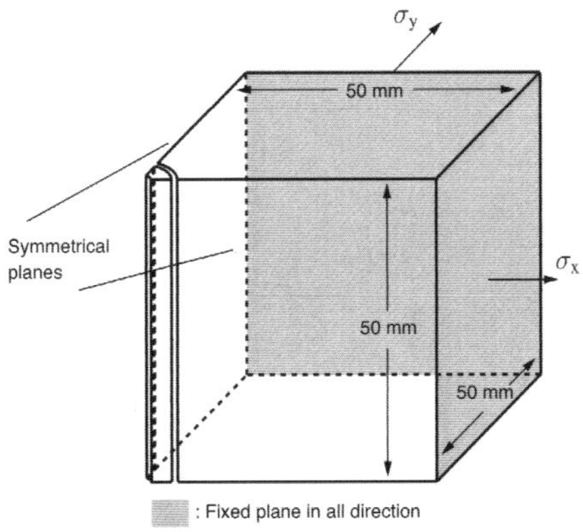

Figure 2. Object used for FE analysis.

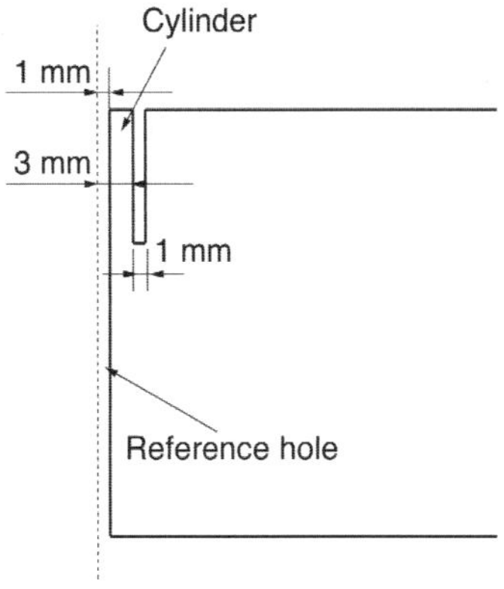

Figure 3. Sizes around the reference hole.

the size of the model is large enough not to interfere with the deformation around the reference hole. Figure 3 shows the sizes of the reference hole. The diameter of the reference hole is 2 mm. The thickness of the cylinder is 2 mm. The thickness of the trepanning is 1 mm. The sizes of the mesh in the radial direction and the through-thickness direction are approximately 0.3 mm and 0.5 mm, respectively. The arc of the reference hole is divided into 18 meshes in the θ-direction. The numerical analysis software is ABAQUS FE code Ver. 6.9. The model change option is used for removing elements. The drilling process and the trepanning process are represented by this operation. The initial residual stress fields are represented by the initial conditions option. Young's modulus, Poisson's ratio and Yield stress used for the analysis are discussed in each of following sections, because the properties change according to the purposes of the analysis.

5. Effects of plastic deformation produced in the drilling process

Stress concentration increases around the reference hole. Plastic deformation does not occur when the stress that exists before the drilling process is low. Therefore, the reference hole deforms elastically after the stress release by trepanning. In contrast, plastic deformation occurs when the stress, which exists before the drilling process, is high. When the stress is higher than one-third of the yield stress, plastic deformation occurs because the stress concentration factor around the hole is 3. When plastic deformation occurs, the trepanning process no longer produces complete stress release. Namely, the relationship between $u_r(z, \theta)$ and the residual stress field differs from that obtained by equation (4). Therefore, in this section, a quantitative evaluation method for evaluating the relationship between the residual stress field and the change in diametric change of the reference hole caused by the plastic deformation is discussed. This method makes it possible to evaluate the effects of the plastic deformation produced in the drilling process.

In the numerical analysis in this section, the material is assumed to have an elastic perfectly plastic body (E: 206 GPa, υ: 0.3, Yield stress: 500 MPa). Trepanning is performed at a time to remove the effects of the trepanning process. Figure 4 shows the relationships between $(d_x)_{plastic}$, $(d_y)_{plastic}$ and the applied stresses in the x direction under the uniaxial tensile stress state obtained by the numerical analysis, where $(d_x)_{plastic}$, $(d_y)_{plastic}$ are the values calculated by subtracting the diametric change obtained by the FE analysis from that obtained by equation (4). These values represent the effects of the plastic deformation produced in the drilling process. Here, the two factors that define these relationships are assumed as follows. The first factor is the maximum stress in the x- and y-axes around the reference hole calculated by the elastic solution. This factor represents the direction of the deformation of the reference hole and the drive force for deformation. The second factor is the von Mises equivalent stress at a position far enough away from the reference hole. This factor represents the ease of

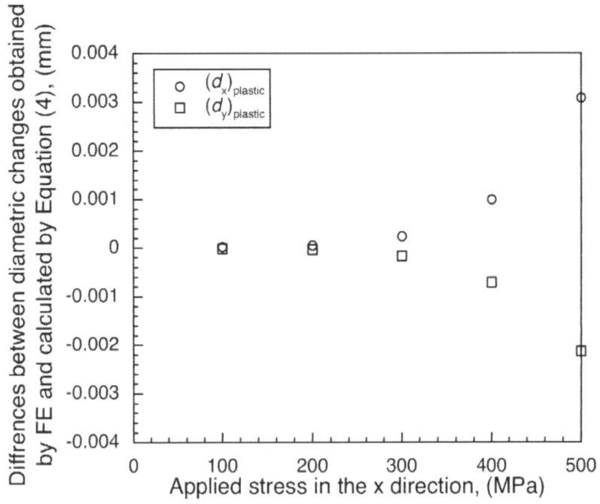

Figure 4. Differences between the diametric changes calculated by equation (4) and by FE analysis.

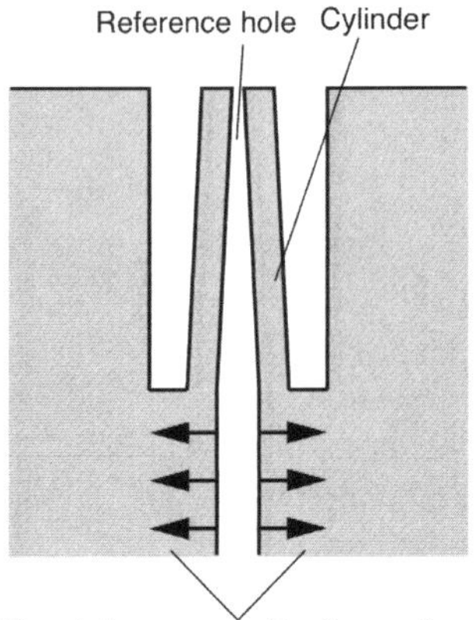

Figure 5. The additional stress caused by the sectional stress release in the trepanning process

expanding the plastic area. Namely, $(d_x)_{plastic}$, $(d_y)_{plastic}$ are given by the unknown functions f_x, f_y as in the following equations:

$$\begin{aligned}(d_x)_{plastic} &= f_x\left(\frac{3\sigma_x}{\sigma_Y}, \frac{\overline{\sigma}}{\sigma_Y}\right) + f_y\left(\frac{-\sigma_x}{\sigma_Y}, \frac{\overline{\sigma}}{\sigma_Y}\right)\\ (d_y)_{plastic} &= f_x\left(\frac{-\sigma_x}{\sigma_Y}, \frac{\overline{\sigma}}{\sigma_Y}\right) + f_y\left(\frac{3\sigma_x}{\sigma_Y}, \frac{\overline{\sigma}}{\sigma_Y}\right),\end{aligned} \tag{5}$$

where σ_Y is Yield stress, and $\bar{\sigma}$ is von Mises stress. f_x, f_y are determined by using the results shown in Figure 4:

$$\begin{aligned}(d_x)_{plastic} &= \alpha\left(\frac{3\sigma_x}{\sigma_Y}\frac{\overline{\sigma}}{\sigma_Y}\right)^2 + \beta\left(\frac{-\sigma_x}{\sigma_Y}\frac{\overline{\sigma}}{\sigma_Y}\right)^2\\ (d_y)_{plastic} &= \alpha\left(\frac{-\sigma_x}{\sigma_Y}\frac{\overline{\sigma}}{\sigma_Y}\right)^2 + \beta\left(\frac{3\sigma_x}{\sigma_Y}\frac{\overline{\sigma}}{\sigma_Y}\right)^2,\end{aligned} \tag{6}$$

where $\alpha = 3.61\times10^{-4}, \beta = -2.69\times10^{-4}$. When a biaxial stress exists, $3\sigma_x$, $-\sigma_x$ are substituted for $3\sigma_x - \sigma_y$, $3\sigma_y - \sigma_x$. Then, it is assumed that $(d_x)_{plastic}$, $(d_y)_{plastic}$ satisfy equation (4) when they are regarded as Δd. $d_{plastic}$, which is the change in diametric change in each angles, is calculated as follows:

$$\frac{d_{plastic}}{d} = -\frac{1}{E'}\left[(1 + 2\cos 2\theta)\sigma_x{'} + (1 - 2\cos 2\theta)\sigma_y{'}\right] - \frac{\upsilon^2}{E}(\sigma_x{'} + \sigma_y{'}), \tag{7}$$

where, $\sigma_x{'}$, $\sigma_y{'}$ are obtained by $(d_x)_{plastic}$, $(d_y)_{plastic}$ and this equation. Namely, $\sigma_x{'}$, $\sigma_y{'}$ are functions of σ_x, σ_y. Therefore, $\sigma_x{'}$, $\sigma_y{'}$ are functions of in-plane stresses σ_x, σ_y, σ_{xy} when σ_x, σ_y are regarded as the principal stresses σ_1, σ_2. Therefore, the relationship between the in-plane residual stress field and the change in diametric change are obtained using equations (6) and (7).

6. Effects of deformation produced in the trepanning process

EDM is performed from one side to the other in the trepanning process as shown in Figure 1. In other words, the process continues to release the sectional stress. The diameter of the reference hole of the stress-unreleased area is affected by the deformation of the stress-released area, as shown in Figure 5. When plastic deformation is produced, the relationship between $u_r(z, \theta)$ and the residual stress field differs from that obtained by equation (4). Therefore, a quantitative evaluation method for evaluating the effect is discussed in this section. The effects at depth z_0 can be seen as the additional stresses $\Delta\sigma_x(z_0)$, $\Delta\sigma_y(z_0)$, $\Delta\sigma_{xy}(z_0)$ produced by deformation of the stress-released area. The additional stresses are applied from an area far enough from the reference hole. These additional stresses can be calculated by subtracting the actual residual stresses from the stresses $(\sigma_x)_{max}$, $(\sigma_y)_{max}$, $(\sigma_{xy})_{max}$ calculated by equation (4) and using the maximum diameter and the final diameter when the object is

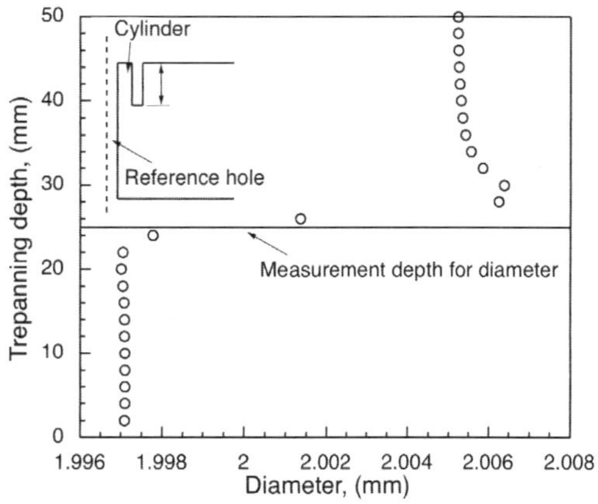

Figure 6. Relationship between diameter and trepanning depth at z= 25 mm.

an elastic body. If the object is an elastic perfectly plastic body, $(\sigma_x)_{max}$, $(\sigma_y)_{max}$, $(\sigma_{xy})_{max}$ are calculated by equations (4) and (7). Here, it is assumed that the additional stresses at depth z_0 can be determined by the in-plane stresses $(\sigma_x)_{basic}(z)$, $(\sigma_y)_{basic}(z)$, $(\sigma_{xy})_{basic}(z)$ $(0>z>z_0)$ calculated by equation (1) using the initial diameter and the final diameter. The effects of deformation of the stress released area are represented by $(\sigma_x)_{basic}$, $(\sigma_y)_{basic}$, $(\sigma_{xy})_{basic}$. Namely, $\Delta\sigma_x(z_0)$, $\Delta\sigma_y(z_0)$, $\Delta\sigma_{xy}(z_0)$ are determined by $(\sigma_x)_{basic}$, $(\sigma_y)_{basic}$, $(\sigma_{xy})_{basic}$. When the diametric changes at 0 and 90 degrees are used, $\Delta\sigma_{xy}(z_0)$, $(\sigma_{xy})_{basic}$ can be ignored. Using the unknown functions g_x, g_y, the following equation can be used to find $\Delta\sigma_x(z_0)$, $\Delta\sigma_y(z_0)$:

$$\begin{aligned} \Delta\sigma_x &= \int_0^{z_0} g_x\left((\sigma_x(z_0\text{-}l))_{basic}, \left(\sigma_y(z_0\text{-}l)\right)_{basic}\right) dl \\ \Delta\sigma_y &= \int_0^{z_0} g_y\left((\sigma_x(z_0\text{-}l))_{basic}, \left(\sigma_y(z_0\text{-}l)\right)_{basic}\right) dl, \end{aligned} \tag{8}$$

where l is the distance from the measurement depth. Then, $\Delta\sigma_x(z_0)$, $\Delta\sigma_y(z_0)$ are not affected by the mechanical properties of the specimens, because the stresses are an additional stress, not an additional strain. Also, $g_x((\sigma_x)_{basic}, (\sigma_y)_{basic}) = g_y((\sigma_y)_{basic}, (\sigma_x)_{basic})$ because g_x, g_y should be symmetrical. As a result, g_x, g_y can be determined under every stress state when g_x, g_y are determined under a uniaxial stress state. By numerical analysis, g_x, g_y can be determined. The material is assumed to have an elastic body (E: 206 GPa, υ: 0.3) in the numerical analysis used in this section. The trepanning process is divided into 25 steps. Figure 6 shows the diametric change in the trepanning process. In this analysis, the applied stress is 300 MPa. It is assumed that $g_x((\sigma_x)_{basic}, 0)$, $g_y(0, (\sigma_x)_{basic})$ are the products of $(\sigma_x)_{basic}$ and a function that outputs the maximum value at l=0 and the minimum value at l=10 by referring to the results. Then,

$$\begin{aligned} g_x((\sigma_x)_{basic}, 0) &= a(10-l)^2\left(\sigma_x(z_0-l)\right)_{basic} \\ g_y(0, (\sigma_x)_{basic}) &= b(10-l)^2\left(\sigma_x(z_0-l)\right)_{basic}, \end{aligned} \tag{9}$$

where $a = -5.44\times10^{-4}, b = 1.77\times10^{-4}$, and

$$\begin{aligned} \Delta\sigma_x &= \int_0^{10} a(10-l)^2\left(\sigma_x(z_0-l)\right)_{basic} + a(10-l)^2\left(\sigma_y(z_0-l)\right)_{basic} dl \\ \Delta\sigma_y &= \int_0^{10} a(10-l)^2\left(\sigma_y(z_0-l)\right)_{basic} + a(10-l)^2\left(\sigma_x(z_0-l)\right)_{basic} dl. \end{aligned} \tag{10}$$

Here, when z_0-l is less than 0, $\sigma_x(z_0\text{-}l)$ and $\sigma_y(z_0\text{-}l)$ should be assumed to be zero because the area does not load the additional stress to the evaluation depth z_0. These equations can only evaluate the additional stress effects under the biaxial stress states. When the x- and y-axes are regarded as principal axes, the additional stress effects can be evaluated under every in-plane stress state.

7. New evaluation procedure

From the results of the above discussion, the evaluation procedure in this study is performed in the following steps.

Step 1 Calculate the stresses $(\sigma_x)_{basic}$, $(\sigma_y)_{basic}$, $(\sigma_{xy})_{basic}$ by equation (4) using the initial diameter and the final diameter.

Step 2 Calculate $\Delta\sigma_x(z_0)$, $\Delta\sigma_y(z_0)$ using equation (10).

Step 3 Calculate the stresses $(\sigma_x)_{max}$, $(\sigma_y)_{max}$, $(\sigma_{xy})_{max}$ using equations (4) and (7) and the maximum diameter and the final diameter.

Step 4 Calculate the actual residual stresses by subtracting the stresses obtained in step 3 from those in step 2.

8. Suitability of the proposed procedure

The numerical analysis is used to discuss the suitability of the proposed procedure. The evaluation results are compared with that obtained by the traditional technique. In this traditional technique, the maximum diametric change is used for Δd to consider the effect of the deformation in the trepanning process. In the other words, this technique ignores the effects of the plastic deformation produced in the drilling process and considers all the deformation produced in the trepanning process as plastic deformation. This technique is called the iDHD technique and is now considered the most precise technique[7]. A material is assumed to have an elastic perfectly plastic body (E: 206 GPa, υ: 0.3, Yield stress: 500 MPa) in the numerical analysis used for this section. The trepanning process is divided into 25 steps. Applied stresses are uniform in the model. The stresses are evaluated at the through-thickness center. Naturally, the evaluation results are changed by the evaluation depth. However, the deformation state $(d\text{-}d_0)/d_0$ is nearly unchanging in area between the depth z=10 mm and z=40 mm, as shown in Figure 7. In this graph, the two deformation states are shown. One is calculated by using the initial diameter in x-axis as d_0. The other is calculated by using the maximum diameter in the x-axis direction as d_0. Namely, the deformation state in the DHD process is almost the same in the area between the depth z=10 mm and z=40 mm. Because the DHD technique is a technique for evaluating the inner stress fields, comparing the results at through-thickness center is enough for discussing the suitability of the proposed procedure.

Figure 8 shows the evaluation results obtained by the proposed procedure. In the graph, σ_x varies and σ_y is 200 MPa. Figure 9 shows the relationships between the evaluation results in the x'-y' system and the applied stresses along the x-axis. The x'-y' system is rotated 30 degrees clockwise from the x-y system. Each figure also shows the results obtained by the traditional technique using equation (1). In addition, Figures 10 and 11 show the errors calculated by applied stresses and evaluation results.
The results indicate that the proposed procedure can evaluate the residual stress fields better than the iDHD technique at all applied stress levels even though the applied stress level is low enough that the plastic deformation does not occur. Figure 12 shows the evaluation results when the material is assumed to have an elastic perfectly plastic body (E: 206 GPa, υ: 0.3, Yield stress: 1000 MPa). Figure 13 shows the errors calculated by applied stresses and evaluation results. The results indicate that the evaluation results obtained by the proposed procedure are in good agreement with the applied stresses when the yield stress is changed.

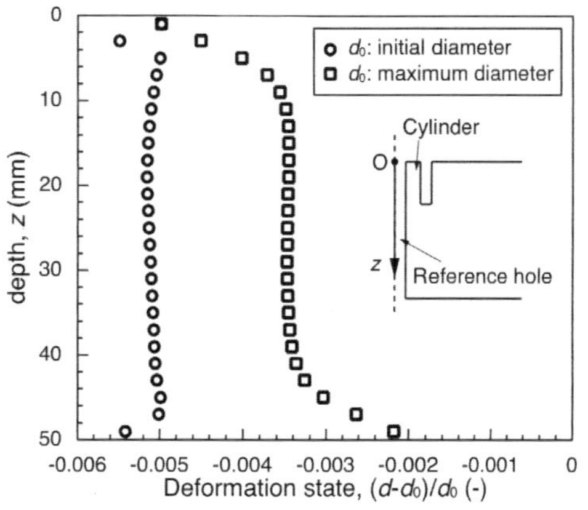

Figure 7. Relationship between deformation state $(d\text{-}d_0)/d_0$ and depth.

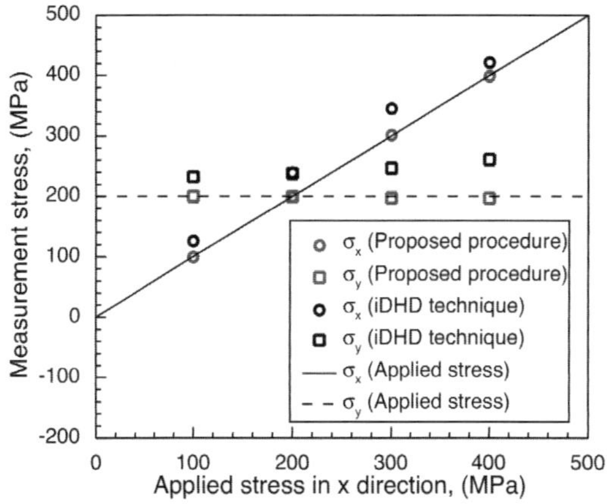

Figure 8. Evaluation results by the proposed procedure.

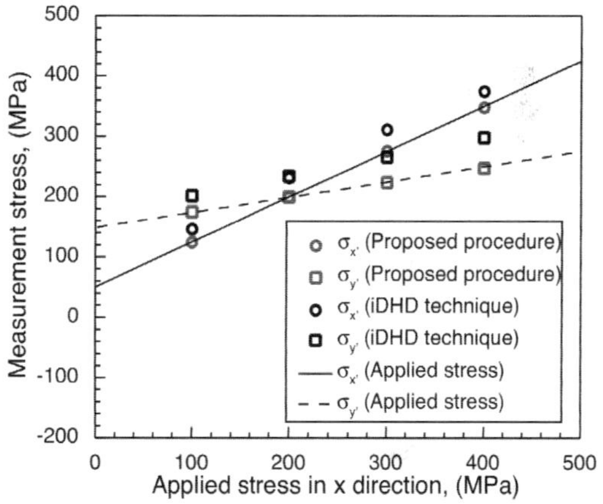

Figure 9. Evaluation results by the proposed procedure (x'-y' system).

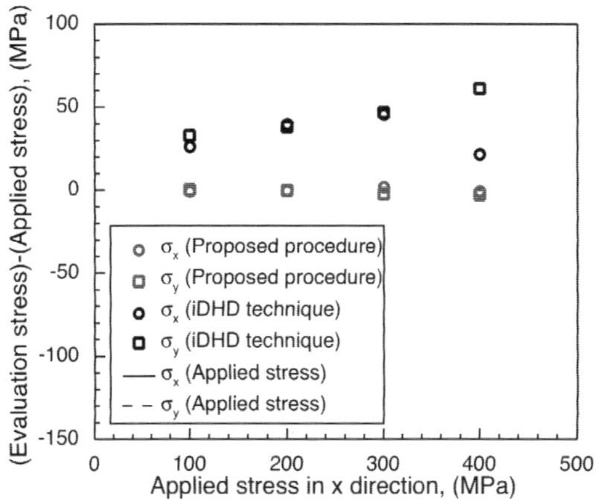

Figure 10. Evaluation errors calculated from (Evaluation stress)-(Applied stress).

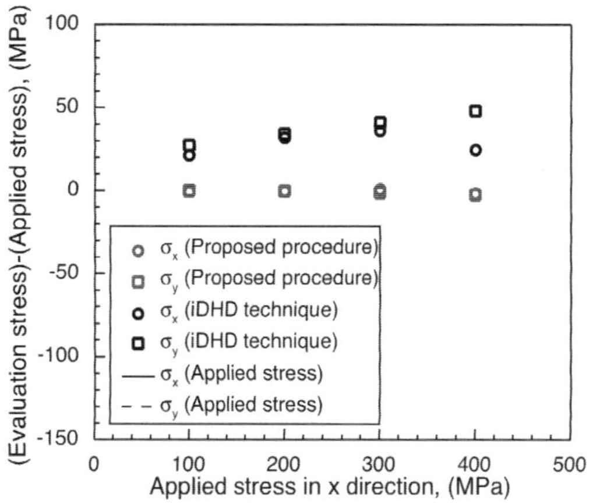

Figure 11. Evaluation errors calculated from (Evaluation stress)-(Applied stress) (x'-y' system).

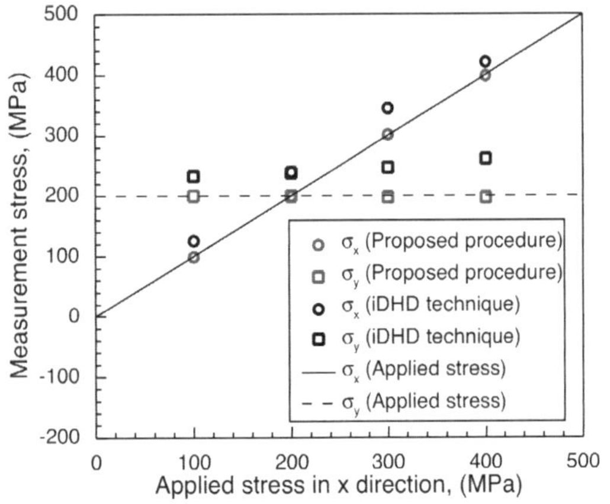

Figure 12. Evaluation results with the proposed procedure (σ_Y=1000).

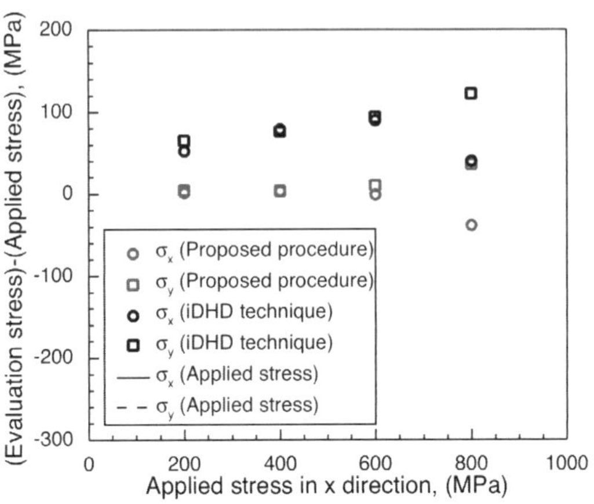

Figure 13. Evaluation errors calculated from (Evaluation stress)-(Applied stress) (σ_Y=1000).

9. Conclusions

A new highly accurate evaluation procedure of internal residual stress fields by the DHD technique was proposed. In addition, the new procedure was applied to evaluate inner stress fields by FE analysis. The results indicated that the proposed procedure evaluates the residual stress fields better than the traditional DHD procedure when the measured object is thick enough that the stress condition can be assumed as the plane strain condition, such as in the model used in this study. Using the proposed procedure, the DHD technique can have wider applicability. In addition, when the procedure for considering the effects of the stress distribution in the through-thickness direction and the effect of the changes in deformation state near the surface, which is not considered in the traditional DHD procedure and the procedure proposed in this study, is proposed, the DHD technique will be able to evaluate the residual stress fields more precisely.

Acknowledgements

This study was supported by Priority Assistance for the Formation of Worldwide Renowned Centers of Research - The Global COE Program (Project: Center of Excellence for Advanced Structural and Functional Materials Design) from the Ministry of Education, Culture, Sports, Science and Technology (MEXT), Japan.

References

[1] Ohata A, Maeda Y, Mawari T, Nishijima S and Nakamura H 1986 Fatigue strength evaluation of welded joints containing high tensile residual stresses *International Journal of Fatigue* **8** 147–50

[2] Park M J, Yang H N, Jang D Y, Kim J S and Jin T E 2004 Residual stress measurement on welded specimen by neutron diffraction *Journal of Materials Processing Technology* **155-156** 1171–77

[3] Leggatt R H, Smith D J, Smith S D and Faure F 1996 Development and experimental validation of the deep hole method for residual stress measurement *The Journal of Strain Analysis for Engineering Design* **31** 177-86

[4] Bouchard P, George D, Santisteban J, Bruno G, Dutta M, Edwards L, Kingston E and Smith D 2005 Measurement of the residual stresses in a stainless steel pipe girth weld containing long and short repair *International Journal of Pressure Vessels and Piping* **82** 299-310

[5] Brown T, Dauda T, Truman C, Smith D, Memhard D and Pfeiffer W 2006 Predictions and measurements of residual stress in repair welds in plates *International Journal of Pressure Vessels and Piping* **83** 809-18
[6] Barber J R 2002 *Elasticity* (Dordrecht: KLUWER ACADEMIC PUBLISHERS) 87-9
[7] Mahmoudi A H, Hossain S, Truman C E, Smith D J and Pavier M J 2009 A New Procedure to Measure Near Yield Residual Stresses Using the Deep Hole Drilling Technique *Experimental Mechanics* **49** 595-604

ECO-MATES 2011 IOP Publishing
Journal of Physics: Conference Series **379** (2012) 012050 doi:10.1088/1742-6596/379/1/012050

Residual stress variation due to piping processes of austenitic stainless steel

R Ihara[1], T Hashimoto[1] and M Mochizuki[1]

[1] Division of Material and Manufacturing Science, Graduate School of Engineering, Osaka University, 2-1 Yamadaoka, Suita, Osaka 565-0871, Japan

E-mail: ihara@mapse.eng.osaka-u.ac.jp

Abstract. In nuclear power plants, stress corrosion cracking (SCC) has been observed near the heat affected zone (HAZ) of the primary loop recirculation pipes made of austenitic stainless steel type 316L. Residual stress is a major cause of SCC. In the joining process of pipes, butt-welding is conducted after machining. Machining is performed to match the inside pipe diameter. Residual stress is generated by both machining and welding. In the case of welding after machining in manufacturing processes of pipes, it appears that residual stress due to machining is varied by the welding thermal cycle. In this study, residual stress variation caused by manufacturing processes was investigated. Residual stress variation was examined by the X-ray diffraction method. The residual stress distribution generated by welding after machining has a local maximum point in the HAZ. The Vickers hardness distribution also has a local maximum point. By the EBSD method, it is clarified that recovery and recrystallization due to welding heat do not occurred in the local maximum point. Residual stress distribution results from the superposition effect of hardening due to machining and welding. The location and value of the local maximum stress are varied by welding conditions. The region of the local maximum stress corresponds to the region where SCC has been observed. Therefore, in addition to a part of the manufacturing processes such as welding or machining, evaluation of all parts of the processes is important to investigate the effect of residual stress distribution on SCC.

1. Introduction

Recently, electric generation in nuclear power plants plays important roles in electrical power supplies and ecology. However, nuclear power plants have safety problems. An important problem is stress corrosion cracking (SCC). SCC has been observed near the heat affected zone (HAZ) of the primary loop recirculation pipes made of low-carbon austenitic stainless steel type 316L [1-3]. SCC is caused by three factors: material, environment and mechanical. The effect of residual stress on SCC is more important for a non-sensitization material such as stainless steel type 316L. In the joining process of pipes, butt-welding and machining are conducted. Machining is performed to match the inside pipe diameter, and provide a smooth surface finishing. Residual stress is generated by both machining and welding. SCC initiates as a transgranular SCC in the machined layer, and SCC propagates through the pipe thickness as an intergranular SCC. In the machined layer, hardness is increased. Moreover, high tensile residual stress is generated [4-5]. It has been proposed that machining has a large influence on SCC initiation [6-9]. However, in a manufacturing process of pipes, welding is often conducted after machining. It would appear that residual stress distribution generated by machining is varied by the welding thermal cycle.

In order to further understand SCC, accurate evaluation of the residual stress is required. To achieve this, not only residual stress generated by only machining or welding but also residual stress variation induced by welding after machining which is conventional in the manufacturing process of pipes is important. In this study, residual stress variation caused by manufacturing processes was investigated.

2. Experimental Procedures

2.1. Manufacturing processes

Machining and bead on plate welding using a Tungsten Inert Gas (TIG) arc were conducted for thin plate specimens made of stainless steel type 316L. Yield stress, tensile strength and elongation are 278 MPa, 540 MPa and 61 %, respectively. The dimensions of the specimens are $150(l) \times 100(w) \times 8(t)$ mm. The schematic illustration of specimen preparation is shown in Figure 1. Welding was conducted after machining. This process was assumed as the manufacturing processes of a welded pipe joint in nuclear power plants. Machining and welding conditions are shown in Table 1. In order to confirm the effect of welding heat input on residual stress variation, two welding currents were used. During welding, argon was used as the shielding gas with a flow rate of 15 *l*/min. Three types of processed specimens were prepared. They are machined specimen, welding specimen and welding specimen after machining. They are named specimen M, specimen W(Q) and specimen M+W(Q), respectively, as shown in Figure 1. The Q provided in parentheses after W is the welding heat input. For example, specimen W(SQ) means a small heat input (I = 120 A), and specimen W(LQ) means a large heat input (I = 200 A).

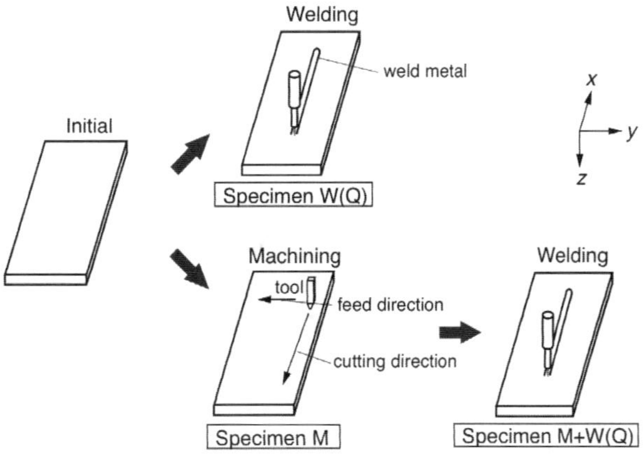

Figure 1. Preparation of each processed specimen.

Table 1. Machining and welding conditions.

Machining		Welding		
		Heat input	SQ	LQ
Cutting speed (m/min)	100	Current, I (A)	120	200
Feed rate (mm/rev)	0.1	Welding speed (mm/s)	2	
Cutting depth (mm)	0.1	Arc length (mm)	3	

2.2. Measurements and observation

Residual stress distributions were measured by X-ray diffraction using the 2D method for the machined or welded surface of each specimen. The measurement device used was a D8 discover instrument with GADDS (Bruker AXS products). Residual stress was calculated from the whole strain of the Debye cone in the 2D method [10]. This method can measure residual stress with good accuracy. The measurement condition is presented in Table 2, and a schematic illustration of the measurement system using a two-dimensional detector is shown in Figure 2. Typically, residual stress measurement of the weld metal is difficult due to the large grains and orientation of the weld metal. For measurement of the weld metal, measurement with multi-axial rocking is efficient [11]. Multi-axial rocking can increase the number of grains contributing to diffraction. The rocking was conducted for the ω axial and Y axial, as shown in Figure 2. To evaluate residual stress distributions, Vickers hardness measurement was performed. The measurement device used was an HMV-FA instrument with an automatic measurement system (Shimadzu Corp.). The indentation load is 490 mN, and the loading time is 15 s. Microstructure observation was performed using a FE-SEM instrument with an EBSD detector (JEOL products) in order to confirm the effect of the welding thermal cycle on the machined layer. The accelerating voltage is 25 kV. In the grain orientation analysis, the measurement step was 0.25 μm.

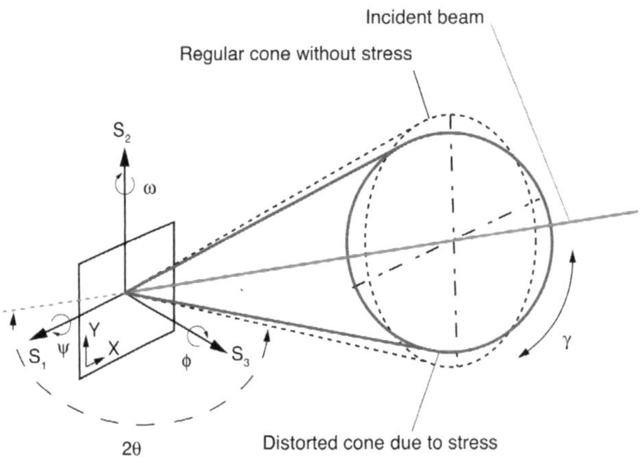

Figure 2. Geometry of the diffraction system for stress measurement.

Table 2. X-ray diffraction conditions for residual stress measurement.

Wave length (nm)	0.2291 (CrKα)
Power (kV, mA)	36, 88
Beam size (mm)	φ 1
Diffraction peak (deg)	128 (220 plane)
Young's modulus (GPa)	209.78
Poisson's ratio	0.278
Measuring time (s)	WM: 80, BM: 30 × 21 frame
Rocking (deg)	ω: ±4, Y: ±0

WM: weld metal, BM: Base metal

3. Results and discussion

3.1. Residual stress measurement

Firstly, residual stress measurement was performed for specimens W(SQ), M and M+W(SQ). Figure 3 shows the comparison of residual stress distributions of each specimen. The *x* direction is the welding and cutting directions, and the y direction is the welding transverse and feeding directions. The dimensions of the weld metal were measured using an optical microscope. The weld metal width is 6.21 mm. Residual stress of specimen W(SQ) have a tensile stress level in the HAZ region. Its stress level around the fusion line in the x direction is nearly the average value of yield stress there. The residual stress decreases with distance from the welding centerline, and changes to the compressive stress. Residual stress in the y direction shows lower tensile stress than the average yield stress in the whole region. These are typical tendency of residual stress distribution due to welding. Residual stress of specimen M was measured for 10 arbitrary points, and the average is shown. The average of σ_x is 618.5 ± 12.4 MPa, and that of σ_y is 381.57 ± 15.3 MPa. Residual stress of specimen M in both directions shows high tensile stress. Residual stress due to machining is caused by large plastic deformation and heat generation during machining. Residual stress distribution of specimen M + W(SQ) in the x direction has a local maximum stress 7 mm from the welded center. The local maximum stress shows over 800 MPa, which far exceeds the value of the yield stress, and shows a higher value than specimen M. Residual stress decreases with distance from the welding centerline. This tendency is the same as specimen W(SQ). Residual stress in the y direction also has a local maximum stress. However, the peak value is lower than that in the x direction, and the location of the stress peak appears outside.

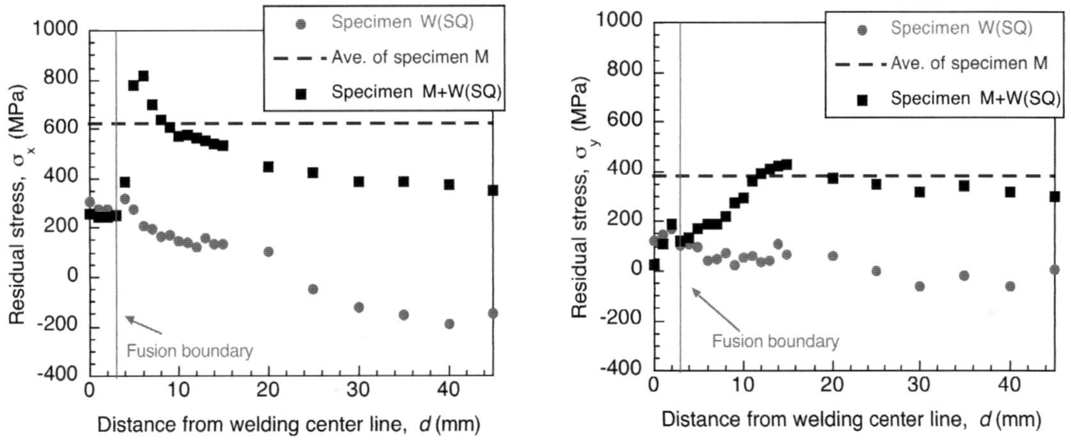

(a) x direction (b) y direction

Figure 3. Residual stress distribution generated by each process.

3.2. Hardness measurement

Vickers hardness measurement was performed to discuss residual stress variation generated by manufacturing processes. Measurement was conducted in the range of 0 to 15 mm from the welding centerline at 0.5 mm intervals. For comparison, measurement of a base metal was conducted for 15 arbitrary points, and this average is used.

Comparison of Vickers hardness distributions is shown in Figure 4. The average of the base metal is 167.2 ± 6.3 HV. Result of Vickers hardness of specimen W(SQ) has a small hardened region near 7.5 mm from the welding center. The other points show almost the same hardness as the average of the base metal. Result of specimen M shows high hardness in the whole region. For this reason, the average of the measured hardness is used, and this average is 328.1 ± 10.5 HV. This results from work hardening induced by machining. Result of specimen M+W(SQ) shows the average of the base metal near the weld metal. However, the hardness has a local maximum value of 370 HV at 8 mm from the welding center. At more than 9 mm, it decreases slightly and becomes the average hardness level of specimen M. Therefore, the local maximum hardness in specimen M+W(SQ) is caused by the superposition effect of hardening due to machining and welding.

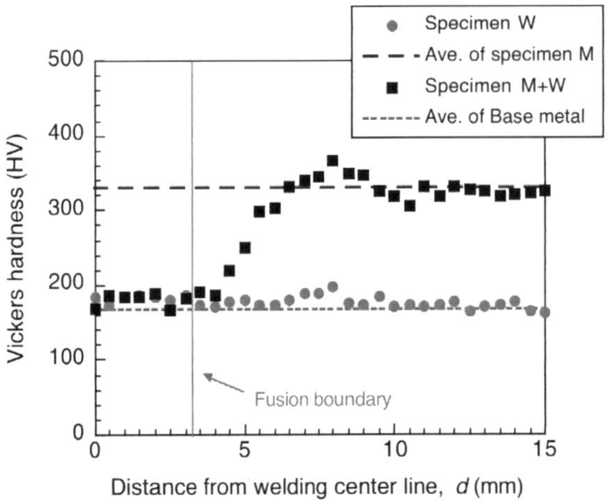

Figure 4. Comparison of Vickers hardness distributions of each processed specimen.

3.3. Microstructure observation

Microstructure observation with the EBSD method was performed for specimen M+W(SQ). Observation was conducted for regions A, B and C near the surface in the *y-z* cross-section, as shown in Figure 5. They are located 3.5, 5.5 and 7.5 mm from the welding centreline. The 3.5 mm location is near the fusion boundary, and the 7.5 mm location is the local maximum point of residual stress and Vickers hardness. The 5.5 mm location corresponds to the intermediate region.

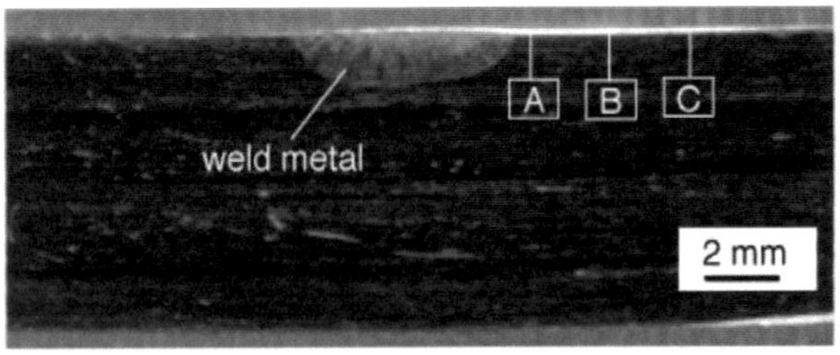

Figure 5. Cross sectional micrograph near the welding centerline.

Results of microstructure observation are shown in Table 3. Table 3 shows the inverse pole figure (IPF) map and image quality (IQ) map superimposed kernel average misorientation (KAM). The IQ map shows reliability. KAM shows misorientation in adjacent pixels excluding grain boundaries. The misorientation parameter like KAM has been used as a barometer of plastic strain distribution [12-13]. The IPF map in region A shows no effect of plastic deformation due to machining, and KAM shows a low value. The IPF map of region B indicates grain rotations and slip lines near the surface. KAM shows a higher value than that of region A. The IPF map in region C shows many grain rotations and slip lines which are deep in comparison with those of other regions. Moreover, random pattern points exist near the surface. These patterns show the difficult points of measurement. KAM in region C shows the highest value. The high value of KAM means that region C has the largest plastic deformation.

From these results, it can be considered that recovery and recrystallization due to the welding thermal cycle occur in the machined layer. A machined layer like region C exists in the whole surface after machining. However, in region A, plastic deformation due to machining disappears completely because of recrystallization. As a result, the IPF map shows smaller grains in comparison with those of other regions. In region A, residual stress of specimen M+W(SQ) shows a similar value to residual stress of specimen W. Moreover, Vickers hardness shows the same value as the base metal. In region B, recovery occurs, and plastic deformation partly disappears. Residual stress and Vickers hardness show higher values than those of only welding. In region C, recrystallization and recovery do not occur, and plastic deformation does not disappear. Although residual stress is calculated by only elastic strain, there is a correlation between plastic strain and elastic strain. Plastic strain indicates the potential of elastic strain, and is related to hardness. The high tensile stress caused by machining mainly contributes to plastic strain. In other words, the effect of machining on residual stress remains only in the hardening region. Moreover, hardening due to welding is generated. This is smaller than that of plastic deformation generated by machining, resulting in a local maximum hardness. Therefore, residual stress distribution has a local maximum point at the region of the local maximum hardness.

Table 3. Comparison of the IPF map and IQ map superimposed KAM at 3.5, 5.5 and 7.5 mm from the welding centerline.

	IPF	IQ+KAM
Region A (3.5 mm from welding center)	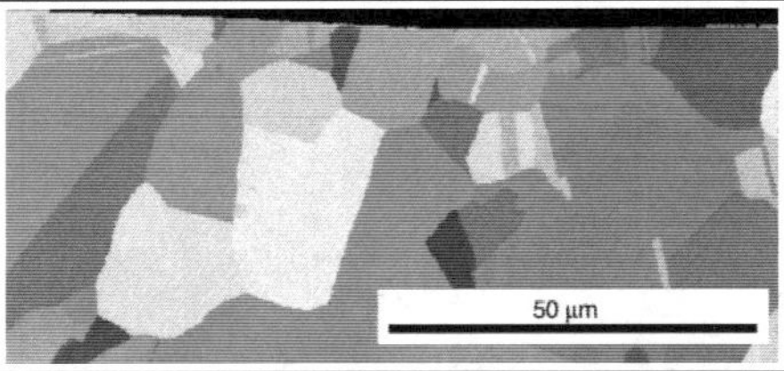	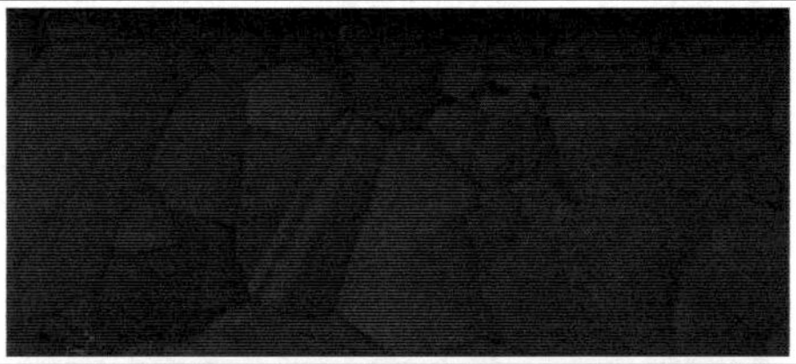
Region B (5.5 mm from welding center)		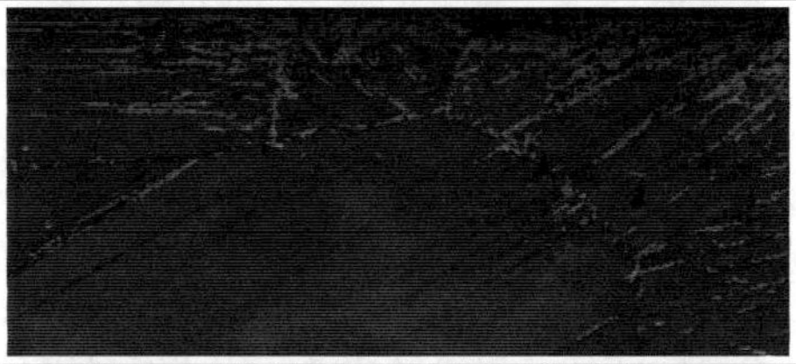
Region C (7.5 mm from welding center)		
Legend	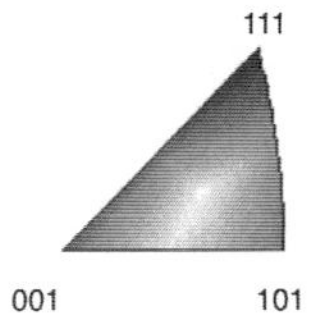	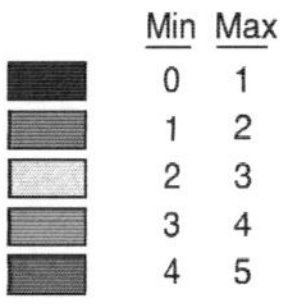

3.4. *Effect of welding condition on residual stress*

Residual stress measurements of specimens W(LQ) and M+W(LQ) were performed in order to clarify the effect of welding heat input. Figure 6 shows residual stress distributions of specimens W(LQ) and

M+W(LQ). The weld metal width with a large heat input is 12.17 mm. The weld metal of specimen W(LQ) has too many coarse and oriented grains. Residual stress measurement of this weld metal using the X-ray diffraction method is very difficult due to insufficient diffraction peaks for analyzing the stress. In this study, residual stress in the weld metal is not important. Thus, the measurement was performed outside the weld metal. Residual stress distributions of specimen M+W(LQ) also have a local maximum point. However, in both directions, locations of the local maximum stress are different from specimen M+W(SQ). In the x direction, the local maximum of the stress is located at 12 mm from the welded centreline. Moreover, the local maximum stress is higher than specimen M+W(SQ). These variations are assumed to be due to difference of temperature distribution during welding. As mentioned above, the local maximum stress is caused by the superposition effect of hardening and microstructure variation such as recovery and recrystallization due to the welding thermal cycle. The hardened region due to welding and microstructure variations is affected by the temperature history. Therefore, relationship between the temperature history and microstructure variation is important in order to evaluate the local maximum stress due to manufacturing processes.

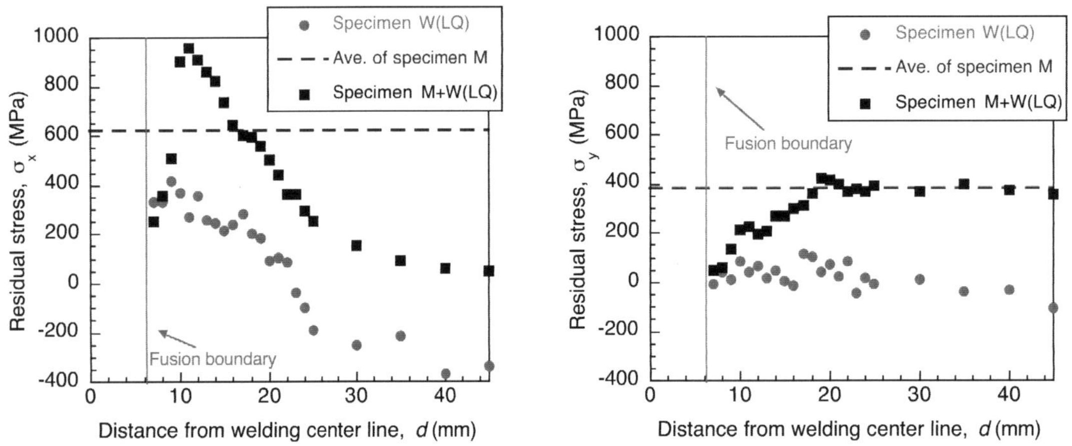

(a) x direction (b) y direction

Figure 6. Effect of welding heat input on residual stress variation generated by each process.

4. Conclusions

Residual stress variation due to manufacturing processes assuming piping of nuclear power plants was investigated using the X-ray diffraction method. In order to make clear residual stress variation, Vickers hardness measurement and microstructure observation with EBSD were performed. The following conclusions are obtained.

(a) Residual stress distribution generated by welding after machining has a local maximum stress in the HAZ.

(b) Vickers hardness distribution also has a local maximum point in the same region, showing the local maximum stress.

(c) In the local maximum region of stress and hardness, recovery and recrystallization due to the welding heat cycle are not observed.

(d) Residual stress has a local maximum point, which results from the superposition effect of hardening caused by machining and welding, in the region where recovery and recrystallization are not observed.

(e) The location and value of the local maximum stress are varied by welding conditions. This is caused by the difference of temperature history.

The location of the local maximum stress corresponds to the region where SCC has been observed. Therefore, in order to evaluate the initiation and propagation of SCC, not only one process but all processes are important.

References

[1] Okamura Y, Sakashita A, Fukuda T, Yamashita H and Futami T 2003 *Trans. SMiRT21* WGO1-1
[2] Suzuki S, Kumagai K, Shitara C, Mizutani J, Sakashita A, Tokuma H and Yamashita H 2004 *J. Maintenology* **3** 65 (in Japanese)
[3] Takamori K, Suzuki S and Kumagai K 2004 *J. Maintenology* **3** 52 (in Japanese)
[4] M'Saoubi R, Outeiro J C, Changeux B Lebrun J L and Dias A M 1999 *J. Materials Processing Technology* **96** 225
[5] Outeiro J C, Umbrello D and M'Saoubi R 2006 *J. Machine Tools & Manufacturing* **46** 1786
[6] Kuniya J, Kasahara S, Anzai H and Fujimori H 2007 *Japan J. Zairyo-to-Kankyo* **56** 22 (in Japanese)
[7] Masuoka T, Mayuzumi M, Arai T and Tani J 2007 *Japan J. Zairyo-to-Kankyo* **56** 93 (in Japanese)
[8] Isselin J, Kai A, Sakaguchi K and Shoji T 2008 *J. Metallurgical and Materials Transactions A,* **39A** 1099
[9] Takeda K, Taniyama A, Kudo T, Uchida H and Mizuki J 2009 *Japan J. Zairyo-to-Kankyo* **58** 228 (in Japanese)
[10] He B B 2003 *J. Powder Diffraction* **18** 71
[11] Hashimoto T, Osawa Y, Hirano S, Mochizuki M and Nishimoto K 2011 *J. Science and Technology of Welding & Joining* **16** 261
[12] Kamaya M, Wilkinson A J and Titchmarsh 2005 *J. Nuclear Eng. and Design* **235** 713
[13] Wilkinson A J, Dingley D J and Meaden G 2009 *J. Electron Backscatter Diffraction in Material Science* 231

Effect of electrode force condition on nugget diameter and residual stress in resistance spot welded high-strength steel sheets

M Iyota[1], Y Mikami[1], T Hashimoto[2], K Taniguchi[3], R Ikeda[3] and M Mochizuki[1]
[1] Division of Materials and Manufacturing Science, Osaka University, 2-1 Yamadaoka, Suita, Osaka 565-0871, Japan
[2] Hashimoto Iron Works Corporation, 7-15 Chikko-hamadera-nishi, Sakai, Osaka 592-8352, Japan
[3] Steel Research Laboratory, JFE Steel Corporation, 1 Kawasaki, Chiba, Chiba 260-0835, Japan

E-mail: iyota@mapse.eng.osaka-u.ac.jp

Abstract. This study examines the effect of the electrode force condition on the nugget diameter and residual stress in spot welded high-strength steel sheets. Numerical simulations of spot welding were performed to examine the nugget diameter and residual stress. The results indicate that adjusting the force profile changes the current density and stress state at the spot welds. Therefore, choosing an appropriate force profile extends the nugget diameter and reduces the residual stress.

1. Introduction

In recent years, global warming has been considered an important issue that must be solved. Therefore, various provisions have been considered, and various statutes have been established to improve the environmental situation. In the automotive industry, the ULSAB (UltraLight Steel Auto Body), ULSAC (UltraLight Steel Auto Closure), ULSAS (UltraLight Steel Auto Suspension), and ULSAB-AVC (UltraLight Steel Auto Body - Advanced Vehicle Concept) projects were promoted from 1994 to 2002. These projects were intended to reduce the environmental load and improve collision safety. Since then, the application of high-strength steel sheet to automobile structural components has been promoted for the same reasons. Now, high-strength steel sheet is being used extensively for automobile bodies and is having a positive impact on the industry and the environment.

Resistance spot welding is a common joining method used for assembling steel sheets. To use high-strength steels in automobiles, it is necessary to ensure the strength of their spot welded joints. A previous report stated that the nugget diameter [1-3] and residual stress [4-6] influence the static and fatigue strengths of spot welded joints. Therefore, generating a large nugget and decreasing the residual stress are crucial for increasing the static and fatigue strengths. Previous studies focused on decreasing the residual stress, and some provisions have already been reported [7-9]. However, these provisions are performed after the spot welding. Therefore, the efficiency of the spot welding is reduced, and the environmental load increases. Instead, it would be very useful to extend the nugget, which would reduce the residual stress without requiring additional processes.

In this study, the nugget diameter and residual stress in spot welded high-strength steel sheets were investigated using coupled electrical-thermal-microstructural-mechanical numerical simulations. Moreover, in order to control the nugget size and residual stress, the effect of the electrode force condition on these characteristics were examined.

2. Numerical simulation

Resistance spot welding is a joining method that utilizes Joule heating. In this process, electric current and electrode force are applied to the sheets. The electrode force influences the current pathway and the contact resistance, and these influence nugget formation. Therefore, they must be taken into account to accurately simulate spot welding. In addition, a phase transformation occurs due to thermal processes above the spot welded area. In a numerical simulation, the change in material properties due to the phase transformation must also be taken into account to accurately predict the residual stress. For this reason, we used a coupled electrical-thermal-microstructural-mechanical numerical simulation to account for not only the effect of the electrode force and the applied current but also the change in the material properties due to the phase transformation [10]. For the spot welding simulations in this study, we used the commercial finite element analysis software SYSWELD [11].

2.1. Analytical model

Figure 1 illustrates the sheets used for the finite element analysis. The sheet was 1.6 mm thick and 62.5 mm wide. To simulate resistance spot welding, an axisymmetric model with a restrained upper electrode edge was used. Figure 2 shows the finite element model. Solid elements were used with an element size of 0.1 × 0.2 mm in the weld area.

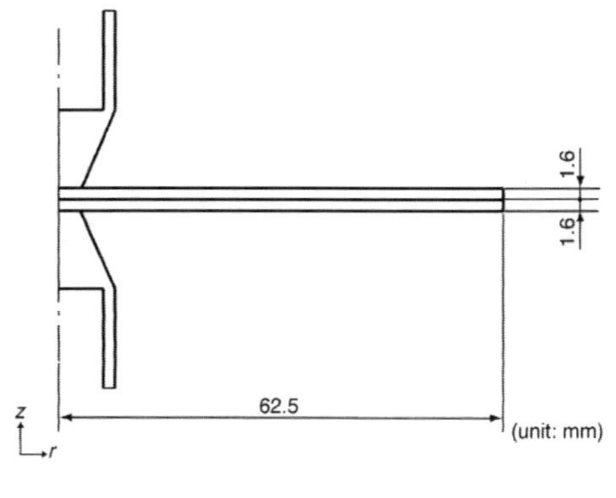

Figure 1. Model used for numerical simulation of resistance spot welding.

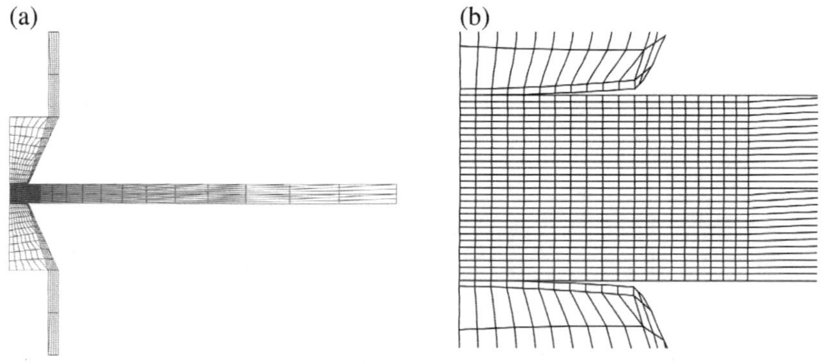

Figure 2. Finite element model used to simulate resistance spot welding: (a) overall view, (b) enlarged view at the spot weld area.

2.2. Material properties

Figure 3 shows the material properties of the high-strength steel sheets used in the resistance spot welding simulation. The physical and mechanical properties were measured experimentally and also obtained from a database [12]. Phase and temperature dependencies were considered for the thermal conductivity, specific heat, yield stress, and thermal strain. Only the temperature dependence was considered for the density and Young's modulus. In the numerical simulation of spot welding, contact resistance is an essential parameter for thermal analysis. The contact resistance depends on the surface state (unevenness, oxide film, etc.), and these resistances disappear in the early stages of applying the electrical current. Therefore, in this study, a contact resistance was assigned to the contact area as a temperature-dependent parameter to imitate the change in contact state during spot welding, as shown in Figure 3(b). When the temperature is low, the value of the contact resistance becomes high, but if the temperature becomes high enough (melting section, etc.), the value of the contact resistance becomes negligible.

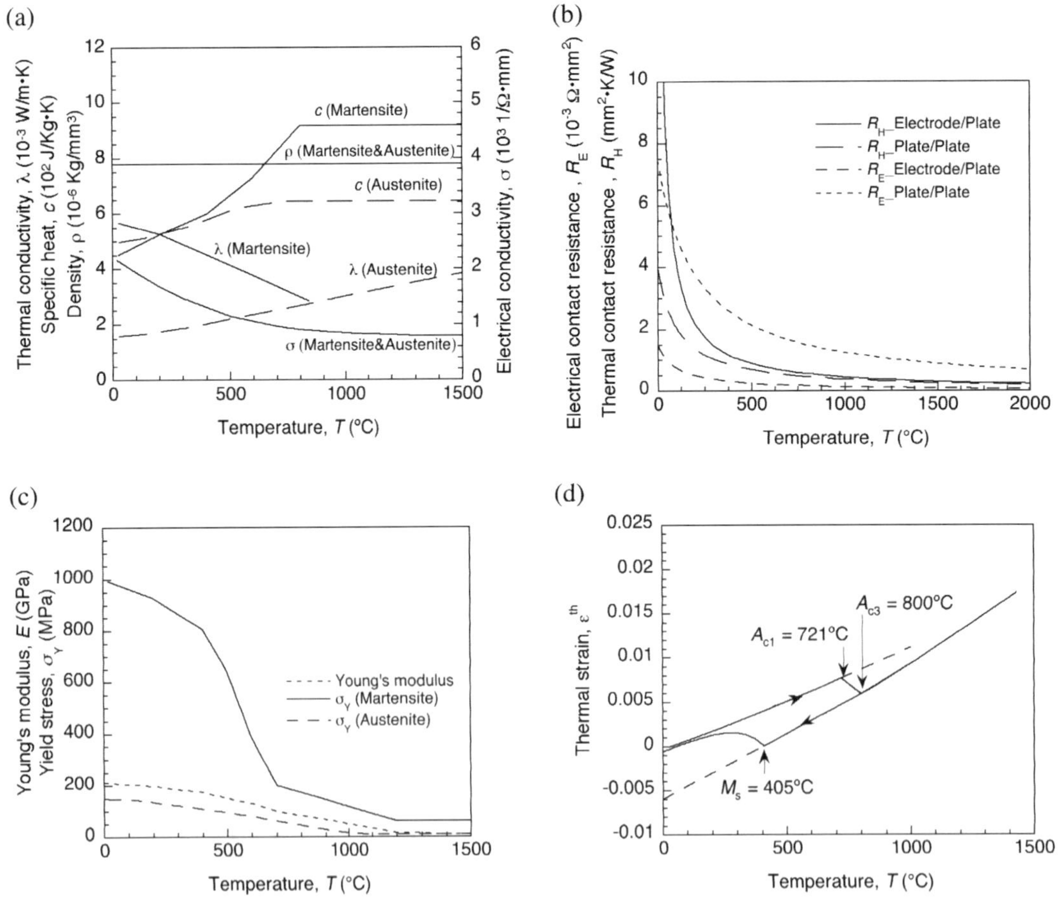

Figure 3. Material properties used in the resistance spot welding simulation: (a) physical properties of the high-strength steel sheets, (b) thermal and electrical contact resistances, (c) mechanical properties of the high-strength steel sheets, (d) thermal strain and transformation properties.

3. Results and discussion regarding nugget diameter

First, a numerical simulation was performed for the basic spot welding condition. Figure 4(a) shows the basic welding condition (hereafter 3.5 kN). The electrical current source was DC, and the profiles of the current and electrode force did not change during spot welding. Figure 4(b) shows the time histories of the nugget diameter and penetration ratio during the spot welding simulation. The nugget diameter and penetration ratio increase until 0.12 s, which is the time at which the current is applied. However, the increases in the nugget diameter and penetration ratio after 0.12 s are small. This phenomenon is caused by changes in the contact area between the sheet and electrode and between each of the sheets. The contact area changes during spot welding, and the current density at the center of the nugget changes with a change in the contact area. In the early stages of spot welding, because the contact area is small, the current density at the center of the nugget is high, and the nugget expands in the thickness and longitudinal directions. However, in the later stages of spot welding, because the contact area is large, the current density at the center of the nugget decreases, and the nugget size does not increase.

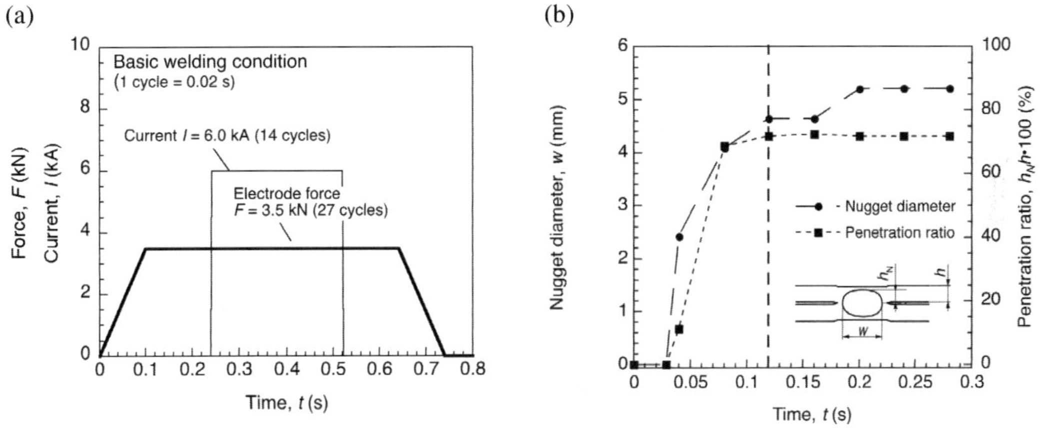

Figure 4. The progression of the nugget diameter and penetration ratio: (a) basic welding condition, (b) time history results for the nugget diameter and the penetration ratio.

Therefore, in order to prevent the decrease in the current density, the welding condition that controls the force profile was applied. This condition is shown in Figure 5(a). The force profile decreases when the growth of the nugget subsides (hereafter 3.5-2.0 kN). Figure 5(b) shows the time history results for the nugget diameter and penetration ratio during spot welding, which can be compared with the results in Figure 4(b). The nugget diameter and penetration ratio increase when the force profile decreases. Figure 6 shows the maximum temperature distribution for each welding condition, and the highest temperature region corresponds to the nugget. The size of the nugget for 3.5-2.0 kN is larger than for 3.5 kN. Figure 7 shows the current density at the center of the nugget. The results for 3.5 kN show the decrease in current density. However, the results for 3.5-2.0 kN show that the current density does not decrease. This is because the contact area becomes small due to the decrease in the force profile. However, the corona bond area for 3.5-2.0 kN is smaller than that for 3.5 kN. As shown in Figure 6, the corona bond is formed between the edge of the nugget and the point of sheet separation. This region represents the weld defect known as expulsion, which causes a decrease in joint strength. Therefore, a corona bond with sufficient width is required.

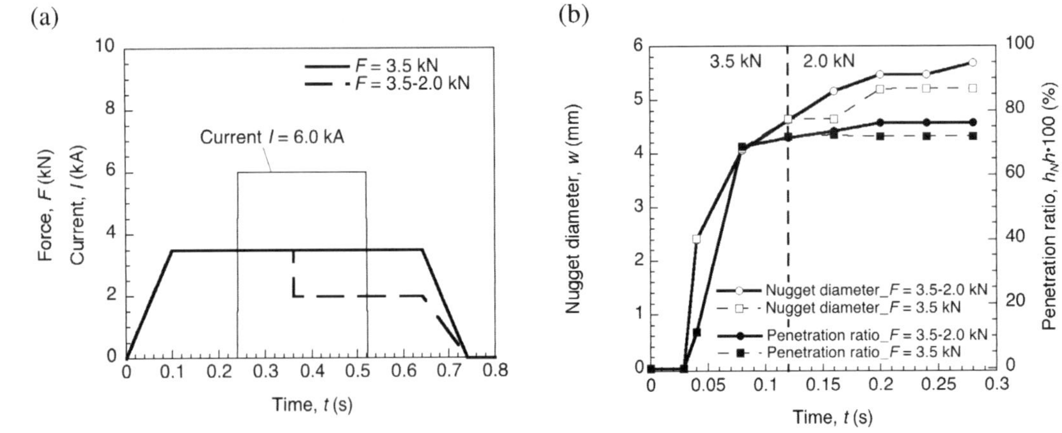

Figure 5. The progression of the nugget diameter and penetration ratio controlled by the electrode force: (a) force controlled condition, (b) time history results for the nugget diameter and penetration ratio for 3.5 kN and 3.5-2.0 kN.

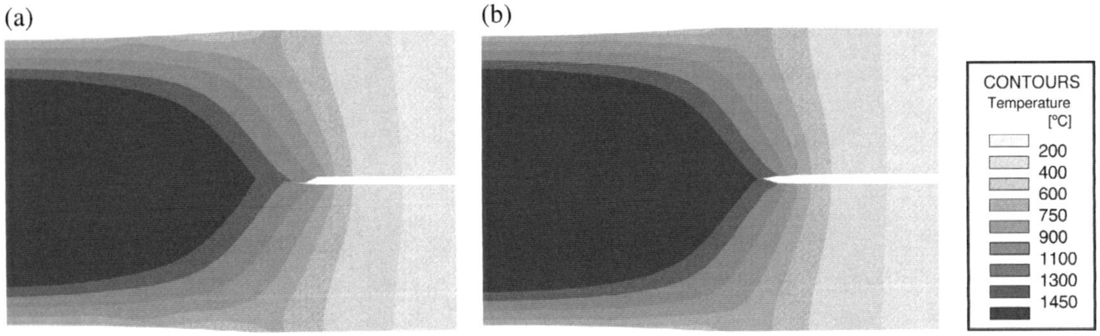

Figure 6. Maximum temperature distributions: (a) result for 3.5 kN, (b) result for 3.5-2.0 kN.

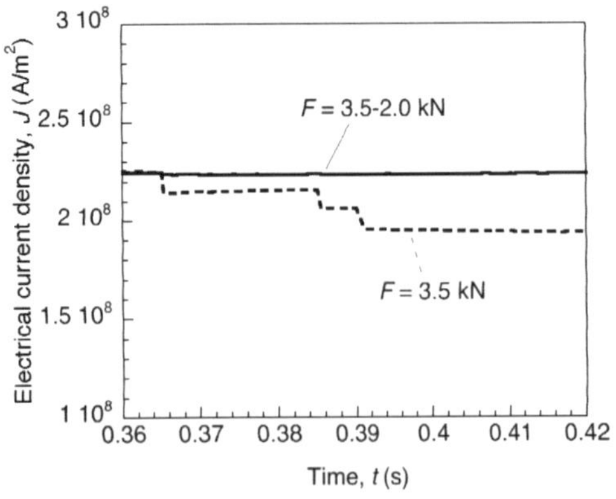

Figure 7. Comparison of current density at the center of the nugget for 3.5-2.0 kN and 3.5 kN.

Therefore, in order to form a corona bond with sufficient width, the welding condition that controls the force profile even further was applied. The considered conditions are shown in Figure 8. When the nugget forms to a sufficient size, the force profile is increased to form a corona bond with sufficient width (hereafter 3.5-2.0-3.5 kN). The results for the maximum temperature distribution are shown in

Figure 9. Figure 9(a) shows the results for 3.5-2.0 kN, and Figure 9(b) shows the results for 3.5-2.0-3.5 kN. The width of the corona bond for 3.5-2.0-3.5 kN is larger than for 3.5-2.0 kN. Therefore, these results show that choosing an appropriate force profile extends the nugget diameter and forms a corona bond with sufficient width.

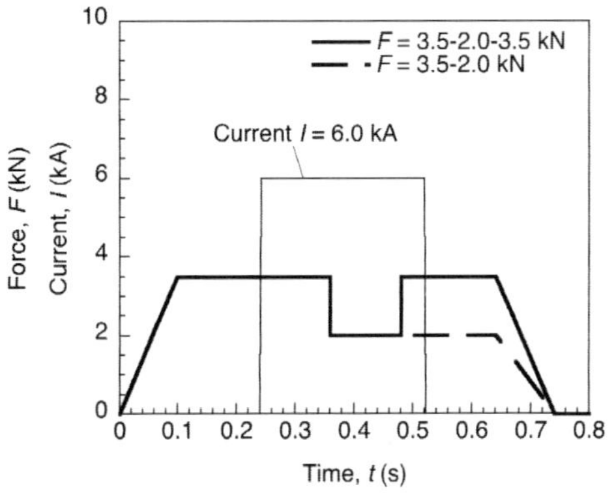

Figure 8. The force controlled condition profile used to form a corona bond with sufficient width.

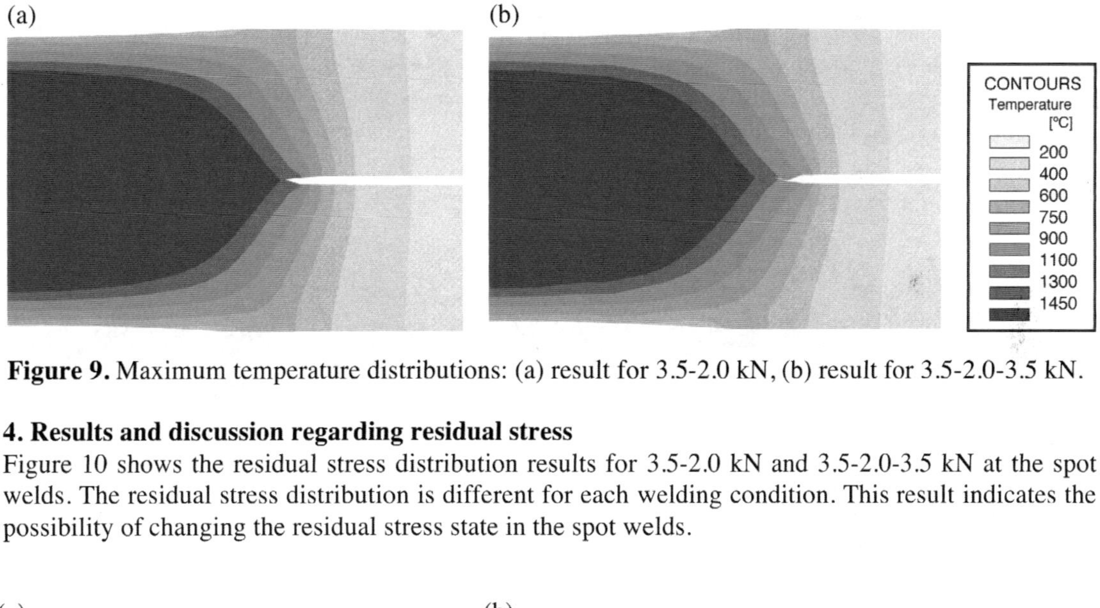

Figure 9. Maximum temperature distributions: (a) result for 3.5-2.0 kN, (b) result for 3.5-2.0-3.5 kN.

4. Results and discussion regarding residual stress

Figure 10 shows the residual stress distribution results for 3.5-2.0 kN and 3.5-2.0-3.5 kN at the spot welds. The residual stress distribution is different for each welding condition. This result indicates the possibility of changing the residual stress state in the spot welds.

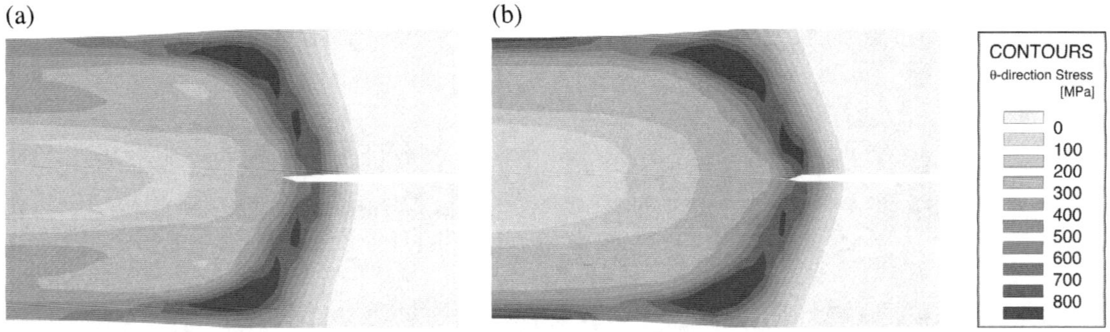

Figure 10. θ-direction residual stress distributions: (a) result for 3.5-2.0 kN, (b) result for 3.5-2.0-3.5 kN.

Therefore, in order to control the residual stress state, the welding conditions that change the force profile were applied. In these conditions, which are shown in Figure 11, the increase in the electrode force is varied (hereafter 3.5-2.0-5.5 kN and 3.5-2.0-7.5 kN).

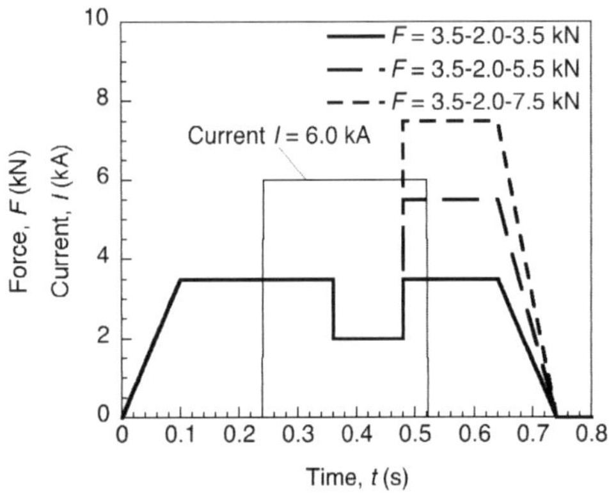

Figure 11. The force controlled condition profiles used to control the residual stress state.

Figure 12 shows the residual stress distribution results for each welding condition. Increasing the changed electrode force tended to decrease the high tensile residual stresses in the heat-affected zone. Figure 13 shows the residual stress distributions at a cross section of the spot welds. Figure 13(a) shows the calculated position of the cross section, and Figure 13(b) shows the residual stress distributions at that cross section. The shapes of the distributions are almost the same, but the peak values of the residual stress decrease compared with 3.5-2.0-3.5 kN and 3.5-2.0-7.5 kN.

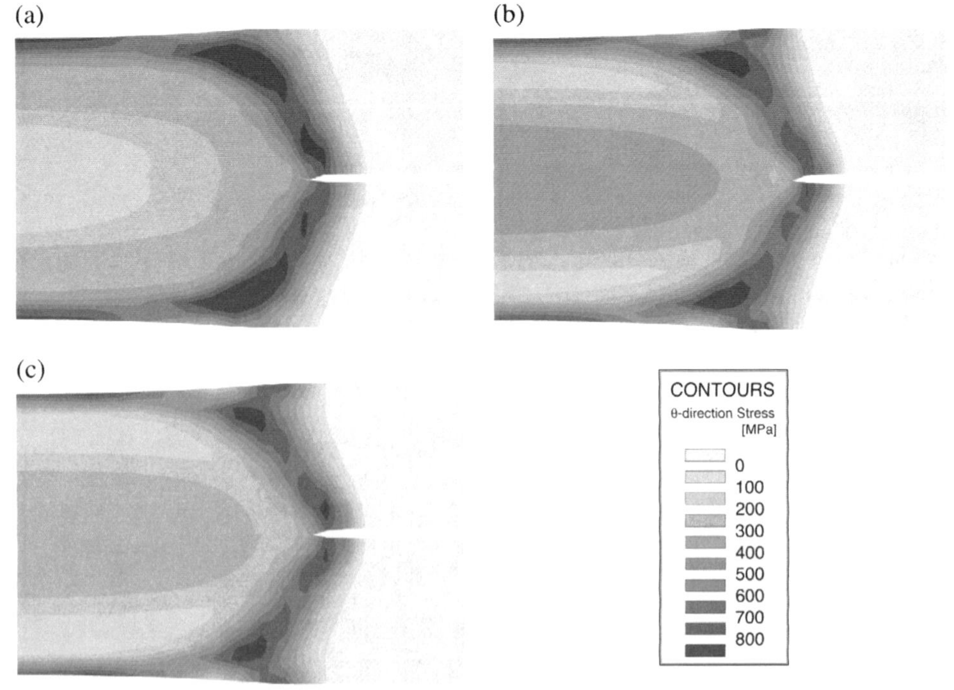

Figure 12. θ-direction residual stress distribution for each condition: (a) result for 3.5-2.0-3.5 kN, (b) result for 3.5-2.0-5.5 kN, (c) result for 3.5-2.0-7.5 kN.

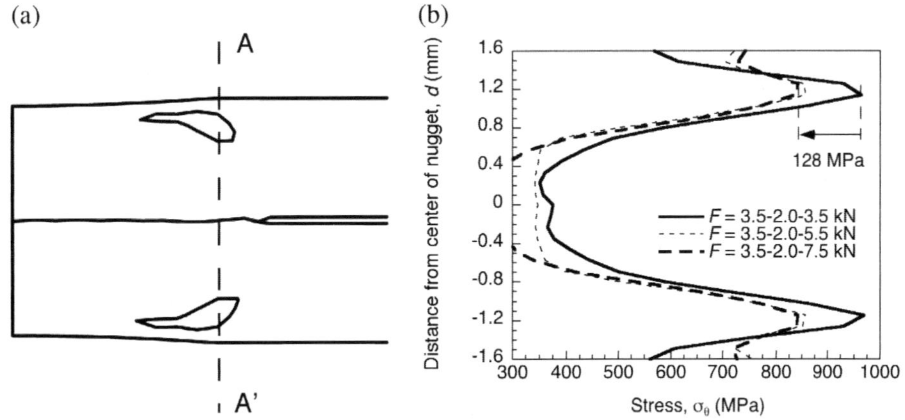

Figure 13. Comparison of the residual stress distribution for each condition: (a) schematic diagram of the measured cross section, (b) residual stress distribution at the A-A' section.

Additionally, it is well known that the residual stress at the corona bond influences the fatigue strength [4-6]. Therefore, the residual stresses at the corona bond in these numerical simulations are shown in Figure 14. The residual stress is tensile, as shown in previous research [13], and increasing the electrode force tends to decrease the tensile stress.

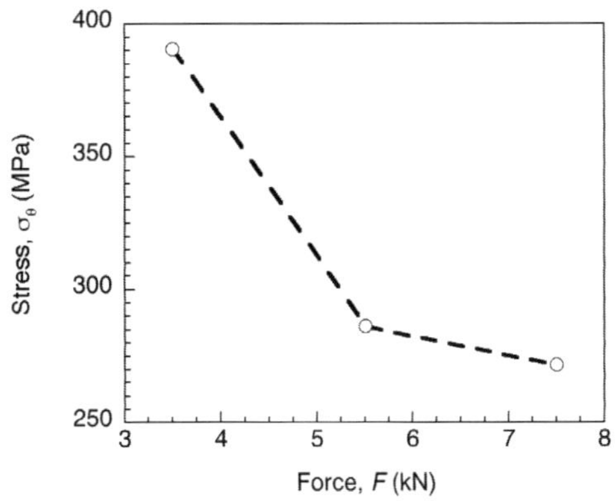

Figure 14. Residual stress at the corona bond for each force condition.

Finally, Table 1 shows a comparison of the nugget size and residual stress between the basic condition and electrode force controlled condition. As shown here, the appropriate force profile, which is considered the mechanism of formation for the nugget and residual stress, extends the nugget diameter and decreases the residual stress at the spot welds.

Table 1. Comparison of the nugget diameter, penetration ratio, and residual stress at the corona bond for the 3.5 kN and 3.5-2.0-7.5 kN force conditions.

	Nugget diameter (mm)	Penetration ratio (%)	Residual stress at corona bond (MPa)
3.5 kN	5.2	71	351
3.5-2.0-7.5 kN	5.7	77	271

5. Conclusions

In this study, the nugget diameter and the residual stress in spot welded high-strength steel were investigated using coupled electrical-thermal-microstructural-mechanical numerical simulations, and the effect of the electrode force profile on these characteristics were examined. The results indicate that an appropriate force profile extends the nugget diameter and decreases the residual stress at the spot welds.

Acknowledgements

This study was partly supported by Priority Assistance for the Formation of Worldwide Renowned Centers of Research – The Global COE Program (Project: Center of Excellence for Advanced Structural and Functional Materials Design) from the Ministry of Education, Culture, Sports, Science and Technology (MEXT), Japan.

References

[1] Kabasawa M, Funakawa Y, Kazuhiro O and Tamura M 1996 *QJJWS* **14**-4 754-61
[2] Sawhill J M, Watabe H and Mitchell J W 1977 *Welding J.* **56**-7 217-24
[3] Vural M, Akkuş A and Eryürek B 2006 *J. Mater. Process. Technol.* **176**-1 127-32
[4] Anastassiou M, Babbit M and Lebrun J L 1990 *Mater. Sci. Eng.* **125**-2 141-56
[5] Yang S Y, Son J K, Cho K S, Hong G S, Kim K S and Mo H K 2001 *STWJ* **6**-6 397-401
[6] Mirsalehi E S and Kokabi H A 2010 *Mater. Sci. Eng.* **527**-23 6359-63
[7] Spitsen R, Kim D, Flinn B, Ramulu M and Easterbrook T E 2005 *J. Manuf. Sci. Eng.* **127**-4 718-23
[8] Hiratsuka K 1971 *J. Jpn. Welding Soc.* **40**-1 13-24
[9] Shinozaki M, Kato T, Irie T and Takahashi I 1982 *TETSU-TO-HAGANE* **68**-9 1444-51
[10] Robin V, Sanchez A, Dupuy T, Soigneux J, Bergheau J M 2002 *Mathematical Modelling of Weld Phenomena 6* 997-1013
[11] ESI Group 2004 *SYSWELD 2004 Reference Manual*
[12] The Japan Society of Mechanical Engineering 1996 *Heat Transfer 4th Edition* (Japan MARUZEN)
[13] Nishiguchi K and Matsuyama K 1987 *Welding in the World* **25**-11/12 222-45

Surface modification of ferritic stainless steel by active screen plasma nitriding

H Nii[1] and A Nishimoto[2]

[1] Graduate School of Science and Engineering, Kansai University
3-3-35 Yamate-cho, Suita, Osaka 564-8680, Japan
[2] Department of Chemistry and Materials Engineering, Kansai University
3-3-35 Yamate-cho, Suita, Osaka 564-8680, Japan

E-mail: akionisi@kansai-u.ac.jp

Abstract. Plasma nitriding is a surface modification process with a low environmental impact. Active screen plasma nitriding (ASPN) is one of the new plasma nitriding technologies, and can eliminate problems related to conventional direct current plasma nitriding (DCPN). In this study, ferritic stainless steel SUS430 samples were treated by ASPN to increase their wear resistance without decreasing their corrosion resistance. ASPN was performed in a nitrogen-hydrogen atmosphere with 25%N_2 + 75%H_2 for 18 ks at 623 K, 673 K, 723 K, 773 K, and 823 K under 600 Pa using an SUS304 screen. When the sample was treated at 673 K by ASPN, the pitting corrosion resistance and wear resistance of its surface were improved because of the formation of the S^{α} phase and a deposited layer containing Ni on the sample surface.

1. Introduction

Nitriding is a surface modification technique that is divided into three types according to the state of the activated nitrogen. These are gas nitriding, salt bath nitriding, and plasma nitriding. Of these three types, plasma nitriding, which is referred to as direct current plasma nitriding (DCPN), offers the following advantages: no environmental pollution, low gas and energy consumption, short treatment time, and high nitrogen potential [1,2]. However, this DCPN process produces a glow-discharge on the component surface, which can cause electrical problems such as the edge effect, hollow cathodic discharge, and arcing. This results in nonuniform properties such as hardness and thickness of the surface layer.

Active screen plasma nitriding (ASPN) offers a solution to these electrical problems. ASPN can eliminate electrical problems because the component is isolated electrically, and the glow-discharge is produced on a screen that is mounted around the isolated component [2-8]. This process can be used to treat ceramics and polymers [5,7].

Among metallic materials, stainless steel is widely used because of its high corrosion resistance. However, stainless steel has poor wear resistance because of its low hardness. Therefore, if stainless steel is applied to sliding parts such as gears and shafts, it is necessary to improve its hardness [9-11]. Most studies related to nitriding of stainless steel have dealt with austenitic stainless steel, but very few have focused on ferritic stainless steel. Ferritic stainless steels do not contain nickel and are inexpensive and eco-friendly materials.

Published under licence by IOP Publishing Ltd

In this study, the ferritic stainless steel sample SUS430 was nitrided by ASPN to increase its wear resistance without decreasing its corrosion resistance.

2. Experimental procedure

The sample material was a ferritic stainless steel SUS430. The chemical composition (mass%) of SUS430 is 16.13% Cr, 0.23% Ni, 0.82% Mn, 0.003% Si, 0.09% C, and the balance is Fe. The sample disk was 20 mm in diameter and 5 mm in thickness. The sample surface was mechanically ground with 150- to 1500-grid SiC, finely polished with a suspension of 0.05 μm alumina, degreased ultrasonically in acetone, and dried in air, before placing it into the nitriding furnace.

ASPN experiments were carried out using a direct-current (DC) plasma nitriding unit (NDK, Inc. Japan, JIN-1S). Figure 1 presents a schematic representation of the ASPN apparatus. A quartz (SiO_2) rod was placed on the cathodic stage in order to construct an isolated stage. The sample was placed on the sample stage and was isolated from the cathodic screen and the anode. The distance between the screen and sample was 10 mm. The screen material used was an austenitic stainless steel SUS304 in order to improve the corrosion resistance of the sample surface by Ni. The shape was an expanded metal mesh with a mesh size of 8 × 3 mm and a thickness of 0.5 mm, formed into a cylinder with a diameter of 77 mm and a height of 100 mm. The screen was thoroughly degreased ultrasonically in acetone and mounted on the cathodic stage around the sample stage.

Nitriding was performed in a nitrogen-hydrogen atmosphere with $25\%N_2 + 75\%H_2$ for 18 ks at 623 K, 673 K, 723 K, 773 K, and 823 K under 600 Pa by the ASPN and DCPN processes. After placing the sample on the sample stage, the chamber was evacuated to ~3 Pa. Nitrogen and hydrogen were then introduced into the chamber and a DC bias voltage was applied, as shown in Table 1.

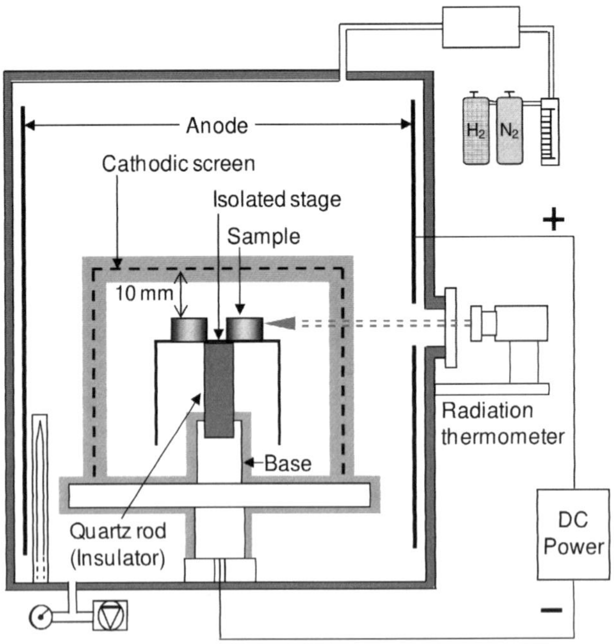

Figure 1. Schematic representation of the ASPN apparatus.

Table 1. Applied electric power of ASPN and DCPN process (W).

	623 K	673 K	723 K	773 K	823 K
ASPN	43	60	109	125	244
DCPN	64	88	150	240	355

After nitriding, the DC supply was switched off, and the sample was cooled to room temperature in the furnace. The nitriding temperature was monitored using a radiation thermometer, as depicted in Figure 1. Then, composition analyses of the sample surface were conducted using electron probe microanalysis (EPMA; JEOL, Japan, JXA-8800). The phase structures on the nitrided surface were determined using theta-2theta X-ray diffraction (XRD; RIGAKU, Japan, RINT-2550V). The entire top surface area of the nitrided samples was analyzed using XRD. Cross sections of each sample were first cut using a low-speed saw and then were polished and chemically etched. The nitrided microstructures of the surface and cross section were examined using an optical microscope and a scanning electron microscope (SEM; JEOL, Japan, JSM-6060LV). The hardness of the surface and the cross sections of the nitrided samples were measured using a Vickers microhardness tester (Matsuzawa, Japan, MXT50) under a load of 0.1 N. Five indentations were performed on each sample, and a 3-point average value (excluding both the maximum and minimum values) was used for the hardness. Wear testing was carried out at room temperature with a pin-on-disk tribometer. The conditions for wear testing were as follows: a running distance of up to 500 m, a wear load of 5 N, a rotation speed of 175 rpm, a wear radius of 6 mm, and a diameter of 6 mm for the alumina ball used for the counter material. After the wear test, the wear loss was calculated using the wear depth and the width of the wear track. A pitting corrosion test was carried out using a potentiostat. The sample was shielded by polytetrafluoroethylene (PTFE), exposing a 6 mm diameter area of the sample surface to the electrolyte. A 3.5 mass% NaCl aqueous solution was used as the electrolytic bath. DC polarization was performed potentiodynamically from -1.0 $V_{Ag/AgCl}$ to 1.5 $V_{Ag/AgCl}$. The anodic polarization curves were recorded at a sweep speed of 1.6 μV/s. On the basis of these polarization curves, the potential corresponding to a current of 1 A/m^2 was selected as the pitting potential.

3. Results and discussion

Figure 2 shows the appearance and EPMA line analysis of the treated sample surfaces. The DCPN sample has a non-uniform surface because of the edge effect. This edge effect occurs because of distortions of the electric field around the corners and edges of the sample, although the sample edge was well heated [12]. On the other hand, the ASPN sample had a uniform surface, and neither the edge effect nor arcing was observed. From the EPMA line analysis, the ASPN sample has a uniform nitrogen intensity distribution, but that of the DCPN sample differs between the edge and the center of the sample surface. In addition, nickel was detected on the ASPN sample. This indicates the presence of nickel from the screen material deposited on the ASPN sample surface.

Figure 3 shows an XRD pattern for the sample treated by ASPN and DCPN. While the phase of the sample surface treated by DCPN was different between the edge and center, the ASPN treated sample had a uniformly nitrided surface. When the sample was treated by ASPN, a γ'-Fe_4N phase was identified on the sample surface, unlike in the case of the DCPN sample. The nitriding mechanism in ASPN mainly involves the deposition and decomposition of activated clusters containing nitrogen and the metal elements of the cathodic screen [13-17]. Therefore, the γ'-Fe_4N phase was formed because of the high concentration of iron nitride, which was present on the screen.

In addition, the S^α phase, which is considered to be a supersaturated solid solution of nitrogen in the ferritic phase, was only observed when the sample was treated below 723 K [18]. When the sample was treated above 723 K, CrN was observed instead of the S^α phase. This indicates that the S^α phase decomposed into the α phase and CrN above 723 K. ($S^\alpha \rightarrow \alpha + CrN$)

Figure 4 shows SEM micrographs of the sample surfaces. On the DCPN sample surface, cauliflower-shaped particles were observed and the surface morphology was different at the edge and center. The surfaces of the ASPN samples have uniform and normal distributions of polygonal particles. It is believed that these correspond to Fe_xN particles from the screen material [13-17].

Figure 5 shows the cross-sectional microstructure of the nitrided samples. While the sample treated by DCPN has a non-uniform nitrided layer on its edge and center, the sample treated by ASPN has a

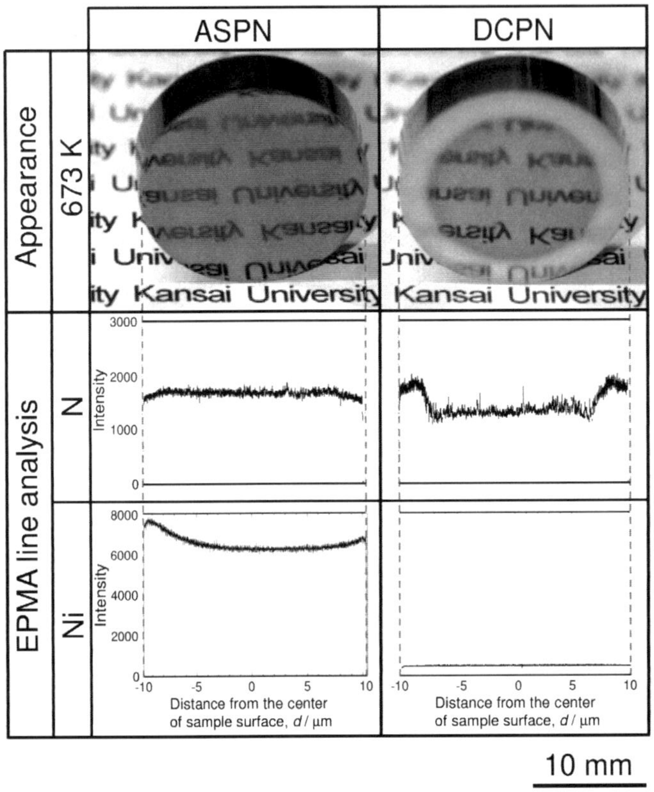

Figure 2. Appearance and EPMA line analysis of samples treated by ASPN and DCPN at 673 K.

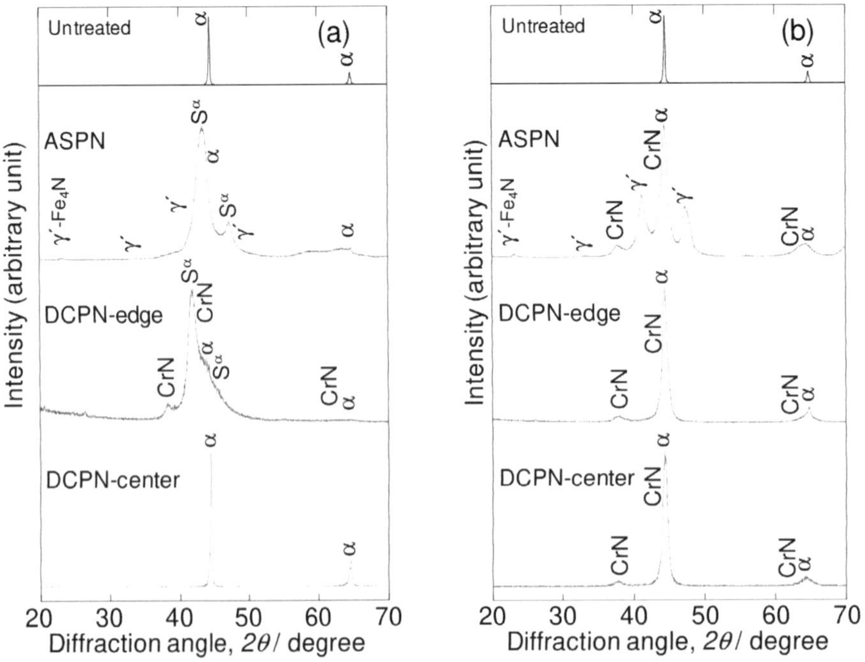

Figure 3. XRD patterns of untreated sample and samples treated at (a) 673 K and (b) 773 K.

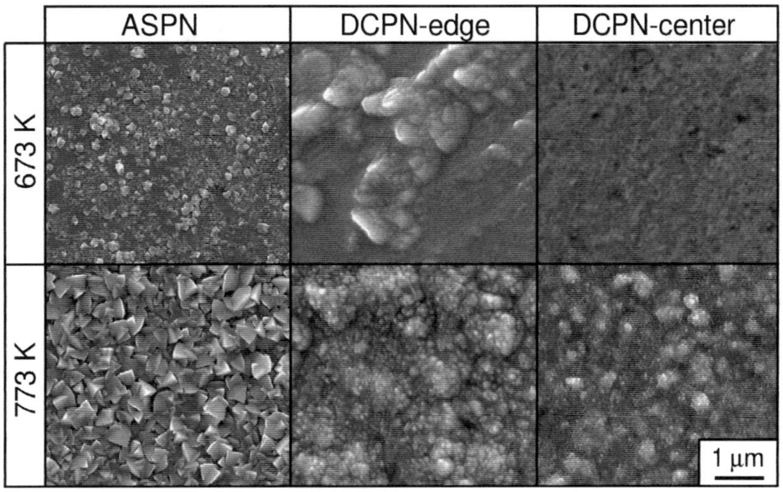

Figure 4. SEM micrographs of samples treated by ASPN and DCPN.

uniform nitrided layer. A white area near the surface was observed when the sample was treated at 673 K, and a black area near the surface was observed when the sample was treated at 773 K. The white area, which was not etched, is considered to be the S^{α} phase. On the other hand, the black area, which was highly etched, is due to the precipitation of CrN and a decrease in the chromium content of the ferritic stainless steel matrix.

Figure 6 shows the cross-sectional hardness distribution. The hardened areas corresponded to the white or black areas observed in the cross-sectional microstructure. The thickness of the hardened areas increased with increasing treatment temperature.

Figure 7 shows the results for wear loss for samples treated by ASPN and DCPN. With an increase in the nitriding temperature, the wear loss for each sample was less than that for the untreated sample. When the sample was treated above 723 K, the wear loss was very small. The decrease in the wear

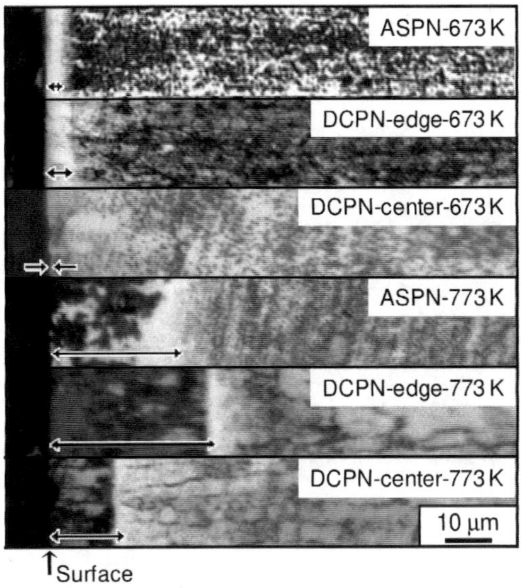

Figure 5. Cross-sectional microstructure of samples treated by ASPN and DCPN.

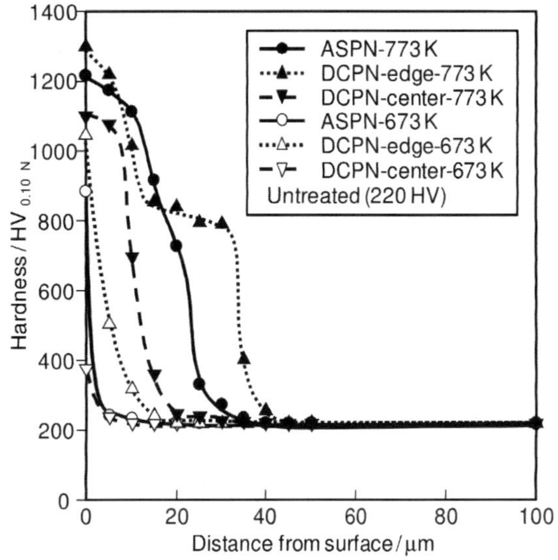

Figure 6. Cross-sectional hardness distribution for samples treated by ASPN and DCPN.

loss corresponds to the change of the surface hardness. Figure 8 shows the surface hardness for samples treated by ASPN and DCPN. With an increase in the nitriding temperature, the surface hardness increased because of either the formation of the S^{α} phase or the precipitation of CrN, as shown in Figure 3. This increase of the surface hardness resulted in a decreased wear loss.

Figure 9 shows the results for the pitting potential in a 3.5 mass% NaCl aqueous solution for samples treated by ASPN and DCPN. While the pitting potentials for DCPN samples were lower than that for the untreated sample (0.184 $V_{Ag/AgCl}$), those for the ASPN samples were higher than those for the untreated and DCPN samples. The ASPN samples have a thin layer on the surface, as shown in Figure 10. However, this thin layer was not observed at the edge and center of the DCPN samples. The layer produced by ASPN was about 200 nm thick, and contained Fe, Cr, Ni, and N. Therefore,

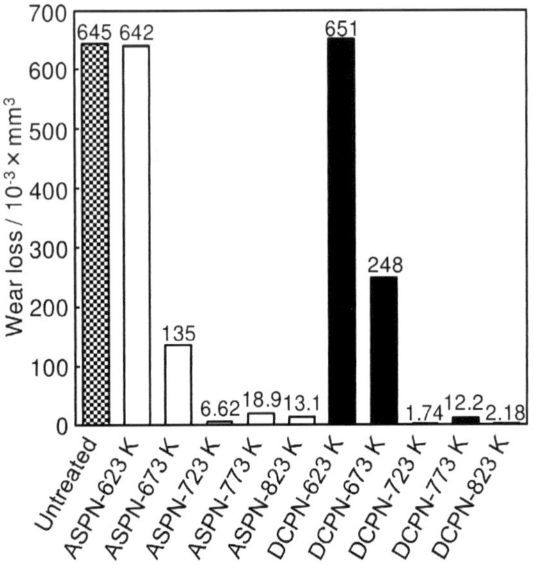

Figure 7. Wear loss for untreated sample and samples treated by ASPN and DCPN.

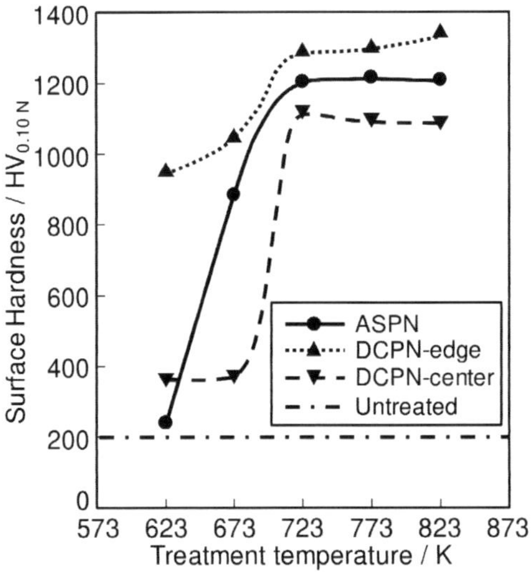

Figure 8. Surface hardness for samples treated by ASPN and DCPN.

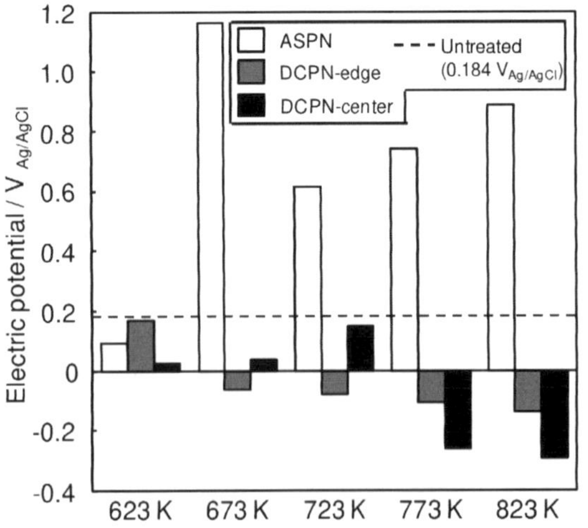

Figure 9. Pitting potential in 3.5 mass% NaCl aqueous solution for untreated sample and samples treated by ASPN and DCPN.

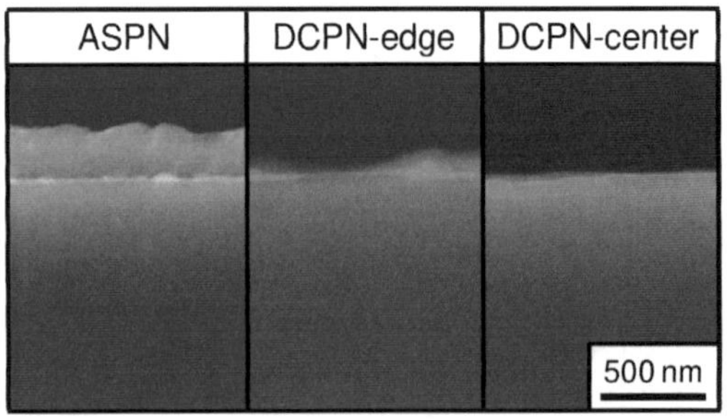

Figure 10. SEM micrographs of the cross section of samples treated by ASPN and DCPN at 673 K.

the layer was assumed to be a deposited layer of the screen material. The high pitting potentials of the ASPN samples were due to a layer containing Ni, which was coated on the sample surface. In particular, the ASPN sample treated at 673 K showed a high pitting potential. This indicates that the presence of the S^{α} phase under the deposited layer plays an important role for the pitting potential, because the corrosion resistance of the nitrided layer depends on the substrate [18].

4. Conclusions

Ferritic stainless steel SUS430 samples were nitrided by the ASPN process using an austenitic stainless steel SUS304 screen to improve their wear resistance without decreasing their corrosion resistance. Although samples treated by DCPN were found to have improved wear resistance, the pitting corrosion resistance values decreased and the nitrided layer was non-uniform. On the other hand, the sample treated at 673 K by ASPN had improved wear resistance, and the pitting corrosion resistance increased significantly because of the formation of the S^{α} phase and a deposited layer containing Ni on the sample surface.

References

[1] Sun Y and Bell T 1991 *Mater. Sci. Eng.* A **140** 419
[2] Nishimoto A, Nagatsuka K, Narita R, Nii H and Akamatsu K 2010 *Surf. Coat. Technol.* **205**(Suppl. 1) S365
[3] Li C X, Bell T and Dong H 2002 *Surf. Eng.* **18** 174
[4] Li C X and Bell T 2004 *Wear* **256** 1144
[5] Li C X and Bell T 2005 *J. Mater. Process. Technol.* **168** 219
[6] Ahangarani Sh, Mahboubi F and Sabour A R 2006 *Vaccum* **80** 1032
[7] Li C X, Dong H and Bell T 2006 J. *Mater. Sci.* **41** 6116
[8] Nishimoto A, Bell T E and Bell T 2010 *Surf. Eng.* **26** 74
[9] Dong H 2010 *Int. Mater. Rev.* **55** 65
[10] Nagatsuka K, Nishimoto A and Akamatsu K 2010 *Surf. Coat. Technol.* **205**(Suppl. 1) S295
[11] Kliauga M and Pohl M 1998 *Surf. Coat. Technol.* **98** 1205
[12] Alves Jr C, da Silva E F and Martinelli A E 2001 *Surf. Coat. Technol.* **139** 1
[13] Asahina S, Hosotani T and Kanayama N 2009 *J.Jpn. Soc. Heat Treat.* **49**(Spec. Iss.) 49
[14] Ahangarani Sh, Sabour A R and Mahboubi F 2007 *Appl. Surf. Sci.* **254** 1427
[15] Li C X 2010 *Surf. Eng.* **26** 135
[16] Zhao C, Li C X, Dong H and Bell T 2006 *Surf. Coat. Technol.* **201** 2320
[17] Li C X and Bell T 2003 *Heat. Treat. Met.* **30** 1
[18] Gontijo L C, Machado R, Casteletti L C, Kuri S E and Nascente P A P 2010 *Surf. Eng.* **26** 265

Comparison of weld residual stress measurement results in low alloy welds between X-ray diffraction and stress relief methods

Asahi Sugahara[1], Shigetaka Okano[1], Tadafumi Hashimoto[1], and Masahito Mochizuki[1]

[1] Graduate School of Engineering, Osaka University, 2-1 Yamada-oka, Suita, Osaka, Japan

E-mail: sugahara@mapse.eng.osaka-u.ac.jp

Abstract. In this study, a comparison of weld residual stress measurements between the X-ray diffraction technique and the stress relief technique is performed with the focus on the effect of the thickness on the stress measurement results. As the difference in the thickness becomes smaller, the difference in the stress measurement results also becomes smaller. At the weld center, where the stress gradient is in the thickness direction, the difference in the thickness is not negligible. In other words, it is concluded that the X-ray diffraction method is advantageous because it can evaluate the unique stress generated only in the surface layer.

1. Introduction

It is important to ensure structural safety and reliability by extending the lifetime of welded structures[1]. An appropriate design concerning the lifetime of a welded structure is a major issue in many industrial products, such as ships, bridges, construction machines, power plants, automobiles, and railway transport systems. For the lifetime assessment of a welded structure, welding-induced metallurgical changes and stress-strain fields should be taken into account. In particular, it is well known that the weld residual stress is an important factor for evaluating the structural soundness of a welded structure, because a variety of fractures, such as brittle fracture, fatigue fracture and corrosion fracture, can be assisted by the tensile residual stress. It is also important to modify the lifetime assessment of an in-service structure by considering age degradation, including corrosion, cracking, material deterioration, and redistribution of the residual stress. Therefore, the demand for a non-destructive measurement of the weld residual stress distribution is increasing.

Various measurement methods of the weld residual stress have been proposed[2]: saw-cutting, center hole drilling (CHD), X-ray diffraction, neutron diffraction, instrumented indentation (IIT), and deep hole drilling (DHD). The X-ray diffraction method has been drawing attention as a non-destructive technique for measuring the residual stress in microscopic detail[3,4,5]. However, the X-ray stress measurement result can be influenced by the microscopic characteristics at the surface layer because the depth of X-ray penetration is within several dozen micrometers[6,7,8]. Since the weld zone has a heterogeneously distributed microscopic material structure and stress distribution, the difference and similarity of the stress measurement results between X-ray diffraction and the conventional stress relief method[9,10,11] should be clarified to appropriately understand the X-ray stress measurement results.

In this study, a comparison of the stress measurement results between the X-ray diffraction method and the stress relief method is performed with the focus on the effect of the thickness. Based on the comparison results, the features of the X-ray stress measurement are discussed.

2. Experimental procedure

2.1. Welded specimens

In this experiment, the specimen is a plate of low alloy steel A533B. The dimensions of the plate are 100 mm in length, 100 mm in width, and 20 mm in thickness, as shown in Figure 1. The chemical composition of this low alloy steel is shown in Table 1. In mechanical property, the yield stress of the low alloy steel A533B is 497 MPa. To measure the residual stress due to welding, the machined layers on the surface at the center of the plate to be welded are removed by electrolytic polishing. In electrolytic polishing, the current density is set to 0.48 A/cm^2 in 10 min, using 5% perchlorate methanol.

In the experiment, bead-on-plate welding is performed on the center of the low alloy steel plate. The welding method is Gas Tungsten Arc (TIG) welding with 100% Ar gas as the assist gas. The welding condition is shown in Table 2.

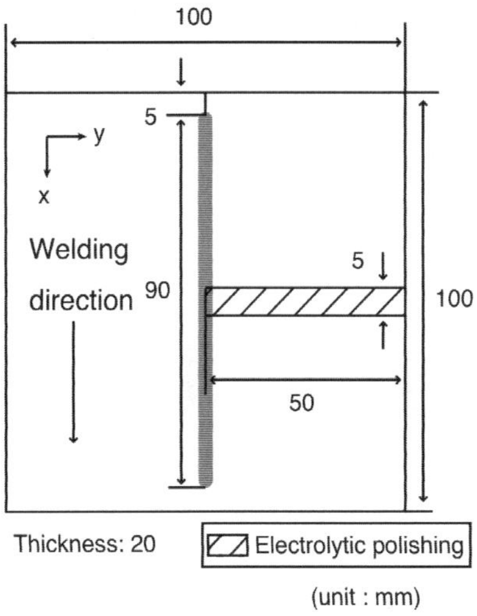

Figure 1. Dimensions of the specimen to be welded.

Table 1. Chemical composition of the low alloy steel used (mass%).

C	Si	Mn	P	S	Ni	Cu	Cr	Mo	Al	Fe
0.12	0.26	1.43	0.06	0.02	0.53	0.02	0.01	0.51	0.038	Bal.

Table 2. Welding condition.

Welding current (A)	200
Welding speed (mm/s)	2
Arc length (mm)	3
Flow rate of shielding gas (L/min)	15

2.2. Stress measurement using the stress relief method

In the conventional stress relief method, the biaxial stresses, σ_x (longitudinal direction to the welding line) and σ_y (transverse direction to the welding line), are calculated using equations (1) and (2), which convert the relieved elastic strain in the specimen by saw-cutting to the original stress before saw-cutting.

$$\sigma_x = \frac{2}{R} \cdot \frac{1}{1-\nu^2} E(\Delta\varepsilon_x + \nu \cdot \Delta\varepsilon_y) \quad (1)$$

$$\sigma_y = \frac{2}{R} \cdot \frac{1}{1-\nu^2} E(\Delta\varepsilon_y + \nu \cdot \Delta\varepsilon_x) \quad (2)$$

where R is the gauge factor (2.08±1%), E is the mechanical Young's modulus (229.9 GPa), and ν is the mechanical Poisson's ratio (0.272). The gauge length of the strain gauge used in the measurement experiment is 1 mm. The measurement locations are shown in Figure 2(a). The dimensions of the broken specimen are 10 mm in length, 10 mm in width, and 20 mm in thickness. To improve the measurement resolution of 5 mm in the transverse direction, two measurement lines with a 10 mm sift along the welding direction are determined because it is known that the weld residual stress is constant in the center of the specimen.

Furthermore, to compare the stress measurement results with the focus on the difference in thickness between the X-ray diffraction method and the stress relief method, the broken pieces are gradually cut into slices from a thickness of 20 mm to 3 mm. In the weld metal, broken pieces 3 mm in thickness are gradually reduced by mechanical polishing to thicknesses of 0.7 mm and 0.3 mm, as shown in Figure 2(b).

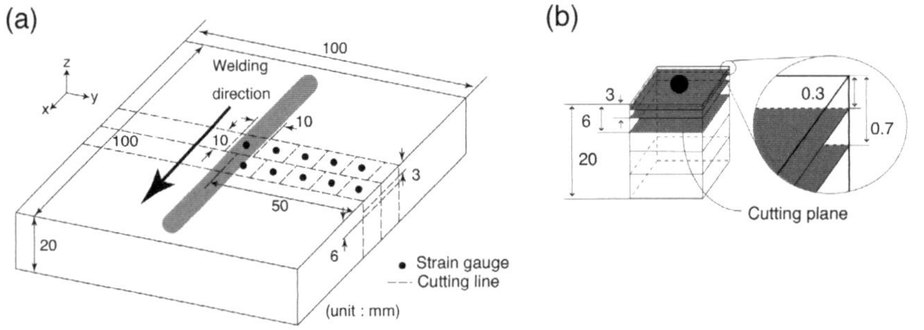

Figure 2. Measurement locations in the stress relief method. (a) Measurement locations, (b) Slices cut along the thickness.

2.3. Stress measurement using the X-ray diffraction method

2.3.1. $sin^2\psi$ method[12]. For the biaxial stress measurement on the welded surface, the $sin^2\psi$ method, which is recommended as the X-ray stress measurement standard, is used. Unless otherwise noted, the X-ray diffraction method in this paper corresponds to the $sin^2\psi$ method. The X-ray stress measurement is performed by the fixed ψ method of the ψ-diffractometer method using an MSF (Rigaku Corporation). The X-ray beam size is set to 4x4 mm in consideration of the size of the coarse grains at the welded zone. The radiation source used is Cr$K\alpha$, the tube voltage is 30 kV, and the tube current is 10 mA. The scattering angle is determined by the half-value breadth method using the 211 reflection of α-Fe. The ψ angles are 7 points equally spaced between 0 and 45 deg. The X-ray elastic constants are the Young's modulus E_{211} of 249.5 GPa and the Poisson's ratio ν_{211} of 0.272. The measurement location is shown in Figure 3. The distance between each measurement location in the X-ray diffraction method is 5 mm, which is the same as that in the stress relief method.

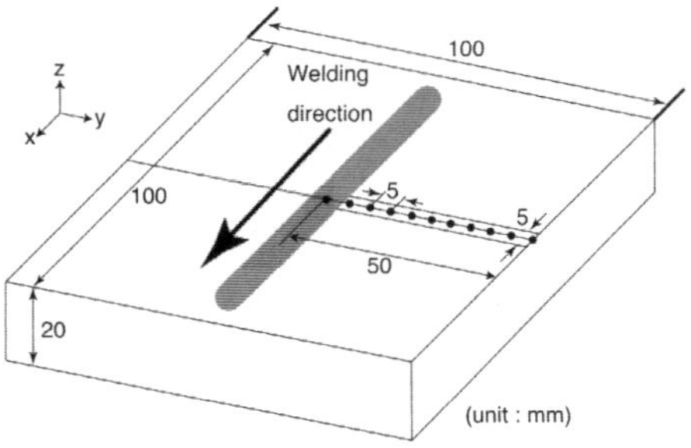

Figure 3. Measurement location for the X-ray diffraction method.

2.3.2. 2D method[13]. The 2D method of D8 Discover with a GADDS (Bruker AXS Corporation) is used to obtain a more detailed residual stress distribution along the thickness of the cross section because the X-ray beam size can converge. It has been shown that the X-ray stress measurement results between the $sin^2\psi$ method and the 2D method are in good agreement[14]. The fluctuation angle is within ±4° in the wavelength distribution direction (ω direction in the diffractometer), so that reciprocal lattice points are on the Ewald sphere of Kα1 and Kα2. In addition, the 211 reflection of α-Fe is used. The number of measurement locations is 21 points in 120 s per 1-Flame with varying φ and ψ. For the X-ray elastic constants, the Young's modulus E_{211} is 249.5 GPa, and the Poisson's ratio ν_{211} is 0.272. Details of the measurement location in the 2D method are shown in Figure 4. As shown in Figure 4, the transverse component of the residual stress σ_y is evaluated along the thickness of the cross section using the specimen, which was cut out from the center of the welded plate.

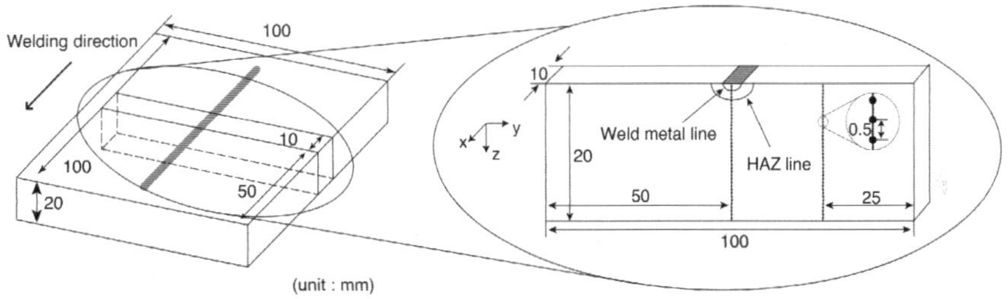

Figure 4. Measurement locations on the cross section thickness.

3. Experimental results and discussion

A comparison of the weld residual stress measurement results between the X-ray diffraction method and the stress relief method is shown in Figure 5. The weld metal is within 3 mm from the center line, the HAZ (Heat-Affected-Zone) is within 3 mm to 5 mm, and the base metal is more than 5 mm from the center line. In the base metal (especially more than 15 mm from the center line), the results are in good agreement. However, an obvious difference appears at the weld center. To discuss the effect of the difference in the thickness on the stress measurement results, broken pieces are gradually cut into slices from a thickness of 20 mm to 3 mm. As shown in Figure 6, the stress measurement results in the base metal are not influenced by the thickness (thickness of the broken pieces), whereas the stress measurement results at the weld center are obviously influenced by the thickness. The results at the weld center are not yet in good agreement.

A broken piece 3 mm in thickness, which is removed from the weld center, is then reduced gradually by mechanical polishing to 0.7 mm and 0.3 mm in thicknesses. The stress measurement results are shown in Figure 7. This figure shows that as the thickness of the broken piece becomes thinner, the difference in the measurement results between the X-ray diffraction method and the stress relief method becomes smaller. It is considered that the X-ray stress measurement result is in good agreement with that of the stress relief method using the specimen of the same thickness between the X-ray diffraction method and the stress relief method.

To discuss the difference in the measurement results shown only at the weld center, the stress distribution along the thickness is measured by the X-ray diffraction method (2D method) on the cross section thickness, as shown in Figure 4. The measurement results of the transverse component of the weld residual stress at the weld center and the base metal are shown in Figures 8(a) and (b), respectively. The measurement result at the point z = 0, which is measured using the $sin^2\psi$ method, is also shown together. At the base metal, no gradient of the stress distribution is shown, whereas an obvious gradient of the stress distribution is shown at the weld center. It is considered that the change of the measured transverse stress from 0 to compressive when the thickness changes from 20 mm to 6 mm, as shown in Figure 6(b), is due to the existence of compressive stresses at the area from the bottom surface (20 mm) to 6 mm, as shown in Figure 8(a). It is concluded that this gradient of stress distribution along the thickness causes the difference in the measurement results at the weld center due to the thickness (thickness of the broken pieces). A drastic change of the stress from compressive to tensile is shown very close to the top surface only at the weld center. This tends to be shown only in the low carbon steel in which the phase transformation occurs. It is considered that the drastic change of the stress is possibly influenced by local re-distribution of the stress due to very rapid cooling only in the thin layer at the surface and the phase transformation. Thus, using the X-ray diffraction method, the unique stress generated only in the surface layer can be evaluated.

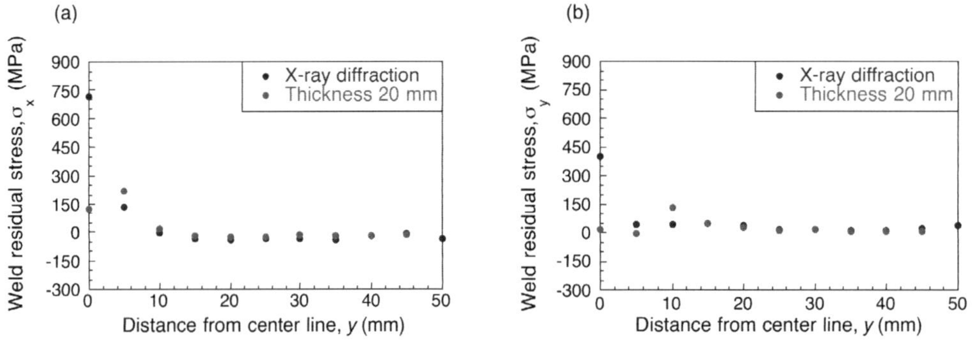

Figure 5. Comparison of the residual stress between the stress relief method and the X-ray diffraction method. (a) Longitudinal component, (b) Transverse component.

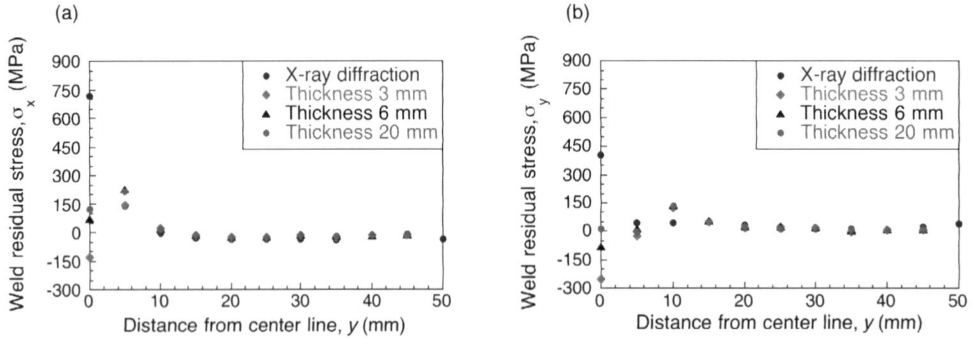

Figure 6. Effect of the thickness on measured residual stress when broken pieces are cut into slices 3 mm in thickness. (a) Longitudinal component, (b) Transverse component.

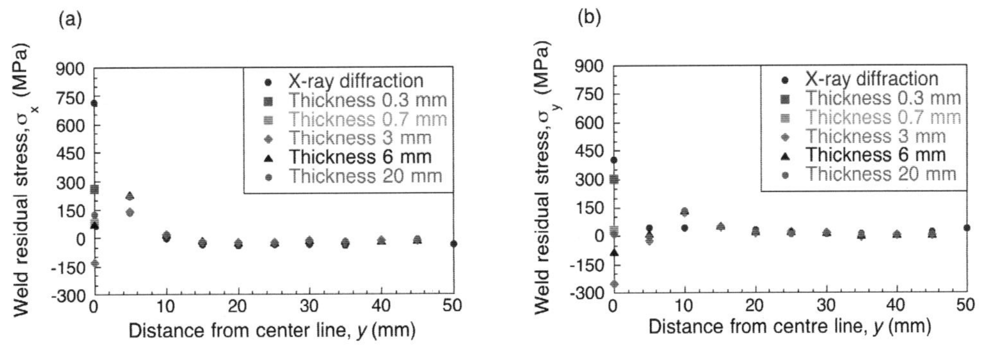

Figure 7. Effect of the thickness on measured residual stress when broken pieces are shaved to 0.3 mm in thickness. (a) Longitudinal component, (b) Transverse component.

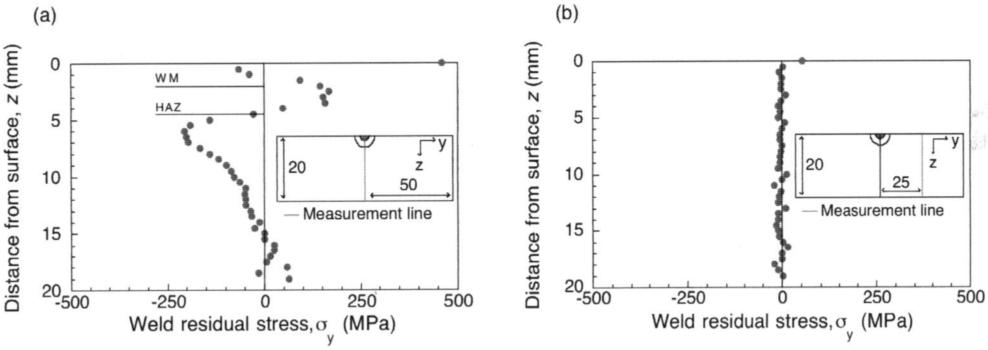

Figure 8. Residual stress distribution along the cross-section thickness of the plate. (a) in the weld metal (y=0 mm from the center line), (b) in the base metal (y=25 mm from the center line).

4. Conclusions

By comparing the weld residual stress measurement results of the X-ray diffraction method and the stress relief method, it is shown that the effects of the thickness on the stress measurement results are not negligible when a stress gradient exists along the thickness. As the difference in the thickness becomes smaller, the difference in the stress measurement results also becomes smaller. In particular, at the weld center, the difference in the stress measurement results between the X-ray diffraction method and the stress relief method tends to become larger, because a stress gradient exists along the thickness. Using the X-ray diffraction method, the unique stress generated only in the surface layer can be evaluated.

Acknowledgements

This study was supported by Priority Assistance for the Formation of Worldwide Renowned Centers of Research - The Global COE Program (Project: Center of Excellence for Advanced Structural and Functional Materials Design) from the Ministry of Education, Culture, Sports, Science and Technology (MEXT), Japan and a Grant-in Aid for Young Scientists (b) (No. 23760096).

References

[1] Sato K, Mukai K, Toyoda M 1979 *Welding engineering* (Science and Engineering, Inc.)
[2] Watanabe M, Sato K 1965 *Welding Mechanics and Applications* (Asakura, Inc.)
[3] Japan Society of Materials Science 2002 *X-ray stress measurement standard*

[4] Tanaka K, Suzuki K, Akiniwa Y 2006 *X-ray evaluation of residual stress –Fundamentals and Applications-* (Yokendo)
[5] Hanabusa T, Fujiwara H 1981 *Residual Stress in Uni-Directionally Deformed Surface Layer and d-sin$^2\psi$ Diagrams by X-Ray Stress Measurements* (Proceedings of the Japan Society of Materials Science) **30** (338) 1095-1101
[6] Goto T, Ohya S 1998 *The Fundamental Principles and Techniques of X-ray Stress Measurement and Current Developments in the Field: 1 The Fundamental Principles and Techniques of X-ray Stress Measurement and Current Development* (Materials) **47**(11) 1188-1194
[7] Tanaka K, Akiniwa Y 1998 *The Fundamental Principles and Techniques of X-ray Stress Measurement and Current Developments in the Field: 2. Microstress and Macrostress* (Materials) **47**(12) 1301-1307
[8] Hanabusa, T 1999 *The Fundamental Principles and Techniques of X-ray Stress Measurement and Current Developments in the Field: 3. Surface Stress and Textured Materials* (Materials) **48**(1) 89-95
[9] Ueda Y, Fukuda K, Nakacho K, Endo S 1975 *A New Measuring Method of Residual Stresses with the Aid of Finite Element Method and Reliability of Estimated Values* (Proceedings of the Japan Society of Naval Architects) **138** 499-507
[10] Ueda Y, Fukuda K, Tanigawa M 1979 *New measuring method of 3-Dimensional residual stresses based of theory of inherent strain* (Proceedings of the Japan Society of Naval Architects) **145** 203-211
[11] Ueda Y, Fukuda K, Fukuda M 1980 *A Measuring Theory of Three Dimensional Residual Stresses in Long Welded Joints* (Proceedings of Japan Welding Society) **49**(12) 845-853
[12] Miyoshi, Y 1989 *X-ray Analysis of Residual Stress in Surface Layer* (Journal of Precision Engineering) **55**(8) 1355-1359
[13] Bob B He 2003 *Introduction to two-dimensional X-ray diffraction* (Bruker Advanced X-ray Solutions, Inc.
[14] Hashimoto T, Osawa Y, Hirano S, Mochizuki M, Nishimoto K 2011 *Accuracy Improvement of X-ray Stress Measurement on GTA Weld of Type 600 Ni Based Alloy* (Materials) **60**(7) 604-609

Numerical simulation of the effect of low-temperature transformation expansion on residual stress in cold cracking test specimens of different restraint factors

Noriyoshi Kubota[1], Yoshiki Mikami[1], Masahito Mochizuki[1] and Kazuo Hiraoka[2]

[1] Division of Materials and Manufacturing Science, Osaka University, 2-1 Yamadaoka, Suita, Osaka 565-0871, Japan
[2] Joining and Welding Research Institute, Osaka University, 11-1 Mihogaoka, Suita, Osaka 565-0047, Japan

E-mail: kubota@mapse.eng.osaka-u.ac.jp

Abstract. Ni-Cr based high-strength weld metals have been developed to reduce preheating processes. In the development process of a welding wire, the evaluation of cold cracking resistance is important. In this study, the effect of low-temperature transformation expansion on the residual stress reduction of welds with different restraint factors in cold cracking test specimens was examined by numerical simulation. The results indicated that the weld metal with low-temperature transformation expansion exhibits the effect of residual stress reduction at a high restraint factor. The reduction decreased for a low restraint factor. Even if the restraint factor changes, the distribution of the restraint factor in the Y-groove weld cracking test is different from that in the H-type restrained weld cracking test. Distributions of residual stresses at the weld root in cold cracking in the Y-groove test and the H-type test have different tendencies. Thus, the difference should be considered when conducting the cold cracking test.

1. Introduction

The use of HT980-class high-strength steels is advantageous because thinner steel plate can be used and a reduced amount of steel and welding consumables leads to a more ecologically beneficial fabrication. In general, cold cracking tends to occur in welds of high-strength steels. Therefore, preheating or post-weld heat treatment processes are essential. However, these processes lead to increased fabrication cost, which limits the applications of HT980-class steels. To realize the use of HT980-class high-strength steels without heat treatment processes, Ni-Cr based weld metals have been developed [1], [2]. Because Ni-Cr welding wire exhibits low-temperature transformation expansion, tensile residual stress, which affects the occurrence of cold cracking, is reduced. In the development of a welding wire, the evaluation of its cold cracking resistance is essential. In cold cracking, the effect of transformation expansion on residual stress is dependent on the restraint factor. It is well known that cold cracking tends to occur for a higher restraint factor at a weld joint [3]. An actual steel structure has varying restraint factor at the weld joint and cold cracking behaviors are changed by the restraint factor. Thus, definitions of the relationship between the cold cracking tendency and the restraint factor at the weld metal are important. In this study, a series of numerical analyses of the residual stress distribution was performed with varying transformation start temperatures of the weld metal in cold cracking test specimens of different restraint factors. The effect of the restraint factor and transformation temperature on the residual stress in the weld metal of high-strength steel is discussed.

2. Numerical simulation method of residual stress considering phase transformation

In this section, a finite element model and material properties are described for the numerical simulation. SYSWELD, the finite element analysis software for heat treatment and welding, was used.

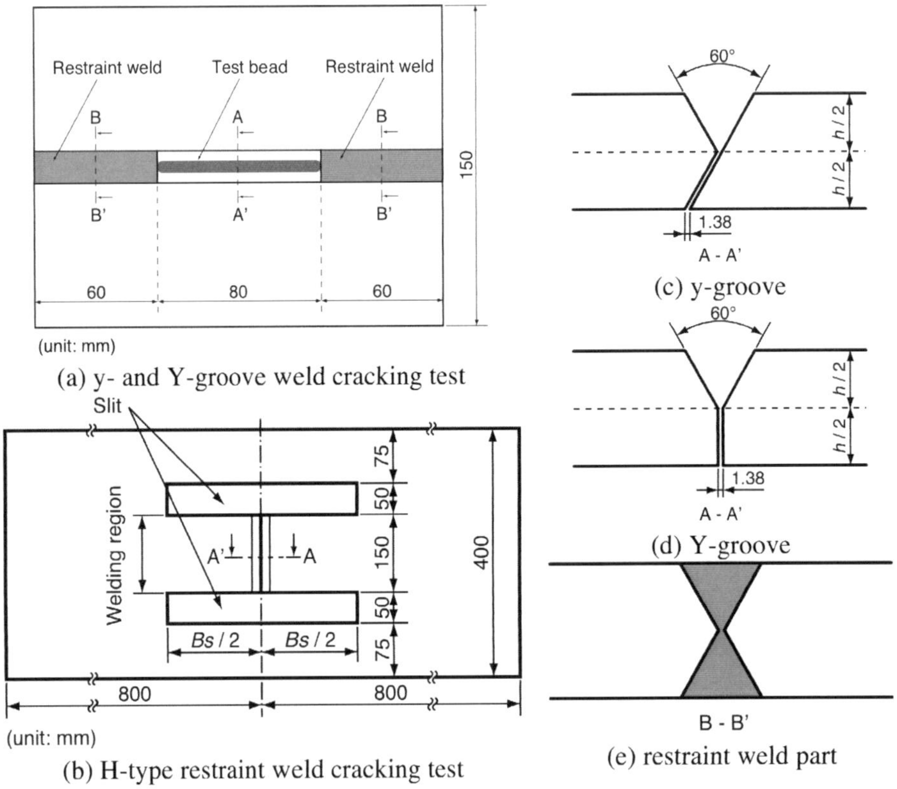

Figure 1. Specimen and dimensions of weld cracking tests.

2.1. *Numerical analysis models and condition*

In this study, a series of numerical analyses with varying transformation start temperatures of the weld metal in cold cracking test specimens of different restraint factors was performed to clarify the effect of transformation expansion on the residual stress. The y- and Y-groove weld cracking test [4] based on JIS Z 3158 (hereinafter called the "y-groove test" or the "Y-groove test") and the H-type restrained weld cracking test [5] with Y-groove (hereinafter called the "H-type test") were used. The shapes and dimensions of these specimens are shown in Figure 1. To change the restraint factor, thickness of the specimens was set to 10, 20, and 25 mm, and in the H-type restrained weld cracking test, the lengths of slits B_s were varied at 30, 50, 140, 300, and 800 mm.

The finite element models used in the simulation are shown in Figure 2. The same finite element mesh divisions around the welds were set for all analysis cases. The Y-groove test and the H-type test are a half model, considering symmetry. The slit in the H-type test was modeled by deactivating finite elements corresponding to the location and size of the slit.

A thermal-elastic-plastic analysis was performed with a moving heat source and consideration of the deposition of the weld metal. The welding speed and heat input were 4.16 mm/s and 1400 J/mm, respectively. The coefficient of heat transfer was 1.5×10^{-5} J/mm^2·°C·s, and the ambient temperature was 30°C.

In addition, the restraint factor that represents the restraint intensity of the weld metal was also calculated. The calculation of the restraint factor was performed as follows: detection of the load-displacement relationship by giving forced displacement to the groove. The slope of the load-displacement curve was defined as the restraint intensity. When the restraint intensity is high, the restraint to the weld metal is considered high. Restraint factors for each specimen were calculated and the relationship between the specimen geometries was discussed.

2.2. *Material properties*

In this analysis, the temperature- and microstructure-dependent material properties of HT980 steel, shown in Figure 3 and Figure 4, were used. The material properties were determined based on the material database [6] [7] as well as the experimental results obtained by the high-temperature tensile test and simulated thermal cycle test. The density used in the simulation was considered to be constant at 7.8×10^{-6} kg/mm^3. The transformation properties were observed at A_{C1} = 650°C and A_{C3} = 850°C for the heating process, where M_s was defined by the Koistinen-Marburger formula [8] at the cooling process. The effect of transformation expansion of the weld metal on the residual stress was evaluated by changing the martensite transformation temperature to 100°C, 218°C, and 418°C. The M_s of the base metal was set to 418°C irrespective of the weld metal.

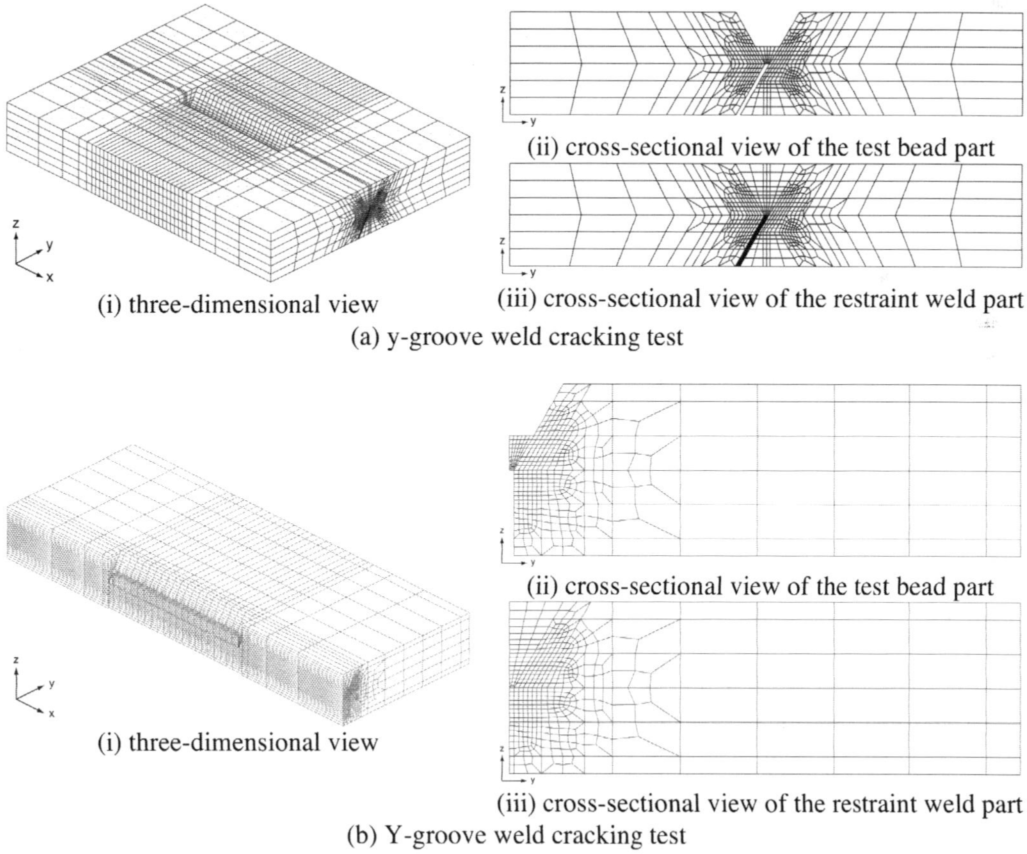

Figure 2. Finite element models.

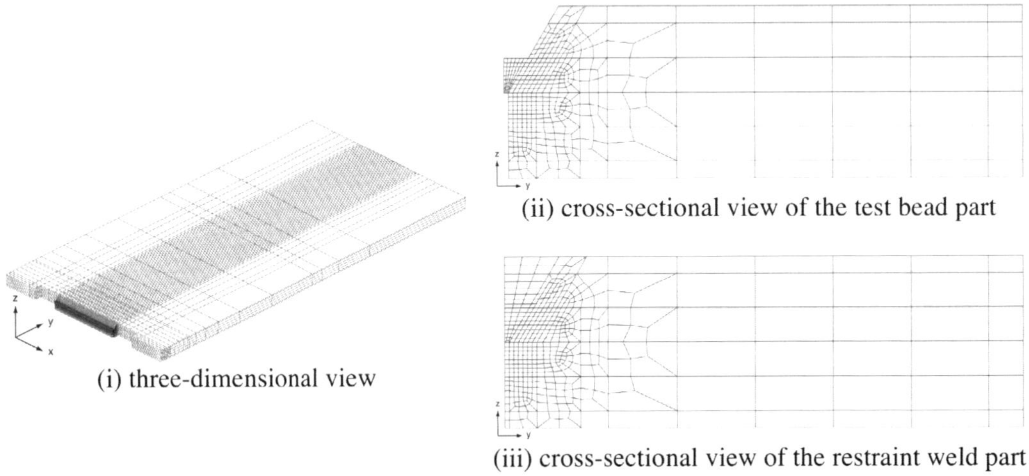

(c) H-type restrained weld cracking test

Figure 2. Continued.

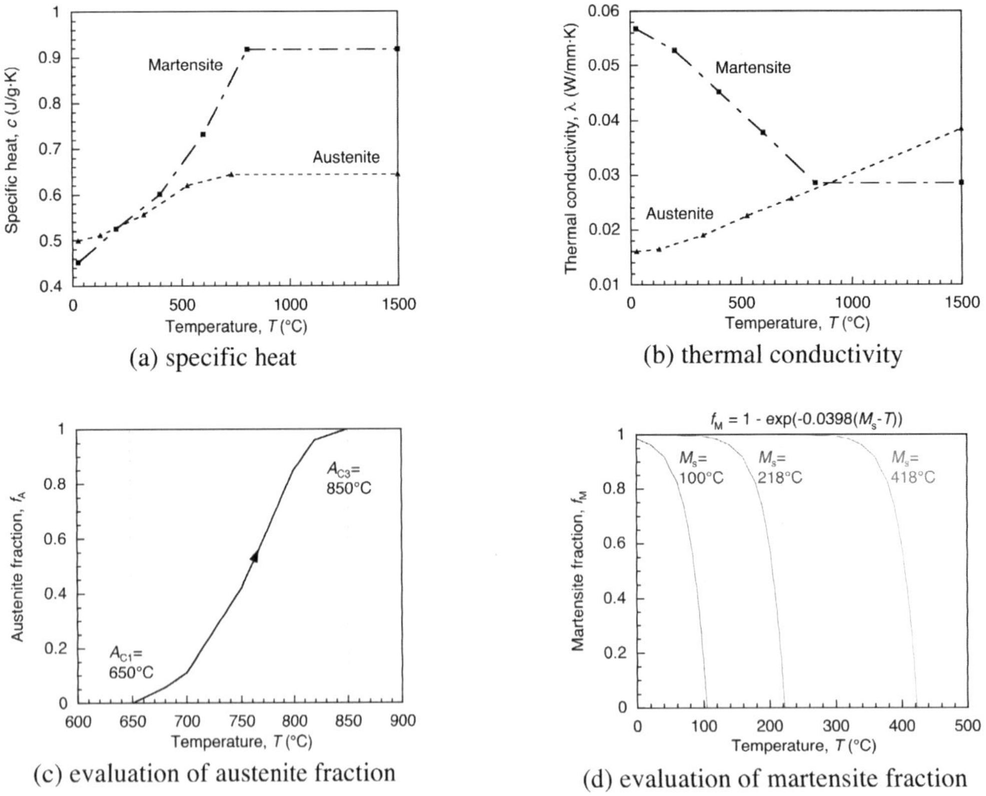

Figure 3. Material properties used in the thermal analysis.

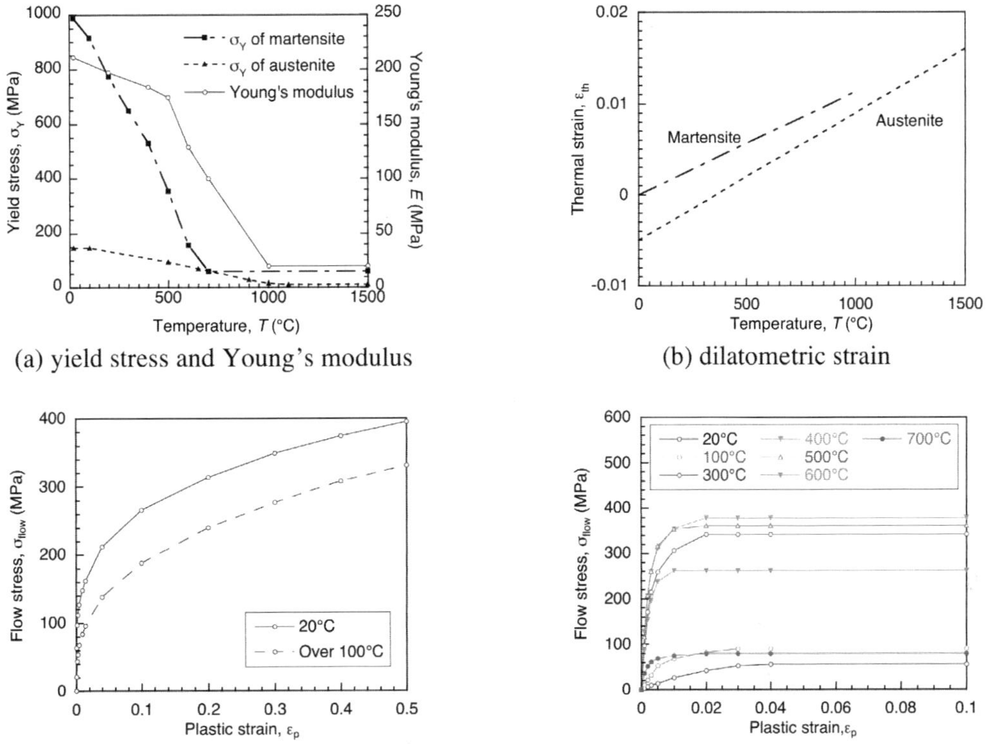

(a) yield stress and Young's modulus (b) dilatometric strain

(c) strain hardening curves of austenite (d) strain hardening curves of martensite

Figure 4. Material properties used in the thermal elastic-plastic analysis.

3. Effect of restraint factor and transformation temperature on restraint stress

The relationship between the restraint stress and the restraint factor is shown in Figure 5. In this study, the restraint stress is defined as the averaged residual stress in the transverse direction σ_y along the throat thickness of the weld metal. Generally, the restraint stress increases with increasing restraint factor. The result of M_s = 418°C, of which the M_s temperature is high, is an example of this relationship. On the contrary, when the M_s temperature decreases, the restraint stress at the high restrain factor deviates from a linear relationship: when the restraint factor is high and M_s decreases, the restraint stress becomes constant. The reduction of the restraint stress from M_s = 418°C to M_s = 100°C originates from the transformation expansion of the weld metal. The results indicated that the transformation expansion of the weld metal is more effective when the transformation temperature is low and the restraint factor is high.

4. Effect of specimen geometry on local stress distribution

To evaluate the occurrence of cold cracking, the stress at the weld root should be discussed. The distribution of residual stresses in the transverse direction at the root along the weld line is shown in Figure 6 and Figure 7. The results of the Y-groove test are shown in Figure 6 and the results of the H-type test are shown in Figure 7. The horizontal axis represents the restraint factor at each location along the weld line. The arrows in the figures indicate the maximum stresses in each M_s case. The residual stresses are almost the same along the weld line in the Y-groove test, shown in Figure 6, although the local stress drops at the edge of the weld metal. The martensitic transformation start temperature M_s clearly affects the residual stress. In contrast, the residual stress at the weld root is distributed from the compressive to tensile values in the H-type test; the critical stress that leads to the occurrence of cold cracking is difficult to determine in such a case. The difference should be taken into consideration when conducting cold cracking tests.

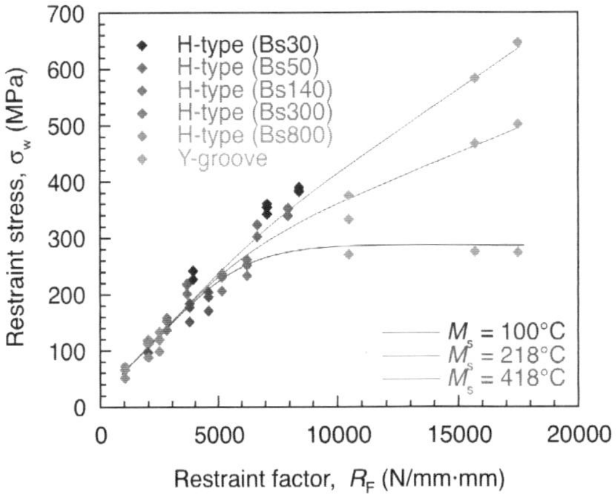

Figure 5. Effect of restraint factor on restraint stress.

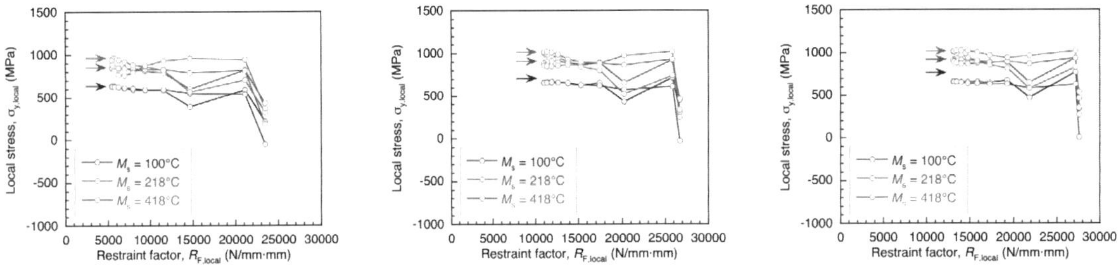

(a) h = 10 mm (b) h = 20 mm (c) h = 25mm

Figure 6. Effect of specimen dimensions on residual stresses distribution in the Y-groove test.

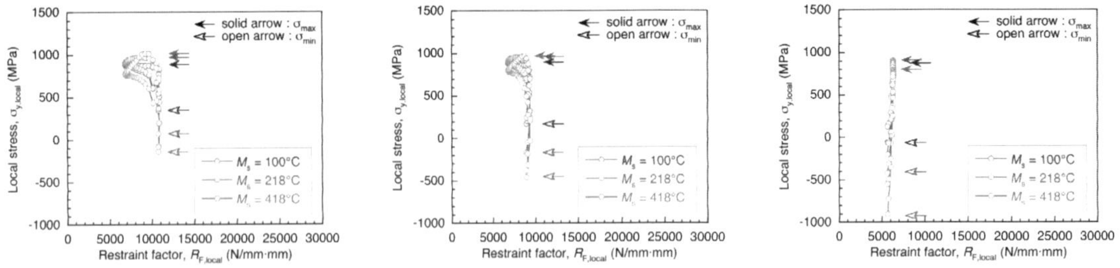

(a) Bs = 30 mm, h = 25 mm (b) Bs = 50 mm, h = 25 mm (c) Bs = 140 mm, h = 25 mm

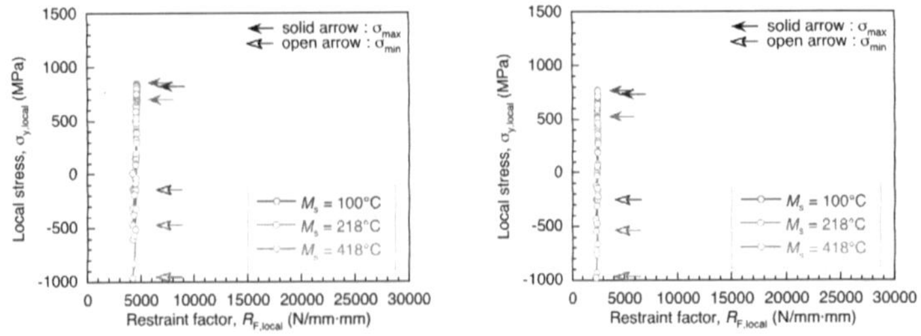

(d) Bs = 300 mm, h = 25 mm (e) Bs = 800 mm, h = 25 mm

Figure 7. Effect of specimen dimensions on residual stresses distribution in the H-type test.

5. Conclusions

In this study, the effect of low-temperature transformation expansion on the residual stress in specimens of different restraint factors was investigated.

1) The weld metal with low-temperature transformation expansion exhibits the effect of residual stress reduction at a high restraint factor.
2) Even if the restraint factor changes, the distribution of the restraint factor in the Y-groove weld cracking test is different from that in the H-type restrained weld cracking test.
3) The following differences should be taken into consideration when conducting cold cracking tests.
    i. In the Y-groove test, the residual stresses are almost the same along the weld line and the M_s temperature clearly affects the residual stress.
    ii. In the H-type test, the residual stresses at the weld root are distributed from the compressive to tensile values. In such a case, the critical stress that leads to the occurrence of cold cracking is difficult to determine.

Acknowledgments

This study was partly supported by the Global COE Program (Center of Excellence for Advanced Structural and Functional Materials Design) from MEXT, Japan and partly by the research activities of the Fundamental Studies on Technologies for Steel Materials with Enhanced Strength and Functions by Consortium of JRCM supported by NEDO.

References

[1] Preprint of the 1st Symposium of Fundamental Studies on Technologies for Steel Materials with Enhanced Strength and Functions Project 2009 pp 1-2 (in Japanese)
[2] Zenitani S and Hayakawa N 2005 Development of New Low Transformation-Temperature Welding Consumable to Prevent Cold Cracking in High Strength Steel Welds *JWS* Vol. 23 No.1 pp 95-102 (in Japanese)
[3] Satoh K, Ueda Y and Kihara H 1973 Recent Trend of Research on Restraint Stresses and Strains for Weld Cracking *JWS* Vol 42 No.1 pp 57-72 (in Japanese)
[4] *Japanese Standards Association* ed Method of y-groove Weld Cracking Test, JIS Z 3158 (in Japanese)
[5] *Japanese Standards Association* ed Method of H-type Restrained Weld Cracking Test JIS Z 3159 (in Japanese)
[6] Material database MATEQ, *JSMS* (in Japanese)
[7] JSME Data Book: Heat Transfer 4th Edition 1997 (in Japanese)
[8] J.B.Koistine and R.E.Marburger 1959 A General Equation Prescribing the Extent of the Austenite-martensite Transfomation in Pure Iron-Carbon Alloys and Plain Carbon Steel *Acta Mettallurgica*, Vol. 7 , pp 59-60

Journal of Physics: Conference Series **379** (2012) 012055 doi:10.1088/1742-6596/379/1/012055

Investigation of factors influencing welding deformation of ship block by inherent strain analysis using idealized explicit FEM

Masashi Hata[1], Shinsuke Itoh[2], Taketoshi Sugihiro[3], Yasuo Yamashita[3], Atsushi Kamio[3], Masakazu Shibahara[2], and Masahito Mochizuki[1]

[1] Graduate School of Engineering, Osaka University, 2-1 Yamada-oka, Suita, Osaka 565-0871, Japan
[2] Graduate School of Engineering, Osaka Prefecture University, 1-1 Gakuen-cho, Naka-ku, Sakai, Osaka, Japan
[3] Sumitomo Heavy Industries Marine & Engineering Co., Ltd., 19 Natsushima-cho, Yokosuka, Kanagawa, Japan

Abstract. It is impossible to avoid welding deformations when steel structures such as ships are assembled. Therefore, quantitative prediction and effective control of residual welding deformation are necessary. However, the current finite element analysis requires very long computational time and memory. To solve this problem, we have developed an inherent strain analysis method using idealized explicit FEM and have applied this proposed method to a ship block model. The simulated results agree well with the measured deformation. Furthermore, the influence of various factors on the welding deformation of targeted ship block is also investigated.

1. Introduction

In ship construction, welding is used to join and assemble components. Welding deformation often causes a deviation from the design. Therefore, corrective work is required. However, since this corrective work relies on the skills of the engineer, it is difficult to estimate the time and the cost. If it was possible to predict the welding deformation accurately, the time and cost of the corrective work could be reduced. Furthermore, if the mechanism of welding deformation is clarified, more effective methods to avoid welding deformation can be developed. Therefore, if welding deformation and corrective work could be accurately predicted, a reduction of total energy is possible in ship construction. However, current methods of predicting welding deformation, such as finite element analysis (FEM), require a long computational time and large memory. Therefore, we developed an inherent strain analysis method using an Idealized Explicit FEM. In this method, the Iterative Substructure Method (ISM) [1] is first used to create a database of inherent strain distributions. Then, using this database, an inherent strain analysis using Idealized Explicit FEM is applied to predict the welding deformation.

Journal of Physics: Conference Series **379** (2012) 012055 doi:10.1088/1742-6596/379/1/012055

2. Analysis method of this study

Generally, inherent strain analysis is used to predict the welding deformation of large welded structures. The computational time for inherent strain analysis is shorter than that for thermal elastic analysis because inherent strain analysis is an elastic analysis. However, to use this method, a database of inherent strain distributions for each joint is necessary. Since the inherent strain distributions must be obtained through a series of experiments, the total time to obtain the results becomes enormous. Therefore, in this study, two improved methods are used. Figure 1 shows the flow of the proposed methods. ISM, which is a high-speed thermal elastic plastic FE analysis, is first used to create the database of inherent strain distributions. Then, by using the database, an inherent strain analysis using Idealized Explicit FEM is applied to predict the welding deformation. The following describes the concepts of the ISM and Idealized Explicit FEM.

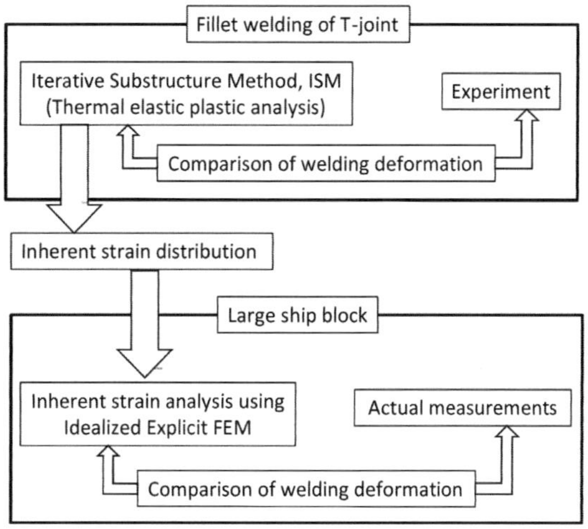

Figure 1. Flow of proposed method.

2.1. Iterative Substructure Method for thermal elastic plastic analysis [1]

In simulations of welding, two characteristics of welding greatly increase the computing time. One is the strong nonlinearity near the welding torch. However, the area that is nonlinear is very limited and small in comparison with the entire structure. Typically, finite element (FE) analysis must be applied as a total nonlinear solution even when the nonlinear region is a very small part of the whole model. The purpose of ISM is to solve these problems. In ISM, the linear and the nonlinear regions are separated. As shown in Figure 2, these regions are called A and B, respectively. The boundary between regions A and B is designated Γ. Most of the computing time in FE analysis is used to solve large-scale simultaneous equations. In welding problems, the equations must be solved (Ns × Ni) times, where Ns is the time step and Ni is the iteration step. As an example, when using Gauss elimination, the computing time for forward elimination and back substitution is proportional to $(DOF)^3$ and $(DOF)^2$, respectively, where DOF is the number of degrees of freedom. If the stiffness matrix is constant, the matrix for forward elimination can be reused. In this case, the computing time becomes almost the same as that for performing the forward elimination one time because the time required by back substitution is markedly shorter than that for forward elimination. The differences between the conventional substructure method and ISM are summarized in Figure 3. In the conventional substructure method, the whole model is divided into regions A and B. In this case, region A changes as region B moves. Thus, the stiffness matrix for region A changes. In contrast, ISM uses the whole region (A+B) and the nonlinear region B. If region B moves, the whole region (A+B) does not change. The stiffness matrix for the whole region (A+B) after forward elimination is saved. In this way, the process can be repeated to solve the displacement in region A without changing the stiffness matrix.

After obtaining the displacement of region A, the displacement in region B is computed by using the displacement of the boundary Γ between regions A and B. Thus, the computing time is reduced by omitting the forward elimination of the stiffness matrix for the whole region (A+B).

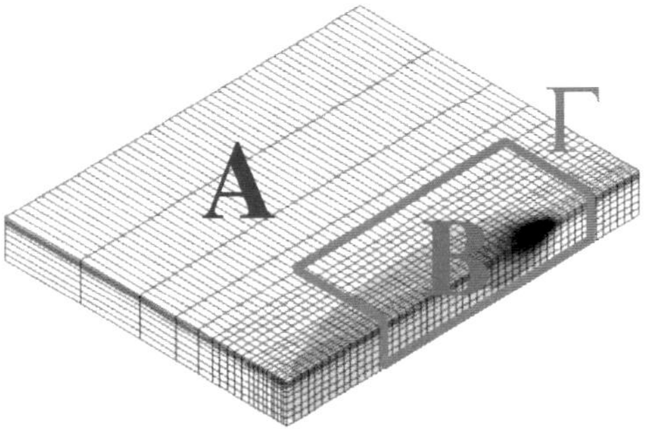

Figure 2. Regions A and B used in ISM. The boundary between A and B is Γ[1].

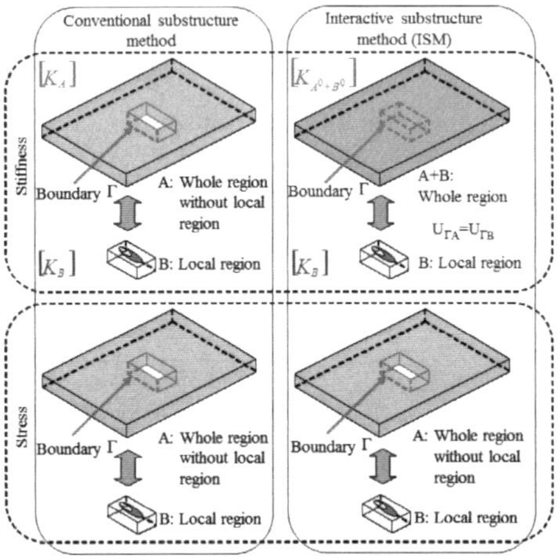

Figure 3. Comparison between conventional substructure method and ISM [1].

2.2. Inherent strain analysis using Idealized Explicit FEM [2]

To simulate the welding deformation, thermal elastic plastic analysis or inherent strain analysis is used. Thermal elastic plastic analysis can simulate transient welding mechanical phenomena. However, if thermal elastic plastic analysis is carried out using implicit static FEM, the computing time and the memory utilization increase in proportion to the square of the number of elements [3]. In contrast, the computing time for inherent strain analysis is much smaller than that for thermal elastic plastic analysis, because inherent strain analysis computes only one step of an elastic analysis by using the inherent strain database. However, the tendency for the computing time and memory to increase is the same as that for thermal elastic plastic analysis if the computation method is implicit static FEM. In this research, Idealized Explicit FEM is used [2]. In this method, a step is divided into hundreds of time steps and the displacements are computed for each. After computing the displacement, the static equilibrium of the whole system is checked. The displacements are computed until a static equilibrium state is obtained. The concept of Idealized Explicit FEM is shown in Figure 4. Generally, in implicit dynamic FEM, the Wilson theta method or the Newmark beta method is used to solve the dynamic problem. To obtain the displacement vector $\{u\}$, very large simultaneous equations have to be solved. In contrast, explicit FEM does not require solving very large matrix equations. The principal theory of explicit FEM can be described as follows [4]. First, the nodal velocity vector $\{\dot{u}\}$ and the nodal acceleration vector $\{\ddot{u}\}$ are discretized by the central difference method:

$$\{\ddot{u}\}_t = (\{u\}_{t+\Delta t} - 2\{u\}_t + \{u\}_{t-\Delta t})/\Delta t^2 \quad (1)$$

$$\{\dot{u}\}_t = (\{u\}_{t+\Delta t} - \{u\}_{t-\Delta t})/(2\Delta t), \quad (2),$$

where t is the time, Δt is the time increment, and $\{u\}_{t-\Delta t}$, $\{u\}_t$, and $\{u\}_{t+\Delta t}$ are the displacement vectors at times $t-\Delta t$, t, and $t+\Delta t$, respectively.

$$[M]\{\ddot{u}\}_t + [C]\{\dot{u}\}_t + \int [B]^T \{\sigma\} dV = \{F\}_t \quad (3)$$

$$\left(\frac{1}{\Delta t^2}[M]+\frac{1}{2\Delta t}[C]\right)\{u\}_{t+\Delta t} = \{F\}_t - \int [B]^T\{\sigma\}dV + \frac{2}{\Delta t^2}[M]\{u\}_t - \left(\frac{1}{\Delta t^2}[M]-\frac{1}{2\Delta t}[C]\right)\{u\}_{t-\Delta t} \qquad (4)$$

Equation (4) is derived by substituting equations (1) and (2) into equation (3).

If the mass matrix $[M]$ and the damping matrix $[C]$ are diagonal matrices, then the displacement vector $\{u\}_{t+\Delta t}$ can be easily obtained from the displacement vectors $\{u\}_t$ and $\{u\}_{t-\Delta t}$. In this formulation, it is well known that the global stiffness matrix is not required and no large simultaneous equations need to be solved. To calculate the displacement, equation (4) is solved. However, for long-duration phenomena such as welding, the computational steps become enormous because of the very short time increment. The short increment is due to the Courant condition [5]. The Courant condition requires the distance travelled by a stress wave in a time increment to be less than the minimum element size. Idealized Explicit FEM overcomes this limitation.

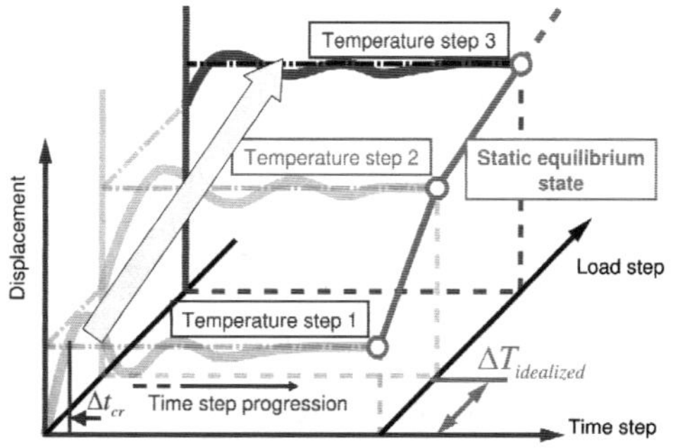

Figure 4. Concept of Idealized Explicit FEM [3].

3. Experiment and simulation for fillet welding of a T-joint

In the proposed method, thermal elastic plastic analysis by ISM is used to create the database of inherent strain distributions. Therefore, in order to verify the adequacy of thermal elastic plastic analysis by ISM and to obtain the database of inherent strain distribution, an experiment and a simulation are performed. In a ship structure, since many stiffeners are attached by fillet welding of a T-joint, in this study, fillet welding of a T-joint is considered. An illustration of the T-joint is shown in Figure 5. The model and welding conditions are shown in Figure 6 and Table 1, respectively. The components are mild steel. The angular distortion is measured at four points, shown in Figure 5. The angular distortion refers to the difference in angle between the base plate and the stiffener after welding. Figure 7 shows a comparison of the angular distortion for the experiment and the simulation. This figure indicates that there are some differences between the experimental and analysis results. These differences are because the experiment was performed simply to reproduce the actual site. However, taking this into account, the ISM simulation corresponds to the experiment.

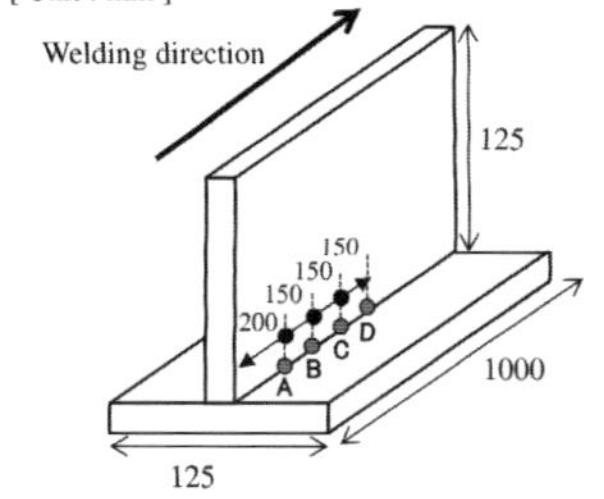

Figure 5. Illustration of T-joint model.

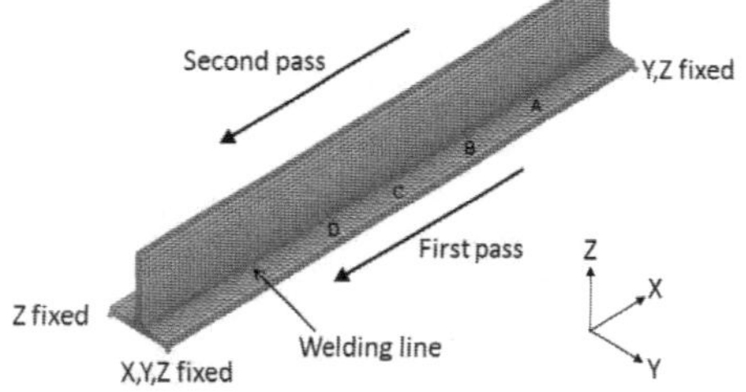

Figure 6. FE mesh divisions of T-joint model.

Table 1. Welding conditions.

Voltage (V)	Current (A)	Welding speed (mm/s)
44.0	290.0	8.0

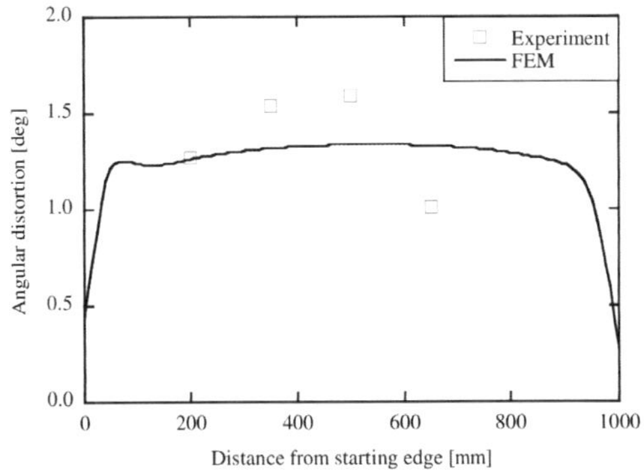

Figure 7. Comparison of angular distortion for experiment and FE analysis using ISM.

4. Effectiveness of the proposed method

In the proposed method, inherent strain analysis using Idealized Explicit FEM is used to predict the welding deformation of a large-scale structure. Therefore, inherent strain analysis using Idealized Explicit FEM and thermal elastic plastic analysis by ISM are compared in order to verify the effectiveness of the former method. This analysis uses the inherent strain distributions from both the welding and straightening by line heating calculated by ISM. In this study, straightening by line heating is performed after welding. The model, welding conditions and heating conditions are shown in Figure 8, Table 1 and Table 2, respectively. The position of line heating and the contour of the Mises plastic strain are shown in Figure 9 and Figure 10, respectively. From Figure 10, since the Mises plastic strain can be regarded as constant, the inherent strain is assumed to be constant for all welding lines in this study. Figure 11 shows a comparison of the deformation of the center section weld line for ISM and inherent strain analysis using Idealized Explicit FEM. As shown in this figure, the simulation using inherent strain analysis using Idealized Explicit FEM corresponds to that using thermal elastic plastic analysis by ISM with high accuracy.

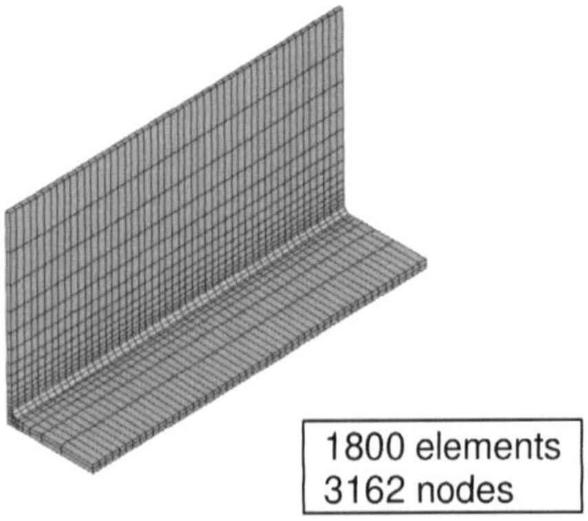

Figure 8. FE mesh division.

Table 2. Line heating conditions.

Heat input (J/mm)	Heating speed (mm/s)
2160	6.0

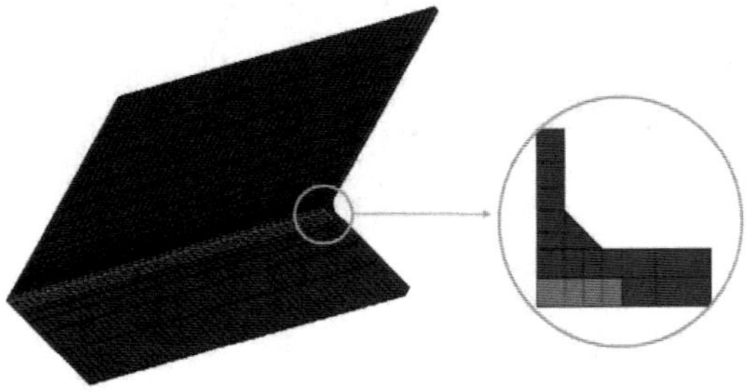

Figure 9. Position of line heating.

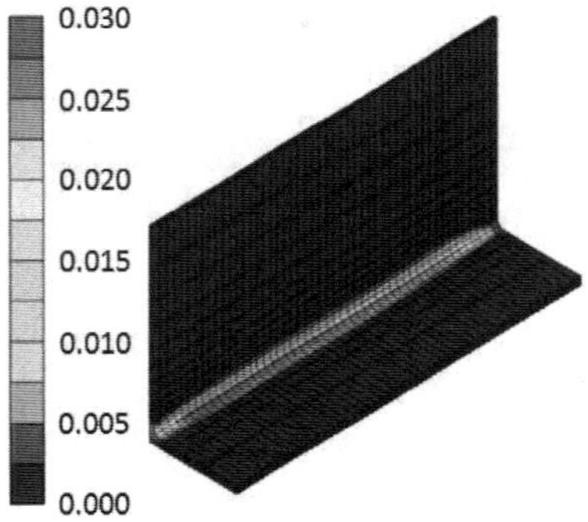

Figure 10. Contour of Mises plastic strain.

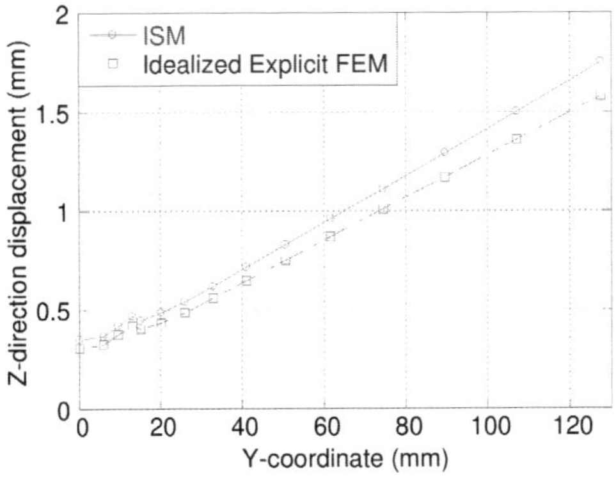

Figure 11. Comparison of deformation of center section weld line for ISM and inherent strain analysis using Idealized Explicit FEM.

5. Inherent strain analysis using Idealized Explicit FEM applied to actual ship block

A ship block is assembled using a large amount of fillet welds. The proposed methods are next applied to the actual ship block shown in the photo in Figure 12. The block is near the engine and the problem of welding and line heating deformation occurs at the aperture. Straightening by line heating is performed after welding. The welding and heating conditions are shown in Table 1 and Table 2, respectively. The edges and the crossed areas are welded manually. The analysis model is shown Figure 13. The total number of nodes and solid elements is 875698 and 614110, respectively. The total number of welding lines in this model is almost 600. Figure 14 and Figure 15 show the contour of X-direction displacement and Y-direction displacement, respectively.

To investigate the influence of the heat input (Q) and the welding and heating velocity (v) on the deformation of the block, the values of each are changed in the simulation. Figure 16 shows the measurement and analysis results for the outline of the ship shape with in-plane deformation. From this figure, it is found that the simulated results correspond to the measurement results. Furthermore, it is found that the deformation at the aperture is strongly affected by the heat input.

Figure 12. Photo of actual ship block.

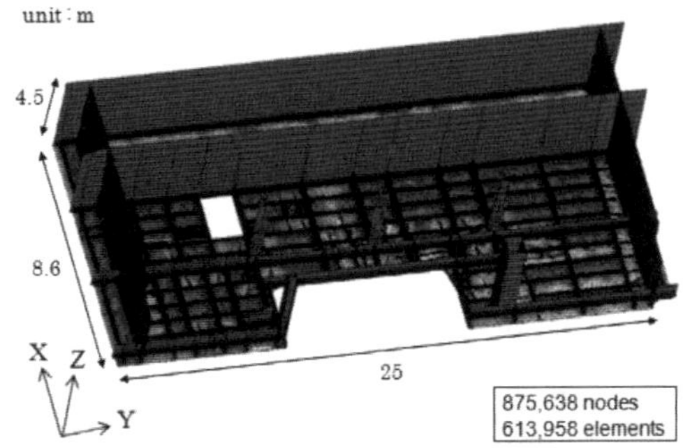

Figure 13. Ship block model.

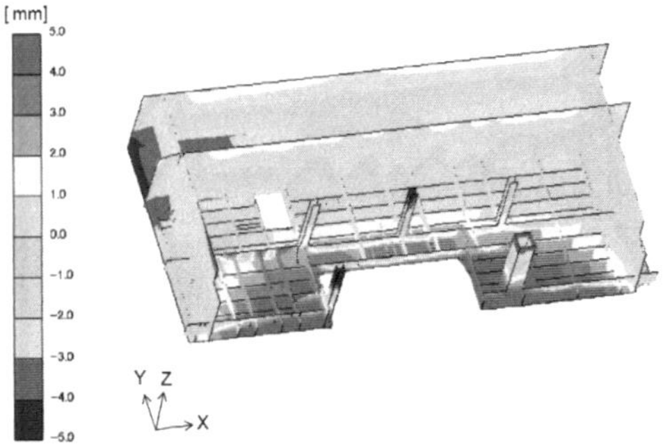

Figure 14. Contour of X-direction displacement.

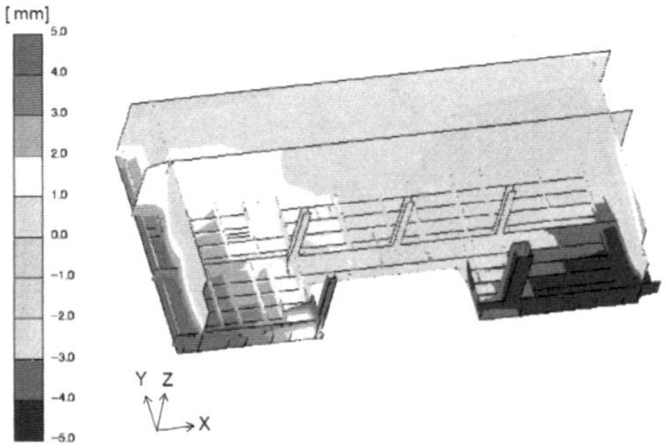

Figure 15. Contour of Y-direction displacement.

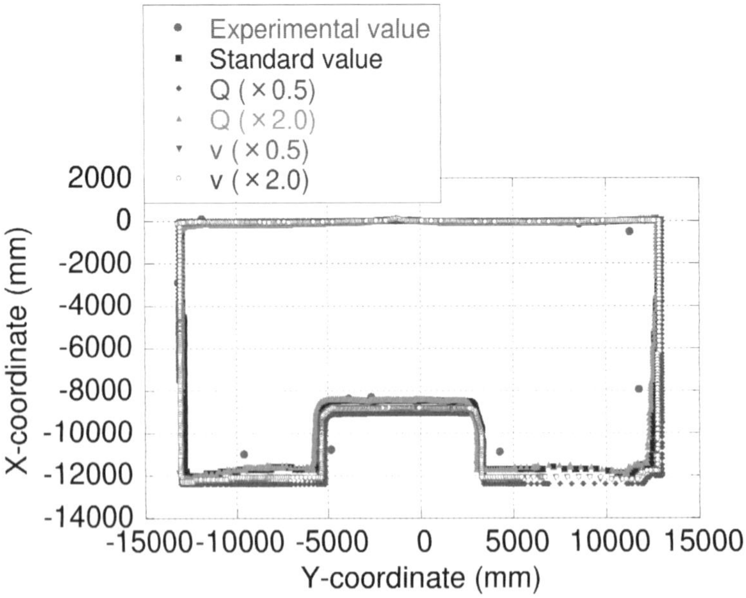

Figure 16. Outline of ship shape with 100 times in-plane deformation.

6. Conclusions

To predict the welding deformation of an actual ship block, an inherent strain analysis method using ISM and Idealized Explicit FEM was developed and applied. The results of the analysis are as follows.
(1) For an actual ship block, the simulated deformation agrees with the experimental results.
(2) The influence of the heat input on the deformation at the aperture is significant.

Acknowledgements

This study was supported in part by Priority Assistance for the Formation of Worldwide Renowned Centers of Research – The Global COE Program (project, Center of Excellence for Advanced Structural and Functional Materials Design) from the Ministry of Education, Culture, Sports, Science and Technology (MEXT), Japan.

References

[1] Murakawa H, Oda I, Ito S, Serizawa H, Shibahara M, Nishikawa H 2005 *IterativeSubstructure Method for Fast Computation of Thermal Elastic Plastic Welding Problems* (Journal of the Kansai Society of Naval Architects) Japan **243** 67-70 (in Japanese)
[2] Shibahara M, Ikusima K, Itoh S, Masaoka K 2011 *Computational Method for Transient Welding Deformation and Stress for Large Scale Structure Based on Dynamic Explicit FEM* (Journal of the Japan Welding Society) **29** (1) 1-9.
[3] Nishikawa H, Oda I, Shibahara M, Serizawa H, Murakawa H 2004 *Three-demensional Thermal-elastic-plastic FEM Analysis for Predicting Residual Stress and Deformation Under Multi-pass Welding* (Proc. Of ISOPE'04) Toulon France CD-ROM
[4] Yagawa M 2004 *KOUZOUKOUGAKU hand book* (MARUZEN) 302
[5] Sugimoto S, Yasuzawa K, Kagawa K 2001 *Elasto-Plastic Collapse Behavior of Plate and Shell Structure Components by Dynamic Explicit FEA* (The Computational Mechanics Conference, The Japan Society of Mechanical Engineers) **11** 183-184

AUTHOR INDEX

Akedo, J 69
Amimoto, M....114
Anno, H....39
Aspera, S M....76
Baba, S....69
Bin, L....53
Cahyanto, W T....76
Castagne, S....224
Chang, S-H....197
Diño, W A....76, 84
Fong, K-S....224
Fujimoto, K....163, 170, 189
Fujimoto, T....163
Fukumoto, S....163, 170, 189
Funahashi, R....69
Halil, A B M....130
Haneda, M....90, 114
Hanzawa, H....100
Hara, K....285
Hashimoto, T....338, 346, 362
Hata, M....376
Hatsuzawa, K....153
Hattori, M....90
Hemthavy, P....136, 142, 269, 278
Heo, Y-M....197
Hidai, H....228
Hiraki, A....10
Hiraki, H....10
Hiraoka, K....369
Hiratsuka, Y....118
Hokazono, M....39
Honda, Y....142
Horikawa, T....95, 100
Hu, A....147
Huang, L....69
Ihara, R....338
Ikeda, R....346
Ikeshoji, T-T....59
Imanaga, Y....118
Ishihara, T....76
Itoh, M....95
Itoh, S....376
Itoh, T....15
Iyota, M....346
Jarfors, A E W....224
Kamio, A....376
Kasai, H....76, 84
Kashiba, Y....163
Katsuyama, S....21
Kencana, W T D....76
Kikuchi, T....250
Kim, G-H....197
Kim, H S....95, 100
Kim, J D....258
Kim, J S....258
Kinoshita, H....250
Kishi, H....76
Kitano, H....328
Kobayakawa, T....228
Kubota, N....369
Kunikata, S....76
Kunisada, Y....84
Kuramoto, D....95
Kuwano, Y....1
Lee, G-A....197
Lim, S C V....224
Liu, J....224
Liu, L....147
Machida, K....95, 100
Maeda, M....10, 122, 130, 178
Maezawa, F....21
Maruo, Y Y....234, 243
Matsumoto, R....285, 291, 297
Matsumura, Y....28
Matsunami, H....189
Matsusaka, S....228
Matsushima, M....170, 189
Matuda, K....90
Mikami, M....33
Mikami, Y....303, 346, 369
Min, I K....258
Minami, M....170
Miura, Y....90
Miwa, K....76
Miyazaki, T....163
Mochizuki, M....303, 328, 338, 346, 362, 369, 376
Morita, N....228
Mugita, N....47
Munetoh, S....47, 53
Muraoka, A....291
Murata, H....104
Nagatomo, Y....47, 53
Nagatsuka, K....309, 318
Nakabayashi, T....39
Nakakohara, Y....47, 53
Nakamura, J....243
Nakanishi, H....76, 84
Nakata, K....309, 318
Nakatani, R....118
Natsuhara, H....15
Nii, H....355
Nishimoto, A....355
Nomura, H....118
Nonomura, S....15
Ohashi, K....297
Ohtsuka, Y....212
Ohyama, Y....178
Okamura, Y....104
Okano, S....328, 362
Okouchi, T....90

AUTHOR INDEX

Okumura, Y 206
Ono, Y 243
Otake, K 250
Ozaki, K 33
Ozawa, M 90, 114
Ren, J 147
Saikubo, A 212
Saisho, M 47, 53
Saito, S 269, 278
Sakai, T 285, 291, 303
Sakaue, M 76
Samir, M F B 278
Sato, H 69
Sechi, Y 309, 318
Sekiguchi, Y 142, 269
Sekimori, N 153
Setsuhara, Y 206
Shibahara, M 376
Shinohara, Y 28
Shiotani, K 163
Shirataki, R 39
Shishihara, Y 189
Shohji, I 153
Shoji, K 59
Soeda, A 170
Sugahara, A 362
Sugihiro, T 376
Sugimoto, T 90
Suzuki, R O 250
Suzuki, S 178
Suzumura, A 59
Tajiri, M 47
Takahara, W 265
Takahashi, K 136, 142, 269, 278
Takahashi, M 170
Takahashi, Y 10, 122, 130, 178
Takenaka, K 206
Tan, M-J 224
Tanaka, M 136, 142
Tanaka, T 21
Taniguchi, K 346
Teranishi, R 47, 53
Tsuda, M 234, 243
Ueda, Y 212
Uetani, T 153
Utsunomiya, H 285, 291, 297
Watanabe, J 153
Wijayanto, Y N 104
Wu, A 147
Wunderlich, W 28
Yamada, H 39
Yamada, T 234
Yamasaki, T 122
Yamashita, Y 376
Yamazaki, T 59
Yan, J 147, 147
Yasuda, K 218
Yokoyama, Y 170
Yuzuriha, H 90
Zhou, Y N 147
Zou, G 147